QUANTITATIVE DYNAMIC STRATIGRAPHY

Timothy A. Cross, Editor

PRENTICE HALL, ENGLEWOOD CLIFFS, NEW JERSEY 07632

Library of Congress Cataloging-in-Publication Data

Quantitative dynamic stratigraphy/Timothy A. Cross, editor.

p. cm.
Includes index.
ISBN 0-13-744749-3
  1. Geology, Stratigraphic—Mathematical models. 2. Sedimentation and deposition—Mathematical models. I. Cross, Timothy Aureal.
QE651.Q26 1990
551.7'01'5118—dc20

89-35124
CIP

Editorial/production supervision: BARBARA MARTTINE
Cover design: BEN SANTORA
Manufacturing buyer: MARY ANN GLORIANDE

The color insert was made possible
by the financial support of Texaco, Inc.

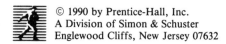 © 1990 by Prentice-Hall, Inc.
A Division of Simon & Schuster
Englewood Cliffs, New Jersey 07632

The publisher offers discounts on this book when ordered
in bulk quantities. For more information, write:

>   Special Sales/College Marketing
>   College Technical and Reference Division
>   Prentice Hall
>   Englewood Cliffs, New Jersey 07632

All rights reserved. No part of this book may be
reproduced, in any form or by any means,
without permission in writing from the publisher.

Printed in the United States of America

10 9 8 7 6 5 4 3 2 1

ISBN 0-13-744749-3

Prentice-Hall International (UK) Limited, *London*
Prentice-Hall of Australia Pty. Limited, *Sydney*
Prentice-Hall Canada Inc., *Toronto*
Prentice-Hall Hispanoamericana, S.A., *Mexico*
Prentice-Hall of India Private Limited, *New Delhi*
Prentice-Hall of Japan, Inc., *Tokyo*
Simon & Schuster Asia Pte. Ltd., *Singapore*
Editora Prentice-Hall do Brasil, Ltda., *Rio de Janeiro*

QDS Philosophy

We dreamed of a world systematic,
Finding patterns would make us ecstatic.
We quantified systems,
And shared algorithms,
While searching for models stochastic.

In the boondocks Tim made us convene,
Flaunting models for all to be seen.
But the lack of consensus,
Just served to incense us,
Teeth-gnashing and venting our spleen.

We waxed philosophic on sandstones
Over Baronet's protests and moans.
Jargoning for position,
(An old geo-tradition),
The clastickers sat chewing their bones.

We brought to Lost Valley our queries
Of Ian's equations and theories.
Somewhere on the road,
Zeitlin wrote the code,
While kissed by the muse of draft beeries.

And finally a model empirical
Which made all the world seem heuristical,
"It's chaotic!" was shouted,
(But still it was touted.)
That it functioned at all is a miracle!

And now my quantitative friends,
We've come to our dynamic end.
Our first QDS,
A raging success,
Go home and continue the trend!

— Katie Joe McDonough and Doug Waples

CONTENTS

Part I—QDS Models in Sedimentary Basin Analysis

Part II—Basin-Scale QDS Models

Part III—Subbasin-Scale QDS Models

Part IV—Stratigraphic Resolution and Model Verification

Part V—Potential Applications of QDS Models

PREFACE

During the past decade, a variety of quantitative geodynamic, thermal (maturation) and fluid-flow models of sedimentary basins have been developed and applied widely in industrial and academic sectors. More recently, sedimentologists and stratigraphers have been developing models that simulate the filling of sedimentary basins at varying scales. Simultaneously, a new generation of diagenetic models is being developed that incorporate fluid-flow and fluid/rock interactions with dynamic simulations of sediment fill. Atmospheric circulation and global climate models are being applied to simulate water circulation and sediment transport in paleooceans. Some structural models that were developed independently are now being incorporated within more integrated basin formation and basin fill models. Petroleum reservoir simulators and groundwater flow models are able to utilize higher resolution, quantitative descriptions of stratigraphic plumbing systems in their simulations of fluid flow.

The history of these developments indicates an increased focus among earth scientists to formulate more exacting and quantitative relations among processes and responses of complex natural systems. It also indicates a progressive expansion in the types of variables, processes and responses included in stratigraphic or sedimentary basin models—there is a trend toward combining models formulated within traditional subdisciplines and integrating them in more comprehensive models of basin evolution. Finally, this history indicates a multi-disciplinary concern with the need to produce more accurate and deterministic descriptions and predictions of stratigraphy—including the petrophysical properties of sedimentary rocks, and the spatial and temporal distributions and geometries of sedimentary rock bodies.

If stratigraphy is the study of layered rocks, incorporating many topics traditionally considered the domains of individual subdisciplines within the earth sciences, then we are in the midst of an evolving form of stratigraphic analysis which has been termed Quantitative Dynamic Stratigraphy. Quantitative Dynamic Stratigraphy is the application of mathematical, quantitative procedures to the analysis of geodynamic, stratigraphic, sedimentologic and hydraulic attributes of sedimentary basins, treating them as features produced by the interactions of dynamic processes operating on physical configurations of the earth at specific times and places.

In February, 1988, about 75 earth scientists assembled at Lost Valley Ranch, near Denver, Colorado, to participate in a four-day wokshop on Quantitative Dynamic Stratigraphy. This workshop was co-sponsored by the American Association of Petroleum Geologists, the Colorado School of Mines, the Gas Research Institute, the National Science Foundation, the U.S. Department of Energy, and the U.S. Geological Survey. Workshop participants, although representing a variety of disciplines, interests and backgrounds, shared a common interest in quantitative approaches to stratigraphic analysis, and a hope that such approaches will provide the vehicle for making more accurate assessments and predictions about the characteristics of sedimentary basin development and fill. The goals of this workshop were:

to summarize the state-of-the-art in quantitative approaches to stratigraphic and sedimentary basin analysis; to identify major types and forms of empirical geological data that are or will be required for constructing, testing and verifying quantitative models; and, to define research directions that will significantly enhance capabilities in predicting temporal and spatial relations of sedimentary facies, stratigraphic architecture, fluid movement, and diagenesis in subsurface strata.

This book, a product of the workshop, summarizes and provides examples of state-of-the-art approaches in quantitative analysis of sedimentary basins. The philosophy of this book is that stratigraphic analysis of sedimentary basins at all scales—from geodynamics, to stratigraphic fill and fluid flow, to sediment transport—is rapidly evolving through employment of quantitative approaches. These approaches, accomplished principally through computer modeling, have the potential of providing more accurate assessment, description and prediction of a variety of sedimentary basin attributes.

Many papers in this book discuss philosophical aspects of numerical simulations of geologic processes and responses, and illustrate how QDS models provide explanations, insight and predictions about behaviors of natural geologic systems. Other papers are reviews or examples of requirements of numerical models for particular applications, or examples of verification studies designed to establish limits on the applications of particular numerical models. The range in topics considered in this book is broad, reflecting the diverse interests of the workshop participants and the diverse applications of QDS models. These topics include philosophies and methods of model construction, concerns about modeling chaotic systems in basin analysis, geodynamic formation and evolution of basins, stratigraphic and sedimentation models, hydrocarbon maturation and sediment diagenesis models, dynamic ocean circulation and climate models, paleogeographic reconstructions using a mass-balance approach, resolution of geologic time from lithostratigraphy, biostratigraphy and chemical stratigraphy, and data bases and methods for evaluating QDS models. Also included are two computer programs that simulate sediment fill and sediment transport.

This book is aimed at college seniors, graduate students and professionals who are interested or engaged in stratigraphic and sedimentary basin analysis. I hope that this collection of state-of-the-art summaries, examples of applications, procedures and cautions about needs for testing and verification, and assessments of future developments of QDS models provide the impetus to, as Katie Joe McDonough and Doug Waples admonish, "Go home and continue the trend!"

Timothy A. Cross

ACKNOWLEDGMENTS

On behalf of the participants of the Quantitative Dynamic Stratigraphy Workshop, I express our gratitude for the financial support of the workshop provided by the Gas Research Institute, the National Science Foundation, the U.S. Department of Energy, and the U.S. Geological Survey. I also thank our two additional sponsors—the American Association of Petroleum Geologists and the Colorado School of Mines. The support of these agencies made this workshop possible and contributed enormously to its success.

I am very appreciative of the time, counsel and guidance provided selflessly by the members of the steering committee at all stages in planning, organizing and completing the workshop: Anthony W. Gorody (Gas Research Institute), Felix M. Gradstein (Geological Survey of Canada), John W. Harbaugh (Stanford University), William W. Hay (University of Colorado, Boulder), Ian Lerche (University of South Carolina), Martin D. Matthews (Texaco E&P Technology Division), and Rudy L. Slingerland (The Pennsylvania State University).

I am grateful for the encouragement, support and massive logistical assistance—for both the workshop and this book—provided by a cadre of graduate students at the Colorado School of Mines: Souad Al Azzawi, Mark Chapin, Mike Gardner, Benito Guerrero, Mark Hanson, Margaret Lessenger, Laird Little, Katie Joe McDonough, Keith Shanley, Mark Sonnenfeld, Tonya Thorn, Dave Valasek, and Dave Witter.

Finally, I thank the following colleagues for their timely, thoughtful and helpful reviews of the papers in this volume. Their conscientiousness is best described as remarkable.

R.S. Anderson
M.R. Baker
J.S. Bridge
D.J. Cant
M.A. Chapin
S.P. Dutton
A.F. Embry
W.E. Galloway
R.K. Goldhammer
P.W. Goodwin
J.W. Harbaugh
W.J. Harrison
W.D. Huff
M.T. Jervey
T.E. Jordan
C.G.C.St. Kendall
G.deV. Klein
D.T. Lawrence
Ian Lerche
M.A. Lessenger

M.D. Matthews
R.K. Matthews
J.A. Nunn
R.H. Osborne
Chris Paola
M.A. Perlmutter
E.P. Poeter
S.B. Romberger
Fred Schroeder
W.J. Schweller
C.A. Shaw
R.L. Slingerland
M.S. Steckler
Randell Stephenson
J.P.M. Syvitski
D.M. Tetzlaff
Julian Thorne
W.L. Watney
D.F. Williams
M. J. Zeitlin

LIST OF PARTICIPANTS AT QDS WORKSHOP

Frederik P. Agterberg (Geological Survey of Canada)
Robert C. Anderson (University of California, Santa Cruz)
Edwin J. Anderson (Temple University)
Stefan Bachu (Alberta Research Council)
Mark R. Baker (University of Texas at El Paso)
Mary Barrett (Mobil Oil Co.)
Eric J. Barron (Pennsylvania State University)
John H. Barwis (Shell Development Co.)
Klaus Bitzer (Universität Freiburg)
Scott Bowman (Rice University)
Samuel W. Butcher (Brown University)
Frank Caruccio (University of South Carolina)
Mark Chapin (Colorado School of Mines)
Sierd Cloetingh (University of Utrecht)
Timothy A. Cross (Colorado School of Mines)
Marc D'Iorio (University of Ottawa)
Walt Dean (U.S. Geological Survey)
Shirley Dutton (Texas Bureau of Economic Geology)
Ashton F. Embry, III (Geological Survey of Canada)
Marc Ericksen (Pennsylvania State University)
Peter B. Flemings (Cornell University)
William E. Galloway (University of Texas at Austin)
Michael Gardner (Colorado School of Mines)
Charles Gilbert (U. S. Department of Energy)
Raymond F. Gildner (Cornell University)
Robert K. Goldhammer (Shell Development Co.)
Peter W. Goodwin (Temple University)
Anthony W. Gorody (Gas Research Institute)
Benito Guerrero (Colorado School of Mines)
Gigi Grace (University of Texas, Dallas)
Felix M. Gradstein (Geological Survey of Canada)
Cedric M. Griffiths (University of Trondheim)
Mark S. Hanson (Colorado School of Mines)
John W. Harbaugh (Stanford University)
Wendy J. Harrison (Colorado School of Mines)
William W. Hay (University of Colorado, Boulder)
Warren D. Huff (University of Cincinnati)
Teresa E. Jordan (Cornell University)
Christopher G.C.St. Kendall (University of South Carolina)
George deV. Klein (University of Illinois)
Dennis R. Kolata (Illinois State Geological Survey)
Robert P. Laudati (Stanford University)

David T. Lawrence (Shell Development Co.)
Ian Lerche (University of South Carolina)
Margaret A. Lessenger (Colorado School of Mines)
Laird D. Little (Colorado School of Mines)
Didier Masson (Pennsylvania State University)
Martin D. Matthews (Texaco E&P Technology Division)
Katie Joe McDonough (Colorado School of Mines)
Kazuo Nakayama (JAPEX)
H. Roice Nelson, Jr. (Landmark Graphics Corp.)
Chuck Norris (University of Illinois)
Christopher Paola (University of Minnesota, Minneapolis)
Martin A. Perlmutter (Texaco E&P Technology Division)
Van Price (University of South Carolina)
Terrence M. Quinn (Brown University)
William C. Ross (Marathon Oil Co.)
Fred Schroeder (Exxon Production Research Co.)
William J. Schweller (Chevron Oil Field Research Co.)
Keith W. Shanley (Colorado School of Mines)
Christopher A. Shaw (University of Colorado, Boulder)
Jack Shelton (Amoco Production Co.)
Rudy L. Slingerland (Pennsylvania State University)
Mark D. Sonnenfeld (Colorado School of Mines)
Mike Steckler (Lamont-Doherty Geological Observatory)
Randell Stephenson (Geological Survey of Canada)
David W. Swetland (Pennsylvania State University)
James P. M. Syvitski (Geological Survey of Canada)
Daniel M. Tetzlaff (Western Atlas International)
Tonya Thorn (Colorado School of Mines)
Julian Thorne (ARCO Oil and Gas Co.)
Harry A. Tourtelot (U.S. Geological Survey)
David W. Valasek (Colorado School of Mines)
Douglas W. Waples (Consultant & Colorado School of Mines)
W. Lynn Watney (Kansas Geological Survey)
Douglas F. Williams (University of South Carolina)
Davin N. Witter (Colorado School of Mines)
Mike Zeitlin (Texaco E&P Technology Division)

Part I

QDS MODELS IN SEDIMENTARY BASIN ANALYSIS

1

QUANTITATIVE DYNAMIC STRATIGRAPHY: A WORKSHOP, A PHILOSOPHY, A METHODOLOGY

Timothy A. Cross [1] *and John W. Harbaugh* [2]

[1]Department of Geology and Geological Engineering, Colorado School of Mines, Golden, CO 80401 USA

[2]Department of Applied Earth Sciences, Stanford University, Stanford, CA 94305 USA

INTRODUCTION

The last decade has seen the development and widespread application of geodynamic and thermal (maturation) models of sedimentary basins. More recently, sedimentologists and stratigraphers have begun to develop a variety of subbasin- to basin-scale models of sediment fill. These range from purely geometrically constrained to (almost) purely dynamic simulations of stratigraphic architecture and facies distributions. Simultaneously, models of fluid flow and fluid/rock interactions (diagenesis) in sedimentary basins are undergoing a second-generation of development. In these models fluid flow is treated dynamically with sediment fill and compaction, in contrast to previous versions where fluid flow was simulated through already formed and filled basins. Atmospheric circulation and global climate models are being applied to simulate water circulation which in turn drives sediment transport in paleooceans. Some types of structural models that were developed independently are now being incorporated within more integrated basin formation and basin fill models. Other forms of models being developed are less restrictive in their traditional disciplinary context. One example is the category of "artificial intelligence" or "expert" systems. These guide or assist users through multiple options using multiple data types in their analysis of geological and geophysical data. Another is fractal analysis which assumes that, regardless of origin, some features of greatly varying spatial scales may be represented simply as progressive spatial expansions of identical geometries. An example might be the description of large-scale reservoir heterogeneity as a multiple expansion of pore-volume/pore-throat geometry.

These developments indicate an increased interest among earth scientists to formulate more exacting and quantitative relations among processes and responses of complex natural systems. Numerical models are an efficient means for effecting this transfer from conceptual understanding to process-response simulations. This history of development also indicates a progressive expansion in the types of variables, processes and responses included in individual models; there is a trend toward developing more complete, integrated models of basin evolution. If stratigraphy is the study of layered rocks, incorporating many topics traditionally considered the domains of individual subdisciplines within the earth sciences, then we are in the midst of an evolving form of stratigraphic analysis which has been termed Quantitative Dynamic Stratigraphy.

Quantitative Dynamic Stratigraphy (1989), T.A. Cross, ed., Prentice Hall, p. 3–20.

In February, 1988, about 75 earth scientists representing a variety of disciplines, interests and backgrounds participated in the Quantitative Dynamic Stratigraphy Workshop at Lost Valley Ranch, near Denver, Colorado. The goals of this workshop were: to summarize the state-of-the-art in quantitative approaches to stratigraphic and sedimentary basin analysis; to identify major types and forms of empirical geological data that are or will be required for constructing, testing and verifying quantitative models; and, to define research directions that will significantly enhance capabilities in predicting temporal and spatial relations of sedimentary facies, stratigraphic architecture, fluid movement, and diagenesis in subsurface strata.

During the workshop, the participants formed seven working groups to address questions and make recommendations related to the workshop goals. Although there was some intentional overlap among questions discussed by the working groups, the titles of the working groups are listed here to indicate the range of issues that each considered. WG 1: Philosophies guiding QDS model development—variables, scales and scope. WG 2: Selection of variables for models of varying scales, scope and purposes. WG 3: State-of-the-art in confirming and refining QDS models with stratigraphic information. WG 4: Future directions and requirements in confirming and refining QDS models with stratigraphic information. WG 5: Interdisciplinary applications of QDS models and their requirements. WG 6: Integration, implementation, and transfer of QDS models. WG 7: Potential impact of QDS on education, science, engineering and public policy.

This paper is distilled from reports written by each working group, augmented by some perspectives that pertain to general issues regarding the philosophy, requirements and directions of Quantitative Dynamic Stratigraphy. In this paper, we summarize the philosophical background of QDS, point out some advantages and problems, forecast QDS's future, and outline what will be required if QDS is to fulfill its promise. This summary is an introduction to this volume, and to the scope, purposes and messages contained in the individual papers.

QDS's NATURE AND PROMISE

Quantitative Dynamic Stratigraphy (QDS) is the application of mathematical, quantitative procedures to the analysis of geodynamic, stratigraphic, sedimentologic and hydraulic attributes of sedimentary basins, treating them as features produced by the interactions of dynamic processes operating on physical configurations of the earth at specific times and places. For example, a QDS model might represent currents of water in sedimentary basins that alternatively erode, transport, and deposit sediment, using equations that are solved repeatedly to represent these processes, with numerical parameters to control their rates.

Although QDS normally requires the construction of computer models that attempt, through simplification of the real world, to represent processes in terms of natural laws and/ or empirical generalizations, it is a philosophically much broader and coherent approach to analysis of sedimentary basins. QDS models are not inherently better than qualitative or conceptual models. Both are sensitive to and limited by the degree to which relations and interdependencies among processes and responses are understood, and by the degree to which initial and boundary conditions and configurations are known. However, the formal description of a natural system in quantitative terms allows for a rigorous, systematic evaluation of the concepts that led to the development of the model. There are numerous advantages of QDS, as illustrated by the following listing:

- QDS allows more complete and effective integration of knowledge from previously separate disciplines;
- it requires more accurate definition of the variables involved in and their operations upon (within specified boundary conditions) the geologic system being simulated;
- it assesses or measures the degree of confidence in descriptions and predictions made by models;
- it may lead to new intuition by revealing interdependencies and feedback mechanisms among process variables and responses that are otherwise obscure or unnoticed;
- it may force the generation of new insights and explanations because model experiments may eliminate concepts or explanations that were previously accepted;
- through real-time graphic simulations, it allows users to visualize and more completely understand interactions among processes and the products produced by their interactions; and,
- it reduces time and effort constraints that manual manipulation of data impose on scientists such that more hypotheses may be developed and tested, and more insights gained about complex geologic systems.

In short, a QDS model represents any quantitative approach that allows an earth scientist to test hypotheses by interacting with a computer terminal at a speed that won't interrupt a train of thought.

In addition to these advantages and the desire to understand and formulate more exacting relations among processes and responses operating in sedimentary basins, there is another overriding impetus for QDS. This is the need to increase the predictive capability, establish confidence levels, and assess the accuracy of sedimentary basin models in general. For example, it is generally acknowledged that there are particular requirements which must be assessed accurately for the discovery of commercial accumulation of hydrocarbons. These requirements include the presence of source, seal, reservoir and conduit facies in appropriate spatial connectedness and geometrical configurations (traps), an appropriate level of maturation, and timing of migration relative to other geologic events. Despite the recognized need to assess these accurately, most geological models are nonquantitative and limited in their predictability, and almost none provides a statement about confidence levels or degree of accuracy of their predictions. Integrated QDS models provide the hope and technology to meet these needs.

The Quest for Natural Laws

QDS models attempt to represent processes in terms of natural laws and sound scientific relationships. Fluid motion in sedimentary basins can be represented by equations that incorporate conservation of fluid (mass) and momentum, relationships so well established that they may be considered natural or fundamental laws and can be employed with confidence. Other relationships, particularly those representing a distillation of empirical observations, may be less fundamental and less broadly applicable, but nevertheless useful under appropriate conditions. Equations for transport of clastic sediment are based on a century of observation and experimentation; we can select alternative equations for sediment transport along the bed of a stream, and these will differ from other alternative equations for sediment transport above the bed. These relationships are largely empirical and have distinct limits in their applicability.

QDS models generally employ mixtures of relationships that span a continuum from "fundamental" to "grossly empirical." This continuum can be divided into four categories for convenience.

1. *Fundamental laws* consist of well established relationships of universal applicability, such as the conservation laws.
2. *First-order approximations* consist of relationships of broad applicability, although their simplifications may only approximate the complexity of actual systems. The elementary diffusion equation is an example, being broadly applicable, even though actual diffusion processes may be more complex and differ according to circumstances.
3. *Empirical relationships* consist of generalizations of more limited applicability, often involving processes that are poorly understood. Functions that relate compaction of clays to overlying load and fluid expulsion are examples.
4. *Gross empirical relationships* are still more limited in their applicability, but can be useful if judiciously employed.

QDS models should be based on fundamental laws insofar as they apply or are known to operate in complex natural systems. Ideally, we could link fundamental relationships so that all major processes and responses are represented, and a QDS model might be a "correct" representation of a natural system. However, it is currently not possible to construct a model of a complex geologic system using only fundamental laws, and such a QDS model is never likely to exist. Instead QDS models employ mixtures of relationships within this continuum. For example, some sediment transport models are predicated on empirical constraints but they also conserve mass, thus linking an empirical relationship with a fundamental law. A more realistic transport model might incorporate more rigorous relationships based on fluid mechanics, but it too would still contain a mixture of relationships.

These mixtures of relationships range widely in applicability and may depend strongly on scale. Although fundamental laws are applicable at all scales, empirical relationships may be applicable only over certain ranges of scale. For example, the diffusion equation may apply fairly accurately at scales of centimeters, but less accurately at larger scales. As a goal, we should strive to incorporate relationships that are drawn increasingly toward the "fundamental" end of the continuum in our quest for models of broadest applicability.

FORWARD AND INVERSE MODELS

Two categories of quantitative models, termed forward and inverse, are applicable at all temporal and spatial scales and to any discipline for any purpose. Forward models make predictions. Inverse models solve for a forward model that best matches predictions and observations (Fig. 1).

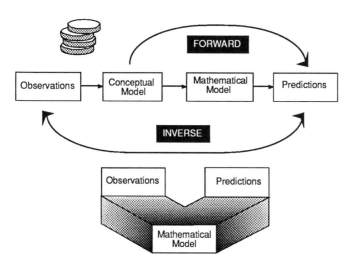

Figure 1. Relations among conceptual, mathematical, forward and inverse models.

Forward models simulate sets of processes and responses—for example open channel flow and sediment transport and deposition—that operate on a system having some specified (assumed) initial condition and configuration—for example, topography, grain-size distribution and dimension. Given an initial configuration of a sedimentary basin (normally some time in the past), a forward model will simulate the evolution of that basin and predict its appearance and character at some later time (normally the present). A model's adequacy depends upon: the degree to which relationships and interdependencies among processes and responses are understood; the knowledge of the initial and boundary conditions; the similarity between the natural system and the set of algorithms, logic statements and mathematical expressions used to represent that system; and, the influence of randomness in the simulated processes.

Forward models are commonly applied in a set of experiments in which the initial and boundary conditions and the process-response variables are progressively adjusted until the simulation matches available observations. Forward models are especially useful for making predictions beyond areas of control or data. If a model result matches the observations available, it may be an accurate model and may be used to simulate the geology between or beyond areas of control. However, because different model experiments (different combinations of initial and boundary conditions and process-response variables) can produce identical simulations, there is no guarantee that a model is accurately representing the real world, even if the simulation predicts the observed. In more formal terms, a model will satisfy the condition of sufficiency if it produces a simulation that matches observations. But the condition of necessity may not also be satisfied in the sense that it is only one of several potential models, each equally reasonable, that might produce identical simulations. Different models producing different simulations may be judged equal in their accuracy and validity even after they are compared with observations, if those observations are uncertain or otherwise less sensitive and of less resolution than model predictions.

Inverse models are the flip side of forward models. Inverse models use present-day observations (truth) to make estimates of the initial and boundary conditions and the combinations of process-response variables that operated through time to produce the observed conditions. Inversion is a mathematically rigorous process of running a forward model multiple times, each time comparing model predictions with observations. The forward model is adjusted progressively such that after each iteration the predictions more closely match the observations (i.e., the differences between predictions and observations are minimized). In addition to calculating process variables through time, inversion can incorporate geologic data to constrain the solution and calculate confidence limits. These constraints may allow accurate solutions when the response is nonunique.

Even though we might wish for simulation models that would run backward—retracing the past from information about the present—genuine process-response models will not run backward because virtually all natural processes are irreversible; water won't run uphill. Irreversibility involves a progressive increase in entropy, or degradation of energy to heat. But even if we were to devise a purely deterministic model that ignores thermodynamics and operates backward, we probably could not preserve enough information that would permit it to run backward for very long.

ATTRIBUTES OF QDS MODELS

All QDS models attempt to simulate some portion of the response to geologic processes. All are based on some combination of theoretical or empirical generalizations—or "concepts"—and commonly incorporate physical, chemical and conservation laws. Geological processes operate over a range of about twelve orders of magnitude in temporal and spatial scales. Temporal scales simulated by QDS models may range from a few seconds (saltation of a sand grain), to years (fluid flow in petroleum reservoirs or groundwater systems), to hundreds of millions of years (thermal history of a basin or simulations of continental drift). Spatial scales may range from microns (pore throats) to hundreds of kilometers (oceanic circulation, tectonic regions and basins).

An ideal basin fill model might simulate everything from the deposition of sand grains from a turbulent eddy to buckling of the lithosphere caused by sediment loading and stress changes. We cannot build such an ideal basin fill model because of our ignorance of geologic processes and responses, their complexity and interdependencies, and computational limitations. It is not even necessary to build such a model, because most applications require only that a model address particular concerns or questions and produce results at an appropriate level of detail. Thus it is possible to limit a particular model to specific scales and constrain it within the initial and boundary conditions of another model or of the real world.

Representations of Processes through Simplifications

What processes should be incorporated in QDS models and how they should be represented? Consider a model for transport and deposition of sediment on a continental shelf. All processes and responses on the shelf cannot be completely represented in any existing QDS process model. Over the shelf as a whole, we cannot represent each fluid element or each sediment grain, nor can we represent the complex motions of turbulent flow of each suspended sediment particle. Our model must be drastically simplified.

We could represent turbulent flow and the movement of sediment particles in generalized form, or we could simplify further and represent flow in nonturbulent form. Still other alternatives are available, but simplification is inevitable. The challenge is to represent the processes involved at scales and levels of detail that are appropriate to the application, with model-building effort and computational capabilities as additional constraints. But, the model should represent the natural systems in some form, incorporating important processes and major interrelationships.

There are three important elements of all QDS models that must be considered in attempting to represent the real world. First is the requirement for representing natural systems through a set of logic statements and mathematical expressions that are simplifications and approximations of the real world as it is understood. Are these assumptions and simplifications valid? Second is the appropriateness of the model dimension(s) selected to simulate the real world. A two-dimensional model (e.g., a 2-D spatial flexure model, or a 1-D spatial/1-D temporal geohistory model) may be an inherently poor approximation of a four-dimensional world and may not behave like the real world. Third is the selection of the ranges in values of variables that operate within the models. In too many instances the possible ranges in values, and the uncertainties of these ranges, are not known and have not been evaluated independently.

Definitions of Purpose, Scales and Boundaries

When a model is devised, a clear statement of the problem being addressed is required at the outset. Assumptions in the model, as well as approximations and limitations, should be clearly stated before the model's development is far along. Assumptions, axioms, approximations, constraints, equations, and logic operations should be consistent with respect to each other, as well with initial values and boundary conditions supplied to the model. Furthermore, time steps and grid spacing in numerical models should be chosen for stability, accuracy, and precision, and should be consistent with respect to the model as a whole.

Temporal and spatial scales of QDS models may range widely, and normally are dictated by the intended application. Regardless of scale, a model's expanse in space and time should be greater than the features of specific interest. This policy helps minimize problems because stable performance in specific areas of interest is more likely if they are not close to model boundaries. Furthermore, if the model is in discrete form, area or space should be represented by a grid in which the spacing of grid nodes is less than dimensions of features of interest. This permits the features to be adequately defined and reduces problems with aliasing. Similarly, time steps should be shorter than variations of interest; a model that represents the geographic distribution of sandbodies formed in response to 100,000-year cyclic variations in sea level should employ time steps much smaller than 100,000 years.

If we could completely limit our models to specific temporal and spatial scales, model construction would be less difficult. But our ability to model is complicated by the dynamic interdependency and feedback between different scales of processes. The problem is to conceptually include the dynamic responses to processes at larger and smaller scales within the model with "intelligent" approximations. Building "intelligent" approximations is difficult because it requires asking questions about the geologic record and providing quantitative explanations where our ignorance and uncertainty are large.

Consequently, it may be necessary or advantageous to create a hierarchy of scales in a single model or in a series of integrated models. For example, a "prospect" level application might incorporate two levels of scale: the evolution of an entire sedimentary basin might be modeled at a coarse scale, while the smaller "prospect" area might be nested within narrower temporal and spatial boundaries and modeled at a finer scale. Where scales are nested, the enclosing larger area dictates the initial and boundary conditions for the smaller area. Whenever multiple levels form a nested sequence, the components should be consistent at all levels, including equations, algorithms, and parameters. Scales may be dictated by available data, and ideally a model's geographic boundaries should coincide with natural boundaries if permitted by the geographic distribution of data.

Representing Responses in Complex Natural Systems

The directionality of historic time complicates our ability to model and understand geologic systems. The stratigraphic record is the result of processes acting before, during and after deposition of sediments, and upon specific configurations of basins that are products of unique histories. For example, the mix of sediment types (texture and mineralogy) supplied depends on previous tectonic, climatic and sedimentary processes in the regional setting of the basin. The physical orientation and types of grains within a particular deposit depends on the fluid flow dynamics at the time of deposition. The final thickness of a deposit depends on the volume of the initial storage space plus compactional and diagenetic processes operating after deposition. The sedimentary record is the result of a complex recursive process. It is difficult to unravel this process because we can only observe the final result.

We know from intuition, experimentation and geologic studies that the complexity of geologic processes and directionality of historic time make QDS model development difficult because so many geologic phenomena are inadequately understood. For any particular response, we may not know which processes are important, nor their relative degrees of importance, values, uncertainties, or interdependencies. We also may not know whether geologic process-response systems are linear or nonlinear. Our knowledge of feedback mechanisms between processes and responses is woefully inadequate. The probability that process interdependencies, feedback mechanisms and response nonlinearity exist in geologic systems, combined with the directionality of historic time, suggest that many geologic systems are chaotic. Consequently, traditional scientific paradigms for building deterministic models are not always appropriate; we need to develop other techniques for building models in these complex historic systems.

Because QDS models normally operate in time, they are strongly influenced by and depend upon knowledge of rates of processes and the effects of those processes on geologic systems. For example, a turbidity current flowing down a submerged slope moves at a rate influenced by the density contrast between the turbidity current and the surrounding water, the size of the turbidity flow, the angle of slope, and the density, sizes and shapes of clastic grains that it transports. The flow rate, however, affects the turbidity current's ability to erode, as does the erodibility of material on the slope. Thus, there are complex interdependencies, with the composition of the turbidity current continually changing as it erodes and deposits. Therefore a simulation model of such a flow will be rate dependent, and will require estimates of rates that may be difficult to obtain and apply.

A century of experimentation and observation has yielded a storehouse of information about rates at which clastic grains are transported. Nonetheless, applying sediment-transport rates measured in flumes to transport rates of turbidity currents is difficult, and applying observed erosion rates to sediments on the sea floor is even more difficult. For example, cohesiveness affects the erodibility of a sediment, but erodibility is also influenced by the cover of benthonic communities and by compaction and cementation, so that measurements of erosion rates must be applied cautiously. Yet, the appropriate rates of processes and responses—in this case erosion, transport and deposition—must be specified in QDS models.

Other rate information is needed, including carbonate production rates as a function of temperature, water depth, chemical and nutrient concentration, presence of suspended detritus, and biogenic processes. Such rates are even more difficult to measure, especially at geologic time scales, and involve complex interdependencies. Some of the same difficulties affect estimates of evaporite deposition rates, although the processes may be less complex.

Knowledge of tectonic rates and responses are also important. For example, isostatic compensation is critical in some QDS process models, and rates of lag and rebound of the crust and mantle may be important, although difficult to estimate. Interactions at plate boundaries, including rifting and subduction, affect sedimentation rates and influence the geometry of sedimentary sequences. The literature on tectonic influences is voluminous, and the challenge is to incorporate appropriate tectonic ideas in formal QDS process models, particularly in three-dimensional models where both lateral and vertical motions are represented.

Rates of post-depositional sedimentary influences are also required in some QDS models. Early compaction of clay and silt affects depth and slopes of the sea bottom, which in turn affect erosion, transportation and deposition. Deep-burial compaction also influences sedimentation, and strongly affects fluid migration and diagenesis. When compaction occurs, the expelled pore fluids affect diagenesis and influence tectonic behavior, such as displacement rates of faults where friction is affected by pore fluids. Thus, rates and interdependencies for compaction, expulsion, and diagenesis should be represented in QDS models, particularly since migrating pore fluids affect permeabilities through solution and cementation, in turn influencing migration and entrapment of hydrocarbons.

Representation of Three Spatial Dimensions

Most traditional geologic displays, such as maps and cross sections, are essentially two-dimensional even though the features portrayed exist in three spatial dimensions plus the dimension of time. QDS modeling is rapidly progressing to three spatial dimensions; dynamic simulation models are now operative that represent sedimentation processes in three dimensions and create three-dimensional bodies of sediment.

Three-dimensional QDS process models can simulate three-dimensional stratigraphic sequences that predict distributions of sedimentary facies and structural features. These predictions are useful in oil and gas exploration and development, in petroleum reservoir management, and in groundwater and waste management studies. The usefulness of 3-D predictions, however, depends upon the model's performance and its scale and level of detail, posing major dilemmas.

Suppose the area of interest, such as a petroleum or mineral prospect, spans four square kilometers (two kilometers by two kilometers), and the prospect occurs in a sedimentary basin that is 200 km in length and breadth. How much of the basin's area should be represented, and how much detail should be represented in the prospect? There is a 10,000-fold difference in the areal expanse of the play with respect to the basin.

Next, suppose that the prospect is represented by a 3-D block of layered rocks in the form of a cube (the block is 2 km deep), with grid nodes every 50 m in all three directions, the grid thus containing $41 \times 41 \times 41 = 68,921$ nodes. A grid of this size is not much of a computational problem. But, if we extend such a grid maintaining the same spacing to the entire basin and retaining the 2-km depth, the larger grid contains $41 \times 4001 \times 4001 = 656,328,041$ nodes, posing a formidable computational challenge even for today's supercomputers. There are two alternatives. A single grid larger than the play but smaller than the basin could be chosen, or two nested grids could be employed, with a coarse grid for the basin and a finer grid for the play. The choice depends partly on computing facilities.

Equations of flow can be represented in three dimensions and can be linked with 3-D stratigraphic simulations, thus providing a grid in which permeability variations can be coupled with flow velocities. Such models have potential for predicting fluid migration routes and entrapment sites for oil and gas. Estimates of rates of expulsion of hydrocarbons from source beds could be linked with migration paths traced through 3-D networks, permitting oil and gas pools to be forecast if the stratal architecture is known in sufficient detail in three dimensions.

Migration paths may be complex, depending on structural configuration as well as variations in sedimentary facies. For example, relatively small variations in sediment texture and fabric may have large influence on variations in permeability, and in turn profoundly affect hydrocarbon migration. Thus, flow simulations through sedimentary sequences, where permeability variations are represented as part of the 3-D stratal architecture, should provide new tools for analyzing fluid movement in the subsurface.

Scales of 3-D flow simulations could range widely, extending over short distances near wells, to reservoirs or regional aquifers, and finally to entire sedimentary basins. Hierarchical nested sequences incorporating progressions of scales may be required.

A goal in the next five years is to develop three-dimensional process models that will represent all major processes operative in sedimentary basins. These processes include erosion, transport, and deposition of clastic sediment, carbonate and evaporite sedimentation, wave activity, redistribution of sediment by wave-induced currents, basin circulation, eustatic changes, compaction, pore-water expulsion, diagenesis, hydrocarbon migration, isostatic compensation, and tectonic activity including faulting and crustal warping. Such models will make large computing demands, and machines operating at 10 to 100 million instructions per second are required, with still faster machines being desirable. Large computer memories are also desirable. Three-dimensional arrays containing as many as one billion elements are desirable for some QDS applications.

Three-dimensional QDS modeling also poses major challenges for graphic display development. Traditional 2-D display procedures are inadequate to represent results of dynamic 3-D models. Fortunately, a new generation of computer-driven graphics workstations is now on the market and is well suited for 3-D geological representations. These workstations employ color, and 3-D representation is accommodated using translucent displays that permit the user to see through multiple layers. Alternatively, workstations employ "slicing" techniques in which 3-D objects are rapidly sliced through and displayed in series of 2-D cross sections. Stereoscopic viewing is also employed.

Model Building Can Enhance Insight

Model building in itself can enhance insight and guide subsequent inquiry. Simulation models can be regarded as experiments in which ideas are defined and progressively redefined, yielding successively improved hypotheses and understanding. Mathematical models require that concepts be expressed as equations and logic operations, generally in computer programs. These tasks add insight because they encourage representation of processes and responses in clear and unambiguous form. Quantification itself, such as adherence to the conservation laws, has strong a influence by requiring that numerical accounting procedures be incorporated in models.

Operation of dynamic models often yields surprises because their performance is different than expected. For example, complex natural systems may be represented with sets of equations that are nonlinear. When run, the models may respond unpredictably, often leading to results that are not intuitive, and their performance cannot be represented by simply "summing" their components. Simple cause-and-effect relationships may not exist, and the value of such models is that they provide insight into interdependencies between components that might otherwise be obscure.

Model development and model operation are elements of the scientific method. Most natural systems, however, are so complex that models always remain interim in form, with any specific model falling short of adequate representation of the natural system. An important role of models, therefore, is to help formulate ideas and stimulate improvement through progressive revision and retesting, with consequent discarding of obsolete ideas and models.

VERIFICATION OF QDS MODELS

These attributes of QDS models demonstrate the necessity of testing, calibrating and validating models at all stages of their development. In assessing the validity of models, the accuracy and uncertainty of the measured values also must be determined, if comparisons with models are to be useful. In some instances, models make predictions or provide results that are identical in type and form to some property that may be measured directly; for example, sediment thickness or lithology. However, in many other instances models make predictions or provide results in a form that is not directly comparable to an observable property; for example, water depth. In these cases, the predicted value must be compared with a measured value of one property that is considered an appropriate proxy for the modeled property; for example, paleontologic or facies information as a proxy for water depth.

Steps in Verification

Verifying QDS models involves five major steps. First, a conceptual model that forms the base for a subsequent mathematical model must be explicitly defined.

Second, the mathematical model must be explicitly defined. Equations and logic operations in computer programs that represent the model must be internally consistent and must parallel the conceptual model on which they are based. The mathematical model can be no better than the conceptual model on which it is based.

Third, when the mathematical model is operative, sensitivity tests must be performed and the model's responses observed while a progression of modest changes are made, modifying only one input variable at a time. If more than one variable is simultaneously changed, the influence of one variable may not be distinguishable from another. Sensitivity tests may reveal interdependencies among processes and responses that were not previously obvious, may provide insight to the model's behavior, and may suggest subsequent modifications to the model. The model can then be progressively modified until it appears to perform satisfactorily over ranges for which the governing input parameters have been tested. When a model has been expanded by incorporation of a new feature, sensitivity tests should be performed with the new feature's influence set at zero, with the intention of ensuring that the expansion itself has not altered the model's performance. Although sensitivity analyses may provide enhanced insight and are an important step in the verification process, satisfactory performance does not constitute verification.

Fourth, models should be adapted to actual geological situations. The user must provide assumptions about the geological setting in which the model is to operate, including the model's geographic boundaries, starting conditions, and rates and effects of processes. For example, a basin experiment might involve specifying the initial topography and/or bathymetry, discharge rates for water and sediment in streams entering the basin, and wind or water current velocity and direction. The possible combinations are unlimited, and numerous experiments may be needed before satisfactory results are obtained. The testing process involves trial-and-error, with repeated comparison of the model's response with the actual situation. Attaining a "satisfactory" comparison still does not complete the verification, for it is unlikely that a unique set of assumptions will yield successive sets of satisfactory approximations.

Prediction and testing constitute the final step. If the model yields predictions in time or space that are geologically reasonable and that match observations, then "verification" for sufficiency—but not necessity—has been attained. The verification process is never final. An essential objective of quantitative modeling is to deepen insight into the underlying conceptual model; model-building is simply a facet of the scientific method.

Stratigraphic Data Required for Verification

Testing and verification of QDS models will require collection of new types and forms of stratigraphic data, as well as collection and reevaluation of existing data.

Three-dimensional QDS models will require information about actual stratigraphic sequences in appropriate three-dimensional detail for comparative purposes. Detailed networks of stratigraphic correlations will be required, which can be based on outcrops, borehole logs and/or intersecting seismic sections. These correlations can be interpolated to yield correlation networks that are fully three dimensional.

Relatively precise stratigraphic correlations are needed. A QDS model synthesizes a stratigraphic sequence in which time-rock relationships are known "perfectly," with detail as fine as specified by the user. For verification, it will be necessary to have information about actual stratigraphic sequences in which the time-rock relationships are known at approximately equivalent detail. Time-rock relationships may be established by sequence-stratigraphic relations, isotopic or faunal ages, marker beds such as bentonites, or other methods that permit temporal resolution greater than is normally attempted and attained. Because QDS models simulate sedimentary facies distributions within discrete time boundaries, the volumes, frequencies of occurrence, and interconnectedness of facies within the previously constructed time framework must also be measured.

Some types of studies will be required to provide information about process variables and responses for QDS models. For example, studies of cyclically bedded stratigraphic sequences with distinctive "rhythms" or "signatures," potentially interpretable with Milankovitch's theories, are needed. Similar studies should be conducted on sequences that exhibit cycles potentially correlatable with worldwide cyclic fluctuation in sea level, such as those represented by Exxon's sea-level curves.

Studies of stratigraphic frameworks at small spatial and temporal scales are needed. These should include detailed analyses of contemporary depositional systems, with emphasis on the rates and frequencies of processes that operate and the sediment bodies that accumulate within them. Since QDS models can yield detailed simulations over time spans as short as a few hours or days, studies of actual deposits forming in equivalent intervals of time are needed. These studies would assist in simulation reconstructions where events are represented in detail over very short periods. For example, studies of stream channels and bars are desirable, as are experimental flume studies; these should be linked with detailed computer simulations. Such studies can also yield dimensions and configurations of individual depositional bodies, as well as sedimentological information including grain size, sorting, bedding, and structural orientation. Fine-scale frameworks in Holocene deposits, such as those of the Mississippi Delta which include detailed correlations based on cores and ^{14}C dates, also can provide linkages between processes that can be observed directly, or linked with historical changes in the past two centuries, including major storms and hurricanes. Fine-scale time frameworks for modern carbonate and evaporite environments are also needed, and studies could be extended in classic regions in Abu Dhabi, Belize, and south Florida. Such studies should emphasize the geochemical and biological aspects—as well as other sedimentologic aspects—of these environments.

POLICIES FOR COMPUTER PROGRAMS

Computers and programs for academic use should be interactive and should graphically display simulation results. Appropriate computers, display equipment, and information links already exist in many geoscience departments, and suitable computer programs are also available, although modifications may be necessary for instructional purposes. New programs are continually developed and many are publicly available, with certain institutions and journals serving as clearinghouses for distribution of programs at little cost.

Programs should be modular, generally consisting of three fundamental components: (1) graphic display (including graphic input and output), (2) numerical calculations, and (3) database manipulation. QDS model code should be structured to reflect these three components, which can be represented by self-contained routines called by a master program or by other subroutines. Within each component, the program code can be further separated into functional subroutines, with each module in the hierarchy fully documented with respect to function, reference, input variables, output variables and any special considerations.

QDS programs should be portable. Networking, portability, data transfer, and model implementation are enhanced by use of common operating system protocols. UNIX is the only current operating system that can be implemented easily and independently of vendor sources; its use is recommended. FORTRAN is preferred for numerical computations, and "C" for graphics and symbolic manipulations, including artificial intelligence applications. These languages are appropriate to problems encountered in QDS models should be widely available in the foreseeable future. FORTRAN continues in wide use and provides a large library of subprograms. "C," although less well known, is growing in use and support, and it is suited for bridging operating system protocols of the future. Where other languages are used, the code should be particularly well documented to facilitate portability and translation.

QDS MODELS IN BASIN ANALYSIS

Many types of basin models have been developed during the last two decades. Most originated in separate disciplines within the earth sciences and were designed to simulate at varying scales the processes and responses operating in natural systems of interest to those disciplines. Only recently have there been attempts to integrate models that cross disciplinary boundaries. The three traditional approaches to modeling the rock properties of sedimentary basins are generally known as geodynamic (including thermal and maturation models) or basin-scale models, sedimentation or small-scale models, and stratigraphic or intermediate-scale models. Other approaches to modeling are concerned with fluid circulation. Examples include atmospheric and oceanic circulation, and fluid flow and fluid/rock interactions within already formed basins—or more recently coupled dynamically with the development and fill of basins.

Sedimentation models

The goal of sedimentation models is to dynamically simulate the generation, transport and deposition of sediment in space and time. These models use physical laws or empirical generalizations, combined with conservation laws, to move and deposit sediment dynamically under uniform or nonuniform flow conditions across dynamically changing depositional surfaces. Current sedimentation models are forward models. Most sedimentation models are concerned with time scales up to 1,000 to 20,000 years, and with spatial scales on the order of tens to thousands of meters. Most simulate sedimentation in topographic basins (i.e., sediments fill an already formed hole), in which case they normally are not coupled with geodynamic models. A few simulate sedimentation in stratigraphic basins, and are coupled in some form with geodynamic models. Some are coupled with atmospheric and/or ocean circulation models which provide the driving mechanism for the transporting fluids. In general, sedimentation models have not been tested against observational data, although some notable exceptions are presented in this volume.

It might appear that sedimentation models provide a more accurate representation of the real world because they are based on more fundamental laws and are concerned with simulating natural systems as those systems actually operate. However, as previously discussed, even modeling the transport of sedimentary grains is an empirical generalization of the ways that turbulent or laminar flow move and deposit those particles. Sedimentation models are neither more fundamental nor better than other types of forward basin models. They are just more appropriate for particular questions and scales of investigation.

Geodynamic Models

Geodynamic models deal with large-scale thermal and mechanical deformation of the crust and lithosphere that produces uplifts and holes to put sediment in. The models are either kinematic or dynamic, and both forward and inverse models are in use. They use mechanics to simulate deformation of crust and lithosphere in the presence of a given stress and temperature field. The properties (e.g., strength, elasticity, thermal conductivity, rheology) and distributions (e.g., crustal thickness and layers) of materials in the crust and lithosphere are either assumed or inverted for by best-fit solutions between data and forward models. Mathematical-mechanical analogs are commonly applied (e.g., the lithosphere as an elastic beam; mantle deformation as fluid flow within a pipe). Most geodynamic models are concerned with time scales of millions of years, and with spatial scales on the order of tens to hundreds of kilometers.

Of the types of basin models, geodynamic models—especially when thermal and maturation models are included in this category—are among the most extensively developed and most widely applied. They are used so routinely that often their results are presumed accurate and precise. Yet, like all models they have their own limitations of accuracy, predictibility and resolution.

Stratigraphic Models

Stratigraphic models use theoretical and empirical approximations from the other two types of models in combination with empirical generalizations to explain or predict the distribution and volumes of lithologies within a basin. They are intermediate in temporal and spatial scales, normally on the order of 1 to 10 million years and a few to tens of kilometers in distance.

At this stage in their development, most stratigraphic models assume that sedimentation and geodynamic equilibrium conditions are reached between time steps of the model. Most model the time-varying response of equilibrium conditions to basin-scale controls of climate, eustasy, tectonics, sediment supply or production, and isostatic compensation. Stratigraphic models contain many of the assumptions, limitations, and potential inaccuracies of the other models. The assumption that equilibrium conditions of different real-world attributes approximate a dynamic system may not be correct. Consequently, information and approximations taken from sedimentation and geodynamic models and used in stratigraphic models must be continually analyzed for conditions of necessity and sufficiency for desired accuracy and precision.

Most often sediment transport and deposition are not simulated dynamically as in sedimentation models. Rather, an empirical generalization, geometric constraint, or inference from sedimentation models—or a combination of these—is employed to constrain the position, geometry and lithology of sediment eroded, transported and deposited during each time step of the model. Lithology estimations may be determined by some type of empirical calibration (e.g., water depths <10 m are sands and water depths >10 m are muds), or through combined diffusion equations that transport two or more synthetic "grain-size populations" at differing rates. Stratigraphic models rely heavily on other empirical generalizations for developing algorithms designed to simulate the real world. Examples include carbonate productivity as a function of one or more variables (e.g., water depth, turbidity, nutrient level), and response of river morphologies and grade to base-level changes.

These inherent disadvantages are advantages in another respect. Because stratigraphic models are partially based on empirical generalizations, they are likely to produce reasonably good approximations of the real world even if they do not simulate all the processes and responses that operate at all scales in the formation and evolution of sedimentary basins. Because stratigraphic models are intermediate in scale to geodynamic and sedimentation models, they are gradually incorporating more of the dynamic elements of the other two. This means that an increasing number of process-response variables will be incorporated in stratigraphic models, and that they will combine some of the best attributes of the other two model types. Another advantage of stratigraphic models is that they make more specific predictions that can be evaluated at a higher resolution and with a greater volume of information. For example, geodynamic models do not make predictions about lithologies or stratigraphic architecture, only about stratal thicknesses. However, as more aspects of geodynamic and sedimentation models are incorporated into stratigraphic models, stratigraphic models will make predictions about lithologies, facies volumes, facies connectedness and stratal architecture, as well as stratal thicknesses.

Outlook for QDS

Potential Applications of QDS Models

The preceding cursory review indicates that QDS models have contributed to understanding the development and evolution of sedimentary basins, particularly at large scales, but that additional potential applications await improvements and new developments in QDS models. Several examples, listed here, were discussed by the working groups and in general sessions.

Basin-scale QDS models that increasingly integrate process-response components of geodynamics, thermal history, tectonics, structure, sedimentology, stratigraphy, fluid flow and fluid/rock interactions is an area of active development. These models will allow users to adjust basin formation and basin fill parameters, such as original topography/bathymetry, sediment influx rates, sea level fluctuations and tectonic influences, and will include simulation of hydrocarbon generation, migration and entrapment in addition to other geological processes. Through such integration of process variables at widely differing scales, we will learn much about the interdependencies and feedback mechanisms among these and other geologic variables. These models should find a ready, interested audience in petroleum and mineral exploration industries.

QDS models of considerably more detail and larger scale, that are appropriately calibrated with the real world, will find numerous applications in oil and gas field development and reservoir management. In contrast to most existing QDS models, these models must portray in three dimensions the volumes, dimensions and geometric arrangements of fluid flow units. Thus, another step in model development and calibration will be required to translate sedimentologic or stratigraphic facies units into petrophysical units that represent strata as an assemblage of pathways, barriers and retardants to fluid flow. To be useful, they must be compatible with and link with reservoir fluid flow simulators.

Similar approaches and requirements also will have direct applications in the fields of groundwater hydrology, contaminant transport, and waste disposal and management. QDS procedures could provide 2-D or 3-D simulations of aquifer, aquiclude and aquitard positions and dimensions, along with estimations of vertical and lateral transmissivities among these units. These models of the "plumbing system" could be linked with dynamic flow models to predict the rates and directions of movement of subsurface waters.

QDS models could be applied to coastal and fluvial engineering problems of stabilization of banks along rivers and shorelines along oceans and lakes. Process models incorporating wave and longshore currents could provide experimental means of analyzing alternative engineering solutions, such as construction of groins and sea walls, and pumping of sand to enlarge beaches. Effects of storms and abnormally high tides also could be simulated. Similarly, the effects of changes in sediment load of streams or shorelines could be predicted, such as produced by damming of streams with consequent reduction in the downstream sediment load.

Impact on Education

QDS will challenge the educational system. Building quantitative models demands skills in mathematics at advanced levels. QDS promotes in geology the historical trend of increasing quantification in all sciences. In addition to mathematics, QDS requires the integration of virtually all geologic fields and makes obvious the need for more rigorous and broad training in the earth sciences and ancillary disciplines. In turn, QDS will be strongly impacted by the products of the educational system; to construct better models, we need to educate better modelers.

QDS has the potential to stimulate and excite students at most levels in education. If simulation models were incorporated in academic courses, geologic processes would be demonstrated in action and students could explore the complex interdependencies among processes and responses. Students would also develop enhanced appreciation of the sensitivity of geologic processes to parameters that regulate them, emphasizing the complexities in analyzing geologic environments, both in the present and the geologic past. Models provide a flexible and powerful mechanism to test, modify and determine limitations of concepts and hypotheses presented in coursework and reading. Simulation would also aid in combining concepts from diverse courses, such as structural geology, solid earth geophysics and sedimentology, which ordinarily are treated separately, but which could be linked with experiments that simulate progressive deformation and fill of sedimentary basins.

Acceptance of QDS

QDS will be accepted to the extent that it is perceived to be useful. Resistance will arise due to lack of understanding, distrust of mathematical procedures, and the belief that numerical models—because they are simplifications of real-world processes and responses—cannot adequately represent the real world. This resistance can be overcome. Training in quantitative procedures should begin at the undergraduate level, so that mathematical "blind spots" are reduced. Articles in major journals could provide exposure to QDS's basic premises, particularly if coupled with case studies of successful applications, which should be publicized and fully described.

Perhaps the biggest danger of all is the risk of overselling the capabilities and applications of QDS models. Only through constant testing and verification procedures, designed specifically to establish the accuracy, confidence limits, resolution and conditions under which a particular model applies, will QDS avoid the almost inevitable crash and disappointment that shadow new approaches to old problems.

ACKNOWLEDGMENTS

We thank all those who attended the QDS Workshop for their enthusiastic participation and their concerted efforts to share their ideas and reach consensus on many of the issues presented in this summary paper. We also thank the sponsors of the Workshop—the American Association of Petroleum Geologists, the Gas Research Institute, the National Science Foundation, the U.S. Department of Energy, and the U.S. Geological Survey—for making it possible.

2

PHILOSOPHIES AND STRATEGIES OF MODEL BUILDING

Ian Lerche
Department of Geological Sciences, University of South Carolina, Columbia, SC 29208
USA

ABSTRACT

The framework of sedimentary basin analysis is used as a vehicle to illustrate the impact of quantitative model building versus data control. Three strategies are examined: "forward" models, "inverse" models and "pseudo-inverse" models. Pragmatic strategies for real case histories are considered using the interwoven aspects of dynamical burial histories, thermal histories, and hydrocarbon generation, migration and accumulation histories. These examples show that the problems of resolution, uniqueness, precision and sensitivity of any quantitative model are tightly connected to the three factors of model assumptions, model parameters, and quality, quantity, noise and distribution of input data. These problems are not unique to quantitative stratigraphic models within basin analysis, but occur in all branches of science. The philosophy and appropriate strategy for examining the sharpness of Occam's razor within the Baconian perception of science for any model in any discipline need to be tailored to allow for the generic problems of information determination catalogued above.

PREAMBLE

The essence of science is model building to account for known data and observations, and to predict new phenomena which are then measured and compared with the model's predictions. When a model ceases to predict observed phenomena, it must be modified, abandoned or replaced such that the perception of science is adjusted to reality. But within this quintessential framework, the philosophies and strategies of model building and model testing are as broad and varied as circumstances and imagination allow.

There is a general understanding that causative agents (sources) are modulated by dynamical processes (filters) to produce end products (outputs). In modeling, the specification of the sources and filters determines the outputs. We mortals have a brief life span compared to the geological time scale, and it is rare when we can observe a system from beginning to end and thus know unequivocally what relations exist among sources, filters and outputs. Instead, usually we are restricted to measuring outputs and inferring filters and sources.

This limitation imposes severe difficulties and constraints on model development. Each model of sources and/or filters depends upon intrinsic assumptions and upon the functional relationships that are assumed to connect one set of physical variables to another. In addition, the functional dependencies among intrinsic and extrinsic variables and parameters are often uncertain, yet they must be provided for or adjusted in models. Determination of these dependencies is one of the major goals of models. The uniqueness, resolution, and sensitivity of a model relative to other competing models then must be assessed by comparing their output behaviors. But effective comparisons of outputs are limited by measurements of resolution, noise, sampling, uniqueness and reproducibility.

Pragmatic strategies for assessing model veracity must be tailored both to the proposed sources and filtering behaviors as well as to the quality, quantity and limitations of data available at the time the model is developed. Various observational and theoretical constraints (for example, requiring that positive variables remain positive, or that mass, momentum and energy be conserved), are often effective aids in designing model strategies.

This paper illustrates by examples the philosophies and strategies that underpin model building. The discipline chosen for illustration is sedimentary basin analysis, although the arguments that are advanced apply equally to any scientific discipline.

INTRODUCTION

Prediction of the amounts and locations of oil and gas in sedimentary basins is the primary goal of sedimentary basin analysis from an oil industry viewpoint. To accomplish this goal we require knowledge of three major components of basin evolution: a dynamical burial history of sediments; a thermal history of the basin; and a hydrocarbon generation, migration, and accumulation history. Unfortunately, we can only observe the end result of the interactive processes that contributed to the basin evolution and, even then, only with limited resolution and limited quality and quantity of data. Our task is to invert observations made today to obtain the dynamical and thermal evolution of the basin. A better determination can then be made of where and when hydrocarbons were produced in the basin, and where they might be now.

Several factors limit the precision of any inversion scheme. Irrespective of either model behavior or intrinsic assumptions from which a model was derived, measurements of input data are limited in their resolution, accuracy, precision, and uniqueness. Even if measurements of input data were perfect, requirements of finite spatial and temporal sampling (discrete samples) imposed by the models further limit resolution.

If we ignore these limitations of data input quality and sampling, there remain three additional sources of ambiguity that limit precision of inversion models. The first includes all intrinsic assumptions inherent to a class of behaviors. As examples, a one-dimensional model is inherently a poor approximation of the real four-dimensional world and may not behave like the real world; an assumed constant, equilibrium-balanced, conductive heat flux ignores likely changes in heat flow or thermal gradients through time; a time invariant porosity with depth function ignores geographic, stratigraphic and temporal changes in subsurface fluid pressures, and therefore porosity, through time.

The second source of ambiguity is the type of behavior proposed or assumed within an intrinsic assumption class. For example it might be assumed that the paleoheat flux of a basin has the particular form of a McKenzie (1978) stretching model with three parameters: time of onset, magnitude "jump" at onset, and thermal decay time scale. Using this assumption, a pseudo-inverse procedure may be employed to determine a value or a family of values for each parameter as precisely as possible. This is termed a pseudo-inverse method because it posits specific forms of the functional behavior. Other models with different functional behaviors and a similar number of parameters might provide equally good fits to the data, and it would not be possible to discriminate among conflicting model behaviors.

Within an intrinsic assumption class the third source of ambiguity arises from a true inverse procedure in which the functional behavior of a required quantity is determined (rather than assumed as in the pseudo-inverse situation) by direct inverse techniques applied to the data. True inverse techniques also must determine the degree to which a functional behavior can be resolved relative to "noise" in the system.

In addition to these philosophies two other end-member classes are available in forward models. One could build an absolutely determined model and calculate a quantity of interest at the present day that evolved from assumed initial conditions. To assess the validity of the forward model, the calculated quantity is then compared in one of two ways with measured values. First, the calculated quantity may be compared directly with observed data of comparable type (e.g., stratal thickness, lithology or porosity). Alternatively, the calculated quantity may be intrinsically unobservable in which case an extra postulate of the quantity's relation to an observable must be made which can never be proven (e.g., TTI to vitrinite reflectance). In both cases, the forward model may be judged acceptable or not depending on how close the calculated quantity matches the observed data. However, even if the calculated and observed are close and the model is judged sufficient, the condition of necessity is not also necessarily proven—that is, the model may be inaccurate (a different model might be a better one) and its calculated quantity may be only fortuitously similar or identical to observed data.

The next section illustrates some of the varying results that arise from applying the above philosophies. We cannot examine all of the consequences, but a representative set will be given to illustrate the consequences of a particular action.

MODELS AND CONSEQUENCES

The requirement of an accurate burial history is interwoven in a fundamental way with all other aspects of basin analysis. For example, the paleotemperature history of a basin is influenced by the paleoheat flux, thermal conductivity and burial depth with time. Both thermal conductivity and burial depth are porosity dependent, and determining porosity variations with respect to depth and time is an essential step in calculating the burial history. Modeling hydrocarbon generation, migration and accumulation depends not only on a chemical formulation of kerogen breakdown, but also on reconstructing the evolution of paleotemperature, pathways for migration, and sealing conditions during a basin's history. These latter three factors also depend on the sedimentary burial history for their resolution.

Burial History in 1-D

A TRUE INVERSE PROCEDURE: ISOSTATIC RECONSTRUCTION   Geohistory (Fig. 1) and burial-history plots (Fig. 2) can be used to predict the age and former thicknesses and positions of stratal units (van Hinte, 1978), estimate the amount of material removed at unconformities by erosion (Fig. 3), and estimate the rate of sediment accumulation as a function of time (Fig. 2). In addition, stratal thicknesses can be used to predict thermal history either from backstripped subsidence or direct thermal indicators (see next section on Thermal History).

The data needed to construct burial curves that model changes in thickness (as in the uncorrected subsidence using present thickness curve $_uR_s$ of Figure 1) are the numerical geologic ages of the layers, their present depths, and the water depth of each layer at its time of deposition. The burial-history plot (Fig. 2) is obtained by linearly interpolating between the depth of the top layer at the present time and the time at which this layer was at the surface. When the top layer is reconstructed to its former position at the surface, and ignoring compaction effects, all lower layers move up at the same rate and their depths are recalculated. The process is repeated until all layers have been moved sequentially to the surface. This is the backstripping process.

In calculating burial histories, it is assumed that all layers were deposited at sea level. By contrast, geohistory calculations incorporate information about paleowater depths of each layer. The geohistory plot (Fig. 1), a plot of the depths of layers with respect to sea level, modifies the burial-history plot by adding the paleowater depth of each layer at the time of deposition.

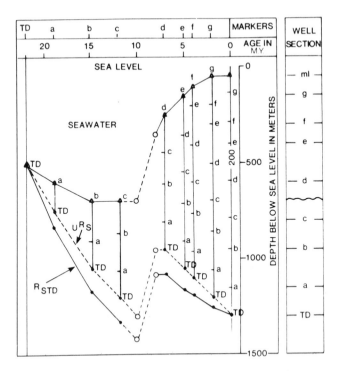

Figure 1. Geohistory diagram for a hypothetical well. Upper curve shows water depth history for location; lower curves shows subsidence and uplift at that location (curve $_uR_s$ = uncorrected subsidence using present thickness; curve R_s = true subsidence using restored thickness). Times TD through c represent period of tectonic subsidence; c through g represent period of uplift (modified from van Hinte, 1978).

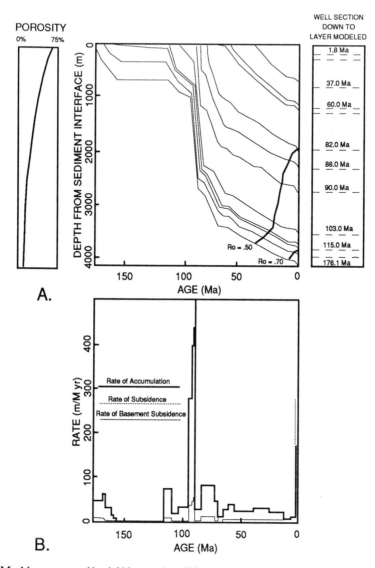

Figure 2. Machine-generated burial history plot of Norway well. $R_o = 0.5$ and $R_o = 0.7$ curves represent Waples' thresholds for onset and maximum generation of oil, respectively. Rates of sediment accumulation, subsidence, and basement subsidence as a function of age are also shown.

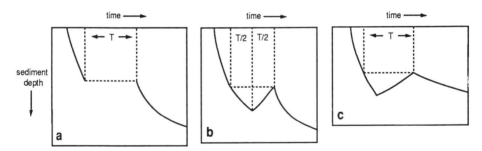

Figure 3. Geohistory diagram for hypothetical well with unconformities handled in different ways. A. Unconformity treated as a nondepositional, hiatal surface without erosion. B. Unconformity treated as a surface representing equal time for deposition and erosion at a fixed rate (taken in this case as the rate prior to the development of the unconformity). C. Unconformity treated as mass deposited at pre-unconformity rate followed by equal mass eroded at post-unconformity rate.

Calculations of burial history and geohistory may be refined by incorporating compaction history, or the changes in rock porosity with respect to depth and time. Van Hinte (1978) illustrated one modeling approach for deriving a compaction history by considering the rate of sediment accumulation, R, with respect to time:

$$R = \frac{T_p(1 - \phi)}{A(1 - \phi_0)} \tag{1}$$

where T_p is the thickness at present porosity (ϕ), A is the time duration, and ϕ_0 is porosity at time of deposition. To model layer thickness as a function of time, it is necessary to model the change in porosity as a function of time or, equivalently, as a function of depth. Van Hinte (1978) illustrated a procedure whereby observed present porosities are calculated backward to arbitrarily chosen initial porosities.

In the absence of porosity data from wells, phenomenological models can be used to relate porosity to depth or load for various lithologies. One such model proposed by Sclater and Christie (1980) supposes an exponential decrease in porosity with depth. The prefactor and exponential decay constant are lithology dependent. Even though actual values of porosity are variable and depend on lithologic variations, effects of overpressure or undercompaction, diagenesis, or a combination of these, they generally decrease with depth. Based on the data of Athy (1930) and Hedberg (1936) for normal pressures, porosity (ϕ) can be represented by

$$\phi = \phi_0 \, e^{-cz} \tag{2}$$

where ϕ_0 is the surface (z = 0) porosity. The constant (c) must be assigned to each lithology. A table of values for ϕ and c is given by Sclater and Christie (1980, Table A1a). Values of ϕ_0 and c from borehole porosity measurements can be obtained by least squares fitting techniques.

An alternate phenomenological model of porosity versus depth was proposed by Falvey and Middleton (1981). In this model, porosity decreases with increasing load on a layer. Falvey and Middleton (1981) argued that in most cases exponential porosity-depth relationships do not correspond to data measured from shallow depths. They formulated a relationship to match empirical observations by assuming the incremental change in porosity ($d\phi/\phi$) is proportional to the change in load (dL) and the void ratio (e). Their expression is

$$\frac{d\phi}{\phi} = -\,k\,e\,dL \tag{3}$$

where $e = \phi/(1-\phi)$, $dL = (1-\phi)\,dz$ with dz equal to the change in depth of a sediment layer due to the change in load, and k is a lithology-dependent constant. Equation (3) has the solution

$$\frac{1}{\phi} = \frac{1}{\phi_0} + k\,z \tag{4}$$

where ϕ_0 is the initial uncompacted porosity and z is the depth.

Well data may be corrected for compaction using Equations (2) and (3) in the following manner. First, the total amount of solid material (h_s) in an interval (z_1 to z_2) is given in the Sclater-Christie model by

$$h_s = \int_{z_1}^{z_2} \left[1 - \phi(z)\right] dz \tag{5}$$

that is,

$$h_s = (z_2 - z_1) - \left(\frac{\phi_0}{c}\right)\left(e^{-c\,z_1} - e^{-c\,z_2}\right) \tag{6}$$

By knowing that the top of a stratal unit was on the surface at a previous time, where $z_1 = 0$, its base may be calculated by solving Equation (6) self-consistently. Similarly, depths and thicknesses of all lower layers can be obtained sequentially and self-consistently. A similar equation applies to the Falvey-Middleton analysis:

$$z_4 - z_3 = h_s + \left(\frac{1}{k}\right)\ln\left(\frac{1/\phi_0 + k\,z_4}{1/\phi_0 + k\,z_3}\right) \tag{7}$$

Given the compaction parameters and values of depth, age, and paleowater depth, a computer program can be written to plot burial-history and geohistory curves such as shown in Figures 1 and 2.

Unconformities within a stratigraphic section disrupt any regular porosity versus time or depth model, regardless of model assumptions or procedures. There are at least three possible ways to model unconformities (Fig. 3). First, they may be considered as a depositional hiatus. Second, they may be treated as a period of deposition followed by erosion in which the thickness of eroded section and the absolute age marking the beginning of erosion are determined from a knowledge of local geology (see van Hinte, 1978). These parameters may be varied to see which model(s) fit observed data most closely (e.g., present-day vitrinite reflectance and/or behavior of velocity or density logs). Third, unconformities may be modeled as a period of deposition followed by erosion in which the thickness of eroded section and the age at the beginning of erosion are calculated by:

$$AGE_E = (R1 \times AGE1 + R2 \times AGE2) / (R1 + R2), \text{ and}$$

Eroded thickness $= R1 \times (AGE_E - AGE1)$

where AGE_E is the age at the beginning of erosion, AGE1 and AGE2 are the observed ages of sediment immediately above and below the unconformity, respectively, and R1 and R2 are the calculated sedimentation rates of the immediately overlying and underlying sediment layers.

If the lithologies and densities of the layers and the depth and thickness of each layer as a function of time are known, it is possible to determine the basement subsidence history by applying Airy (1855) isostasy calculations (Bomford, 1971). Steckler and Watts (1978) discussed the procedures involved in backstripping (removing) successive layers, replacing them by water, and allowing underlying layers to rebound isostatically. The basement subsidence (Y) in terms of the water depth at time of sediment deposition, W_d, and sediment thickness, S, is given by (see also Figure 4):

$$Y = S \frac{(\rho_m - \rho_s)}{(\rho_m - \rho_{w'})} + W_d - \Delta_{SL} \frac{(\rho_m)}{(\rho_m - \rho_w)} \qquad (8)$$

where ρ_m is the average mantle density, ρ_w is the average water density, ρ_s is the average sediment density, Δ_{SL} is the change in elevation of mean sea level, and Y is the depth to basement without sediment and water loads and represents the subsidence caused by tectonic effects. The terms W_d and Δ_{SL} must be determined by paleobathymetric analysis.

The major assumption in this isostatic reconstruction is that the observed present-day porosity variation with depth has been constant through time. This type of reconstruction does not incorporate effects of fluid overpressuring, despite observations from many basins demonstrating that overpressuring occurs (Hunt, 1979).

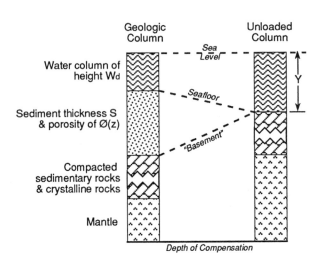

Figure 4. Schematic diagram of reconstructed (loaded) sedimentary section and backstripped (unloaded) sedimentary section. Parameters are defined in text (from Steckler and Watts, 1978).

PSEUDO-INVERSE PROCEDURE: FLUID FLOW/COMPACTION   To incorporate overpressuring, fluid flow and permeability variations through time, another procedure—a pseudo-inverse model—is required. In this procedure, present-day observations are used to constrain the limits on the variables employed in the equations thought to describe compaction histories.

Sediment compaction is the result of fluid expulsion caused by increasing overburden. Fluid movement through sediment is described by Darcy's equation:

$$\frac{\partial (\rho_f e)}{\partial t} = \frac{\partial}{\partial \xi} \left(K_z \rho_f \mu^{-1} (1 + e)^{-1} \frac{\partial P}{\partial \xi} \right) \tag{9}$$

where e is the void ratio of the rock, $\rho_f (\mu)$ is the density (viscosity) of the fluid, K_z is the vertical permeability of the rock, and P is the fluid pressure in excess of hydrostatic. The fully compacted depth coordinate, ξ, represents the net thickness of the rock such that when a new layer of sediment is deposited, underlying layers do not move in ξ. Darcy's law, expressing flow as a function of permeability and pressure gradient, can then be applied correctly. The physical depth, z, is related to ξ by $d\xi/dz = (1-\phi)$, where the porosity, ϕ, is related to the void ratio, e, by $e(1-\phi) = \phi$.

Excess fluid pressure, P, of Equation (9) is represented as a function of frame pressure, P_f, by

$$P = \int_{\xi}^{\xi_*} g (\rho_s - \rho_f) \, d\xi - (P_f - P_{f*}) \tag{10}$$

where g is gravitational acceleration, ρ_s is the solid density, P_f is the frame pressure (a prescribed function of void ratio, e, for each lithology), and $_*$ denotes the appropriate value on the sediment surface ξ_* ($P = 0$ on ξ_*).

All these variables are time-dependent and related to the sediment accumulation rate. Solving Equation (9) determines the changes of sediment thickness, porosity, permeability, pressure and fluid flow rate with respect to time and depth. Observations constrain the current values for comparison with model predictions (Fig. 5). Two numbers conventionally are varied until the constraints of Figure 5 are met. These numbers, A and B, occur in the quasi-empirical equations of state for shales as:

$$K_z = K_* (e/e_*)^A \text{ and} \tag{11a}$$

$$P_f = P_{f*} (e/e_*)^{-B} \tag{11b}$$

Empirical evidence from many wells suggests $A = 3 \pm 2$ and $B = 3 \pm 3$ (Cao et al., 1985). This procedure is a pseudo-inverse model, because the system is constrained by adjusting A and B until all thicknesses, porosity variations, fluid pressure variations, and permeabilities with depth are in mimimum least squares discord with present-day observations. Figures 6-9 show the results of applying this procedure to the Navarin basin COST #1 well. Figure 9 also shows the resulting burial history.

A true inverse procedure is not necessarily preferred over a pseudo-inverse procedure, particularly if the true inverse method is more restricted in its intrinsic assumptions and allowed behavior than the pseudo-inverse procedure. Similarly, any 2-D or 3-D model must be constrained either in functional form or in parameter values by observed present-day data related to the system's evolution.

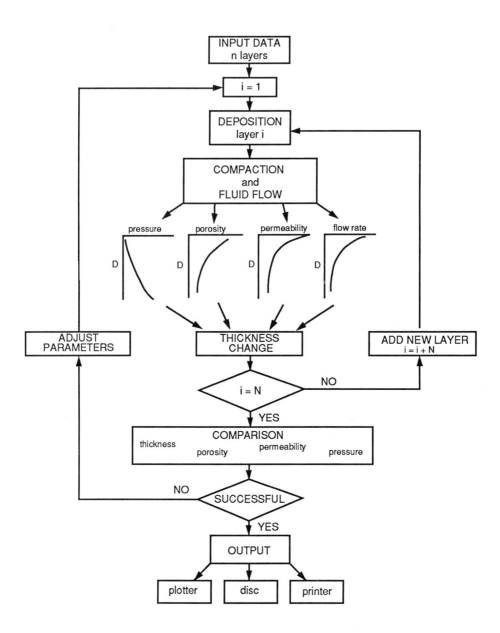

Figure 5. Flow diagram for the dynamic evolution model.

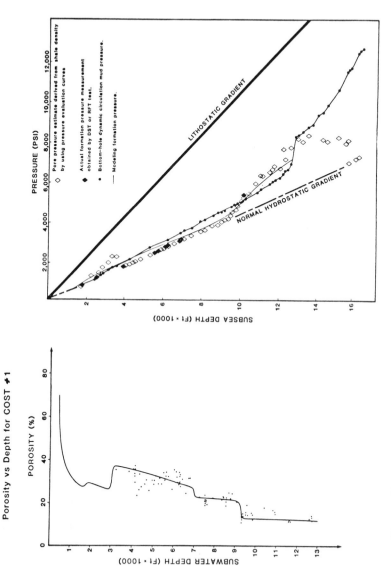

Figure 6. Navarin basin COST #1 well. (Left): Profile of porosity measurements with depth and superimposed curve from 1-D fluid-flow/compaction modeling. (Right): Fluid pressure with depth and superimposed curve from 1-D fluid-flow/compaction modeling.

31

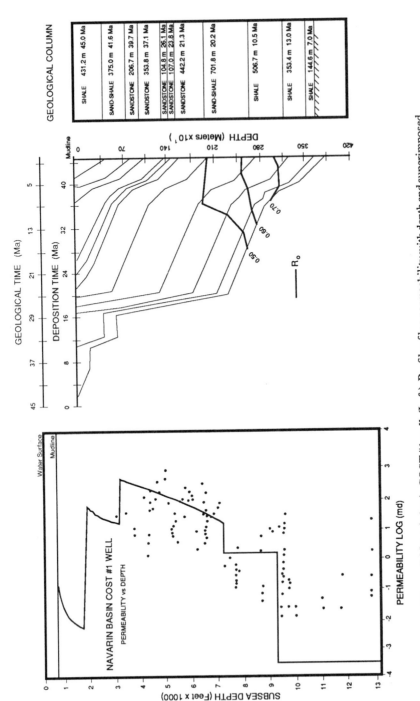

Figure 7. Navarin basin COST #1 well. (Left): Profile of log permeability with depth and superimposed curve from 1-D fluid flow/compaction modeling. (Right): 1-D Burial history of Navarin basin COST #1 well using fluid-flow/compaction code and superimposed curves of constant vitrinite reflectance. At far right is the stratigraphic column. Note the agreement of present day formation thicknesses predicted on the burial history curve with those observed in the stratigraphic column.

32

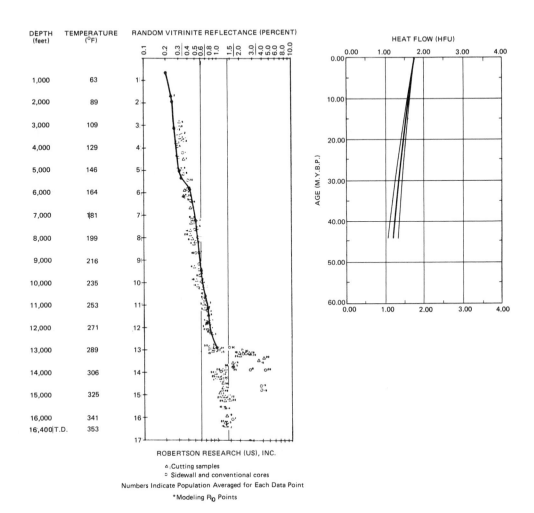

Figure 8. Navarin basin COST #1 well. (Left): Profile of vitrinite reflectance measurements with depth and superimposed best determined curve from pseudo-inverse, modeling. (Right): Best one parameter linear heat flux (dark line) and uncertainty curves (faint lines) with time produced by pseudo-inversion scheme.

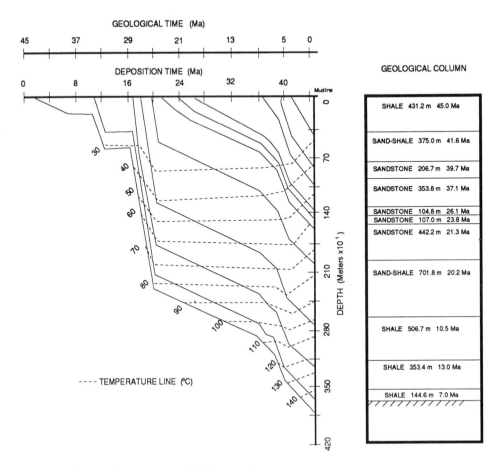

Figure 9. Navarin basin COST#1 well. Isotemperature curves with time based on best linear heat flux from Figure 8, superimposed on burial history curve.

Thermal History

To assess the thermal history of sediments we need to determine how their temperatures changed with time. Four factors influence the temperature history of sediment layers: conductive transport; convective transport; focusing and defocusing by heterogeneous thermal conductivity variations; and basement heat flux, $Q(t,x)$, variations with time, t, and spatial location, x. In principle, if basement heat flux through time is known, then the remaining three factors can be determined if the burial history also is known. Two end-member models may be constructed: forward models with parameters to be determined, and pseudo-inverse models based on present-day observations.

FORWARD MODELS OF PALEOHEAT FLUX   McKenzie (1978) argued that rapid stretching of continental lithosphere produces thinning and passive upwelling of hot asthenosphere. This is associated with block faulting and subsidence. After the heating event the lithosphere thickens by cooling, and the surface slowly subsides, typically without faulting. The exponentially decreasing rate of subsidence and heat flow depend only on the amount of stretching which can be estimated numerically. In the McKenzie model, the initial subsidence (S_i) is given by

$$S_i = a \left((\rho_0 - \rho_c)\left(\frac{t_c}{a}\right)\left(1 - \alpha T_1\left[\frac{t_c}{2a}\right]\right) - \frac{\alpha T_1 \rho_0}{2}\right)\left(\frac{1 - 1/\beta}{[\rho_0 (1 - \alpha T_1) - \rho_w]}\right) \qquad (12)$$

where a is the thickness of the lithosphere, t_c is the initial thickness of the continental crust, ρ_0 is the density of the mantle, ρ_c is the density of the continental crust, ρ_w is the density of sea water, α is the thermal expansion coefficient of the mantle, T_1 is the temperature of the asthenosphere, and β is the crustal stretching factor.

For large values of time and small values of β, the surface elevation, $E(t)$, is given by

$$E(t) = \frac{4\,a\,\rho_0\,\alpha T_1}{\pi^2(\rho_0 - \rho_w)} e^{-t/\tau} \qquad (13)$$

where

$$\tau = a^2/\pi^2 \kappa \qquad (14)$$

and κ is the thermal diffusivity. The heat flow, $Q(t)$, is given by

$$Q(t) = \left(\frac{K\,T_1}{a}\right)\left[1 + \left(\frac{2\beta}{\pi}\right)\sin\left(\pi/\beta\right)e^{-t/\tau}\right] \qquad (15)$$

For small values of β and large values of time, a straight line fit which best approximates Equation (13) for the subsidence is given by

$$Y = E_0\,r\left[\left(\frac{2t}{e\,\tau}\right)^{1/2} - 2\,e^{-1/2} + 1\right] \qquad (16)$$

where $E_0 = 4a\rho_0\,\alpha T_1 / \pi^2(\rho_0 - \rho_w)$; $r = (\beta/\pi) \sin(\pi/\beta)$; and $\tau = 62.8$ m.y. The subsidence predicted by the McKenzie model is similar in form to that predicted by the Turcotte-Ahern model (1977).

In the McKenzie model, prior to the time of stretching, the heat flow is given by the leading coefficient, KT_1/a, in Equation (15). Sclater and Christie (1980) noted that the present-day heat flow (approximately 0.8×10^{-2} cal/m² • sec) predicted by the McKenzie model is lower than present measurements of heat flow in the North Sea (approximately 1.5 $\times 10^{-2}$ cal/m² • sec). Sclater and Christie modified the McKenzie model by adding another heat flow component of 0.8×10^{-2} cal/m² • sec from the decay of radioactive elements within the crust. Another modification was proposed by Keen et al. (1981) in which the effect of thermal blanketing by low thermal conductivity sediments is included, reducing the rate of lithospheric cooling. This modification is thought to be particularly important for studies of older rifted margins, such as the North Sea, where sediment thicknesses on the order of 15 km (49,000 ft) occur. Royden and Keen (1980) and Hellinger and Sclater (1984) proposed a modified stretching model encompassing two layers to explain uplift and erosion on the flanks of a rift basin.

Royden (1980) proposed another model that provides an exponential subsidence compatible with observational data. This model assumes fracturing of the continental lithosphere and its intrusion by dikes from the mantle. Replacement of light crustal rocks by denser ultramafic materials results in initial subsidence and avoids the general problem of uplift. If γ_d is the fraction of lithosphere composed of dike material intruded from the asthenosphere, then the heat flow, $Q(t)$, and surface elevation, $E(t)$, are given by

$$Q(t) = \left(\frac{T_m K}{l} \right) \left(1 + 2\gamma_d \sum_{n=1}^{\infty} e^{-n^2 \pi^2 \kappa t / l^2} \right) \quad \text{and} \tag{17}$$

$$E(t) = \left(\frac{\alpha l \rho_m T_m}{(\rho_m - \rho_w)} \right) \left(\frac{4 \gamma_d}{\pi^2} \right) x \sum_{n=0}^{\infty} \left[\frac{1}{(2n+1)^2} \right] e^{\left(-(2n+1)^2 \pi^2 \kappa t / l^2 \right)} \tag{18}$$

where T_m is the temperature at the base of the mantle, l is the thickness of the lithosphere, κ is the thermal diffusivity of the lithosphere, α is the thermal expansion coefficient of the mantle, ρ_m is the density of the mantle, ρ_w is the density of water, and K is the thermal conductivity.

Subsidence may be computed from the surface elevation, and since subsidence also is inferred from the burial history, the discordance between model predictions and observations can be minimized by parameter adjustment. These adjusted parameters are then used directly in the posited heat flux models. These examples show how model strategies may be mixed for increased effectiveness; the subsidence components are treated as pseudo-inverse models, but the heat flux components are treated as forward models.

These theories of basin formation rely on thermal contraction of the crust to produce subsidence. An alternative explanation, proposed by Falvey (1974) and expanded by Middleton (1980) and Falvey and Middleton (1981), suggests that the initial stage of subsidence is caused by deep crustal metamorphism in basins of relatively young or unaltered crust. Falvey (1974) demonstrated that elevation of the geotherm during continental rifting may produce subsidence along the margins due to deep crustal metamorphism. He also suggested that some intracratonic basins may have similar origins. The deep crustal metamorphism mechanism implies a subsidence of:

$$\frac{\rho_m \left(\rho_2 - \rho_1 \right)}{\rho_1 \left(\rho_m - \rho_s \right)} L \left(t \right) \qquad (19)$$

where $L(t)$ is the thickness of greenschist facies rocks metamorphosed to amphibolite facies at time t, ρ_1 is the density of the greenschist facies, ρ_2 is the density of the amphibolite facies, ρ_s is the density of sediment, and ρ_m is the density of the upper mantle. Falvey (1974) suggested that the metamorphism driving basin subsidence occurs predominantly at the greenschist-amphibolite facies boundary. The model entails a period of heating followed by a period of cooling, with subsidence occurring in two stages. The initial subsidence, which occurs during the later part of the heating period, is attributed to deep-crustal metamorphism and continues as long as crustal temperatures increase. As temperatures decline during the subsequent cooling period, subsidence continues due to thermal contraction of the lithosphere. The time of transition between the heating phase and the cooling phase is t_1.

The downward movement, $l(t)$, of the earth's surface due to metamorphism is offset by the tendency for uplift, $h(t)$. Total displacement, taking these two factors into account and neglecting erosion, is given by

$$S \left(t \right) = \left(l(t) - h \left(t \right) \right) \qquad (20)$$

for $l(t) < h(t)$, and by

$$S \left(t \right) = \left(\frac{\rho_m}{\rho_m - \rho_s} \right) \left(l \left(t \right) - h \left(t \right) \right) \qquad (21)$$

for $l(t) > h(t)$, where $-S(t)$ is uplift, ρ_m is the density of the mantle, ρ_s is the density of the sediment, and $\rho_m/(\rho_m - \rho_s)$ is the isostatic loading factor. Subsidence due to thermal contraction is given by

$$S \left(t \right) = \left(\frac{\rho_m}{\rho_m - \rho_s} \right) \left(h \left(t_1 \right) - h \left(t \right) + h \left(t - t_1 \right) \right) \qquad (22)$$

where t_1 is the time of transition between the heating and cooling phases.

The distinguishing features of the Falvey-Middleton metamorphism model are that heat flow increases to the point of breakup and then it decreases. This contrasts with the other models in which heat flow decreases after the time of stretching. The heat flow from the metamorphism model is on the order of half that of the crustal extension model.

These geologic models for basin formation predict different heat flows through time. If we are to use these models quantitatively to predict heat flow, it is necessary to determine the optimum set of parameters for a given model for specific wells. Even then there are major differences among predicted histories, all consistent with inferences based on present-day data. Figure 10 compares the McKenzie (1978) and Falvey-Middleton (1981) models of heat flux tied to subsidence. The models have a discordance in heat flux that varies by a factor two.

Subsidence histories described by these models may be evaluated by comparison with inferred subsidence from a burial-history model. The large discordances in reconstructing estimates of paleoheat flux are caused by the different model postulates that connect modeled basement heat flux changes with the dependent subsidence (elevation).

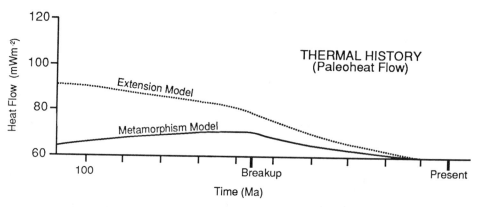

Figure 10. Comparison of relatively high heat flow resulting from McKenzie crustal extension model derived using stretching factor of $\beta = 1.8$, and low heat flow from Falvey deep crustal metamorphism model (see Falvey and Middleton, 1981).

INVERSE AND PSEUDO-INVERSE MODELS OF PALEOHEAT FLUX   Dissatisfaction with forward heat flux models has focused attention on inverse methods applied to measured thermal indicators. If the chemistry and/or physics of the kinetic evolution of a thermal indicator is known, and if the burial history also is known, then the evolution of a thermal indicator from deposition to the present day depends solely on the temperature variation along the burial path of the sediments. By connecting the temperature to basement heat flux through a conductive or conductive-convective model, and by measuring the thermal indicator today at numerous depths, paleoheat flux through time may be recovered by inversion. Detailed mathematical developments of inversion procedures for several different thermal indicators have been presented elsewhere (e.g., vitrinite reflectance—Toth et al., 1981; Lerche et al., 1984: sterane and hopane isomers and sterane aromatization—McKenzie and MacKenzie, 1983; Armagnac et al., 1988; Lerche, 1988: Ar^{39}/Ar^{40} variations—Harrison and Be, 1983; Lerche, 1988a: pollen translucency—Grayson, 1975; McKenna and Lerche, 1989: apatite fission tracks—Gleadow, Duddy and Lovering, 1983; Huntsberger and Lerche, 1987; Lerche, 1988a,b).

Thermal indicator inversion schemes for one-dimensional systems are constructed with the following logic. In a one-dimensional burial history variations of all quantities can occur only with depth and time. Therefore the temperature, $T(z,t)$, varies only with depth, z, below the sediment-water interface and with time, t. Since the position of a sedimentary stratum changes with time we reference the ith layer by its present position, z_i. If the sedimentation, subsidence, and compaction history of the basin are known, the position of any stratum through time is given by the function $z = \zeta(t; z_i)$ and the time of sedimentation of any stratum is given by the function $\tau_i = \tau(z_i)$. Since pressure, $p(z,t)$, also affects the physical and chemical evolution of the sediments, we will assume that it, too, is known.

A thermal indicator, $TI(z)$, is then related to the temperature history of the stratum in which it is measured

$$TI(z) = \int_{\tau(z_i)}^{t_{present}} FT\left[\left(\zeta(t;z_i), t\right), p\left(\zeta(t;z_i), t\right)\right] dt \tag{23}$$

where F is a known function.

If several different thermal indicators, $TI_j(z_i)$, j=1,... N, are measured at several depths, $z_i = 1, \ldots M$, can we reconstruct the temperature history, T(z,t)? The temperature field is a function of position and time, and because temperature must obey the heat flow equation, the problem may be transformed into one with a one-dimensional unknown. The heat flow equation may be written as

$$LT(z,t) = 0 \tag{24}$$

where L, the linear differential operator of the heat flow equation which includes terms for thermal conductivity, K, heat capacity, and source is assumed known. Then the temperature is completely determined by an initial condition (for example sediment is added to the basin with a given temperature), and two boundary conditions (for example, the temperature at the top of the basin is at the known value, $T_s(t)$, and that the heat flux through the bottom of the basin is $Q(t) = -k(\partial T/\partial z)$). Thus we can replace the problem of determining the entire temperature field with the problem of determining the heat flux, Q(t), at one depth in the basin. The unknowns are reduced from a two-dimensional function to a one-dimensional function. At any depth in the basin the temperature history, T(z,t) is described by:

$$T(z,t) = T_s + Q(t) \int_0^z \frac{dz'}{K(z')} \tag{25}$$

This equation applies if the thermal recovery time through the sedimentary overburden is shorter than the time scale marking changes in the basement heat flux, and if convective heat transfer is either minimal or accounted for in dealing with real basins.

Inserting T(z,t) into Equation (23) expresses the thermal indicator TI(z) in terms of the one unknown, Q(t). By measuring TI(z) at numerous depths, it is possible to invert to obtain directly the unknown heat flux variation Q(t). As shown elsewhere (Lerche, 1988a,b), the details change depending on the individual thermal indicators used, but the logic stands. Figure 11, from McKenna and Lerche (1989), shows the results of such an inversion procedure using both pollen translucency and vitrinite reflectance measurements.

HYDROCARBON MODELS—FORWARD AND PSUEDO-INVERSE   To determine the amounts of hydrocarbons that have been generated and accumulated and the locations in which they occur, the following information is needed. (1) The amount of oil and gas generated per square kilometer in each source rock in every part of the basin. (2) The timing of hydrocarbon formation, development of impermeable seals, and folding and faulting so traps can be identified. (3) The amounts of oil and gas that have been expelled from and how much is still retained in the source rock(s). (4) Evaluation of the ultimate oil and gas reserves of the sedimentary basin. Only a quantitative approach, allowing a computation of the amount of oil and gas generated in any place in the basin as a function of time, can provide this information. However, because many of the intermediate steps in petroleum and ga°

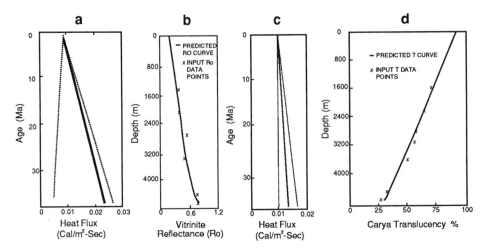

Figure 11. Heat flux variations with time in well Pan Am #27 based on pseudo-inversion of vitrinite reflectance data, and pollen translucency data. A. Heat flux calculated from vitrinite reflectance data shown in B. C. Heat flux calculated from *Carya* translucency data shown in D. Note the common domain of overlap of both indicators in predicted heat flux behavior. Also note the high degree of fit of observed versus predicted behavior with depth for both indicators.

generation from kerogen degradation are unknown, such a scheme has to be phenomenological; that is, parameters have to be determined by least squares fitting to basin and laboratory data.

A mathematical model of petroleum generation incorporating geologic time and reaction kinetics was introduced by Tissot (1969) and discussed more completely by Tissot and Espitalie (1975). The model is based on kinetics of kerogen degradation and uses the general scheme of first-order chemical kinetics. Kerogen is a macromolecule composed of polycondensed nuclei bearing alkyl chains and functional groups. As burial depth and temperature increase heteroatomic bonds are successively broken, roughly in order of increasing rupture energy. The reactions generate heavy heteroatomic compounds, carbon dioxide, and water, then progressively smaller molecules, and finally hydrocarbons. At the same time, the residual kerogen becomes progressively more aromatic and evolves toward a carbon residue.

The validity of a phenomenological model depends on two factors: (1) the validity of the physics and chemistry involved in the calculation, and (2) the values of the parameters chosen. Tissot and Welte (1978) compared model results with volumes of petroleum generated in various basins. They claimed a quadratic deviation lower than 10^{-2} and a correlation coefficient better than 0.9. The same set of constraints given by Tissot and Welte (1978, Table V.4.1) is said to be sufficient to account for all conditions of kerogen degradation including: (a) evolution at relatively low temperatures (50°-150° C; 122°-302°F) over a time of 10 to 400 m.y., (b) artificial evolution through laboratory experiments (180°-250°C; 356°-482°F), and (c) high-temperature (400°-500°C; 752°-932°F) retorting of oil shales.

This model assumes *in situ* generation. The constants were determined assuming constant geothermal gradients and ignoring the effects of differential primary and secondary migration, and thermal evolution after accumulation. Strictly speaking, the changing thermal gradients with depth and time require a recalibration of values for greater validity and applicability.

To apply the phenomenological model of Tissot and Welte (1978) to predict changes in hydrocarbon content of rocks with respect to time and temperature, it is necessary to assume first-order chemical reaction kinetics for the breakdown of kerogen to hydrocarbons. These reactions are governed by

$$\frac{dx_i}{dt} = k_i x_i \tag{26a}$$

$$k_i = A_i e^{-E_i/k_B T(t)} \tag{26b}$$

where x_i is the number density of component i, k_i is the reaction rate for this reaction, k_B is Boltzmann's constant, A_i is the reaction rate in the limit of high temperature, E_i is the activation energy, and $T(t)$ is the temperature as a function of time acting on the element x_i.

The Tissot and Welte scheme considers three types of initial kerogens which differ in their hydrogen:carbon ratio. Type 1 (algal) has a high liquid-generating potential, type 2 (amorphous) has an intermediate liquid-generating potential, and type 3 (woody) has a low liquid-generating potential. Each type is assumed to have six different bond types that have strengths or activation energies in the range 10 to 80 kcal/mole. The amount of organic material (x_i) reacting in the ith reaction (breaking of the ith bond type) for i = 1 to 6 is specified for each kerogen type. Calibrations for the formation of gas have been made from laboratory experiments on hydrocarbon cracking. Consideration of a single reaction with an activation energy of 80 kcal/mole was convenient to account for gas generation in the deep parts of sedimentary basins. Tables of the activation energies (E_i), the reaction constants (A_i), and the initial values (x_i) for each kerogen type are given by Tissot and Welte (1978, Table V.4.1).

To calculate the quantity of hydrocarbons produced by potential source rocks, it is necessary to use the above calculations and measurements or estimates of the percentage of organic material in those source rocks. Nakayama and Van Siclen (1981) detailed the methodology of this system to work out a hydrocarbon budget for a basin. To make predictions about the hydrocarbon generation and accumulation history of a basin, it is also necessary to determine the temperature as a function of time. Tissot and Welte (1978) assumed the heat flux was constant and used constant geothermal gradients in their models. The phenomenology of the model's kinetic structure is not dependent on this point, but parameter determination is very much dependent on the constancy of temperature gradient.

Two problems are apparent. Is the structure of the kinetic model adequate? How do we determine the best kinetic parameters for a given model? Other kinetic schemes have been proposed (e.g., Cao, Glezen and Lerche, 1986; Braun and Burnham, 1987) that also are consistent with limited observational data. At this time, kinetic models are rudimentary and either can be handled in a forward modeling sense or a pseudo-inverse sense, but we cannot yet distinguish between many competing models. The generation parameters are determined from measurements made after the produced hydrocarbons have moved from their source "kitchen" to reservoirs, confounding the parameter determination even more.

SENSITIVITY, NOISE, RESOLUTION AND UNIQUENESS   In discussing the forward, pseudo-inverse and true inverse philosophies for analyzing burial, thermal and hydrocarbon histories, three underlying considerations were taken for granted, either explicitly or implicitly. First, we assumed that any input data required (e.g., stratigraphic ages, formation thicknesses, porosity variations with depth, thermal indicator measurements, present day heat flux) were accurate (no error) and continuous with depth (no finite sampling interval problem). Neither assumption is true in general. Second, we assumed that, apart from the parameters or functional forms we were attempting to determine, all other parameters and functional forms were known precisely (e.g., thermal conductivity, permeability scale values, mantle density). This assumption is likely invalid. Third, we assumed that the intrinsic underlying assumptions of a model were accurate enough to describe the real world (e.g., 1-D, 2-D, 3-D, conductive heat flux only, no diagenesis, hydrocarbon kinetic model forms). But this is known to be perhaps the most invalid point of any model. Indeed, even the models connecting heat flux variations to basement subsidence vary amongst themselves quite markedly and all meet constraints of present-day data (Sleep, 1971; Royden et al., 1980; McKenzie, 1978; Falvey and Middleton, 1981; Turcotte and Ahern, 1977).

Each of these problems is associated with uniqueness, resolution, sensitivity, precision and signal versus noise, in forward, pseudo-inverse and true inverse models (Cao, 1987). This area would seem to be one requiring a more detailed investigation and recognition than it has so far received to date.

Perhaps the best that can be hoped for in our attempts to determine hydrocarbon accumulations is to bracket least probable, average, and most probable situations with confidence intervals so that we can at the least quantify precisely the level of our lack of determinism. And even that would be a major step forward.

DISCUSSION AND CONCLUSION

We have examined the philosophies and practices of forward and inverse models in sedimentary basin analysis. In general forward models suffer from being presented independently of the observed data, but have the advantage of guiding thought about dynamical sedimentary behavior. Forward models often conflict one with another, and occasionally have internal inconsistencies as well. True inverse procedures are often more difficult to obtain. They usually honor some components of data sets, but may be so restrictive in their underpinning assumptions as to omit the dominant physical and geological behaviors of the system. Pseudo-inverse models would seem to represent a compromise between forward models and true inverse models. On the one hand they permit determination of a finite number of parameters for a given functional shape, but they also contain the capability of allowing an extended set of functional shapes should the data warrant such an extension.

All models suffer from the vagaries of data noise and sampling, from assumptions concerning parameter values and functional behaviors within a model framework, and from intrinsic model assumptions. Measures of determination which can provide resolution, uniqueness and precision of dynamical evolution in the face of such irresolution of information would seem to be needed rather urgently.

These types of problems are not unique to sedimentary basin analysis but are generic problems faced by any model of any process in any area of science.

ACKNOWLEDGMENTS

This work was supported by the Industrial Associates of the Basin Analysis Group at the University of South Carolina. Comments by John Harbaugh and Margaret Lessenger on a previous version of this paper are much appreciated.

REFERENCES CITED

Airy, G.B., 1855, On the computation of the effect of the attraction of mountain-masses as disturbing the apparent astronomical latitude of station of geodetic surveys: Philosophical Transactions of the Royal Society of London, v. 145, p. 101-104.

Armagnac, C., Kendall, C., Kuo, C., Lerche, I. and Pantano, J., 1988, Determination of paleoheat flux from vitrinite reflectance data and from sterane and hopane isomer data: Journal of Geochemical Exploration, v. 30, p. 1-28.

Athy, L.F., 1930, Density, porosity and compaction of sedimentary rocks: American Association of Petroleum Geologists Bulletin, v. 14, p. 1-24.

Beaumont, C., 1981, Foreland basins: Geophysical Journal of the Royal Astronomical Society, v. 65, p. 291-329.

Bomford, G., 1971, Geodesy (second edition): London, Oxford Press, p. 441-443.

Braun, R.L., and Burnham, A.K., 1987, Analysis of chemical reaction kinetics using a distribution of activation energies and simpler models: Journal of Energy Fuels, v. 1, p. 153-161.

Cao, S., 1987, Sensitivity analysis of 1-D fluid flow model for basin analysis [Ph.D. thesis]: Columbia, South Carolina, University of South Carolina.

Cao, S., Glezen, W.H., and Lerche, I., 1985, Fluid flow, hydrocarbon generation and migration: A quantitative model of dynamical evolution in sedimentary basins. Proceedings Offshore Technology Conference Paper OTC 5182, v. 2, p. 267-276.

Falvey, D.A., 1974, The development of continental margins in plate tectonic theory: Australian Petroleum Exploration Association Journal, v. 14, p. 95-106.

Falvey, D.A., and Middleton, M.F., 1981, Passive continental margins: Evidence for a prebreakup deep crustal metamorphic subsidence mechanism: 26th International Geological Congress, Colloque C3.3, Geology of Continental Margins, Supplement to v. 4, p. 103-114.

Gleadow, A.J.W., Duddy, I.R., and Lovering, J.F., 1983, Fission track analysis: A new tool for the evaluation of thermal histories and hydrocarbn potential: Australian Petroleum Exploration Association Journal, v. 23, p. 93-102.

Grayson, J.F., 1975, Relationship of palynomorph translucency to carbon and hydrocarbons in clastic sediments, in Alpern, B., ed., Petrographie de la matiere organique des sediments, relations avec la paleotemperature et le potential petrolier: Paris, Centre National de la Recherche Scientifique, p. 261-273.

Harrison, M.T., and Be, K., 1983, $^{40}Ar/^{39}Ar$ age spectrum analysis of detrital microclines from the southern San Joaquin basin, California: An approach to determining the thermal evolution of sedimentary basins: Earth and Planetary Science Letters, v. 64, p. 242-256.

Hellinger, S.J., and Sclater, J.G., 1983, Some comments on two-layer extensional models for the evolution of sedimentary basins: Journal of Geophysical Research, v. 88, p. 8251-8269.

Hunt, J.M., 1979, Petroleum geochemistry and geology: San Francisco, W.H. Freeman Co., 617 p.

Huntsberger, T.L., and Lerche, I., 1987, Determination of paleo heat-flux from fission scar tracks in apatite: Journal of Petroleum Geology, v. 10, p. 365-394.

Keen, C.E., Beaumont, C., and Boutilier, R., 1981, Preliminary results from a thermo-mechanical model for the evolution of Atlantic-type continental margins: 26th International Geological Congress, Colloque C3.3, Geology of Continental Margins, Supplement to v. 4, p. 123-128.

Lerche, I., 1988a, Inversion of multiple thermal indicators: Quantitative methods of determining paleoheat flux and geological parameters, I. The theoretical development for paleoheat flux: Mathematical Geology, v. 20, p. 1-36.

Lerche, I., 1988b, Inversion of multiple thermal indicators: Quantitative methods of determining paleoheat flux and geological parameters, II. The theoretical development for chemical, physical and geological parameters: Mathematical Geology, v. 20, p. 73-96.

Lerche, I., Yarzab, R.F., and Kendall, C.G.St.C., 1984, The determination of paleoheat flux from vitrinite reflectance data: American Association of Petroleum Geologists Bulletin, v. 68, p. 1704-1717.

MacKenzie, A.S., and McKenzie, D., 1983, Isomerization and aromatization of hydrocarbons in sedimentary basins formed by extension: Geological Magazine, v. 120, p. 417-470.

McKenna, T., and Lerche, I., 1989, Pollen translucency as a thermal maturation indicator: Journal of Petroleum Geology (submitted).

McKenzie, D., 1978, Some remarks on the development of sedimentary basins: Earth and Planetary Science Letters, v. 40, p. 25-32.

Middleton, M.F., 1980, A model of intracratonic basin formation, entailing deep crustal metamorphism: Geophysical Journal of the Royal Astronomical Society, v. 62, p. 1-14.

Nakayama, K. and Van Siclen, D.C., 1981, Simulation model for petroleum exploration: American Association of Petroleum Geologists Bulletin, v. 65, p. 1230-1255.

Royden, L., and Keen, C.E., 1980, Rifting process and thermal evolution of the continental margin of eastern Canada determined from subsidence curves: Earth and Planetary Science Letters, v. 51, p. 343-361.

Royden, L., J.G. Sclater, and Von Herzen, R.P., 1980, Continental margin subsidence and heat flow: Important parameters in formation of petroleum hydrocarbons: American Association of Petroleum Geologists Bulletin, v. 64, p. 173-187.

Sclater, J.G., and Christie, P.A.F., 1980, Continental stretching: An explanation of the post-mid-Cretaceous subsidence of the central North Sea basin: Journal of Geophysical Research, v. 85, p. 3711-3739.

Steckler, M.S., and Watts, A.B., 1978, Subsidence of the Atlantic-type continental margin off New York: Earth and Planetary Science Letters, v. 41, p. 1-13.

Tissot, B., 1969, Premieres donnees sur les mecanismes et la cinetique de la formation du petrole dans les sediments (First data on the mechanism and kinetics of the formation of petroleum in sediments): Revue de l'Institut Francais du Petrole, v. 24, p. 470-501 (in French).

Tissot, B., and Espitalie, J., 1975, L'evolution thermique de la matiere organique des sediments: Applications d'une simulation mathematique (Thermal evolution of organic material in sediments: Applications of a mathematical simulation): Revue de l'Institut Francais du Petrole, v. 30, p. 743-777 (in French).

Tissot, B., and Welte, D.H., 1978, Petroleum formation and occurrence: New York, Springer-Verlag, p. 500-521.

Toth, D.J., Lerche, I, Petroy, D.E., Meyer, R.J., and Kendall, C.G.St.C., 1981, Vitrinite reflectance and the derivation of heat flow changes with time, in Bjorney, M., ed., Advances in organic geochemistry: Proceedings of the 10th International Meeting on Organic Geochemistry, Bergen, Norway, p. 588-596.

Turcotte, D.L., and Ahern, J.L., 1977, On the thermal and subsidence history of sedimentary basins: Journal of Geophysical Research, v. 82, p. 3762-3766.

van Hinte, J.E., 1978, Geohistory analysis—Application of micropaleontology in exploration geology: American Association of Petroleum Geologists Bulletin, v. 62, p. 201-222.

3

Predictability and Chaos in Quantitative Dynamic Stratigraphy

Rudy Slingerland
Department of Geosciences, The Pennsylvania State University, University Park, PA 16802 USA

Abstract

Models of quantitative dynamic stratigraphy are usually nonlinear equations representing forced, dissipative systems, and as such they are susceptible to a rich mathematical behavior only recently recognized. Even a simple nonlinear dynamical system such as the Lorenz equations describing Rayleigh-Benard convection, used here as an example, contains periodic, slightly aperiodic, and seemingly random solutions called chaotic, depending upon the Rayleigh number. Systems of this type display a sensitive dependence upon initial conditions making prediction in its present sense impossible. Some periodicities that arise are likely to be explained by external causes when in fact, they are due to nonlinear coupling. In a positive light, these nonlinear dynamics may explain some of the complexities in the stratigraphic record.

Introduction

Over the last two decades a quiet revolution has occurred in the science and mathematics of nonlinear dynamical systems. What was once a backwater topic of research, for the most part ignored by physicists after Poincaré, now has its own journals, conferences, centers for nonlinear studies, and even its own toys, such as Space Balls. The reasons are several—the diminishing returns of particle physics, advances in computers and numerical analysis, for example—but two others seem especially noteworthy. first, we are now at the stage where the interesting problems are the more difficult nonlinear ones. To practice reductionist science with its linearized models is to throw out the baby with the bathwater. Thus an alternative scientific approach, termed "analysis by synthesis" by Hut and Sussman (1987), has arisen wherein one constructs and solves nonlinear mathematical models on the computer. Of a set of models, the model configuration that best accounts for the observations is assumed to be the correct one. This is the technique of quantitative dynamic stratigraphy (QDS) as presented elsewhere in this volume. Second, as scientists in as diverse fields as meteorology and population ecology began constructing nonlinear models, they discovered a rich mathematical behavior (see Crutchfield et al., 1986, and Stewart and Thompson, 1986, for reviews, and Gleick, 1987, for a popular account). Even the simplest of deterministic equations generated periodic, slightly aperiodic, and random solutions, the latter being called chaos. This rich behavior drew the interest of mathematicians, making the study of nonlinear dynamical systems fashionable once again. The result has been a deeper understanding of such enigmas as the transition to turbulence in fluids (Feigenbaum, 1980; Hofstadter, 1981).

There is a certain irony in the revolution however. The possibility of chaotic behavior in even simple nonlinear deterministic systems makes analysis by synthesis all the more difficult. Over a certain range of initial conditions, the solutions may be well behaved, settling down to a fixed point or simple orbit in the state space. In other, *a priori* unknowable ranges, the solutions may be chaotic. And, in either case, small differences in the initial conditions may produce great differences in the solutions, thus magnifying small errors in the initial conditions. Prediction becomes impossible.

Two questions arise then—how do we recognize chaos, and are quantitative dynamic stratigraphy models susceptible to it? If the answer to the latter is yes, an additional question follows—what are the stratigraphic implications of a chaotic solution? The remainder of this article addresses these questions, although no straightforward answers are presented. A review of the Lorenz model of Rayleigh-Benard convection provides an analogue for recognizing chaos and I will attempt to summarize the few known necessary conditions for chaotic behavior of a system. Some comments on stratigraphic implications close the discussion.

An Example of Nonlinear Behavior and Chaos

Classical Rayleigh-Benard convection serves as an ideal example of nonlinear behavior, because it is simple enough to allow for intuitive understanding and because it is quite well studied. Rayleigh-Benard convection is one possible mode of fluid circulation deep within sedimentary basins where it could contribute to the origin of diagenetic pressure seals of gas reservoirs.

When a fluid is heated uniformly from below and cooled uniformly from above, heat is first transported vertically by conduction with no apparent fluid motion. After some time, and for temperature differences above a critical minimum, cylindrical rolls develop, convecting heat by fluid transport. As demonstrated in Shirer (1987), the smallest effective model of these system states through time is the two-dimensional equation set of Lorenz (1963),

$$dx/dt^* = -px + py$$

$$dy/dt^* = -xz + rx - y$$

$$dz/dt^* = xy - bz$$

where x is proportional to the intensity of convective motion, y is proportional to the temperature difference between ascending and descending currents, z is proportional to the distortion of the vertical temperature profile from linearity, and t^* is dimensionless time. The coefficient p is the *Prandtl number*, taken as 10 in the original calculations; r is the normalized *Rayleigh number* (equal to 24.7 at the onset of steady convection), taken to be 28; and b is a constant equal to 8/3. For an initial condition, Lorenz chose (0,1,0), a slight departure from the state of no convection.

Our intuition tells us that at this slightly supercritical normalized Rayleigh number, r, fluid flow should commence at $t^* > 0$ and evolve into a steady state convection. Lorenz found otherwise. The temporal behavior of solutions to this small system of equations is complicated for this and certain other r in the range $24.7 < r < 215$. This is illustrated by plotting y, the difference in fluid temperatures on the rising and falling limbs of a cell (Fig. 1). It grows with time up to a t^* of about 30 when warm fluid is at the top of the cell. Then y decreases and by $t^* = 50$ changes sign, signifying that the overly vigorous flow of the cell has caused the warm fluid originally at the bottom of the cell at $t^* = 0$ to continue over the top of the cell and descend; likewise the cold fluid ascends. The resulting buoyant forces cause the motion to cease and reverse direction at $t^* = 60$. For $85 < t^* < 1650$ the fluid motion matches our intuition in that it is quasi-steady around one solution with a fixed mean value of y (also x and z). It oscillates however, and the oscillations increase in amplitude until $t^* = 1650$ after which the motion is quite irregular. The motion is sometimes clockwise and sometimes counterclockwise with no apparent long-term periodicity.

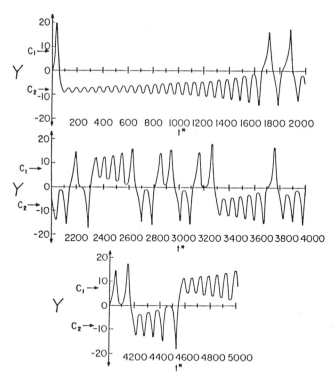

Figure 1. Times series solution of the Lorenz convection model for $p = 10$, $b = 8/3$, and $r = 28$. Variable y is proportional to the difference in fluid temperatures on the rising and falling limbs of the convection cell and t^* is dimensionless time. At $t^* = 1650$ the motion becomes chaotic (modified from Nese, 1985).

Is the motion chaotic? To answer this question, consider a more revealing graphical representation—a plot of the solutions in their state space, an abstract construct whose coordinates are the dependent variables of the system. A system that proceeds from some initial condition to a steady state solution would be represented by a trajectory from an initial point to a single steady state point. Because many systems end up at the same steady state solution regardless of initial conditions, a steady state point is said to attract nearby trajectories or *orbits* and is called an *attractor*. There may be many attractors in a state space, each with its own *basin of attraction*. Other systems may not come to rest in the long term; rather they may cycle periodically thorough a sequence of states in a *periodic orbit*. The associated attractors are called *limit cycles* (see Crutchfield et al., 1986; May, 1976; and Stewart and Thompson, 1986).

The graph of the Lorenz equations in state space (Fig. 2) was constructed by solving the equations at timesteps so infinitesimal as to produce a line. The trajectory loops around one stationary solution and then another, returning near to itself but never duplicating an individual orbit. Solutions such as these are called *chaotic*. They are random in the sense that no predictions about future states can be made, yet they arise from a completely deterministic system. Attractors of this type are called *chaotic* or *strange* attractors (see Devaney, 1986, p. 50, for formal definitions). Interestingly, they have a fractal nature; an infinite number of points show a self-similar detail at all levels of magnification.

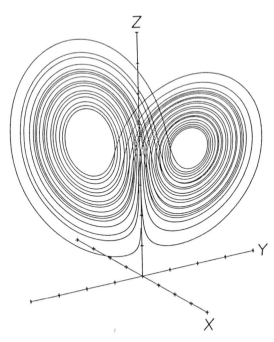

Figure 2. Plot of solutions to the Lorenz convection model in state space for the same conditions as Figure 1 but starting after t* = 1650. The trajectory loops around two stationary solutions but never repeats itself, thereby defining a strange attractor (modified from Nese, 1985).

To grasp just how chaotic this behavior is, consider the experiment performed by Crutchfield et al. (1986) illustrated in Figure 3. Solutions of the equations are shown in state space at selected times for each of 10,000 initial conditions, so close together they appear as one dot at $t^* = 0$. The solutions spread out through time to cover the entire attractor, dramatically illustrating sensitive dependence on initial conditions and the unpredictability of future states. This sensitivity has become known as the butterfly effect, from Lorenz's (1979) address entitled, "Predictability: Does the flap of a Butterfly's Wings in Brazil Set Off a Tornado in Texas?"

To summarize, the above example illustrates that some forced nonconservative hydro-dynamical systems may exhibit quasi-periodic behavior over the short term with no periodicity in the forcing. Over the longer term they may show chaotic behavior, depending upon the magnitude of the coefficients. The chaos is unpredictable, sensitive to initial conditions, yet bounded, and recurrent, producing a fractal geometry.

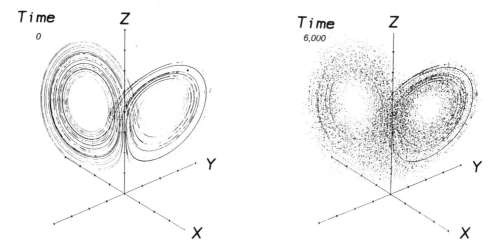

Figure 3. The Lorenz strange attractor of Figure 2 with additional solutions overlaid to illustrate sensitive dependence to initial conditions. At $t^* = 0$ a point represents 10,000 initial conditions that lead to the 10,000 solutions presented at $t^* = 6,000$ (modified from Crutchfield et al., 1986).

ARE QDS MODELS SUSCEPTIBLE TO CHAOS?

Although there is no formal mathematical answer to this question, there are some general guidelines we can use. These guidelines, which will be considered in turn, are:

1. Classification of the model's system of equations with respect to physical type, dimension, degree of coupling among equations, and degree of nonlinearity;
2. Values of the Lyapunov dimension and correlation exponent; and
3. Sensitivity to initial conditions.

Classification

A QDS model can be classified first according to physical type. By this is meant whether it is an open or closed organizational, kinematic, or dynamical model, and if the latter, whether it is forced and whether it is dissipative. Most of the QDS models presented in this volume are open and dynamical in that they receive mass from outside and go beyond the geometrical relations of the kinematic model to include evolution of the state variables with forces also considered. Most are forced in that they are fed energy by a boundary condition, and most are dissipative in that they lose energy through friction. It is now understood that open, forced and dissipative dynamical systems such as these are susceptible to chaotic solutions (Shirer, 1987). This arises because of the competition between the forcing and dissipative processes. Also, it has been argued on thermodynamic grounds that systems closed to their environment with respect to mass transport should not exhibit instability (Feinberg, 1980).

Dimension or number of degrees of freedom seems to be an important consideration for chaotic behavior. The Lorenz attractor disappears in a three-dimensional convection model (although new chaotic attractors appear), probably because turbulence is three-dimensional and Lorenz's two-dimensional model is hunting for a stable solution that only is available with another degree of freedom (Shirer, 1987). This raises the issue, long debated in population ecology, of whether more complex systems are more stable or less stable than simpler systems. In this usage, complex means both more variables and greater degree of coupling among variables. The general conclusion from qualitative stability analysis of partially specified systems is that progressively more complex systems are likely to be progressively less stable (Levins, 1974). However, Shaw (1987, p. 1653) concluded the opposite: "Computer experiments show that the coupling together of complex systems often increases . . . the degree of order in the composite system."

Finally, one might suspect that the degree of nonlinearity may determine whether a system exhibits a chaotic attractor. No general rules seem to exist on the subject, however (H. N. Shirer, personal communication, 1988).

Lyapunov Dimension and Correlation Exponent

The Lyapunov Dimension and Correlation Exponent are thought to measure the number of dimensions necessary to specify the region of the attractor in the state space. For example, if there are $N = 3$ equations in a dynamical system, then the state space has three axes corresponding to the three state variables, and the largest dimension possible for an attractor is 3 or generally, N. This would be a volume in the state space that attracts or traps trajectories orbiting near it. Similarly, the point and limit cycle attractors mentioned earlier would have dimensions of 0 and 1, respectively. This information becomes useful in the present context because the dimensions of strange attractors are usually nonintegers, a reflection of the folded, fractal structure of the chaotic solution sets. Thus, determining whether a model will exhibit chaotic behavior reduces to determining the dimensions of its attractors.

The Lyapunov dimension was defined by Kaplan and Yorke (1979) as a function of the Lyapunov exponents of an attractor. In the interests of brevity and because the correlation exponent is easier to calculate, the Lyapunov dimension will not be discussed further here; see Nese (1987) for details.

The correlation exponent v, was defined by Grassberger and Procaccia (1983a,b) as a measure of the local structure of an attractor. It is conjectured to be related to the Lyapunov dimension and can be calculated from a time series of one component of the dynamical system. Let $Y_j, j = 1,2,...n$, be n points on an attractor residing in N-dimensional state space. The points may be obtained from a QDS model, for example, as a times series $Y_j = Y(t+jT)$ of a dependent variable, where T is a fixed time increment. A point Y_k is selected and all the distances $\|Y_j-Y_k\|$ of this point from the remaining n-1 points are calculated. This procedure is repeated for all the Y_k points on the attractor and a correlation integral C(L) is computed as

$$C(L) = \lim_{n \to \infty} \frac{1}{n^2} \sum_{\substack{j,k = 1 \\ \text{when } j \neq k}}^{n} H\left(L - \| Y_j - Y_k \|\right)$$

where H is the Heaviside function (if $L - \|Y_j - Y_k\| > 0$, then $H = 1$, otherwise, $H = 0$) and L is a fixed distance measured from Y_k. This is equivalent to calculating the density of points on the attractor within a range of distances L from Y_k, and then finding the average of this density over all values of k. In general one expects that

$$C(L) \propto L^v$$

where v, the correlation exponent should be 1 if the attractor is a line, 2 if a surface and so on up to N, the dimension of the state space. In the latter case the data points are totally uncorrelated, i.e., random. Operationally, v is determined by finding the slope of the line when $\ln[C(L)]$ is plotted against $\ln[L]$.

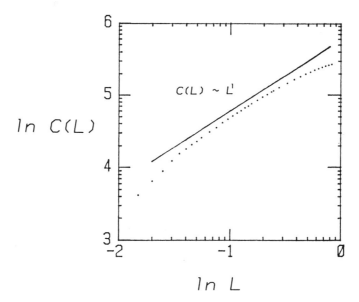

Figure 4. Graph of the correlation integral versus L computed using 4,000 points on the periodic attractor of the Lorenz system when $p=10, b=8/3$, and $r=300$. The slope of the points is approximately one, indicating the attractor is a line, and therefore nonchaotic.

As an example, consider a plot of this type for the Lorenz system (Fig. 4) when the normalized Rayleigh number r, is 300 (Nese, 1985). At this r the attractor is a stable limit cycle, meaning that the solutions have settled down to stable oscillations and the graph of the state space is a loop. By the above reasoning we expect v to be 1, a value closely approximated by the slope of the data points in Figure 4. For the case discussed above when r = 28 and the attractor is chaotic, Grassberger and Procaccia (1983a,b) calculated a v of 2.05 + 0.01, a fractal dimension as expected. Thus, it may be possible to examine a QDS model's output for chaotic behavior by examination of its correlation exponent.

Sensitivity to Initial Conditions

Probably the most straightforward method for determining a model's susceptibility to chaotic behavior is to test it for sensitivity to initial conditions. A chaotic attractor is strongly suggested if for similar (but not identical) initial conditions the solutions show early similar behavior that diverges with time.

It appears then, that as a class, QDS models could be susceptible to chaotic behavior, especially when integrated over long time periods. Solutions can be inspected for chaos, however, and this possibility should be considered along with other more common explanations such as numerical instability.

IMPLICATIONS OF CHAOS TO STRATIGRAPHY

There are several implications of chaotic solutions to QDS models. First, they make prediction difficult. Without chaos we expect the final configuration of sedimentary facies, for example, to be only weakly affected by slight changes in the initial conditions. With chaos, any configuration within the region of the attractor is possible. Second, as Shaw (1987) pointed out, there need be no unique causative periodic forcing required to explain apparent periodicities in the rock record. They can arise from the nonlinear coupling as sets of interacting resonances, much as in the Lorenz model (Fig. 2). The seemingly periodic repetitive successions of lithologies, currently interpreted as a Milankovitch signal, should be examined with this in mind. Finally, and in a more positive vein, certain complexities of the stratigraphic record may now have an explanation in chaos theory.

CONCLUSIONS

This paper has attempted to define typical behaviors of nonlinear dynamical systems, particularly chaos. It has explored the extent to which quantitative dynamic stratigraphic models may be susceptible to chaos, and has suggested some implications for stratigraphy. There is every reason to believe that some QDS models will contain chaotic attractors in their state spaces; indeed, this situation may be necessary if we are ever to explain complexities of the stratigraphic record.

ACKNOWLEDGMENTS

H. N. Shirer and Lee Kump played larger roles than they know in the gestation of these ideas, and the reviewers, Julian Thorne, Ian Lerche, and Tim Cross, did their best to eliminate my wooly thinking.

REFERENCES CITED

Crutchfield, J.P., Farmer, J.D., Packard, N.H., and Shaw, R.S., 1986, Chaos: Scientific American, v. 255, no. 6, p. 46-57.

Devaney, R.L., 1986, An introduction to chaotic dynamical systems: Menlo Park, Benjamin/Cummings, 320 p.

Feigenbaum, M.J., 1980, Universal behavior in nonlinear systems: Los Alamos Science, v. 1, p. 4-27.

Feinberg, M., 1980, Chemical oscillations, multiple equilibria, and reaction network structure, *in* Stewart, W.E., et al., eds., Dynamics and modelling of reactive systems: New York, Academic Press, p. 59 130.

Gleick, J., 1987, Chaos—Making a new science: New York, Viking, 352 p.

Grassberger, P., and Procaccia, I., 1983a, Characterization of strange attractors: Physical Review Letters, v. 50, p. 346-349.

Grassberger, P., and Procaccia, I., 1983b, Measuring the strangeness of strange attractors: Physica, v. 9D, p. 189-208.

Hofstadter, D.R., 1981, Metamagical Themas—Strange attractors: Mathematical patterns delicately poised between order and chaos: Scientific American, v. 245, no. 5, p. 22-43.

Hut, Piet, and Sussman, G.J., 1987, Advanced computing for science: Scientific American, v. 257, no. 4, p. 144-153.

Kaplan, L.P., and Yorke, J.A., 1979, Chaotic behavior of multidimensional difference equations, *in* Peitgen, H.O., and Walther, H.O., eds., Functional differential equations and the approximation of fixed points: New York, Springer-Verlag, Lecture Notes in Mathematics, v. 730, p. 228-237.

Levins, R., 1974, The qualitative analysis of partially specified systems: New York Academy of Sciences Annals, v. 231, p. 123-138.

Lorenz, E.N., 1963, Deterministic nonperiodic flow: Journal of the Atmospheric Sciences, v. 20, p. 130-141.

Lorenz, E.N., 1979, Predictability: Does the flap of a butterfly's wings in Brazil set off a tornado in Texas? Address at the annual meeting of the AAAS, Washington, D.C., December 29th.

May, R.M., 1976, Simple mathematical models with very complicated dynamics: Nature, v. 261, p. 459-467.

Nese, J.M., 1985, Phase space structure and dimension of attractors of finite spectral models [M.S. thesis]: University Park, Pennsylvania, Department of Meteorology, The Pennsylvania State University, 180 p.

Shaw, H.R., 1987, The periodic structure of the natural record and nonlinear dynamics: EOS (American Geophysical Union Transactions), v. 68, p. 1651-1665.

Shirer, H.N., 1987, ed., Nonlinear hydrodynamic modeling: a mathematical introduction: Berlin, Springer-Verlag, Lecture Notes in Physics, 546 p.

Stewart, H.B., and Thompson, J.M., 1986, Nonlinear Dynamics and Chaos: Chichester, Wiley, 376 p.

4

LIMITS TO THE PREDICTIVE ABILITY OF DYNAMIC MODELS THAT SIMULATE CLASTIC SEDIMENTATION

Daniel M. Tetzlaff
Atlas Wireline Services, Western Atlas International, Inc., P.O. Box 1407, Houston, Texas 77251 USA

ABSTRACT

An important practical application of models that simulate processes in the geologic past is predicting the present configuration of geologic features. This use has inherent limitations that arise from three interdependent factors: adequacy of the model in representing the actual processes, knowledge of initial conditions and boundary conditions, and randomness in the simulated processes. A three-dimensional computer model, SEDSIM, illustrates these factors.

A model's adequacy in representing geologic processes depends on the degree to which the processes are understood and on the computer power available for simulating them. Although both are limited at present, there appears to be no limit to their future development. Quantitative information about initial and boundary conditions, athough necessary for operating a simulation model, often is unavailable. A common approach to place limits on these boundary conditions is to perform repeated experimental runs in which the conditions are progressively adjusted so that the simulated results match available observations. The development of automatic adjusting procedures could greatly facilitate the process.

Randomness, in a broad sense, can appear even in models that are defined to be completely deterministic. Processes such as fluid turbulence are deterministic because they are governed by precise physical laws, but their behavior can be predicted only within certain limits. These limits cannot be reduced by increasing the precision of the initial and boundary conditions. Such processes have been called "chaotic," and they abound in nature. They may place an absolute limit to the predictive ability of dynamic models.

INTRODUCTION

The simulation of geologic processes serves three main purposes: (1) to help achieve a better understanding of the physical processes represented, (2) to understand present configurations of geologic features by simulating how they were formed, and (3) to predict the effect of geologic processes acting in the future for engineering or environmental management purposes. This paper focuses on the second purpose, and illustrates the principles, methods and limitations of developing models for such simulations by examples. The examples use

Quantitative Dynamic Stratigraphy (1989), T.A. Cross, ed., Prentice Hall, p. 55–65.

the computer program SEDSIM (Tetzlaff, 1987) to simulate sedimentary processes and responses. The principles they represent, however, apply generally to all process models that simulate accumulation of clastic sediments.

SEDSIM simulates free-surface flow in two horizontal dimensions, taking into account flow depth, but using vertically averaged flow parameters. The flow equations are approximated by means of a "marker-in-cell" or "particle-cell" method (Harlow, 1964; Hockney and Eastwood, 1981), which assumes that the fluid consists of a large number of small elementary volumes moving over a fixed grid. Erosion, transport, and deposition of clastic sediment are handled by SEDSIM using a modification of the Meyer-Peter and Muller (1948) formula.

SEDSIM is lengthy to execute but extremely flexible. It can simulate steady or unsteady flow in a variety of clastic environments (terrestrial, marine, or mixed). The program can handle up to four sediment grain sizes and their mixtures. Sedimentary deposits are represented in three dimensions and can be displayed graphically as maps, sections, and perspective views. Successions of displays show the evolution of the simulated systems through time.

Adequacy of Geologic Process Models

The representation of an actual physical system by a model may involve two types of approximations: a reduction in the number of spatial dimensions, and a simplification of the physical processes involved. The ways in which physical processes may be represented and the consequences of simplifications employed are specific to each model (Abraham et al., 1981), and are not considered here. However, the representation of space and time and the limitations of those representations are common to all sedimentation models and are discussed.

Representation of Space

The number of spatial dimensions represented in a model is a key factor in the balance between model realism and computer power limitations. Before the availability of digital computers, most geologic models were limited to one spatial dimension. With computers, however, they became two dimensional (2–D). With the present state of computer hardware and software, the number of spatial dimensions is ceasing to be a limitation, and three-dimensional (3-D) dynamic models are becoming more common.

Two-dimensional vertical or horizontal sedimentation models are most commonly used today. They present results in a manner that is familiar to geologists, and their computational demands are much less than those of 3–D models. Yet, they can simulate processes whose responses are difficult to predict intuitively or analytically.

One limitation of 2–D models is their potential failure to accurately represent the real world. Since at least one dimension is not represented, the simulated process(es) may be assumed erroneously to lack change in the missing dimension(s). As an example, Figure 1 compares the 2–D output, in the form of synthetic stratigraphic cross sections, of a 2–D and a 3–D model that simulate a prograding delta. Both models assume a steady source of water and sediment provided by a river flowing into a depositional basin. Figure 1A was produced by modifying algorithms in the 3–D program SEDSIM so that it would run in only two dimensions with no flow perpendicular to the plane of the figure. The simulated deposits are relatively homogeneous and the sequence is monotonously coarsening upwards. When SEDSIM is run in three dimensions (Fig. 1B), it produces alternating layers of coarse and fine sediment. The figure shows a similar coarsening-upwards trend of the entire deposit, but with large vertical and lateral variations in grain size. These variations are caused because the fluvial channels and the main centers of deposition in the 3–D model are allowed to shift laterally, whereas in the 2–D model they are not. Although both models display output in the same 2–D form, the differences are due to model algorithms that operate in either two (Fig. 1A) or three (Fig. 1B) dimensions.

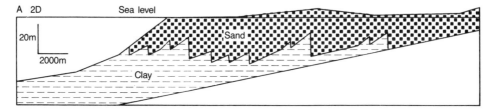

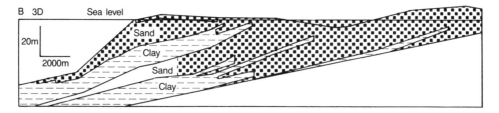

Figure 1. Comparison of the 2-D output of two models simulating deltaic deposition. Both models operate under similar conditions and numerical methods, but the upper model (A) involves only two dimensions, ignoring flow across the section shown, whereas the lower model (B) is three dimensional.

The appropriateness of a model with less than three spatial dimensions also may depend on the scale of the system being investigated. For example, although individual streams may flow perpendicular to the plane of the cross section in Figure 1B, the net regional flow and sediment transport are parallel to the section. If the geologic system under consideration is sufficiently larger than the scale of lateral movement of rivers within the delta system, then a 2–D model might provide a sufficiently realistic simulation of that system. This is because the effects of flows that are perpendicular to the plane of the simulation may average out, and the 2-D model may be an adequate (although not necessarily accurate) representation of the system that is of concern to the investigator.

A similar scale effect involving the vertical dimension must be considered when running SEDSIM. The SEDSIM program uses vertically-averaged flow parameters, and thus does not fully represent the vertical dimension when simulating flow. Therefore, SEDSIM cannot simulate small sedimentary features, such as ripples, because vertical flow variations are important at a scale of a few centimeters. However, if features of larger scale are the focus of the modeling, the small-scale vertical-flow variations become unimportant and SEDSIM can adequately simulate them.

Representation of Time

Time usually is represented in sedimentation models as a series of discrete steps. Processes that occur relatively slowly, such as changes in heat flow or crustal subsidence by thermal contraction, can be represented with adequate resolution using long time steps in a single model. Similarly, processes that occur relatively quickly, such as river meandering or avulsion, can be represented using short time steps in a single model. A major problem with time representation in models that attempt a relatively detailed reconstruction of the geologic past is representing the operation of short-term processes through longer periods of geologic time, without requiring excessively long computations.

SEDSIM utilizes two methods to span this range of temporal scales and combine short-term and long-term processes more efficiently. The first method, termed *time-extrapolation*, is employed for sedimentary processes that are relatively steady; that is, their intensity changes slowly with time. The method consists of running the model over a relatively short period of simulated time (e.g., a few days), and recording changes that occur due to erosion or deposition during that period. These changes are then extrapolated over a longer period of time (e.g., a year). The combined results establish a new base for successive iterations of the two-step algorithm. The extrapolated time cannot be made too large, or unrealistic results will ensue.

The second method, termed *time-skipping*, can be used for unsteady or punctuated processes. Some sedimentary processes, such as turbidity currents or river floods, operate only during brief periods of time and are inactive during longer, intervening intervals. In the time-skipping approach, the processes are modeled only during the brief intervals they operate, and calculations are omitted for the intervening periods. Turbidity currents are ideal examples provided that pelagic sedimentation between flows is negligible. The model can run for the few hours of simulated time during which a specific turbidity flow occurs. It then simulates the next flow, which might occur many years later, without simulating the interval between the two events.

River floods also are highly unsteady. Although they may last only a few days, they often are responsible for most of the sediment eroded or deposited by a river during a year. floods are more difficult to simulate using time-skipping. first, floods of different intensities occur at different time intervals. Second, although the steady processes that occur between floods may produce small changes, they usually are not negligible. By superimposing time-extrapolation and time-skipping in a simulation model, this difficulty is circumvented.

Most limitations that affect a model's realism ultimately reside in two causes: the degree to which we understand the processes involved, and the computer capability available to run the model. The usual limiting factor is computer capability. Many of the limitations in a model's adequacy to represent sedimentary processes could be reduced or eliminated through the use of more powerful computers. Fortunately, the present rate of increase in computer capabilities promises tremendous advances in the next decade.

KNOWLEDGE OF PAST CONDITIONS

Previous geologic conditions must not be ignored in order to operate a sedimentary model because unlike a real sedimentary system, a simulated system is isolated from its "surroundings" by space and time boundaries. Conditions at these boundaries must be provided as input to the model, and can be divided into (1) spatial boundary conditions (relationships between variables specified at the model's geographic boundaries), and (2) initial conditions (values of variables specified at the outset of a simulation experiment). Both conditions limit the ability of a model to accurately predict the configuration of sedimentary deposits because it is usually difficult to establish what they were in the past for the system being simulated.

Spatial Boundary Conditions

Conditions that specify the interactions between a simulated system and the outside world are imposed at the spatial boundaries. Boundary conditions must be defined because a simulated system generally cannot be treated as being closed, and external variables that affect the system must be specified. Even the earth as a whole cannot be treated as a closed system because variations in solar radiation, for example, may indirectly but significantly affect sedimentation through climatic changes. By specifying the boundary conditions, the model is prevented from evolving independently of external influences or constraints.

A boundary condition that must be specified in most sedimentation process models is how fluid and sediment are treated at the edges (spatial limits) of the simulated area. In rudimentary models where sedimentation is affected only by conditions upstream, the model may work properly without rigorous definition of boundary conditions along the lateral edges. But more realistic models are affected by the same problems along their edges as are physical models of sedimentation. For example, if flow is unrestricted along the edges (i.e., water and sediment are removed as they reach the edges), a "waterfall effect" results in which flow and erosion increase drastically at outflow points (Fig. 2A). By contrast, if flow is prevented across the boundary sediment tends to accumulate near the edges (Fig. 2C) because flow velocities become very low.

Through experimentation with SEDSIM using these two end-member as well as intermediate conditions, a satisfactory solution to this boundary problem was achieved. The solution allowed free outflow of water, but prohibited erosion and deposition of sediment along the edges (Fig. 2B). However, the simulated deposits still differ from those that would form if the boundaries were more distant. Figure 3 shows the "wedge effect" that occurs near the boundaries as a result of decreasing sedimentation rates toward the edges.

Figure 3 also shows how the locations and characteristics of fluid and sediment sources may affect the behavior of a sedimentation process model. In SEDSIM, the use of flow rates or sediment input rates that are slightly inappropriate will cause a dynamic disequilibrium in which net erosion or deposition occurs near the inflow, creating holes or mounds, respectively, near the sources (Fig. 3). The appearance of such artificial features indicates the use of inappropriate boundary conditions at the sources. Thus, although SEDSIM requires correct flow and sediment input rates, these parameters may be adjusted by trial and error if they are not precisely known.

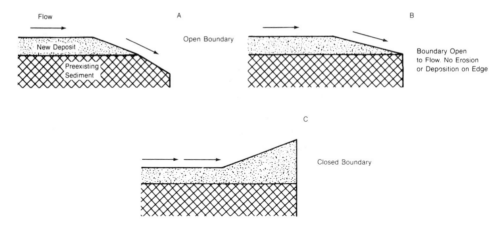

Figure 2. Alternative procedures for representing outflow conditions in sedimentation simulation models. (A) Open boundary results in excessive flow and rapid erosion along the boundaries. (B) Boundary that is open to flow but at which no erosion or deposition is permitted results in wedge effect that has least influence within simulated area. (C) Closed boundary results in excessive deposition.

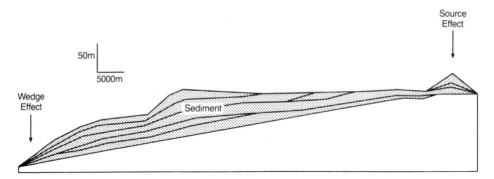

Figure 3. Boundary conditions that affect predictive ability of both sedimentation process models, and physical sedimentation experiments. Wedging of beds occurs along the edge of the simulated area, and a sediment mound forms near the source.

Initial Conditions

In geology, our knowledge of the initial conditions of a model used to predict present configurations of sedimentary deposits usually is limited. Paleogeographic and paleoenvironmental conditions can be estimated only from the observed sedimentary record. An ideal solution, when initial conditions are not known, might be to devise a model that would run backwards. The model would be supplied with information about present geologic configurations, and would reproduce in reverse the processes that generated them.

Initially, it might seem theoretically possible to devise such a model. Many physical processes are symmetric with respect to time. For example, given the position and velocity of a ball on a pool table, we can trace its movement backwards for a few seconds with the same accuracy as we can trace its forward path. Friction has a significant effect, however, and information about the ball's past history is progressively lost. When friction is considered, we must know the ball's position and velocity with greater accuracy if we are to retrace its movement backwards. When the ball comes to rest, we can no longer deduce its past history. Similarly, geologic features do not retain all the information about how they were formed. No matter how detailed our inverse models might be, they cannot retrace the past using only information from the present.

Pseudoinverse Modeling

One possible way to circumvent the problem of incomplete knowledge of initial conditions in forward models is to use pseudoinverse modeling (see also Lerche, this volume). Pseudoinverse modeling is a trial-and-error procedure in which a model's input is progressively adjusted in successive computer experiments, in an attempt to achieve a closer match between simulated and actual geologic features. Pseudoinverse modeling can be tedious and expensive when each experiment involves extensive computations. It is also partly subjective because intuition often is used in deciding how to adjust the input to improve the results.

To reduce subjectivity in pseudoinverse modeling, an automatic or interative minimization procedure may be employed. First, a function is defined that represents, as a single number, the error between the simulation's results and the observed features. Then the combination of input parameters that minimizes the error function must be determined. Either standard minimization procedures (Himmelblau, 1981; Dennis and Schnabel, 1983), or procedures that do not require knowledge of the derivatives of the error function (Brent, 1973) may be used.

The error function may be difficult to define, and the choice in its construction depends on the characteristics of the simulated features that are considered most important. For example, a function that accounts only for differences in overall shaliness between simulated and actual deposits might yield deposits with the desired average shaliness, but shale distribution could differ significantly between simulated and actual deposits. By contrast, a function that is too detailed, such as an index reflecting point-by-point correlation between observed and simulated deposits will complicate the minimization procedure.

Preliminary results with simple sedimentary systems, simulated by SEDSIM, suggest an error function (E) for sand/shale sequences. The expression determined most appropriate is the sum of the squares of the differences between the moments of vertical shaliness distribution, integrated over the simulated area:

$$E = \int_A \sum_{i=1}^{n} \left[M_i(p) - m_i(p) \right]^2 dp$$

where A is the area of the simulated system, and n is the maximum order of moments considered. $M_i(p)$ is the shaliness moment of order i for observed system at point p, defined as:

$$\int_{-1}^{1} z^i\, S(z)\, dz$$

where z is the depth scaled so that -1 represents the bottom of the sequence and +1 represents the top, $S(z)$ is the observed shaliness at level z above point p, $m_i(p)$ is the shaliness moment of order i for the simulated system at point p (defined as M_i, but for simulated rather than observed shaliness), and p is the point in the horizontal plane. The value of n, a positive integer, can be selected to define the detail of model adjustment. A value of +1 will lead to agreement in average shaliness only; higher values of n will take into account vertical variations in shaliness with progressively greater detail.

RANDOMNESS

A simulation model is deterministic when every state is unequivocally determined by any preceding state. A probabalistic model, on the other hand, is one that contains elements of chance, and its future performance cannot be predicted except in a statistical sense. As sedimentary systems are better understood, their representation with deterministic models will become more common, but there may be absolute limits to the ability of deterministic models in reproducing actual systems.

Pseudorandom Models

A deterministic dynamic model can be regarded as a function (f) that links the initial state of a system with its state after a given period of time. For example, the initial system could be represented by a specified topographic surface, a point on that surface where a river flows into the area, and rates of water and sediment discharge. After a simulated period of time a deterministic model unequivocally leads to a specific new state.

The function f may be continuous or discontinuous. An example of a discontinuity would be a ridge or watershed in the area simulated. A series of experiments in which the point source of the river is moved a short distance each time with respect to its position in the preceding experiment would show a "jump" difference in the model's response when the source crosses the ridge.

Alternatively, the function f may be discontinuous not just at a few points, but at all points over a region within the set of specified initial conditions. An example is shown in Figure 4, which displays the results of two experiments with the SEDSIM program. Both experiments simulate 20 turbidity flows, and the state of the system is shown after the 20th flow. The only difference between the two experiments is that in the first of the 20 flows represented, the initial position of a single fluid element was changed. The difference is minute, considering that hundreds of fluid elements are represented. Yet, the result after the 20th run is significantly different. The flows emerged from their respective canyons in different directions, and deposition occurred on different sides of the simulated fans. The large differences in the final states result from the small initial difference in flow conditions, which was amplified by turbulence, erosion, and deposition. The difference between the final states in the two experiments cannot be reduced significantly by reducing the difference between the initial states, unless the initial states are made absolutely identical. Even though SEDSIM is deterministic, its behavior is remarkably similar to that of a probabalistic model.

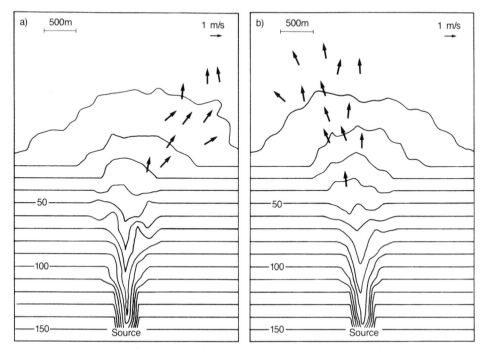

Figure 4. Contour maps comparing the results of two computer experiments simulating 20 turbidity flows. The experiments were identical, except for a very small difference in position of the first flow. These results differ appreciably due to pseudorandom or chaotic behavior. The contours are shown in meters above arbitrary datum.

A model, such as SEDSIM, in which f is discontinuous at every point, can be called a pseudorandom model. Given this definition, pseudorandom models include "chaotic" models, which after a certain period of time, evolve within a bounded set of states, even though the states are never repeated (Mayer-Kress, 1986; Schuster, 1975; Slingerland, this volume). Many properties of chaotic models are also applicable to the broader category of pseudorandom models.

If we use a pseudorandom model to predict on which side of a turbidity fan deposition is going to occur after 20 flows, we need to know the initial conditions exactly. Knowing the conditions with arbitrarily high precision is not sufficient. If the system is truly random, even exact knowledge of initial conditions is not sufficient for such predictions. Pseudorandom models differ from random models only when initial conditions are known exactly, a situation that is unattainable in real-world applications. From a purely theoretical standpoint, however, pseudorandom models are deterministic.

As discussed previously, it may be impossible to interpret past histories from present configurations because information about the past is lost, and many alternative possible past states may have led to the present state. The inverse is also true. Information about the initial state is not sufficient to predict the future because an incompletely known initial state could lead to many possible future states (Fig. 5).

Pseudorandom models such as SEDSIM can still be useful for interpretation if used cautiously, employing the following steps (Fig. 5): (1) make an "observation" consisting of all the available geologic data of the geologic features in the area of interest, (2) use pseudoinverse modeling to obtain a probability distribution of all the initial states that could have produced the observed features, (3) operate the model in a forward direction to produce a probability distribution of all the final states that the possible initial states could have generated, and (4) limit or weight this probability distribution in accord with the observations used initially. The resulting probability distribution of present states may be the best that can be obtained using simulation models.

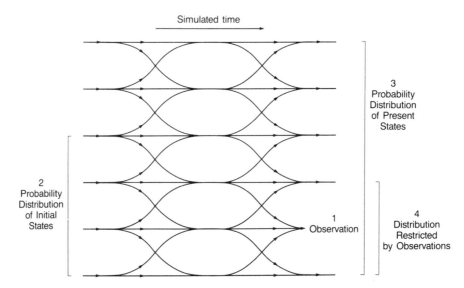

Figure 5. Schematic representation of the pseudorandom simulation modeling procedure. The possible initial states are at the left, and the states after a given period of time are at the right. Pseudorandomness causes the paths to diverge, while a loss of information causes the paths to merge. The ensuing "shuffling" allows only probabilistic predictions to be made, using the steps indicated by numbers 1 through 4.

Such a procedure may seem awkward, but it parallels the scientific method as commonly employed in geology: (1) observe geologic features, (2) propose several possible models that could be responsible for the observed features, (3) infer the configurations of geologic features these conceptual models could have produced, and (4) discard the configurations that do not fit observations. finally, the remaining possible configurations may be used in predicting geological features at locations where there are no observations. Using simulation models, it is possible to use this conceptual method in a rigorous quantitative manner.

Conclusions

The ability of a simulation model to predict the present configuration of geologic features is limited by several factors. Some limitations depend on the computer power available to perform simulations, but they will probably be overcome as computers become more powerful. However, there are other absolute limitations to the ability of models to predict present features, or to "postdict" features produced in the past. Even if our models were totally realistic, we might be forced to adopt a probabilistic approach, in which statistical distributions of possible geologic configurations are incorporated. We may be unable to do any better because of limitations imposed by loss of information and pseudorandomness.

Acknowledgments

I wish to thank Atlas Wireline Services for supporting my attendance at the Workshop on Quantitative Dynamic Stratigraphy and for encouraging the presentation of this paper. I also thank Prof. John W. Harbaugh of Stanford University, who directed the development of computer program SEDSIM. Finally, I thank James P.M. Syvitski of the Geological Survey of Canada, Timothy A. Cross of the Colorado School of Mines, and J. W. Harbaugh for their constructive reviews of an earlier version of this paper.

References Cited

Abraham, G., van Os, A.G., and Verboom, G.K., 1981, Mathematical modeling of flows and transport of conservative substances: requirements for predictive ability, *in* Fischer, H.B., ed., Transport models for inland and coastal waters, Proceedings of a symposium on predictive ability: New York, Academic Press, p. 1-31.

Brent, R.P., 1973, Algorithms for minimization without derivatives: Englewood Cliffs, New Jersey, Prentice Hall, Series in Automatic Computing, 195 p.

Dennis, J.E., Jr., and Schnabel, R.B., 1983, Numerical methods for unconstrained optimization and nonlinear equations: Englewood Cliffs, New Jersey, Prentice Hall, 358 p.

Harlow, F.H., 1964, The particle-in-cell method for fluid dynamics, *in* Alder, B., ed., Computational physics: New York, Academic Press, v 3, p. 319-343.

Himmelblau, D.M., 1972, Applied nonlinear programming: New York, McGraw-Hill, 498 p.

Hockney, R.W, and Eastwood, J.W., 1981, Computer simulation using particles: New York, McGraw-Hill, 540 p.

Mayer-Kress, G., 1986, ed., Dimensions and entropies in chaotic systems: New York, Springer-Verlag, 382 p.

Meyer-Peter, E., and Muller, R., 1948, Formulas for bedload transport: Stockholm, Proceedings, Third Meeting of the International Association of Hydraulic Resources, p. 39-64.

Schuster, R.E., 1975, Deterministic chaos, an introduction: New York, VCH Publishers, 228 p.

Tetzlaff D.M., 1987, A simulation model of clastic sedimentary processes [Ph.D. thesis]: Stanford, California, Stanford University, 345 p.

5

Choice of Lithology Identifiers for Modeling Purposes in Quantitative Stratigraphy—A Role for Petrophysics in Stratigraphy

Cedric M. Griffiths
University of Trondheim, Division of Petroleum Engineering and Applied Geophysics,
S P Andersens v.15, N-7034 Trondheim, Norway

Abstract

Effective modeling needs a precise language. Geological language often lacks precision, but its flexibility has served traditional interpretations and applications well. If we are to move from a descriptive stratigraphy to one that makes quantitative predictions, we must find a way of increasing the precision of language while maintaining much of the flexibility of current terminology. An example is the inclusion of temporal terms in stochastic models.

Sandstone is a common geological term that is of limited use in modeling due to the need to add so many qualifying statements or terms before it becomes unambiguous. This imprecision makes forward modeling easy (but untestable), and inverse modeling very difficult.

Petrophysical analysis provides a relatively accurate "snapshot" of a lithology that can be incorporated in both vertical and areal stochastic models derived from seismic and well data. Such models are being used today for reservoir description and basin evaluation in the oil industry. Regardless of the increased accuracy that petrophysically-defined lithologies may provide, they often lack the genetic component that enables development of predictive temporal and spatial models. The challenge is to derive precise genetic features from petrophysics to improve the predictive capacity of stratigraphic models.

There are many problems in developing a terminology for petrophysical attributes that can have predictive potential in modeling studies. This paper examines some of the requirements for such a language, and suggests a systematic approach to the problem, using syntactic pattern recognition and both classical and petrophysical descriptors.

Introduction

In many fields of geology there is increasing pressure on geologists and geophysicists to make quantitative predictions about vertical and lateral variations in rock properties. In making predictions it is useful to consider several hypotheses that can be tested against available data and ranked according to degree of correlation or certainty. Hypothesis testing is one impetus for quantitative modeling.

Quantitative Dynamic Stratigraphy (1989), T.A. Cross, ed., Prentice Hall, p. 67–86.

67

However, geologists have been slow to establish formal hypothesis tests in stratigraphy. One possible reason for this is that the language of geological expression does not lend itself to precise communication and testing. In the field of clastic sedimentology, for example, grain size is the most commonly used classification base. Yet, as Miall (1984, p. 179) has shown, if grain size alone is used in vertical profile analysis, it is not possible to predict lithosome shape because similar changes in grain size through vertical profiles may be produced by different processes and, therefore, may be associated with different types of sedimentary bodies. As an example, similar coarsening- and thickening-upward profiles may be generated by prograding alluvial fans, river-dominated deltas, wave-dominated deltas, barrier islands, storm-dominated shorelines, or submarine fans.

At the level of a single sample it is necessary to include both textual and structural qualifiers to a term before the level of ambiguity is reduced sufficiently such that hypothesis testing would stand a chance of success. Miall's (1978) fluvial facies scheme is an example of such a combined terminology, but even in this scheme there is considerable ambiguity that would make the units difficult to use in a predictive way. Unit 'Sh,' for example, includes all sandstone grain-size subclasses with horizontal lamination that may be produced by both high and low flow velocity regimes.

Another common classification scheme, derived from principal component and/or cluster analysis, provides a multivariate terminology that includes grain size, chemical and structural elements (e.g., Imbrie and Purdy, 1962; Park, 1974). The descriptive terminology produced has limited utility, and the number of major terms generally is limited to less than ten for psychological reasons related to human short-term memory capacity (see Griffiths, 1982).

Grain size is commonly used as a classifier in siliciclastic rocks for several reasons: it is easy to measure in the field (at least the modal grain size); it may be related directly to the energy of the depositing medium; it exerts a significant control on rock properties; and siliciclastic rocks are composed of grains bound together in one way or another. However, if we have seismic or petrophysical data and we wish to model the three-dimensional development of stratigraphic units and their relationships, then grain size alone is not sufficient because it has an ambiguous relationship to larger-scale properties.

There are many problems in developing a terminology for petrophysical attributes that can have predictive potential in modeling studies. This paper examines some of the requirements for such a language, and suggests a systematic approach to the problem that uses both classical and petrophysical descriptors.

Syntax Analysis

The formal study of languages has been boosted recently by the need to provide fast and efficient compilers for computer languages and by the drive towards natural language computer communications. At present, computer systems are not very good at handling ambiguity, and computing languages have a limited vocabulary and well-defined, rigid structures. A common way of representing this structure, or syntax, is the Backus-Naur-Form (BNF) notation, first developed to define the syntax of Algol 60 (Chomsky, 1965). A language can be considered as having two components. The simplest is its grammar or syntactic rule system that determines whether or not a legal phrase has been generated. For example, "Geologists do it with hammers." is a legal English sentence. A simple substitution syntactic analyser or "parser" could quickly recognize that a sentence had been generated, as it also would with the phrase "Hammers do it with geologists." The BNF grammar for these phrases would appear as follows.

<SENTENCE>::=<NOUN.PHRASE> <VERB.PHRASE–PREPOSITIONAL.PHRASES>
<SENTENCE>::=<SENTENCE> <CONJUNCTION> <SENTENCE>
<NOUN.PHRASE>::=<DETERMINER> <ADJECTIVES–NOUN> I <PLURAL> I <PRONOUN>
<ADJECTIVES–NOUN>::=<ADJECTIVE> <ADJECTIVES–NOUN>
<ADJECTIVES–NOUN>::=<NOUN>
<VERB.PHRASE–PREPOSITIONAL.PHRASES>::=<VERB.PHRASE–PREPOSITIONAL.PHRASES>
        <PREPOSITIONAL.PHRASE>
<VERB.PHRASE–PREPOSITIONAL.PHRASES>::=<VERB.PHRASE>
<VERB.PHRASE–PREPOSITIONAL.PHRASES>::=<VERB> <PREPOSITIONAL.PHRASE>
<VERB.PHRASE>::=<VERB>
<VERB.PHRASE>::=<VERB> <NOUN.PHRASE>
<PREPOSITIONAL.PHRASE>::=<PREPOSITION> <NOUN.PHRASE>
<PREPOSITIONAL.PHRASE>::=<NOUN.PHRASE> <PREPOSITION> <NOUN.PHRASE>
<DETERMINER>::=a I the I this I that
<ADJECTIVE>::=strong I big I small
<NOUN>::=geologist I hammer
<PRONOUN>::=it
<PLURAL>::=geologists I hammers
<VERB>::=does I do
<PREPOSITION>::=to I of I with
<CONJUNCTION>::=and

These rules state that a sequence of noun phrase followed by a verb phrase–prepositional phrases forms a sentence in English. Each of the components to the right of the '::=' symbol can be substituted by the symbol to the left. A selection of symbols which may be substituted for each other are separated by 'I.'

The BNF notation describes a simple context-free hierarchical structure that combines "terminal" and "nonterminal" symbols. Terminal symbols are the terminology of the language, the means of communication. Nonterminal symbols describe progressively higher levels of structure developed according to defined rules. The symbols '<' and '>' are used to delineate nonterminal higher-order symbols that substitute for either the terminal or lower-order nonterminal symbols.

These rules can be used for both forward modeling and pattern recognition. The forward modeling capability arises because a series of legal alternatives can be generated from any high-level structure. The difference between the two can be illustrated by considering the use of sedimentation models in seismic modeling. What type of legal facies sequence could be expected given an anticipated sedimentary environment? Using the relevant syntax for a particular sedimentary environment, a set of legal synthetic facies sequences can be generated which, after production of synthetic seismograms, can be tested against the observed seismic traces. This is the forward modeling capability offered by syntactic analysis. In this context it allows for more complex sequence structures than the approach suggested by Sinvhal and Khattri (1983, 1984).

As a pattern recognition tool the grammatical rules for different environments can be applied to a petrophysical or grain-size data set, and the degree of success in parsing the data sequence is an indication of the conformity of the data with the theoretical model embodied in the syntax. Such a pattern recognition mode applied to a data set from a known sedimentary environment also can test the sufficiency of the grammatical rule set.

The second and more complex part of a language is its semantics. This is exemplified by the phrase "time flies like an arrow" which can be interpreted in several different ways. It is syntactically correct according to the rules of English, but it is ambiguous without further information.

Syntactic analysis is applicable to more domains than the structure of natural and computer languages. In its general sense it is a pattern recognition tool that can be used to identify relationships within sequences. It has been used to identify chromosome shapes (Ledley, 1964, 1965), chemical bond patterns (Watanabe, 1985), and image shapes (Gaglio, Morasso, and Tagliasco, 1982; Gonzales, 1983, p. 363).

Stratigraphic Syntax

In a geological context, many features are governed by natural physical or chemical laws, some of which are reasonably well known. In the field of stratigraphy and basin analysis there are some additional principles that explain or synthesize common stratigraphic elements, such as Walther's Law of Facies Succession, the Law of Superposition of Strata, and Gressly's Facies concept. Each of these stratigraphic laws or principles has been the subject of much discussion (for example, Krumbein and Sloss, 1963, p. 299; Miall, 1984, p. 133 and 277; Davis, 1983, p. 106; Reading, 1978, p. 4; Visher, 1984, p. 37).

The Law of Superposition of Strata, if taken in the most strict sense, may be flawed at small scales; for example, load structures or fluidization of sediment locally invert normal stratigraphic superposition. But if we accept deviations at this scale, then it can still be considered true that in a vertical sequence successive layers of undeformed strata are younger upwards. The facies concept has been interpreted in various ways, but in this paper it will be used in Gressley's original sense of being a collection of physical properties that are distinguishable from those of neighboring rocks at a chosen scale (Krumbein and Sloss, 1963, p. 317). In this sense it is close to Serra's (1977) electrofacies concept. Walther's law has potential as a predictive tool if used selectively, and may be used to form the basis of a syntactic rule set. For any given depositional system at a specified scale, there will be a set of facies relationships such that a traverse in a given direction relative to the direction of transport or the direction of progradation would be expected to encounter a given sequence of facies with a known probability.

In stratigraphy we thus have a set of terminal symbols (facies) which, if they can be mapped unambiguously within a present-day depositional system, also can be identified using the same syntax from borehole sequences if there were no intervening periods of erosion. As an example, the following sequence of facies and facies associations for carbonate shelf deposits is taken from Serra (1985, p. 144; after Wilson, 1975). Wilson identified nine major facies groups that may be representative of a horizontal sequence from deep to shallow water environments as shown below.

<1>::=black shale | evaporite | thin–limestone | <1> <1>

<2>::=marl | fossiliferous–limestone | <2> <2>

<3>::=fine–limestone | chert | <3> <3>

<4>::=breccia | <4> <4>

<5>::=massive–carb | limesand <5> | <5> limesand | <5> <5>

<6>::=oolites | <6> <6>

<7>::=variable–carb | clastics | <7> <7>

<8>::=dolomitic–limestones | dolomites | <8> <8>

<9>::=dolomite | anhydrite | red–beds | <9> <9>

<basin>::=<2> | <1> <basin> | <basin> <3> | <basin> <basin>

<slope>::=<5> | <4> <slope> | <slope> <slope>

<platform>::=<9> | <8> | <7> <platform> | <6> <platform> | <platform> <platform>

<carb.shelf>::=<basin> <slope> <platform>

The expression '<1> <1>' implies that association (nonterminal symbol) '<1>' can repeat indefinitely and equates to the phrase '<1> *' in strict BNF terminology. This example indicates that an environmental interpretation is possible provided that the rules can be specified with some degree of confidence and that the terminal symbols are not ambiguous.

The syntactic approach has the potential to include quite complex facies relationships in a formal pattern, provided that a set of unambiguous terminal symbols can be identified. In this sense it is more powerful than first- or second-order stochastic modeling and, as will be shown subsequently, it is possible to use stochastic information while building the grammar. Second, the syntactic approach allows inclusion of alternative facies in the set of terminal symbols. This is potentially important in modeling situations where one facies may substitute for another in a vertical or lateral succession according to some empirically or dynamically defined substitution probability. Finally, building a set of syntactic rules for a depositional system is also a useful discipline for testing the adequacy and uniqueness of the available terminology.

Terminal Symbols

In both the professional literature and textbooks, there is a state of terminal disarray concerning choices of facies names and characteristics, and even basic building blocks for sequence descriptions, for most depositional environments. Grain-size profiles, which normally do not include information about grain-size sorting, are most commonly used because of ease of representation and similarity to a weathered rock outcrop (e.g., Miall, 1984, p. 86). This causes problems when comparing maps with columnar sections, as the profile information is not easily translated to map view. There is no sorting information in profiles, the (often hand drawn) sedimentary textures and fabrics do not have standard shapes, and it can be difficult to compare sections drawn by different people.

An example of this can be seen in the two fluvial facies schemes published by Cant and Walker (1976, p. 111-114) and Miall (1978). In these schemes, lithofacies 'F' of Cant and Walker, described as "cross-laminated sandstones . . ., and alternating cross-laminated sandstones and mudstones . . .," would seem to lie in the same position in the sequence as a combination of Miall's 'Sr' (very fine to coarse sand with ripple marks of all kinds) and 'Fl' facies (sand, silt and mud with fine laminations and very small ripples). Distinguishing unambiguously between 'Sr' and 'Fl' would apparently be a problem on the basis of either grain size or the existence of ripples at a given scale, whereas 'F' is a case of a homogeneous unit defined in terms of higher-order structure, that is, the oscillation between two lower-order lithofacies.

SYNTACTIC PATTERN RECOGNITION AND GEOLOGICAL FACIES

The formal application of syntactic analysis to geological sequences faces two main problems. The first is the choice of terminal symbol, and the second is the derivation of a consistent, formal syntax.

Choice of terminal symbols often is limited by the data available, for example grain-size and structural data from cores and sections, petrophysical (including dipmeter) data from oil wells, and impedance data or signal properties from seismic sections. An example of terminal symbols using petrophysical data is a numerical lithology based on a subdivision of M-N-gamma space (Griffiths, 1983), where 'M' and 'N' are the slope of the density slowness and density neutron crossplots, respectively (Burke et al., 1969). M-N-gamma space is subdivided as shown in Figure 1 to give 125 different possible lithostates or terminal symbols with "hard" class edges. This numerical lithology satisfies some of the requirements of a repeatable petrophysical facies set that has an approximate relationship to mineralogy if not to grain size. The choice of numbers within the numerical lithology cube is such that the plotted section has a similar appearance to a traditional stratigraphic section. High gamma shales have low rock numbers, whereas clean quartz and calcite units have high rock numbers. Increasing the 'M' value raises the rock number in units of five, whereas increasing the 'N' value raises the rock number in unit steps. By subdividing velocity-density space in a similar manner a different numerical lithology can be created that is related to porosity and mineralogy. It also would be possible to use results from a standard log analysis package to erect a numerical lithology index based on, for example, relative shale volume, matrix grain density and porosity.

The next problem is one of deriving a syntax from the spatial relationships of rock units such that an environmental interpretation can be derived directly from petrophysical data. Common to all four-dimensional (x,y,z,t) facies relationships is the potential existence of at least two levels of variation or oscillation. Beerbower (1964) applied the terms "autocyclic" and "allocyclic" to different types of oscillation or repetition of facies in sedimentary sequences. Autocyclic oscillations were considered to originate from the normal energy or environmental fluctuations along the borders of two neighboring facies at any scale. Thus facies boundaries are not static, they can oscillate about any point on the margin, and any vertical sequence would record these variations as a cyclicity over time. By contrast, allocyclic oscillations are controlled from outside the immediate environment, for example by tectonic movement, climatic change, or eustatic change.

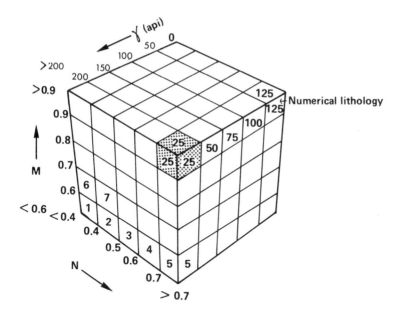

Figure 1. Numerical lithology derivation in M-N gamma space.

Autocyclic oscillations may be produced in two ways. The first is an artifact of the classification process and results in oscillations of facies within a gradational sequence. If boundaries are arbitrarily placed within a continuous sequence, such as a grain-size and sorting distributions, one would anticipate neighboring classes to occur as a function of random variation about the boundary. For example, a single coarsening-upward sequence might display oscillations of medium to coarse to medium to coarse grain-size classes. The second is chaotic in nature and results from instability in or disjunct nature of sedimentation processes. This type of oscillation is typified by the association between basinal muds and slumped sand from an unstable delta front.

Terminal symbols should be incorporated in a syntactic rule set to express both types of autocyclic oscillation, whereas the allocyclic variation should be treated by higher-order nonterminal symbols. Allocyclic oscillations are difficult to treat quantitatively and predictively on the scale of a few hundred meters. By contrast, and in comparison with autocyclic oscillations, they may be modeled more easily on the seismic and basin scale and, therefore, are of more use in stratigraphic correlation.

In discussing conceptual models for deltas, Elliott (1986, p. 114) noted that there is a causal link between hinterland and coastal climate, tectonics, topography, and basinal and fluvial energy regimes to the nature of the delta regime, delta morphology and delta facies pattern. The principal processes involve water transport at various energy levels and directions, and gravity. The main autocyclic gradational features are thus controlled by variations in hydrodynamic energy in three dimensions, and a given flow strength will oscillate about a given point in space. We may anticipate that the grain size transported by the energy at a given point will vary with time. At the very lowest level, terminal symbols should reflect this grain-size oscillation. We can examine this feature quantitatively by looking at a transition matrix where the classes are grain size derived.

The upward transition matrix (UTM) shown in Table 1 was derived from a deltaic sequence in the Jurassic of the North Sea shown in Figure 2. The abbreviations used in the following tables and syntaxes are: s=silt; vf=very fine sand; f=fine sand; m=medium sand; vc=very coarse sand; c=coarse sand; mud=mudstone; lst =limestone; cc=coal. This matrix shows autocyclic oscillations from silt to very fine sand, very fine to fine sand, fine to medium sand, medium to coarse sand, and coarse to very coarse sand. It also shows that mud to silt cannot be considered gradational autocyclic in this sequence, and thus silt to mud may represent either a chaotic autocyclic transition or an allocyclic transition, or that silt and mud are identified differently in different contexts but actually represent the same facies. There are also transitions of more than one grain-size class, such as silt to fine sand, and fine to coarse sand.

The syntax for this type of oscillation is shown below as derived from the UTM.

<ao1>=s vf,vf s

<ao2>=s f,f s

<ao3>=f vf,vf f

<ao4>=vf mud,mud vf

<ao5>=f m,m f

<ao6>=m cc,cc m

<ao7>=m c,c m

<ao8>=cc mud,mud cc

<ao9>=c vc,vc c

Chaotic autocyclic oscillations would not necessarily be identifiable from the transition matrix alone, but in this case the "substitutability" matrix may be of some help. Substitutability analysis is discussed by Davis (1973, p. 288) and Davis and Cocke (1972). Developed originally for the study of word substitution in written text, it identifies states that have the same relationship to other states either above or below, and both above and below. For example, in the case of

"a b c d a b e d a b c d a b e d a b c d a b e d a b c d a b e d"

the upward transition matrix is

	a	b	c	d	e	
a	0	8	0	0	0	8
b	0	0	4	0	4	8
c	0	0	0	4	0	4
d	7	0	0	0	0	7
e	0	0	0	4	0	4

and the upwards substitutability matrix (states tend to have a common successor) is

	a	b	c	d	e
a	1.00	0.00	0.00	0.00	0.00
b	0.00	1.00	0.00	0.00	0.00
c	0.00	0.00	1.00	0.00	1.00
d	0.00	0.00	0.00	1.00	0.00
e	0.00	0.00	1.00	0.00	1.00

and the downwards substitutability matrix (states tend to have a common predecessor) is

	a	b	c	d	e
a	1.00	0.00	0.00	0.00	0.00
b	0.00	1.00	0.00	0.00	0.00
c	0.00	0.00	1.00	0.00	1.00
d	0.00	0.00	0.00	1.00	0.00
e	0.00	0.00	1.00	0.00	1.00

i.e., state 'e' is a perfect substitute for state 'c' in relation both to transitions from, and transitions to states 'b' and 'd.'

It seems likely that the substitutability matrix could be used to identify chaotic autocyclic phenomena. High scores for state pairs on the downward substitutability matrix identify consistent environmental oscillations that possibly are due to chaotic instability in depositional processes. In other words, once a state is in existence its transition to one of two (or more) possible other states depends on the nature of an environmental trigger. Such oscillations may be exemplified by the migration of distributary channels, or the erosion or silting up of a mouth bar complex.

Allocyclic events may be marked by anomalous lithologies within the context of the system being examined, for example, by the existence of a marine limestone or a volcanic ash, or a marine shale. In relation to transition matrices, such events only have significant probabilities in large data sets and they may be found in the relationship between rankings on the upward and downward substitutability matrices. In vertical section, lithostates that have a higher than average score for upward substitutability, combined with lower than average scores (row totals) on the downward substitutability matrix have some of the properties of sequence tops. That is to say that few other states have the same predecessor states, and many other states have similar overlying states. Lithostates with the opposite relationship have properties that one would associate with sequence bases, that is, few other states have the same successors, and many other states share the same predecessor.

BRENT GROUP, NORTHERN NORTH SEA

As a demonstration of some of these concepts we will consider a syntactic analysis of the Middle Jurassic Brent Group. This group of sands, shales, coals, occasional limestones and calcareous sands forms a productive reservoir unit in many parts of the North Sea, and has been described as consisting of several delta lobes with interspersed organic-rich embayments and marine incursions (Norwegian Petroleum Society, 1987). A section showing both core-logged lithologies and numerical lithologies is shown in Figure 2.

The upward transition matrix (UTM) for this group in a cored North Sea well is shown in Table 1. Tables 2 and 3 show the upward (USM) and downward (DSM) substitutability matrices, respectively, derived from Table 1. Several autocyclic pairs are identified on the UTM (Table 1), for example, vf-s, f-s, f-vf, vf-mud, f-m, cc-m, mud-cc, vc-c, c-m. The DSM (Table 3) shows that silt and mud, silt and medium sand, and coal and coarse sand score relatively high and could be indicative of an autocyclic-chaotic relationship. That is, these pairs of lithostates often have common predecessor states. From an examination of both the

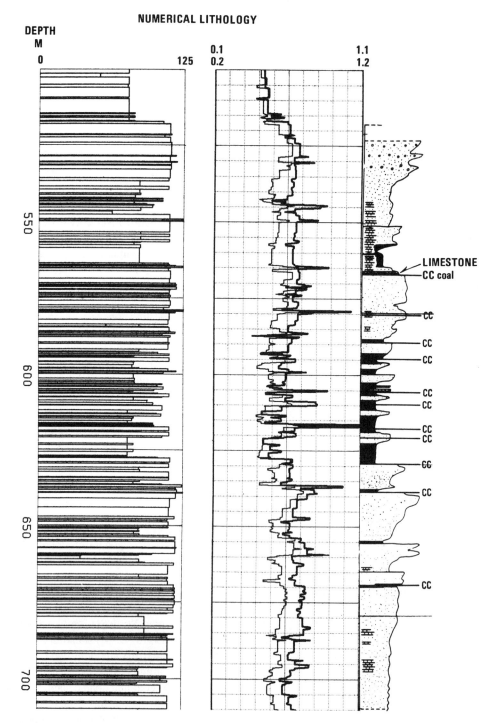

Figure 2. Middle Jurassic deltaic sequence from a North Sea well. 'M' and 'N' values after Burke, Campbell and Schmidt (1969).

DSM and UTM, the autocyclic chaotic syntax could be described as follows.

cc <ac1> s::=cc s

f <ac2> mud::=f mud

vc <ac3> mud::=vc mud

lst <ac4> mud::=lst mud

vf <ac5> cc::=vf cc

f <ac6> cc::=f cc

c <ac7> cc::=c cc

f <ac8> c::=f c

Table 1 The upward transition matrix (UTM) from the Brent section of a North Sea well. Based on grain-size classes.

	mud	s	vf	f	m	c	vc
mud	0	0	4	1	0	0	0
silt	0	0	3	3	0	0	0
v. fine	2	2	0	7	0	0	0
fine	1	2	3	0	5	2	1
medium	0	0	0	2	0	3	0
coarse	0	0	0	1	2	0	5
v.coarse	1	0	0	0	1	3	0

Table 2 The upwards substitutability matrix (USM) derived from Table 1 data.

	s	vf	f	m	cc	vc	c	mud	lst	row total
s	1.00	0.65	0.31	0.26	0.46	0.00	0.13	0.69	0.50	4.00
vf	0.65	1.00	0.15	0.43	0.43	0.08	0.19	0.26	0.84	4.02
f	0.31	0.15	1.00	0.37	0.72	0.52	0.44	0.51	0.10	4.13
m	0.26	0.43	0.37	1.00	0.24	0.50	0.20	0.50	0.26	3.75
cc	0.46	0.43	0.72	0.24	1.00	0.53	0.20	0.39	0.62	4.58
vc	0.00	0.08	0.52	0.50	0.53	1.00	0.11	0.00	0.21	2.94
c	0.13	0.19	0.44	0.20	0.20	0.11	1.00	0.14	0.13	2.53
mud	0.69	0.26	0.51	0.50	0.39	0.00	0.14	1.00	0.14	3.63
lst	0.50	0.84	0.10	0.26	0.62	0.21	0.13	0.14	1.00	3.79

Table 3 The downwards substitutability matrix (DSM) derived from Table 1 data.

	s	vf	f	m	cc	vc	c	mud	lst	row total
s	1.00	0.43	0.62	0.69	0.36	0.13	0.35	0.75	0.24	4.56
vf	0.43	1.00	0.30	0.53	0.52	0.10	0.27	0.36	0.23	3.75
f	0.62	0.30	1.00	0.08	0.42	0.12	0.18	0.55	0.26	3.53
m	0.69	0.53	0.08	1.00	0.37	0.50	0.54	0.51	0.24	4.47
cc	0.36	0.52	0.42	0.37	1.00	0.25	0.60	0.18	0.51	4.21
vc	0.13	0.10	0.12	0.50	0.25	1.00	0.08	0.05	0.00	2.23
c	0.35	0.27	0.18	0.54	0.60	0.08	1.00	0.42	0.59	4.02
mud	0.75	0.36	0.55	0.51	0.18	0.05	0.42	1.00	0.53	4.36
lst	0.24	0.23	0.26	0.24	0.51	0.00	0.59	0.53	1.00	3.60

Finally, comparing the USM (Table 2) and DSM rankings it appears that satisfying the sequence-top criteria would be: very coarse sandstone (vc), limestone (lst), fine sandstone (f), and very fine sandstone (vf). Satisfying the sequence-base criteria would be: mud, coarse sand (c), medium sand (m), and silt (s).

Appendix I lists the syntax reflecting these structures. The data sequence as logged from the core was, from base to top—

s vf f m cc m f vc c vc mud vf f m c vc m cc s f m
cc mud f cc f mud cc mud vf mud cc mud vf mud cc
lst mud vf s f s vf s vf cc vf f cc vf f c cc c f c m
c m cc m lst f s f m f vf f vf f vf f m c vc c vc c vc.

The corresponding interpretation is shown in Figure 3.

Petrophysical Lithostates

If we then take the same well, but look at the petrophysically derived terminal symbols only, then the matrices shown in the appended tables (Appendices IV, V and VI) would be generated, and we could derive a syntax as shown in Appendix II. The numerical lithology sequence for the same interval as the cored section above is shown in Appendix III. The interpretation is shown in Figure 3.

As can be seen by a comparison of the petrophysical and grain-size derived interpretations in Figure 3, the petrophysical grammar has identified many of the delta front lobes that

Figure 3. Comparison between environmental interpretations based on grain size and petrophysically based numerical lithologies.

were recognized in the core, but confusion between limestone and diagenetically calcite-cemented sandstone can be seen. The organic sequence has not been identified from the petrophysical section due to the association of silt units with calcite-cemented sandstones rather than coals.

In the case described above, the second-order syntax is the same in both the petrophysical and core cases. This is not necessary, and work in progress uses the petrophysical data alone in the same way that the sequence boundary criteria were developed. The greater resolution available in practice from a 125 state petrophysical lithology, rather than the core logger's five to nine states, gives the possibility to develop and test more intricate environmental syntax than is available with grain size alone, but it is probable that dipmeter data should be included before a reliable environmental syntax can be developed using petrophysics.

DISCUSSION

There is a wealth of literature on the subject of lithology and paleoenvironmental prediction from subsurface data. Reviews by Cant (1984) and Serra (1985) point out many of the problems. Ambiguity exists in the relationship between a Gressly-type facies (or electrofacies) and an environment of deposition, suggesting that facies order is at least as important as the facies identification itself. The use of first-order sequence data in environmental description and prediction has been widely studied since the early 1960s, and has been essentially nondeterministic in nature. Recent applications are seismic modeling (e.g., Sinvhal et al., 1984) and stochastic shale determination in reservoir engineering (e.g., Haldorsen and Lake, 1984).

The use of first-order stochastic modeling implies that the occurrence of a given facies can be adequately predicted from the preceding facies alone. However, investigations using substitutability matrices, the observation of several orders of cyclicity in sea level, and the existence of similar facies types at different positions in several depositional environments implies that a deeper level of structure is necessary to adequately predict facies occurrence away from the borehole. Syntax analysis offers one way of combining both stochastic and deterministic information in a formal manner. Whether a context-free grammar is adequate for such purposes is open to question, and it may be that contextual information and semantic structures are necessary before some of the ambiguities in interpretation can be removed.

CONCLUSIONS

1. A geological syntax can be developed on the basis of both classical and petrophysical states.
2. Sequences of lithostates can be parsed using such a grammar to give an environmental interpretation from both classical and petrophysical lithologies. This could provide one way of integrating core and log information.
3. A petrophysically-based lithology with structure described by a formal syntax can provide a genetic component that is otherwise lacking from such measurements. The syntax can be developed either empirically from field observations or from a theoretical understanding of depositional environments as in dynamic stratigraphic modeling.

4. Resolving the ambiguity in the interpretation of petrophysically-derived facies may need a semantic element in addition to the context-free grammar illustrated here.

5. The direct application of Walther's Law and formal syntax analysis to electrofacies will be enhanced by the direct mapping of lateral electrofacies variation in the field. Wagoner and McKague (1986) have shown, for example, that it is feasible in some circumstances to map a sedimentary environment in terms of physical properties such as matrix density and moisture content.

6. Changes in petrophysical signature due to diagenesis and compaction must be taken into account when building the syntax. This can partly be done by careful design of the numerical lithology or electrofacies itself, but it is possible to include sequential information in the form of modifications to the depositional grammar.

ACKNOWLEDGMENTS

This work was partly supported by the Norwegian Continental Shelf and Petroleum Technology Institute.

REFERENCES CITED

Beerbower, J.R., 1964, Cyclothems and cyclic depositional mechanisms in alluvial plain sedimentation: Kansas Geological Survey Bulletin 169, v. 1, p. 31-42.

Burke, J.A., Campbell, R.L., Jr., and Schmidt, A.W., 1969, The litho-porosity crossplot: The Log Analyst, v. 10, p. 25-43.

Cant, D.J., 1984, Subsurface facies analysis, in Walker, R.G., ed., Facies models (second edition): Geoscience Canada Reprint Series 1, p. 297-310.

Cant, D.J., and Walker, R.G., 1979, Development of a braided-fluvial facies model for the Devonian Battery Point Formation: Canadian Journal of Earth Sciences, v. 13, p. 102-119.

Chomsky, N., 1965, Aspects of the theory of syntax: Cambridge, Massachusetts, MIT Press, 251 p.

Davis, J.C., 1973, Statistics and data analysis in geology: New York, John Wiley and Sons, 550 p.

Davis, J.C., and Cocke, J.M., 1972, Interpretation of complex lithological successions by substitutability analysis, in Merriam, D.F., ed., Mathematical models of sedimentary processes: New York, Plenum Press, p. 27-52.

Davis, R.A., 1983, Depositional systems: A genetic approach to sedimentary geology: New Jersey, Prentice Hall, 560 p.

Elliott, T., 1986, Deltas, in Reading, H.G., ed., Sedimentary environments and facies (second edition): Oxford, Blackwell Scientific Publications, p. 113-154.

Gaglio, S., Morasso, P., and Tagliasco, V., 1982, Syntactic techniques in scene analysis: New York, Crane Russak and Co., Digital Systems for Industrial Automation, v. 1, no. 2-3, p. 241-258.

Gonzales, R.C., and Wintz, P., 1983, Digital image processing: Reading, Massachusetts, Addison-Wesley Publishing Co., 576 p.

Griffiths, C.M., 1982, Pigeonholes and petrography: Computer applications in geology III, Geological Society of London Miscellaneous Paper 15, p. 81-102.

Griffiths, C.M., 1983, An M-N-Gamma crossplot as the basis of a numerical lithostratigraphy with geotechnical implications: Procedings of International Symposium on Recent Advances in Quantitative Stratigraphic Correlation, Kharagpur, Dec. 12-13, 1983.

Haldorsen, H.H., and Lake, L.W., 1984, A new approach to shale management in field-scale models: Society of Petroleum Engineering Journal, v. 24, p. 447-457.

Imbrie, J., and Purdy, E.G, 1962, Classification of modern Bahamian carbonate sediments: American Association of Petroleum Geologists Memoir 1, p. 253-272.

Krumbein, W.C., and Sloss, L.L., 1963, Stratigraphy and sedimentation (second edition): San Francisco, Freeman, 660 p.

Ledley, R.S., 1964, High-speed automatic analysis of biomedical pictures: Science, v. 146, p. 216-223.

Ledley, R.S., et al, 1965, FIDAC: Film input to digital automatic computer and associated syntax-directed pattern recognition programming system, *in* Tippett, J.T., et al., eds., Optical and electro-optical information processing: Cambridge, Massachusetts, MIT Press, 780 p.

Miall, A.D., 1973, Markov chain analysis applied to an ancient alluvial plain succession: Sedimentology, v. 20, p. 347-364.

Miall, A.D., 1984, Principles of sedimentary basin analysis: New York, Springer-Verlag, 490 p.

Miall, A.D., editor, 1978, Fluvial sedimentology: Canadian Society of Petroleum Geologists Memoir 5, 859 p.

Norwegian Petroleum Society (NPF) Staff, editor, 1987, Geology of the Norwegian oil and gas fields: An atlas of hydrocarbon discoveries: London, Graham and Trotman Ltd.

Park, R.A., 1974, A multivariate analytical strategy for classifying palaeo-environments: Journal of Mathematical Geology, v. 6, p. 333-352.

Serra, O., 1977, Méthode rapide d'analyse faciologique par diagraphies différées (Rapid method of facies analysis from logs): Society of Professional Well Log Analysts, 5th European Logging Symposium Transactions, Paris, France, Paper 9, 15 p.

Serra, O., 1985, Sedimentary facies from wire-line logs: Schlumberger, 211 p.

Sinvhal, A., Khattri, K.N., and Awasthi, A.K., 1984, Seismic indicators of stratigraphy: Geophysics, v. 49, p. 1196-1212.

Visher, G.S., 1984, Exploration Stratigraphy: Tulsa, Oklahoma, Pennwell Press, 350 p.

Wagoner, J.L., and McKague, H.L., 1986, Variation of physical properties of alluvium in an arid basin: Sedimentary Geology, v. 47, p. 53-68.

Watanabe, S., 1985, Pattern recognition: Human and mechanical: New York, John Wiley and Sons, 592 p.

Appendix I: Brent section grain size based syntax.

f <seq.bound> c=f c

vf <seq.bound> c=vf c
vf <seq.bound> m=vf m

cc <seq.bound> c=cc c

lst <seq.bound> mud=lst mud
lst <seq.bound> c=lst c
lst <seq.bound> m=lst m
lst <seq.bound> s=lst s

vc <seq.bound> mud=vc mud
vc <seq.bound> m=vc m
vc <seq.bound> s=vc s

<s>=s
<s>=<s> <s>

<cs>=vc,c
<cs>=<cs> <cs>

<org>=cc,mud
<org>=<org> <org>

<fines>=m,f,very_fine_sand

<fines>=<fines> <fines>

<calc>=lst
<calc>=<calc> <calc>

<C.up>=<fines> <cs>,<s> <cs>
<C.up>=<C.up> <C.up>

<Org.seq>=<org> <fines>,<fines> <org>,<Org.seq> <org>
<Org.seq>=<Org.seq> <Org.seq>

<Delta.front.lobe>=<cs>,<C.up> <Delta.front.lobe>
<Delta.front.lobe>=<fines> <Delta.front.lobe>

<Embayments>=<fines> <org>,<fines> <Org.seq>,<Embayments> <Embayments>

<Transgression>=<calc> <seq.bound>

Appendix II: Brent section numerical lithology based syntax.

87 <seq.bound> 124=87 124
87 <seq.bound> 100=87 100
87 <seq.bound> 106=87 106

118 <seq.bound> 124=118 124
118 <seq.bound> 100=118 100
118 <seq.bound> 106=118 106

107 <seq.bound> 124=107 124
107 <seq.bound> 113=107 113
107 <seq.bound> 100=107 100

88 <seq.bound> 124=88 124
88 <seq.bound> 100=88 100
88 <seq.bound> 106=88 106

<cs>::=14,19,39,44,64,69,89,94,95,114,119,120
<cs>::=<cs> <cs>

<cc>::=22,23,24,25,47,48,49,50,72,73,74,75,97,98,99,100,122,123,124,125
<cc>::=<cc> <cc>

<org>::=1,2,3,4,5,26,27,28,29,30,51,52,53,54,55
<org>::=<org> <org>

<fines>::=76,77,78,79,80,101,102,103
<fines>::=<fines> <fines>

<s>::=6,7,8,9,10,31,32,33,34,35,56,57,58,59,60
<s>::=81,82,83,84,85,106,107,108,109,110
<s>::=<s> <s>

<calc>::=12,13,18,37,38,42,62,63,67,68,87,88,92,93,112,113,117,118
<calc>::=<calc> <calc>

<C.up>::=<fines> <cs>,<s> <cs>
<C.up>::=<C.up> <C.up>

<Org.seq>::=<org> <fines>,<fines> <org>,<Org.seq> <org>
<Org.seq>::=<Org.seq> <Org.seq>

<Delta.front.lobe>::=<cs>,<C.up> <Delta.front.lobe>
<Delta.front.lobe>::=<fines> <Delta.front.lobe>

<Embayments>::=<fines> <org>,<fines> <Org.seq>,<Embayments> <Embayments>

<Transgression>::=<calc> <seq.bound>

Appendix III: Numerical lithology sequence for Brent section of a North Sea well.

82 107 112 87 113 108 113 112 87 88 113 112 107 113 112 107 113 118 113 108 113 112 118 93 118 118 119 118 119
118 119 118 113 114 112 113 118 113 88 37 49 99 119 113 108 113 108 113 118 113 114 119 120 119 113 124 107 82 87 113 112 11
3 114 107 82 107 113 108 113 114 107 106 107 106 107 82 107 106 107 82 107 113 108 107 108 107 82 107 118 112
106 107 108 113 88 113 112 87 112 114 112 118 113 112 113 108 113 114 113 119 118 113 114 112 114 118 113 112 87 112 113
112 113 112 113 123 113 88 113 124 118 107 113 108 113 112 113 114 119 113 118 119 114 113 114.

Appendix IV: The upward transition matrix for the Brent section of a North Sea well. Petrophysically derived numerical lithologies.

	82	107	112	87	113	108	88	118	93	119	114	37	49	99	120	124	106	97	100	123	Row Totals
82	0	5	0	1	0	0	0	0	0	0	0	0	0	1	0	0	0	0	0	0	7
107	5	0	2	0	4	2	0	1	0	0	2	0	0	0	0	0	3	0	0	0	17
112	0	2	0	4	8	2	1	2	0	0	2	0	0	0	0	0	1	0	0	0	19
87	0	0	2	0	2	0	1	0	0	0	0	0	0	0	0	0	0	0	0	0	5
113	0	0	11	0	0	8	3	5	2	1	8	0	0	0	0	2	0	0	0	1	39
108	0	2	0	0	8	0	0	0	0	0	0	0	0	0	0	0	0	0	0	0	10
88	0	0	0	0	3	0	0	1	0	0	0	1	0	0	0	0	0	0	0	0	5
118	0	1	2	0	7	0	0	1	0	4	0	0	0	0	0	0	0	0	0	0	15
93	0	0	0	0	0	0	0	2	0	0	0	0	0	0	0	0	0	0	0	0	2
119	0	0	0	0	3	0	0	4	0	0	1	0	0	0	1	0	0	0	0	0	9
114	0	2	3	0	3	0	0	1	0	2	0	0	0	0	0	0	0	0	0	0	11
37	0	0	0	0	0	0	0	0	0	0	0	0	1	0	0	0	0	0	0	0	1
49	0	0	0	0	0	0	0	0	0	0	0	0	0	1	0	0	0	0	0	0	1
99	0	0	0	0	0	0	0	0	0	1	0	0	0	0	0	0	0	0	0	0	3
120	0	0	0	0	0	0	0	0	0	1	0	0	0	0	0	1	0	0	0	0	1
124	0	1	0	0	0	0	0	2	0	0	0	0	0	0	0	0	0	0	0	0	3
106	0	4	0	0	0	0	0	0	0	0	0	0	0	0	0	0	0	0	0	0	4
97	0	0	0	0	0	0	0	0	0	0	0	0	0	0	0	0	0	0	1	0	1
100	1	0	0	0	0	0	0	0	0	0	0	0	0	0	0	0	0	0	0	0	1
123	0	0	0	0	1	0	0	0	0	0	0	0	0	0	0	0	0	0	0	0	1

Appendix V: Upwards substitutability matrix for numerical lithologies derived from petrophysical data. Brent section, North Sea well.

	82	107	112	87	113	108	88	118	93	119	114	37	49	99	120	124	106	97	100	123
82	1.00	0.00	0.28	0.00	0.00	0.23	0.00	0.11	0.00	0.00	0.37	0.00	0.19	0.00	0.00	0.43	0.96	0.19	0.00	0.00
107	0.00	1.00	0.50	0.52	0.33	0.51	0.51	0.46	0.09	0.40	0.45	0.00	0.00	0.00	0.00	0.12	0.00	0.00	0.65	0.52
112	0.28	0.50	1.00	0.55	0.16	0.86	0.80	0.71	0.15	0.68	0.56	0.00	0.00	0.00	0.00	0.28	0.21	0.00	0.00	0.83
87	0.00	0.52	0.55	1.00	0.49	0.65	0.65	0.63	0.24	0.38	0.77	0.00	0.00	0.00	0.00	0.00	0.00	0.00	0.00	0.67
113	0.00	0.33	0.16	0.49	1.00	0.00	0.00	0.10	0.33	0.32	0.42	0.00	0.00	0.03	0.06	0.26	0.00	0.00	0.00	0.00
108	0.23	0.51	0.86	0.65	0.00	1.00	0.94	0.83	0.00	0.56	0.65	0.00	0.00	0.00	0.00	0.11	0.24	0.00	0.00	0.97
88	0.00	0.51	0.80	0.65	0.00	0.94	1.00	0.81	0.00	0.56	0.56	0.00	0.00	0.00	0.00	0.00	0.00	0.00	0.00	0.97
118	0.11	0.46	0.71	0.63	0.10	0.83	0.81	1.00	0.00	0.48	0.78	0.00	0.00	0.27	0.47	0.05	0.12	0.00	0.00	0.83
93	0.00	0.09	0.15	0.24	0.33	0.00	0.00	0.00	1.00	0.54	0.00	0.00	0.00	0.00	0.00	0.63	0.00	0.00	0.00	0.00
119	0.00	0.40	0.68	0.38	0.32	0.56	0.56	0.48	0.54	1.00	0.33	0.00	0.00	0.00	0.00	0.63	0.00	0.00	0.00	0.58
114	0.37	0.45	0.56	0.77	0.42	0.65	0.56	0.78	0.00	0.33	1.00	0.00	0.00	0.22	0.38	0.17	0.38	0.00	0.00	0.58
37	0.00	0.00	0.00	0.00	0.00	0.00	0.00	0.00	0.00	0.00	0.00	1.00	0.00	0.00	0.00	0.00	0.00	0.00	0.00	0.00
49	0.19	0.00	0.00	0.00	0.00	0.00	0.00	0.00	0.00	0.00	0.00	0.00	1.00	0.00	0.00	0.00	0.00	1.00	0.00	0.00
99	0.00	0.00	0.00	0.00	0.03	0.00	0.00	0.27	0.00	0.00	0.22	0.00	0.00	1.00	0.58	0.00	0.00	0.00	0.00	0.00
120	0.00	0.00	0.00	0.00	0.06	0.00	0.00	0.47	0.00	0.00	0.38	0.00	0.00	0.58	1.00	0.00	0.00	0.00	0.00	0.00
124	0.43	0.12	0.28	0.00	0.26	0.11	0.00	0.05	0.63	0.69	0.17	0.00	0.00	0.00	0.00	1.00	0.45	0.00	0.00	0.00
106	0.96	0.00	0.21	0.00	0.00	0.24	0.00	0.12	0.00	0.00	0.38	0.00	0.00	0.00	0.00	0.45	1.00	0.00	0.00	0.00
97	0.19	0.00	0.00	0.00	0.00	0.00	0.00	0.00	0.00	0.00	0.00	0.00	1.00	0.00	0.00	0.00	0.00	1.00	0.00	0.00
100	0.00	0.65	0.00	0.00	0.00	0.00	0.00	0.00	0.00	0.00	0.00	0.00	0.00	0.00	0.00	0.00	0.00	0.00	1.00	0.00
123	0.00	0.52	0.83	0.67	0.00	0.97	0.97	0.83	0.00	0.58	0.58	0.00	0.00	0.00	0.00	0.00	0.00	0.00	0.00	1.00

Appendix VI: Downwards substitutability matrix for numerical lithologies derived from petrophysical data. Brent section, North Sea well.

	82	107	112	87	113	108	88	118	93	119	114	37	49	99	120	124	106	97	100	123
82	1.00	0.00	0.17	0.00	0.26	0.24	0.00	0.14	0.00	0.00	0.00	0.00	0.00	0.00	0.00	0.00	0.93	0.00	0.00	0.00
107	0.00	1.00	0.08	0.43	0.40	0.00	0.00	0.11	0.13	0.22	0.06	0.00	0.00	0.39	0.00	0.12	0.09	0.00	0.00	0.00
112	0.17	0.08	1.00	0.00	0.16	0.95	0.90	0.68	0.08	0.37	0.90	0.00	0.00	0.00	0.00	0.95	0.16	0.00	0.00	0.93
87	0.00	0.43	0.00	1.00	0.51	0.00	0.00	0.27	0.08	0.00	0.00	0.00	0.00	0.14	0.00	0.00	0.31	0.00	0.00	0.00
113	0.26	0.40	0.16	0.51	1.00	0.06	0.04	0.29	0.46	0.47	0.23	0.26	0.00	0.00	0.00	0.09	0.42	0.00	0.00	0.00
108	0.24	0.00	0.95	0.00	0.06	1.00	0.88	0.71	0.00	0.20	0.93	0.00	0.00	0.00	0.00	0.87	0.23	0.00	0.00	0.97
88	0.00	0.00	0.90	0.00	0.04	0.88	1.00	0.68	0.00	0.19	0.87	0.00	0.00	0.00	0.00	0.81	0.00	0.00	0.00	0.90
118	0.14	0.11	0.68	0.27	0.29	0.71	0.68	1.00	0.00	0.15	0.81	0.00	0.00	0.00	0.00	0.63	0.22	0.00	0.00	0.70
93	0.00	0.13	0.08	0.08	0.46	0.00	0.00	0.00	1.00	0.83	0.00	0.00	0.00	0.00	0.00	0.00	0.00	0.00	0.00	0.00
119	0.00	0.22	0.37	0.00	0.47	0.20	0.19	0.15	0.83	1.00	0.20	0.00	0.00	0.00	0.00	0.37	0.00	0.21	0.21	0.21
114	0.00	0.06	0.90	0.00	0.23	0.93	0.87	0.81	0.00	0.20	1.00	0.00	0.00	0.00	0.00	0.86	0.08	0.00	0.00	0.96
37	0.00	0.00	0.00	0.00	0.26	0.00	0.00	0.00	0.00	0.00	0.00	1.00	0.00	0.00	0.00	0.00	0.00	0.00	0.00	0.00
49	0.00	0.00	0.00	0.00	0.00	0.00	0.00	0.00	0.00	0.00	0.00	0.00	1.00	0.00	0.00	0.00	0.00	0.00	0.00	0.00
99	0.00	0.00	0.00	0.14	0.00	0.00	0.00	0.00	0.00	0.00	0.00	0.00	1.00	1.00	[A0.0	0.00	0.00	0.00	0.00	0.00
120	0.00	0.00	0.00	0.00	0.00	0.00	0.00	0.56	0.00	0.00	0.12	0.00	0.00	1.00	1.00	0.00	0.00	0.00	0.00	0.00
124	0.00	0.12	0.95	0.00	0.09	0.87	0.81	0.63	0.00	0.37	0.86	0.00	0.00	0.00	0.00	1.00	0.00	0.00	0.00	0.89
106	0.93	0.09	0.16	0.31	0.42	0.23	0.00	0.22	0.00	0.00	0.08	0.00	0.00	0.00	0.00	0.00	1.00	0.00	0.00	0.00
97	0.00	0.00	0.00	0.00	0.00	0.00	0.00	0.00	0.00	0.21	0.00	0.00	0.00	0.00	0.00	0.00	0.00	1.00	1.00	0.00
100	0.00	0.00	0.00	0.00	0.00	0.00	0.00	0.00	0.00	0.21	0.00	0.00	0.00	0.00	0.00	0.00	0.00	1.00	1.00	0.00
123	0.00	0.00	0.93	0.00	0.00	0.97	0.90	0.70	0.00	0.21	0.96	0.00	0.00	0.00	0.00	0.89	0.00	0.00	0.00	1.00

Part II

BASIN-SCALE QDS MODELS

6

THE ROLE OF THE THERMAL-MECHANICAL STRUCTURE OF THE LITHOSPHERE IN THE FORMATION OF SEDIMENTARY BASINS

Michael S. Steckler
Lamont-Doherty Geological Observatory, Palisades, New York 10964 USA

ABSTRACT

The thermal and mechanical properties of the lithosphere exert important controls on the formation of sedimentary basins. The strength of the lithosphere is best understood as controlled by the geotherm and the crustal thickness and composition, and can be estimated using yield stress envelopes. The spatial variation of strength influences the location of lithospheric deformation, and the geometry of the resulting sedimentary basin. The shape of foreland basins, for example, directly reflects the lithospheric flexural rigidity. For extensional basins, the position of the mantle thinning controls the site of eventual lithospheric rupture, and the initial crustal failure generally occurs within ~100 km. The large heterogeneity in the strength of the crust determines the form of the rift's surface expression. The zone of crustal rifting will be wider for the case of thick orogenic crust than for cratonic regions. The flexural rigidity of lithosphere during rifting is not well understood. Uncertainties in lithospheric thermal models and backstripping techniques constrain the maximum resolution of extension estimates in rift basins to ~10%, but neglect of physical properties can cause errors exceeding 50%.

INTRODUCTION

Sedimentary basins are one expression of deformation processes on a lithospheric scale. The thermal-mechanical properties of the lithosphere, along with tectonic deformation processes, are the key factors that determine the physical shape of sedimentary basins and the stratigraphic pattern of the sediments that fill them. To produce quantitative models of basin stratigraphy requires an understanding of the dynamics of basin-forming processes. Although our understanding of lithospheric dynamics is incomplete, many variations in shape, subsidence histories and stratigraphic fill among different basins can be explained with our present knowledge of the thermal and mechanical structure of the lithosphere.

Quantitative modeling of the development of sedimentary basins and correct interpretation of the observed tectonics depend upon several factors including: 1) the ability to accurately estimate the tectonic subsidence throughout a basin; 2) an understanding of the dynamics of lithospheric deformation; and 3) the uncertainty in the lithospheric parameters used to calculate lithospheric deformation and to scale the resulting vertical movements. The

Quantitative Dynamic Stratigraphy (1989), T.A. Cross, ed., Prentice Hall, p. 89–112.

extent to which these factors can be incorporated into quantitative models differs for foreland basins versus extension basins.

Formulations of quantitative geodynamic models are in transition. As occurred in the history of plate tectonics investigations, there is a shift in research emphasis from determining kinematics to understanding the dynamic processes controlling basin formation. This shift reflects an increase in the knowledge of lithospheric-scale processes relevant to basin formation and evolution. Although complete comprehensive models of basin formation cannot yet be calculated, quantitative comparisons using present models are very valuable to our understanding of factors that control basin formation.

Most studies of basin subsidence have not adequately assessed uncertainties either in the data analyses or in the models used to interpret results of these data analyses. Thus, scientists have sometimes overestimated the accuracy of geodynamic models, consequently overinterpreting results, or sometimes dismissed worthwhile methodologies of investigation as being too unconstrained to be of value. The limits of both data and model accuracies must always be kept in mind.

This paper assesses the current understanding of processes involved in basin formation at the lithospheric scale. In doing so, I have at times necessarily incorporated a biased view based on my research experience. The first part of this paper summarizes and evaluates present knowledge of the thermal and mechanical structure of the lithosphere. The second part considers lithospheric deformation that creates basin subsidence, and investigates its dependence on lithospheric structure. I also illustrate some of the diverse attributes observed among basins that can be explained by quantitative models.

LITHOSPHERIC STRUCTURE

Thermal Lithosphere

Postrift subsidence of extensional basins is due to the cooling and thermal contraction of thinned lithosphere, similar to the subsidence of oceanic lithosphere along oceanic ridges. In fact, thermal models of basin formation generally have been calibrated against oceanic ridge subsidence (Sleep, 1971; McKenzie, 1978). In young sea floor, the basement depth is proportional to the square root of its age. Observations of ocean-floor bathymetry at greater ages (>80 m.y.) show a deviation from this proportionality. This flattening has been modeled by Parsons and Sclater (1977) as due to the cooling of a lithospheric plate with an equilibrium thickness of 125 km. The asthenospheric temperature at the base of the plate is estimated to be ~1300°-1350°C. The cooling-plate model effectively predicts the mean depth of the oceanic sea floor (Fig. 1). The subsidence of basins is related to the perturbation of the lithosphere from this equilibrium state.

Despite the success of the cooling-plate model, the bathymetry data contain some deviations from calculated thermal subsidence. The mean scatter for older sea floor is 130 m in the North Pacific and 230 m in the North Atlantic, or about 3-6% of the total subsidence (Fig. 1; Parsons and Sclater, 1977; Renkin and Sclater, 1988). In addition, there are numerous locations where the subsidence differs significantly (Trehu, 1975; Cochran, 1986). Many of these variations can be explained by modest (<100°C) variations in the temperature at the

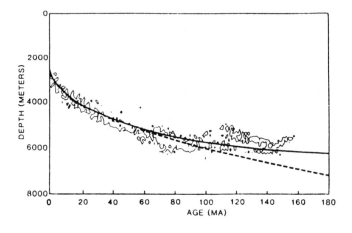

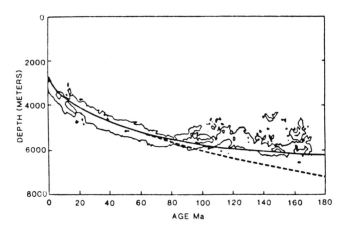

Figure 1. Subsidence of the North Atlantic (top) and North Pacific Oceans (bottom). The contours enclose two-thirds of the depths around the mode. The dashed and thick solid lines show the empirical subsidence curves for the cooling of a half-space and the cooling-plate model, respectively (after Renkin and Sclater, 1988).

base of the lithosphere over 1000 km-wide zones (Cochran, 1986; Hayes, 1988). These factors represent the limit of predictability of the cooling-plate model, and should be considered when interpreting sedimentary basin subsidence.

The flattening of the ocean-floor bathymetry with age, consistent with the lithosphere approaching an equilibrium thickness, requires the addition of heat to the base of the lithosphere. The cooling-plate model does not describe the origin of this extra heat. It was thought to be due to the onset of a convective instability at ages greater than ~70 m.y. originating in the thermal boundary layer at the base of the lithosphere (Parsons and McKenzie, 1978; Houseman and McKenzie, 1982). Recent observational (Haxby and Weissel, 1986) and theoretical results (Buck, 1987) show that small-scale convective instabilities develop in the thermal boundary layer at or near the ridge crest and do not cause the deviation from √ age subsidence at older sea floor. Robinson and Parsons (1988) proposed a larger-scale convective instability is responsible for the flattening.

Heestand and Crough (1981) suggested that the flattening of the age-depth curve is due to heat transferred to the lithosphere by hotspot swells. They showed a dependence of ocean-floor bathymetry on distance from hotspots. If this is indeed the case, then the mean flattening may be the result of a stochastic process and may not necessarily apply to any given segment of lithosphere. This has important consequences for extensional basin subsidence models as it implies considerable variability in the subsidence of basins at old ages.

The thermal structure of continents is less well known than that of the oceans. There had been considerable controversy over whether heat flow data indicate that continental lithosphere beneath shields is thicker than oceanic lithosphere beneath old oceans (Sclater et al., 1980; Chapman and Pollack, 1975). However, now there is general acceptance that cratonic lithosphere is thicker. Inclusions in diamonds indicate that material from ~200 km depth has remained fixed to the South African craton for over one billion years (Boyd et al., 1985). Analyses of seismic travel times indicate that continental cratons are thicker than old oceanic lithosphere (Lerner-Lam and Jordan, 1987). The flexural rigidity of cratonic regions loaded in foreland basins reaches values equivalent to 100 km-thick elastic plates (Karner and Watts, 1983; Lyon-Caen and Molnar, 1983, 1985). This requires temperatures of less than 700°-800°C at these depths, also supporting the existence of thick continental lithosphere. Thus, although thermal models of sedimentary basins generally assume an initial lithospheric thickness of 125 km, the actual initial conditions and basin subsidence histories could differ significantly in some settings (Jarvis, 1984).

Mechanical Lithosphere

The mechanical lithosphere, which can support large stresses for tens of millions of years, is several times thinner than the thermal lithosphere. The strength of the lithosphere in bending, as measured by its flexural rigidity, usually is determined by examining the isostatic response to loads. It is commonly expressed as the thickness of an elastic plate of equivalent flexural rigidity (T_e). The T_e of oceanic lithosphere has been found to increase with age approximately to the depth of the 450°C isotherm (Watts, 1978).

The actual rheology of the lithosphere is more complex than an elastic plate overlying a weak fluid. Bodine et al. (1981) showed that the empirical variation of T_e is due to the temperature dependence of ductile flow in the mantle. Although the strength of the uppermost part of the lithosphere is controlled by lithology-independent brittle failure, the strength at greater depths is controlled by thermally activated creep in olivine (Goetze and Evans, 1979; Kirby, 1983). This produces a roughly diamond-shaped variation in the strength of the oceanic lithosphere with depth, commonly referred to as the yield stress envelope (Figure 2a). Olivine retains substantial long-term strength at temperatures up to 700°-800°C. The lithosphere above this level comprises the mechanical thickness of the lithosphere.

The T_e of unstressed lithosphere would be its mechanical thickness. However, as the lithosphere is loaded, its flexural rigidity decreases (Bodine et al., 1981; McNutt, 1984). Under intense bending, compression or extension, the T_e can be reduced substantially

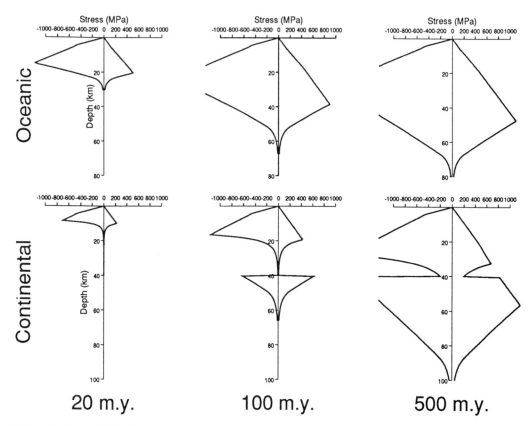

Figure 2. Shape of yield stress envelopes (YSE) for oceanic and continental lithosphere for several thermal ages. Both the overall strength of the lithosphere and its flexural rigidity vary with the size of the YSE. The calculations assume 5 and 40 km-thick crust and 125 and 175 km-thick lithosphere for the ocean and continent, respectively. The lower crust uses an anorthosite rheology (Kirby, 1983) and the mantle is approximated by an olivine rheology (Goetze, 1978).

(McAdoo and Sandwell, 1985; Buck, in press). Beneath seamounts and at trenches, the T_e of oceanic lithosphere is estimated empirically as increasing with depth to the 450°-500°C isotherm. Some of the scatter in the effective T_e values can be attributed to the effects of variable load size or in-plane stresses. McNutt and Menard (1982) and McNutt (1984) suggested that T_e is underestimated due to reheating of the lithosphere beneath seamounts, and that 600°C is a better average value.

The rheology of continental lithosphere is more complex than that of oceanic lithosphere due to the presence of a thick crust. The yield strength of the materials composing the lower continental crust is considerably less than that of the mantle. Thus, yield stress envelopes for the continents (Fig. 2b) commonly exhibit a weak zone in the lower crust, sometimes referred to as a "jelly sandwich" rheology. As a result of this more complex rheology, the strength of continental lithosphere is generally less than oceanic lithosphere for an equivalent geotherm. For very cold, thick lithosphere, T_e is controlled by the rheology of olivine, as in oceanic lithosphere (Karner et al., 1983). However, as the geotherm increases, the isotherm representing T_e can decrease by a factor of two (Kusznir and Karner, 1985). This is because the layer supporting flexural stresses shifts from the stronger mantle, as in the oceans, to the weaker continental crust.

The yield stress envelope can be used to estimate not only flexural rigidity, but also the overall strength of the lithosphere. The area of the yield stress envelope is an approximate measure of total force needed to produce failure of the entire lithosphere and thereby allow large-scale extension or compression. The main factors controlling the strength of the lithosphere are the geotherm, the thickness and composition of the crust, and the sediment thickness. The strength increases with the lithospheric thickness but decreases with larger crustal thicknesses. For the same geothermal gradient, therefore, oceanic lithosphere will be stronger than continental. However, cratons with large lithospheric thicknesses are stronger than oceanic regions (Fig. 3).

This variability in continental strength and its subsequent evolution during basin formation must be a major factor in the development of sedimentary basins. For the case of the early stages of continental rifting, there has been a considerable effort to examine the evolution of lithospheric strength during extension. In a series of papers, England (1983, 1986) and Sawyer (1985, 1986) debated whether lithospheric strength decreases or increases,

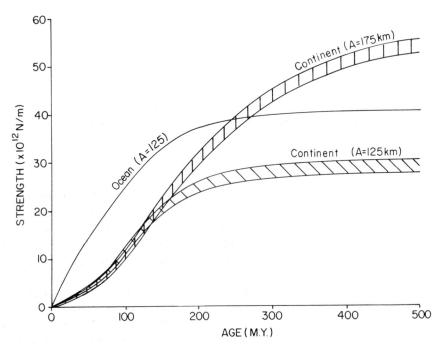

Figure 3. Plots of the strength of lithosphere in extension versus age. The strength of the lithosphere is approximated by the area of the extensional side of the YSE. Values for the equilibrium lithospheric thickness are shown. The widths of the continental lithosphere curve represent results for a range of mafic, lower crustal rheologies.

shutting off further extension, during rifting. Kusznir and Park (1987) and Lynch and Morgan (1987) demonstrated how the changing strength of a rift undergoing extension depends on the initial crustal thickness, the geothermal gradient and the rate of extension. Figure 4a shows contours of the composite strength of the lithosphere as a function of geotherm and crustal thickness; the lithosphere weakens as the crustal thickness or geothermal gradient increases. Depending on the initial conditions and rate of extension, the trajectory of rifting lithosphere can cause it to move to a state that is either weaker or stronger, thereby promoting or hindering continued deformation at that site (Fig. 4b).

Lithospheric strength can be estimated for a variety of different tectonic regimes. Calculations of strength variations across continental margins (Vink et al., 1984; Steckler and ten Brink, 1986) indicate a propensity for rifting just inboard of the hinge zone at continental

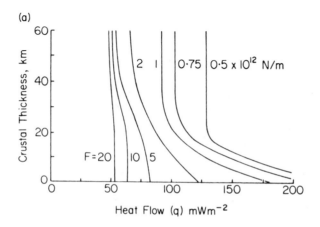

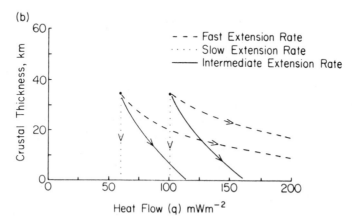

Figure 4. a) Plot of contours of the strength of the lithosphere from YSEs as a function of crustal thickness and heat flow. b) Trajectories of the lithosphere in crustal thickness-heat flow space for different extension rates and initial conditions. Depending on the initial conditions and extension rate, the evolving lithosphere can undergo either strain hardening or weakening (after Kusznir and Park, 1987).

margins, creating thin continental slivers. Similarly, old rifts are more likely to be reactivated along their flanks than at the thinner crust within the rift. The heterogeneity in composition of continental crust and the widely different strengths of these materials (Kirby, 1983) suggest that there may be significant lateral variation in the strength of even adjacent regions within continents. The sensitivity of deformation to lithospheric strength must play a major role in the development of sedimentary basins.

FORELAND BASINS

Variations in form or shape of sedimentary basins reflect the diversity of the thermal and mechanical structure of the underlying lithosphere. Foreland basins are primarily mechanical basins formed by the loading and depression of the lithosphere in front of mountain belts (Price, 1973; Beaumont, 1981; Karner and Watts, 1983; Jordan, 1981). The width and depth of foreland basins directly reflect the flexural rigidity and therefore the thermal and mechanical state of the underlying lithosphere. Broad and shallow foreland basins are developed on thick, cold lithosphere, whereas deeper and narrower basins are developed on thinner, hotter lithosphere. In addition, a recent study by McNutt et al. (1988) indicates that the dip of the underthrust plate, the curvature of the mountain belt thrust front, and the length of the mountain belt all depend on the flexural rigidity of the lithosphere.

The cause of foreland basin subsidence generally is attributed to the overthrusting of convergent systems. The subsidence of foreland basins is commonly episodic, reflecting discontinuities in motion of convergent systems and in development of mountain-belt loads. Unlike extensional basins, there is no later thermal subsidence phase. The main thermal effect distant from the collision zone is the thermal blanketing by the sediment fill. Heating of sediment and underlying lithosphere contributes to slight uplift and erosion between thrusting events, accentuating the discontinuous nature of foreland basin subsidence (Kominz and Bond, 1986).

The weight of the mountain topography at overthrust belts is not the only load involved in generating foreland basins. A number of studies have shown that the topographic mass at many collision belts is insufficient to cause the entire depression of the foreland basin (Jordan, 1981; Karner and Watts, 1983; Royden and Karner, 1984). Karner and Watts (1983) proposed the existence of "hidden," or subsurface, loads at the suture zone marked by a Bouguer gravity high to explain the size of foreland basins. They suggested that the hidden load could be related to the overriding of an earlier passive margin. Stockmal et al. (1986) modeled the subduction of a continental margin and confirmed that a thick overthrust belt and foreland basin could be accommodated with little topographic relief. However, not all subsurface loads can be associated with subduction of a passive margin. In the Apennines and Carpathians (Royden and Karner, 1984), the position of the required subsurface load corresponds to the Tyrrhenian and Pannonian extensional basins which postdate the collisions. The additional forces required in these mountain belts, as in the Himalaya (Lyon-Caen and Molnar, 1985), may be due to forces in the subducted slab.

The Anadarko basin of Oklahoma is a clear example of the existence of static subsurface loads in a flexural loaded basin (Fig. 5). This 6 km-deep flexural basin was produced by the

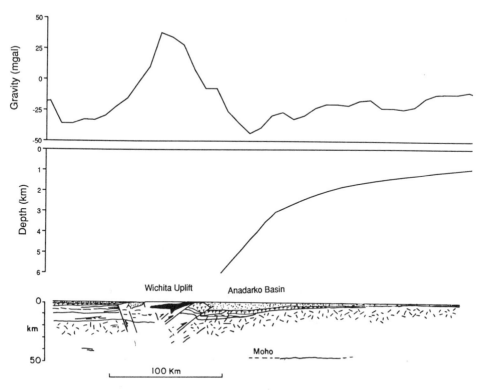

Figure 5. Cross section of the Anadarko basin and Wichita uplift. The dotted pattern on the structure section shows the distribution of the Upper Mississippian-Permian sediments caused by the compression and overthrusting of the Wichita uplift. The T-pattern and black represent the sediments and igneous bodies of the earlier southern Oklahoma aulacogen, respectively. The flexural deflection shape of the unconformity at the base of the Anadarko basin sediments is shown in more detail in the center. The gravity anomaly above illustrates the coupled gravity high over the mountain-belt load with an accompanying broad gravity low over the foreland basin that characterizes flexural systems. The lack of significant topographic relief in the region implies the existence of subsurface loads, related to the underlying southern Oklahoma aulacogen.

Pennsylvanian-Permian overthrusting of the Wichita Mountains which has no present-day topographic relief. Thus, the entire load creating the Anadarko basin is in the subsurface. The core of the uplift is covered by synrift volcanics of the southern Oklahoma aulacogen (Gilbert, 1983). Compression of the Late Precambrian-Cambrian aulacogen resulted in the overthrusting of the heavy rift crust which had been made more dense by crustal thinning and igneous activity (Brewer et al., 1983). This uplifted block acts as the subsurface load for the Anadarko basin to the north. In the absence of subsurface loads, erosion of the mountain belt would have caused the rebound and erosion of the foreland as well.

Subsidence of Extensional Basins

During the formation of large-scale extensional basins and passive margins, the crust and lithosphere are thinned. McKenzie (1978) explained the subsidence in these basins with a simple, one-dimensional kinematic model. The vertical isostatic movements resulting from basin formation are due primarily to a balance of subsidence caused by thinning of the crust, and uplift caused by the advection of heat into the lithosphere. Following rifting, continued subsidence results from the conductive cooling of the heated lithosphere. The thermal, postrift subsidence, in particular, has been studied extensively at sedimentary basins and continental margins.

Backstripping

To investigate the subsidence of extensional basins, sedimentary sections are analyzed by backstripping (Watts and Ryan, 1976; Steckler and Watts, 1978). Backstripping attempts to isolate the tectonic component of subsidence by correcting for sediment compaction and loading by sediment plus water. The tectonic subsidence may then be compared to theoretical subsidence curves to estimate the degree of extension in a basin.

Sawyer et al. (1982a,b) estimated uncertainties in backstripping studies as ±5%. Steckler et al. (1988) compared different backstripping results of the same well and confirmed this estimate. Bond and Kominz (1984) utilized a range of porosity-depth curves in backstripping Paleozoic sections to determine the uncertainties in the compaction corrections. The extension models have been calibrated using oceanic ridge subsidence models. As discussed earlier, ridge subsidence also has a variability of ~5% away from hotspots or other perturbations. Thus, estimates of fractional thinning of the lithosphere have at best an accuracy of about ±10%.

A number of physical processes are neglected both in McKenzie's (1978) extension model and by one-dimensional backstripping. These include lateral conduction of heat (Steckler and Watts, 1980), finite duration of rifting (Cochran, 1983; Alvarez et al., 1984), thermal blanketing by the sedimentary cover (Lucazeau and LeDouaran, 1985; Stephenson, this volume), and flexural isostasy (Steckler, 1981; Beaumont et al., 1982; Watts et al., 1982). Consideration of many of these factors that modify the subsidence of extensional basins can only be approximated in backstripping. Thus, comparison of tectonic subsidence obtained from backstripping with one-dimensional theoretical curves can yield estimates that are significantly in error (Steckler, 1981; Bond and Kominz, 1984).

Forward models of basin evolution are necessary to accurately determine the distribution of extension within a basin. In this way processes such as thermal blanketing, in which the sediment loading and thermal cooling of the lithosphere are coupled, can be accommodated. As an example, Figure 6 shows the subsidence across a continental margin calculated using a two-dimensional model including lateral conduction and flexural effects. These factors significantly distort the shape and/or amplitude of subsidence curves. Airy backstripping of the sedimentary thicknesses produced by the model can yield estimates for the extension that are in error by over 50%. Estimates of extension calculated without such consideration of the regional structure of the basin and forward modeling are not reliable.

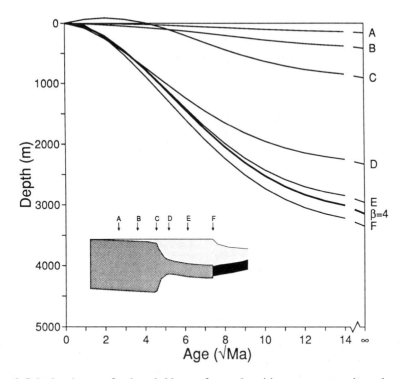

Figure 6. Calculated curves for the subsidence of several positions across a continental margin. The geometry of the margin and locations are shown on the inset. The model included lateral as well as vertical heat flow and flexure due to both the cooling of the lithosphere and the sediment load. The subsidence curves were generated by Airy backstripping of strata. Sites A-C are on unthinned continental crust and D-F on crust thinned by a factor of 4. Large errors in estimates of extension factors result when two-dimensional physical properties are neglected in backstripping.

Uplift and Heating at Rifts

The uniform extension model (McKenzie, 1978) is a simple kinematic model that illustrates the origin of basin subsidence. However, investigations of passive margins have found that the uniform extension model underestimates the amount of heat advected by extension (Royden and Keen, 1980; Hellinger and Sclater, 1983) and cannot explain the size of rift-flank uplift (Hellinger and Sclater, 1983; Jarvis, 1984; Watts and Thorne, 1984). The two-layered extension model (Royden and Keen, 1980) provided a method of addressing this problem within a kinematic framework.

Subsequently, fluid dynamic models of rifts that incorporate the rheologic properties of mantle rocks have indicated that the lateral temperature gradients introduced by rifting were capable of inducing significant flow in the mantle (Buck, 1984, 1986; Keen, 1985; Fig. 7). This small-scale convective flow is a direct result of temperature structures established by

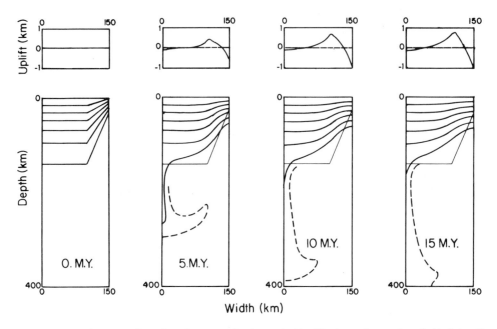

Figure 7. Development of small-scale convection beneath rifts. The lower boxes show half of the rift which is assumed symmetrical at 0, 5, 10 and 15 m.y. after rifting. The fluid in the boxes has a temperature- and pressure-dependent olivine rheology. The isotherms within the lithosphere are shown at 200°C intervals. The 1250°C isotherm (dashed line) was included to indicate the convective flow. The initial position of the 1200°C isotherm is repeated as a light line to show the sweeping away of the material beneath the rift flank. The upper boxes show the topography developed from conduction and convection.

lithospheric thinning and is not predicted by purely kinematic models. Small-scale convection advects additional heat into a rift beyond that which can be introduced by the extension alone. This results in thermal uplift of the rift flanks on the order of 1000 m (Fig. 7). Evidence of such uplift over zone a few-hundred kilometers wide has been determined at the Gulf of Suez (Steckler, 1985), where significant prerift uplift can be discounted (Chenet and Letouzey, 1983; Steckler, 1985; Garfunkel, 1988).

Watts and Thorne (1984) needed to include additional heating of the lithosphere at the landward parts of the Baltimore Canyon segment of the U.S. east coast passive margin to explain the Mesozoic coastal plain stratigraphy. Celerier (1986) and Steckler et al. (1988) reached similar conclusions at other sites along this margin. Steckler et al. (1988) suggested that these results indicate that small-scale convection was active at this margin during rifting, and that it is probably common whenever substantial horizontal temperature gradients are induced at the base of the lithosphere.

Small-scale convection also may have a major effect on the production of melt at rifts (Mutter et al., 1988). Basaltic melts produced at both oceanic ridges and rifts are primarily due to adiabatic depressurization melting (McKenzie, 1983). Local convective circulation at rifts can cycle undepleted mantle through shallow depths, thereby increasing the volume of melt which can be produced at a rift (Mutter et al., 1988). The vigor of small-scale convective flow and of melt production will depend on the lateral temperature gradients produced by rifting. The flow will be the most pronounced for narrow, rapid rifting events. This may be able to explain the existence of thick basalt layers and seaward-dipping reflectors at abruptly rifted margins such as the Voring Plateau.

An alternative explanation for variations in extent of igneous activity at margins is minor variations in asthenospheric temperature (White et al., 1987). The volume of melt generation is sensitive to minor changes in the temperature conditions at rifts (McKenzie and Bickle, in press). Variability in mantle temperatures, similar to those estimated for oceanic ridges, can explain the range of melting observed at rifted margins for broad regions of volcanism such as the North Atlantic margins near the Iceland hotspot (White et al., 1987). However, in the northwest Australian margin, there is a sharp change from the broadly extended continental Exmouth Plateau with little melt to the abrupt margin at the Cuvier basin with seaward-dipping reflectors immediately to the south (Mutter et al., 1988). This argues strongly for control of melt production by the rifting geometry and small-scale convection as well as regional variations in asthenospheric temperatures.

The continued existence of substantial uplifts at several older passive margin and rifts (e.g., southeastern Australian highlands, Brazilian continental margin, Broken Ridge; Weissel and Karner, in press) indicate processes other than thermal expansion are involved. Thickening of the crust by underplating or igneous activity is probably of major importance. Flexure may be another control of large uplifts at rift flanks. The increase of flexural rigidity with time following rifting can help permanently "lock in" the topography of uplifted rift flanks (Karner, 1985). Furthermore, Vening-Meinesz (1950) showed that isostatic unloading at normal faults can produce large footwall uplifts. However, calculations of the magnitude of flexural uplifts at rifts using the Vening-Meinesz (1950) method predict uplifts larger than some authors consider reasonable (Hellinger and Sclater, 1983).

There is still considerable debate over whether there is flexural or local isostasy during rifting and, if so, what the flexural rigidity would be. The low amplitude of gravity anomalies at extensional basins argue for small flexural rigidities (Karner and Watts, 1982; Barton and Wood, 1984). As a result, it has been argued that during extension and active normal faulting the flexural rigidity of the lithosphere is negligible (White and McKenzie, 1988; Fowler and McKenzie, in preparation). At the least, T_e at extensional sedimentary basins apparently corresponds to a lower temperature than for oceanic lithosphere (Beaumont et al., 1982; Barton and Wood, 1984; Watts, 1988).

Despite this there is considerable evidence of flexural effects at rifts. These include footwall uplift (Jackson and McKenzie, 1983), and the association of rift-flank uplift with the polarity of fault blocks (Rosendahl, 1987). Kusznir et al. (1987) argued Airy isostasy implies considerable distortion of detachments below fault blocks that is not supported by deep seismic-reflection profiles. Bell et al. (1988) identified gravity anomalies at the Newark and Gettysberg basins predicted by flexural compensation models of extension along detachments. Weissel and Karner (in press) argue that permanent rift-flank uplift is in part due to a flexural response to the tectonic denudation caused by rifting. In formulating the flexural response to tectonic denudation, they found that the amount of rebound depends on the depth

Hanley Library
University of Pittsburgh
Bradford Campus

to a reference surface which they equate to detachments. They calculate rift-flank uplifts ranging from 500 to 5000 m depending on the depth of a detachment surface. The controversy over the flexural strength of the lithosphere during rifting may be due in part to the neglect (and therefore implicit assumptions concerning) of the depth of their isostatic reference surface.

GEOMETRY OF EXTENSIONAL BASINS

In recent years, geologists investigating extended terranes have realized that the geometry of extensional faulting can be as complex as has been observed in thrust belts. The recognition of low-angle normal faults, in particular, has altered the perception of the way the lithosphere extends. The asymmetric structure of most rift basins was rediscovered and the geometries of conjugate passive margins have been interpreted in terms of simple-shear extension models (Gibbs, 1984; Wernicke, 1985; Lister et al., 1986). However, direct evidence of

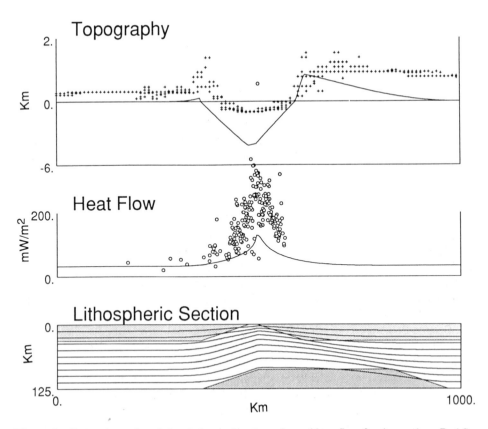

Figure 8a. Comparison of model and observed topography and heat flow for the northern Red Sea. Results for simple-shear extension along a detachment dipping at 15°. The lithospheric sections show the resulting crust in the light shading and the asthenosphere advected into the section in the darker shading with representative isotherms superimposed.

asymmetric detachment structures still is limited to upper and middle levels of the crust. The evidence for simple-shear extension through the subcrustal lithosphere comes from asymmetries of uplift and volcanism at rifts and passive margins (Wernicke, 1985; Lister et al., 1986).

Whole-lithosphere simple shear was first proposed to explain the differences in volcanism between the Arabian and African sides of the Red Sea (Wernicke, 1985). Buck et al. (1988) investigated this asymmetry using a kinematic thermal model of pure- and simple-shear extension (Fig. 8). They found that simple-shear extension could not account for the elevation difference on opposite sides of the rift, and suggested that the difference was due to flexural loading by the Zagros on an Arabian plate broken by the Red Sea rifting. Passive extension by simple- and pure-shear models do not account for the local rift-flank uplift; additional heating, such as that generated by small-scale convection, is required. More importantly, simple-shear extension shifts much of the heat introduced by rifting to one side of the Red Sea where it is distributed over the broad zone of the detachment. As a result, the lithosphere beneath the rift remains relatively cool. In fact, the model temperatures remained

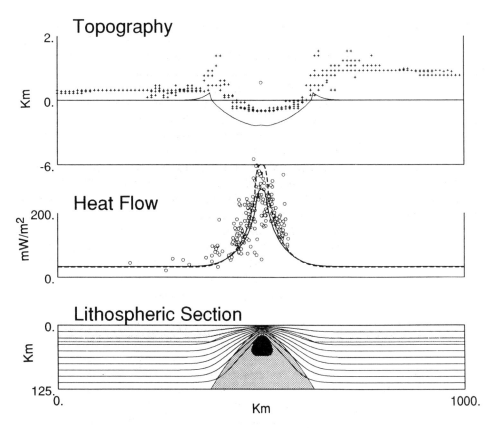

Figure 8b. Results for pure shear extension on a rift zone narrowing from 100 to 20 km. Lithospheric section uses same conventions as in Figure 8a with the black locating mantle undergoing partial melting. Neither model can match the observed rift-flank uplift and only the pure-shear model significantly thins the lithosphere and generates high heat flow at the rift axis.

cool enough to never generate any melting or volcanism. The heat-flow values predicted by the whole-lithosphere simple-shear models underestimate the observed heat-flow values in the northern Red Sea by a factor of three (Fig. 8a).

Of the pure-shear models, cases in which the zone of active extension either expanded or remained constant with the opening of the rift also could not reproduce the heat flow or induce melting for the conditions at the northern Red Sea. Only a pure-shear model with a narrowing zone of extension (Fig. 8b) could match the observed heat flow and generate partial melting of the upwelling mantle.

These results strongly suggest that the later stages of continental rifting are dominantly pure shear. Even if rifting is initiated along detachments, it evolves to a pure-shear geometry as extension progresses. Furthermore, it suggests that as rifting continues the width of the region of active extension decreases. Indeed, oceanic ridges can be considered as pure-shear extension over a narrow zone (Jarvis, 1984).

Rheologic controls on extension

The narrowing of rifts during the final stages of extension is a consequence of the rheology of the lithosphere. The dependence of rock strength on temperature focuses the extensional strain on the hottest, and therefore weakest, part of the rift. Similarly, the initial location of lithospheric rifting will be controlled by regional strength variations. Rifts tend to follow local zones of weakness and avoid cold, cratonic areas (Sykes, 1978).

Braun and Beaumont (1987) constructed a dynamic model of extension in which the strength of the lithosphere was rheologically controlled. They found that the initial width of the extending region varied with depth and was inversely proportional to the local lithospheric strength. For example, rifting in the Central Atlantic initially extended over a broad zone in the weak upper crust, forming numerous Mesozoic rift basins along North America and West Africa. Many of these reactivate Appalachian thrusts (Ratcliffe and Burton, 1985; Swanson, 1986). However the eventual continental margin underwent large amounts of extension concentrated into a narrow zone seaward of the present hinge zone.

Braun and Beaumont (1987) also found the sharp contrast in strength at the Moho, which changes the width of active extension, caused the development of subhorizontal detachment zones accommodating the differential movement of the crust and mantle. Dunbar and Sawyer (1988), using a similar rheologic model, extended lithosphere in which initial weak zones in the crust and mantle were offset. This geometry induced a subhorizontal detachment in the weak lower crust connecting the two regions of extension.

The model experiment by Dunbar and Sawyer (1988) shows the relative importance of the crust and mantle in rifting. Most of the strength in the lithosphere lies within the subcrustal mantle. As rifting progresses, the zone of mantle thinning yields the most weakening of the lithosphere and determines the eventual site of rupture. Thus, while the initial location of rifting is determined by preexisting weaknesses throughout the lithosphere, the deep-seated mantle thinning dominates the evolution of the rift. This is corroborated by studies of the Red Sea (Buck et al., 1988) who found that at breakup, the lithospheric thinning has a pure-shear form and is centered near the Red Sea axis. Even if a rift initiates with a simple-shear geometry, both of these studies suggest it will tend towards pure shear with time.

Figure 9 shows a reconstruction of Africa and South America with the South Atlantic Ocean closed (Cande et al., in press). The shades depict the present-day elevations of South America and Africa. The highest topography along the reconstructed margin does not lie symmetrically astride the ocean/continent boundary, but rather shifts from one side to the other. The highest relief facing the margin occurs in Brazil and southern Africa, despite the greater elevation of the entire African continent that developed in the Tertiary (Bond, 1978). The jump in peak elevations occurs at an offset in the coastlines of both continents. This high topography along the Brazilian and southern African margins was likely the locus of greatest heating and igneous activity (underplating?) during rifting. Thus, the uplifted zone forms a straighter line than the location of the final crustal break between the continents, causing the uplift to shift from one side to the other. One interpretation of these observations is that the uplifted zone marks the position of mantle thinning that controlled the site of rifting, and that the crustal rifts utilized nearby zones of crustal weakness. That the crustal rifts deviate by up to 100 km to either side, or approximately the thickness of the lithosphere, is not unexpected.

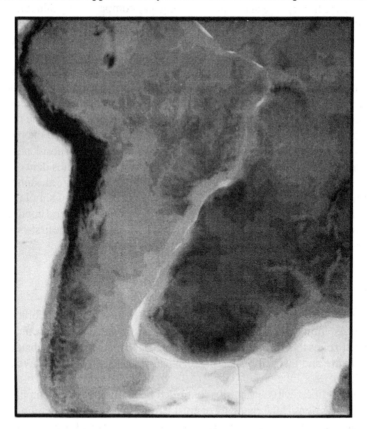

Figure 9. Reconstruction of the South Atlantic generated by rotating Africa back against South America. The image displays the present-day topography contoured at a 300 m interval for low elevations. The contour interval increases to 500 m for elevations above 1500 m and 1000 m for depths below than -1000 m. Note the antisymmetric topography along the South American and African continental margins.

Discussion

The initial state of the lithosphere has an important control on the development of sedimentary basins. Determining the crustal and lithospheric thickness at the start of basin formation is necessary for understanding geodynamic processes. This may be made more difficult because the site of deformation may be concentrated at an anomalous zone of weakness.

Still, large-scale variations can be determined. Prior to rifting, the continental lithosphere bordering both the South Atlantic Ocean and the Red Sea was probably thick because these areas had been tectonically quiescent since the Pan-African (500-700 Ma). Both of these continental margins are relatively straight and simple, and reconstructions indicate that the zones of initial rifting were narrow. This is consistent with the results of the rheological models presented above. A stronger lithosphere yields a narrower initial rift. It will also favor weakening of the lithosphere and, as a result of rifting, promote continued extension.

The North and Central Atlantic, by contrast, were sites of repeated orogeny during the Paleozoic, and the lithosphere at the time of rifting was comparatively thinner. Lithospheric thickness at the start of extension was probably about 125 km, as assumed in previous analyses (Steckler and Watts, 1978; Royden et al., 1980). The crust, however, was riddled with major compressional structures and may have been thickened. Initial rifting in the Central Atlantic occurred over a broad zone hundreds of kilometers wide extending from the basins of the Newark Supergroup in North America to their equivalents in Africa. Many of these reactivated Paleozoic thrusts (Swanson, 1986). Rifting extended over a long time (>50 m.y.) before it became focused into the highly thinned crust seaward of the hinge zone.

The resulting sedimentary basins and continental margins differ in the pattern of both synrift, isostatic subsidence and postrift, thermal subsidence. Synrift subsidence is controlled by both the net crustal thinning and the advection of heat. Postrift subsidence is dominated by lithospheric cooling. Their distribution is linked to the evolving lithospheric structure during rifting and will not precisely correspond to the predictions of simple models (e.g., McKenzie, 1978; Lister et al., 1986). Nonetheless, factors such as lateral conduction of heat and flexure help reduce local variability within rifts and contribute to the first-order successes of simple, geometric, rifting models.

SUMMARY AND CONCLUSIONS

Uncertainties in determining tectonic subsidence by backstripping and calibration of oceanic thermal models limits the maximum accuracy of estimates of crustal and lithospheric thinning in rift zones to ±10%. Oversimplification of the physics of rift subsidence, such as neglect of flexure and thermal blanketing, can cause these estimates to have actual errors exceeding 50%.

Flexural rigidities, while primarily controlled by temperature, are complicated within continents by variations in the rheology of crustal rocks. The isotherm defining the effective elastic thickness decreases when ductile failure of the lower crust occurs. The variation in flexural rigidity and overall lithospheric strength can be explained by changes in the shape of the yield stress envelope.

The shape and style of foreland basins are controlled by the lithospheric flexural rigidity. Subsurface loads are a major component of the mechanics of flexural basins.

The flexural rigidity during lithospheric failure is not well understood. Flexural effects are present in extensional rifts but the rigidity may be less than estimated by simple models.

The style of extensional failure in the crust is strongly controlled by preexisting fabrics. This is because the heterogeneity in crustal strength is large. A thickened orogenic crust will tend to fail over a wide (100s of km) zone. Asymmetric styles of upper crustal failure are the general observation in rifts.

The major zone of crustal thinning generally overlies the zone of thinning in the mantle lithosphere plus or minus the ~100 km thickness of the lithosphere. Together with thermal modeling of extension, this suggests that the subcrustal lithospheric extension is better explained by pure-shear mechanisms.

ACKNOWLEDGMENTS

Roger Buck, Bernie Coakley, Tim Cross, Garry Karner, Jeff Nunn, and Randell Stephenson reviewed the manuscript and provided helpful discussions. I would like to thank Steve Cande and Bill Haxby for their assistance in making the image of South Atlantic topography. This work was supported by National Science Foundation grants EAR-85-18537 and OCE-88-00703, and by Gas Research Institute contract 5087-260-1546. Lamont-Doherty Geological Observatory contribution number 4365.

REFERENCES

Alvarez, F., Virieux, J., and Le Pichon, X., 1984, Thermal consequences of lithosphere extension over continental margins: The initial stretching phase: Geophysical Journal of the Royal Astronomical Society, v. 78, p. 389-411.

Barton, P., and Wood, R., 1984, Tectonic evolution of the North Sea basin: Crustal stretching and subsidence: Geophysical Journal of the Royal Astronomical Society, v. 79, p. 987-1022.

Beaumont, C., 1981, Foreland basins: Geophysical Journal of the Royal Astronomical Society, v. 65, p. 291-329.

Beaumont, C., Keen, C.E., and Boutilier, R., 1982, On the evolution of rifted continental margins: Comparison of models and observations for Nova Scotian margin: Geophysical Journal of the Royal Astronomical Society, v. 70, p. 667-715.

Bell, R.E., Karner, G.D., and Steckler, M.S., 1988, Early Mesozoic rift basins of eastern North America: The role of detachments during extension: Tectonics, v. 7, p. 447-462.

Bodine, J.H., Steckler, M.S., and Watts, A.B., 1981, Observations of flexure and the rheology of the oceanic lithosphere: Journal of Geophysical Research, v. 86, p. 3695-3707.

Bond, G.C., 1976, Evidence for Late Tertiary uplift of Africa relative to North America, South America, Australia and Europe: Journal of Geology, v. 86, p. 47-65.

Bond, G.C., and Kominz, M.A., 1984, Construction of tectonic subsidence curves for the early Paleozoic miogeocline, southern Canadian Rocky Mountains: Implications for subsidence mechanisms, age of breakup, and crustal thinning: Geological Society of America Bulletin, v. 95, p. 155-173.

Boyd, F.R., Gurney, J.J., and Richardson, S.H., 1985, Evidence for a 150-200-km thick Archean lithosphere from diamond inclusion thermobarometry: Nature, v. 315, p. 387-389.

Braun, J., and Beaumont, C., 1987, Styles of continental rifting: Results from dynamical models of lithospheric extension, in Beaumont, C., and Tankard, A., eds., Sedimentary basins and basin-forming mechanisms: Canadian Society of Petroleum Geology Memoir 12, p. 241-258.

Brewer, J.A., Good, R., Oliver, J.E., Brown, L.D., and Kaufman, S., 1983, COCORP profiling across the southern Oklahoma aulacogen: Overthrusting of the Wichita Mountains and compression within the Anadarko basin: Geology, v. 11, p. 109-114.

Buck, W.R., 1984, Small-scale convection and the evolution of the lithosphere [Ph.D. thesis]: Cambridge, Massachusetts Institute of Technology, 256 p.

Buck, W.R., 1986, Small-scale convection induced by passive rifting: The cause for uplift of rift shoulders: Earth and Planetary Science Letters, v. 77, p. 362-372.

Buck, W.R., 1987, Analysis of the cooling of a variable viscosity fluid with application to the earth: Geophysical Journal of the Royal Astronomical Society, v. 89, p. 549-577.

Buck, W.R., 1988, Flexural rotation of normal faults: Tectonics (in press).

Buck, W.R., Martinez, F., Steckler, M.S., and Cochran, J.R., 1988, Thermal consequences of lithospheric extension: Pure and simple: Tectonics, v. 7, p. 213-234.

Cande, S., LaBrecque, J.L., and Haxby, W.B., 1988, Plate kinematics of the South Atlantic: Chron 34 to Present: Journal of Geophysical Research (in press).

Celerier, B., 1986, Models for the evolution of the Carolina Trough and their limitations [Ph.D. thesis]: Cambridge, Massachusetts Institute of Technology, 206 p.

Chapman, D.S., and Pollack, H.N., 1975, Global heat flow: A new look: Earth and Planetary Science Letters, v. 28, p. 23-32.

Chenet, P.Y., and Letouzey, J., 1983, Tectonics of the area between Abu-Durba and Gebel Mezzazt (Sinai, Egypt) in the context of the evolution of the Suez rift: Bulletin Centre Recherche Exploration-Production, Elf-Aquitaine, v. 7, p. 201-215.

Cochran, J.R., 1983, Effects of finite extension times on the development of sedimentary basins: Earth and Planetary Science Letters, v. 66, p. 289-302.

Cochran, J.R., 1986, Variations in subsidence rates along intermediate and fast spreading mid-ocean ridges: Geophysical Journal of the Royal Astronomical Society, v. 87, p. 421-454.

Dunbar, J.A., and Sawyer, D., 1988, Preexisting lithospheric weaknesses and the relative timing of volcanism and graben formation in continental rifting: EOS (Transactions of the American Geophysical Union), v. 69, p. 465.

England, P.C., 1983, Constraints on extension of continental lithosphere: Journal of Geophysical Research, v. 88, p. 1145-1152.

England, P.C., 1986, Comment on "Brittle failure in the upper mantle during extension of continental lithosphere" by Dale S. Sawyer: Journal of Geophysical Research, v. 91, p. 10487-10490.

Garfunkel, Z., 1988, Relation between continental rifting and uplifting: Evidence from the Suez rift and northern Red Sea: Tectonophysics, v. 150, p. 33-49.

Gibbs, A.D., 1984, Structural evolution of extensional basin margins: Journal of the Geological Society of London, v. 141, p. 609-620.

Gilbert, M.C., 1983, Timing and chemistry of igneous events associated with the southern Oklahoma aulacogen: Tectonophysics, v. 94, p. 439-455.

Goetze, C., 1978, The mechanisms of creep in olivine: Philosophical Transactions of the Royal Society of London, Series A, v. 288, p. 99-119.

Goetze, C., and Evans, B., 1979, Stress and temperature in the bending lithosphere as constrained by experimental rock mechanics: Geophysical Journal of the Royal Astronomical Society, v. 59, p. 463-478.

Haxby, W.F., and Weissel, J.K., 1986, Evidence for small-scale convection from Seasat altimeter data: Journal of Geophysical Research, v. 91, p. 3507-3520.

Hayes, D.E., 1988, Age-depth relationships and depth anomalies in the southeast Indian Ocean and South Atlantic Ocean: Journal of Geophysical Research, v. 93, p. 2937-2954.

Heestand, R.L., and Crough, S.T., 1981, The effect of hotspots on the ocean age-depth relation: Journal of Geophysical Research, v. 86, p. 6107-6114.

Hellinger, S.J., and Sclater, J.G., 1983, Some comments on two-layer extensional models for the evolution of sedimentary basins: Journal of Geophysical Research, v. 88, p. 8251-8269.

Houseman, G., and McKenzie, D.P., 1982, Numerical experiments on the onset of convective instability in the Earth's mantle: Geophysical Journal of the Royal Astronomical Society, v. 79, p. 2961.

Jackson, J.A., and McKenzie, D.P., 1983, The geometrical evolution of normal fault systems: Journal of Structural Geology, v. 13, p. 189-193.

Jarvis, G.T., 1984, An extensional model of graben subsidence—The first stage of basin evolution: Sedimentary Geology, v. 40, p. 13-31.

Jordan, T.E., 1981, Thrust loads and foreland basin evolution, Cretaceous, western United States: American Association of Petroleum Geologists Bulletin, v. 65, p. 2506-2520.

Karner, G.D., 1985, Thermally induced residual topography within oceanic lithosphere: Nature, v. 318, p. 527-532.

Karner, G.D., and Watts, A.B., 1982, On isostasy at Atlantic-type continental margins: Journal of Geophysical Research, v. 87, p. 2923-2948.

Karner, G.D., and Watts, A.B., 1983, Gravity anomalies and flexure of the lithosphere at mountain ranges: Journal of Geophysical Research, v. 88, p. 10449-10477.

Karner, G.D., Steckler, M.S., and Thorne, J., 1983, Long-term mechanical properties of the continental lithosphere: Nature, v. 304, p. 250-253.

Keen, C.E., 1985, The dynamics of rifting: Deformation of the lithosphere by active and passive driving forces: Geophysical Journal of the Royal Astronomical Society, v. 80, p. 95-120.

Kirby, S.H., 1983, Rheology of the lithosphere: Reviews of Geophysics and Space Physics, v. 21, p. 1458-1487.

Kominz, M.A., and Bond, G.C., 1986, Geophysical modeling of the thermal history of foreland basins: Nature, v. 320, p. 252-256.

Kusznir, N., and Karner, G.D., 1985, Dependence of the flexural rigidity of the continental lithosphere on rheology and temperature: Nature, v. 316, p. 138-142.

Kusznir, N.J., and Park, R.G., 1987, The extensional strength of the continental lithosphere: Its dependence on geothermal gradient, crustal composition and thickness, in Coward, M.P., Dewey, J.F., and Hancock, P.L., eds., Continental extensional tectonics: Geological Society of London Special Publication 28, p. 35-52.

Kusznir, N., Karner, G.D., and Egan, S., 1987, Geometric, thermal and isostatic consequences of detachments in continental lithosphere extension and basin formation, in Beaumont, C., and Tankard, A., eds., Sedimentary basins and basin-forming mechanisms: Canadian Society of Petroleum Geology Memoir 12, p. 185-204.

Lerner-Lam, A.L., and Jordan, T.H., 1987, How thick are the continents?: Journal of Geophysical Research, v. 92, p. 14007-14026.

Lister, G.S., Etheridge, M.A., and Symonds, P.A., 1986, Detachment faulting and the evolution of passive continental margins: Geology, v. 14, p. 246-250.

Lucazeau, F., and LeDouaran, S., 1985, The blanketing effect of sediments in basins formed by extension: A numerical model. Application to the Gulf of Lion and Viking graben: Earth and Planetary Science Letters, v. 74, p. 92-102.

Lynch, H.D., and Morgan, P., 1987, The tensile strength of the lithosphere and the localization of extension, in Coward, M.P., Dewey, J.F., and Hancock, P.L., eds., Continental extensional tectonics: Geological Society of London Special Publication 28, p. 53-66.

Lyon-Caen, H., and Molnar, P., 1983, Constraints on the structure of the Himalaya from an analysis of gravity anomalies and a flexural model of the lithosphere: Journal of Geophysical Research, v. 88, p. 8171-8191.

Lyon-Caen, H., and Molnar, P., 1985, Gravity anomalies, flexure of the Indian plate, and the structure, support and evolution of the Himalaya and Ganga basin: Tectonics, v. 4, p. 513-538.

McAdoo, D.C., and Sandwell, D.T., 1985, Folding of oceanic lithosphere: Journal of Geophysical Research, v. 90, p. 8563-8569.

McKenzie, D.P., 1978, Some remarks on the development of sedimentary basins: Earth and Planetary Science Letters, v. 40, p. 25-32.

McKenzie, D.P., 1984, The generation and compaction of partially molten rock: Journal of Petrology, v. 25, p. 713-765.

McKenzie, D.P., and Bickle, M., 1988, The volume and composition of melt generated by extension of the lithosphere: Journal of Petrology (in press).

McNutt, M.K., 1984, Lithospheric flexure and thermal anomalies: Journal of Geophysical Research, v. 89, p. 11180-11194.

McNutt, M., and Menard, H.W., 1982, Constraints on yield strength in the oceanic lithosphere derived from observations of flexure: Geophysical Journal of the Royal Astronomical Society, v. 59, p. 363-394.

McNutt, M., Diament, M., and Kogan, M.G., 1988, Variations of elastic plate thickness at continental thrust belts: Journal of Geophysical Research, v. 93, p. 8825-8838.

Mutter, J.C., Buck, W.R., and Zehnder, C.M., 1988, Convective partial melting, 1: A model for the formation of thick basaltic sequences during the initiation of spreading: Journal of Geophysical Research, v. 93, p. 1031-1048.

Parsons, B., and Sclater, J.G., 1977, An analysis of the variation of ocean-floor bathymetry and heat flow with age: Journal of Geophysical Research, v. 82, p. 803-827.

Parsons, B., and McKenzie, D., 1978, Mantle convection and the thermal structure of the plates: Journal of Geophysical Research, v. 83, p. 4485-4496.

Price, R.A., 1973, Large-scale gravitational flow of supracrustal rocks, southern Canadian Rockies, in deJong K.A., and Scholten, R., eds., Gravity and tectonics: New York, Wiley and Sons, p. 491-502.

Ratcliffe, N.M., and Burton, W.C., 1985, Fault reactivation models for the origin of the Newark basin and studies related to eastern U.S. seismicity, in Robinson, G.P., and Froelich, A.J., eds., Proceedings of the second U.S. Geological Survey workshop on the early Mesozoic basins of the eastern United States: U.S. Geological Survey Circular 946, p. 6-45.

Renkin, M.L., and Sclater, J.G., 1988, Depth and age in the North Pacific: Journal of Geophysical Research, v. 93, p. 2919-2935.

Robinson, E.M., and Parsons, B., 1988, Effect of a shallow low-viscosity zone on small-scale instabilities under the cooling oceanic plates: Journal of Geophysical Research, v. 93, p. 3469-3479.

Rosendahl, B.R., 1987, Architecture of continental rifts with special reference to East Africa: Annual Reviews of Earth and Planetary Sciences, v. 15, p. 445-503.

Royden, L., and Karner, G.D., 1984, Flexure of lithosphere beneath Apennine and Carpathian foredeep basins: Evidence for an insufficient topographic load: American Association of Petroleum Geologists Bulletin, v. 68, p. 704-712.

Royden, L., and Keen, C.E., 1980, Rifting process and thermal evolution of continental margin of eastern Canada determined from subsidence curves: Earth and Planetary Science Letters, v. 51, p. 343-361.

Royden, L., Sclater, J.G., and von Herzen, R.P., 1980, Continental margin subsidence and heat flow: Important parameters in formation of petroleum hydrocarbons: American Association of Petroleum Geologists Bulletin, v. 64, p. 173-187.

Sawyer, D.S., 1985, Total tectonic subsidence: A parameter for distinguishing crust type at the U.S. Atlantic continental margin: Journal of Geophysical Research, v. 90, p. 7751-7769.

Sawyer, D.S., 1986, Reply to comment on "Brittle failure in the upper mantle during extension of continental lithosphere": Journal of Geophysical Research, v. 91, p. 10491-10492.

Sawyer, D.S., Swift, B.A., Sclater, J.G., and Toksöz, M.N., 1982a, Extensional model for the subsidence of the northern United States Atlantic continental margin: Geology, v. 10, p. 134-140.

Sawyer, D.S., Swift, B.A., Toksöz, M.N., and Sclater, J.G., 1982b, Thermal evolution of the Baltimore Canyon Trough and Georges Bank Basins, in Watkins, J.S., and Drake, C.L., eds., Studies in continental margin geology: American Association of Petroleum Geologists Memoir 34, p. 743-764.

Sclater, J.G., Jaupart, C., Galson, D., 1980, The heat flow through oceanic and continental crust and the heat loss of the earth: Reviews of Geophysics and Space Physics, v. 18, p. 269-311.

Sleep, N.H., 1971, Thermal effects of the formation of Atlantic continental margins by continental breakup: Geophysical Journal of the Royal Astronomical Society, v. 24, p. 325-350.

Steckler, M.S., 1981, The thermal and mechanical evolution of Atlantic-type continental margins [Ph.D. Thesis]: New York, Columbia University, 261 p.

Steckler, M.S., 1985, Uplift and extension at the Gulf of Suez—Indications of induced mantle convection: Nature, v. 317, p. 135-139.

Steckler, M.S., and ten Brink, U.S., 1986, Lithospheric strength variations as a control on new plate boundaries: Examples from the Arabian Plate: Earth and Planetary Science Letters, v. 79, p. 120-132.

Steckler, M.S., and Watts, A.B., 1978, Subsidence of the Atlantic type continental margin off New York: Earth and Planetary Science Letters, v. 42, p. 1-13.

Steckler, M.S., and Watts, A.B., 1980, The Gulf of Lion: Subsidence of a young continental margin: Nature, v. 287, p. 425-429.

Steckler, M.S., Watts, A.B., and Thorne, J.A., 1988, Subsidence and basin modeling at the U.S. Atlantic passive margin, in Sheridan, R.E., and Grow, J.A., eds., The Atlantic continental margin, U.S.: Geological Society of America, The Geology of North America, v. I-2, p. 399-416.

Stephenson, R., 1988, Beyond first-order thermal subsidence models for sedimentary basins?, in Cross, T.A., ed., Quantitative dynamic stratigraphy: Englewood Cliffs, New Jersey, Prentice-Hall, (this volume).

Stockmal, G.S., Beaumont, C., and Boutilier, R., 1986, Geodynamic models of convergent margin tectonics: Transition from rifted margin to overthrust belt and consequences for foreland-basin development: American Association of Petroleum Geologists Bulletin, v. 70, p. 181-190.

Swanson, M.T., 1986, Preexisting fault control for Mesozoic basin formation in eastern North America: Geology, v. 14, p. 419-422.

Sykes, L.R., 1978, Intraplate seismicity, reactivation of preexisting zones of weakness, alkaline magmatism and other tectonism postdating continental fragmentation: Reviews of Geophysics and Space Physics, v. 16, p. 621-688.

Trehu, A.M., 1975, Depth versus age$^{1/2}$: A perspective on mid-ocean rises: Earth and Planetary Science Letters, v. 77, p. 287-304.

Vening Meinesz, F.A., 1950, Les Grabens Africans resultant de compression ou de tension dans la croûte terrestre?: Institute du Royal colonial Belge Bulletin, v. 21, p. 539-552.

Vink, G.E., Morgan, W.J., and Zhao, W.-L., 1984, Preferential rifting of continents: A source of displaced terranes: Journal of Geophysical Research, v. 89, p. 10072-10076.

Watts, A.B., 1978, An analysis of isostasy in the world's oceans, 1. Hawaiian-Emperor seamount chain: Journal of Geophysical Research, v. 83, p. 5989-6004.

Watts, A.B., 1988, Gravity anomalies, crustal structure and flexure of the lithosphere at the Baltimore Canyon trough: Earth and Planetary Science Letters, v. 89, p. 221-238.

Watts, A.B., and Ryan, W.B.F., 1976, Flexure of the lithosphere and continental margin basins: Tectonophysics, v. 36, p. 25-44.

Watts, A.B., and Thorne, J., 1984, Tectonics, global changes in sea-level and their relationship to stratigraphic sequences at the U.S. Atlantic continental margin: Marine and Petroleum Geology, v. 1, p. 319-339.

Watts, A.B., Karner, G.D., and Steckler, M.S., 1982, Lithospheric flexure and the evolution of sedimentary basins, *in* Kent, P., Bott, M.H.P., McKenzie, D.P., and Williams, C.A., eds., The evolution of sedimentary basins: Philosophical Transactions of the Royal Society of London, v. 305A, p. 249-281.

Weissel, J.K., and Karner, G.D., 1988, Flexural uplift of rift flanks due to tectonic denudation of the lithosphere during extension: Journal of Geophysical Research (in press).

Wernicke, B., 1985, Uniform-sense normal simple shear of the continental lithosphere: Canadian Journal of Earth Sciences, v. 22, p. 108-125.

White, N.J., and McKenzie, D.P., 1988, Formation of the "steer's head" geometry of sedimentary basins by differential stretching of the crust and mantle: Geology, v. 16, p. 250-253.

White, R.S., Spence, G.D., Fowler, S.R., McKenzie, D.P., Westbrook, G.K., and Bowen, A.N., 1987, Magmatism at rifted continental margins: Nature, v. 330, p. 439-444.

7

Beyond First-Order Thermal Subsidence Models for Sedimentary Basins?

Randell Stephenson
Geological Survey of Canada, Institute of Sedimentary and Petroleum Geology, 3303–33
Street NW, Calgary, Alberta, T2L 2A7 Canada

Abstract

The general characteristics of sedimentary basins at rifted or sheared continental margins are explained by models of crustal subsidence driven by the thermal contraction of an anomalously hot lithosphere. Conventional local isostatic and simple elastic-plate models of sedimentary loading on a thermally perturbed lithosphere may be inadequate because of the important interactions of thermal and mechanical forces. Several important thermo-mechanical processes have the tendency to reduce and/or retard the thermal subsidence mechanism. These include lithospheric thickening (or thinning) due to the basal phase change, accompanied by the introduction (or removal) of latent heat; the displacement of hot asthenospheric material by sedimentary loading; the insulating effect of the thickening, low thermal conductivity, sedimentary sequence atop the cooling, thickening thermal lithosphere; and the regional isostatic compensation of the thermal body forces, given regional compensation of the sediment load. Regional compensation of the thermal forces includes the implicit effects of thermal bending moments. The spatial interactions of these processes can produce significant stratigraphic variations within sedimentary basins. Furthermore, the ongoing relaxation of thermal and mechanical lithosphere stresses resulting from the basin subsidence mechanism needs to be considered as a source of stratigraphic cyclicity within basins.

Introduction

There is almost universal agreement that crustal subsidence resulting from the conductive cooling of a thinned and heated lithosphere is responsible for the postrift, first-order development of many of the Earth's continental margin and intracratonic sedimentary basins. McKenzie (1978) presented a simple and concise quantitative formulation of the transient lithosphere temperatures and resulting crustal heat flux and subsidence patterns predicted by such a mechanism and his paper forms the benchmark for numerous subsequent studies of the regional development of specific sedimentary basins (e.g., Bond and Kominz, 1984). McKenzie's formulation, in which lithospheric thinning results from pure-shear extension, was based on the assumption that the resulting anomalous temperature field decays by transient conduction of heat in the vertical direction and subsidence occurs according to local isostasy. The complicating effects on basin subsidence analyses of adopting a regional—or flexural—instead of a local isostatic model and of incorporating lateral heat transfer in the lithosphere are moderately well known (e.g., Cochran, 1983). But there are some other fundamental processes inherent to the postrift, thermal subsidence model that also require assessment for a fuller understanding of the relationship of the tectonic driving mechanism and the internal architecture of extensional sedimentary basins.

A more rigorous thermo-mechanical model of postrift sedimentary basin subsidence is examined here. It comprises a sedimentary basin overlying a thinned, rheologically layered (elastic-viscoelastic) lithosphere in turn overlying an inviscid asthenosphere. Active sediment deposition, at a specified rate, occurs within the overlying basin and the thermal state of the sedimentary basin is fully coupled with that of the lithosphere. Temperatures in the model are governed by the effects of vertical and horizontal thermal conduction such that the lithosphere/asthenosphere boundary is defined as a (partial) melt isotherm or phase-change boundary that migrates vertically depending on the transient thermal state. The potential buoyancy effects of a compositionally less dense asthenospheric partial melt (e.g., Brown and Beaumont, 1988) are not included in the calculations nor are the origins of the partial melt phase beneath extensional basins explicitly considered (i.e., Mutter et al., 1988). Vertical deformations of the lithosphere result from the purely mechanical effects of sediment loading as well as from changes in the ambient temperature field. The temperature anomalies contribute to these deformations not only by setting up body forces but also by creating thermal in-plane forces and associated bending moments. The model allows explicit consideration of: the effect of a partial liquidus/solidus boundary (the lithosphere/asthenosphere boundary) on the heat budget of the cooling lithosphere; the thermal equilibrium changes in the lithosphere resulting from the overlying, relatively lower thermal conductivity, sedimentary layer; the thermo-elastic consequences of rheological layering in the lithosphere, including the generation of bending moments resulting in uplift of a cooling lithosphere capped by a relatively strong, elastic layer; and how these various processes interact to complicate the internal structure of thermally subsiding sedimentary basins.

A schematic representation of the model is shown in Figure 1 with some of its physical parameters. The mechanical responses of such a rheologically layered lithosphere to conductive cooling, and to loading of overlying water and sediment layers, were quantified by Nakiboglu and Lambeck (1985) and Lambeck and Nakiboglu (1981) respectively. Their solutions were combined and modified slightly to deduce the appropriate regional compensation model for three-dimensional sedimentary basin applications in Stephenson et al. (1987) and for two-dimensional sedimentary loads at rifted continental margins in Stephenson et al. (in press). The upper elastic layer of the two-layered lithosphere model is the cooler, more competent part of the lithosphere with a thickness equal to its long-term effective elastic thickness. The base of the underlying viscoelastic layer is defined by the thermal thickness of the lithosphere. It is assumed that the two layers are perfectly coupled such that the vertical and horizontal deformations are continuous across the interface. Temperatures in the models are computed using a finite difference solution to the two-dimensional transient heat conduction equation for a moving lower (phase) boundary case following Hastaoglu (1986) and using generally simple, temperature boundary conditions (Ozisik, 1980). All of the calculations that follow were made using those values for model parameters listed in Table 1. The basic input variable to the model calculations is β, the lithosphere stretching parameter as defined by McKenzie (1978).

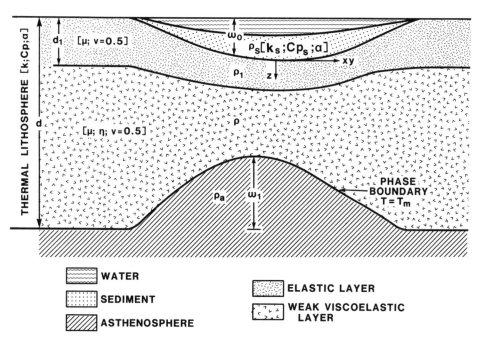

Figure 1. Schematic representation of the coupled thermo-mechanical model of sedimentary basin subsidence. Displacement of the upper and lower lithosphere boundaries in the model are labelled as ω_0 and ω_1, respectively; other symbols and parameters are listed in Table 1.

Table 1 Thermal and mechanical parameters used in model calculations.

Parameter	Definition	Value
ρ_1	Mean crustal density	2.7×10^3 kg m^{-3}
ρ	Lower lithosphere density	3.2×10^3 kg m^{-3}
ρ_a	Asthenosphere density	3.2×10^3 kg m^{-3}
ρ_s	Sediment density	2.5×10^3 kg m^{-3}
k	Thermal conductivity of lithosphere	3.1 Wm^{-1} K^{-1}
k_s	Thermal conductivity of sediments	1.3/2.0/5.2 Wm^{-1} K^{-1}
C_p	Specific heat capacity of lithosphere	1210 J kg^{-1} K^{-1}
C_{ps}	Specific heat of sediment	900 J kg^{-1} K^{-1}
T_m	Melting point temperature	1200°C
a	Linear thermal expansion coefficient	1.03×10^{-5} K^{-1}
L	Latent heat of melting	4.2×10^5 J kg^{-1}
d	Thickness of thermal lithosphere	125 km
μ	Effective rigidity of lithosphere	0.39×10^5 MPa
d_1	Effective elastic thickness of lithosphere	0/20/40 km
ν	Poisson's ratio	0.5
η	Newtonian viscosity	$\leq 10^{23}$ Pa s

BASIN-WIDE THERMAL SUBSIDENCE

Figure 2 shows thermal subsidence and lithosphere thickness as functions of the square root of time for several values of β for the present model. The basins are assumed to be in a state of local isostasy and, as no sediments are included, they are water filled only. The degree of partial melt in the asthenosphere beneath the stretched lithosphere is taken to be 5% so that the value of L used in the computations is 5% of the value listed in Table 1, following the method of Oldenburg (1975). Geophysical evidence for significant volumes of partially melted material in the asthenosphere beneath oceanic ridges and beneath relatively young oceanic lithosphere is ample (e.g., Solomon, 1973). That partial melt exists extensively elsewhere in the asthenosphere, especially in relation to zones of continental rifting (e.g., Parker et al., 1984), is probable.

Figure 3 shows how variations in the degree of partial melt affect the thickening of the lithosphere. The model also accounts implicitly for the introduction of equal amounts of heat

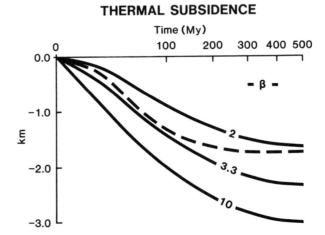

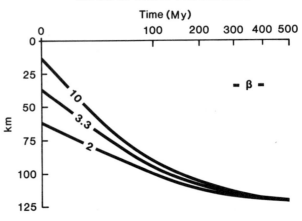

Figure 2. Thermal subsidence (top) and lithosphere thickness (bottom) for various βs as functions of $t^{1/2}$, the former compared to equivalent β=2 McKenzie (1978) model (dashed curve). No sedimentation; local isostasy.

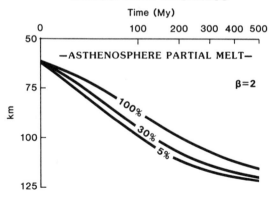

Figure 3. The influence of variations in percent asthenospheric partial melt on lithosphere thickness as a function of $t^{1/2}$. $\beta=2$; no sedimentation; local isostasy. Because these basins are water-filled only, the equilibrium thicknesses eventually attained by the lithosphere will be the same for all three curves (d=125 km).

to the base of the stretched and to the base of the unperturbed lithosphere by mantle conduction or convection such that the thickness of the unperturbed lithosphere remains constant. That is, old, cold, unperturbed lithosphere is assumed to have a finite thickness, as suggested by seismological and other observations from North America (Kono and Amano, 1978).

Superimposed on the top half of Figure 2 (dashed curve) is the equivalent thermal subsidence predicted by the model of McKenzie (1978) for $\beta=2$ calculated using the present physical parameters where applicable. Because McKenzie's model does not incorporate into its heat budget the processes noted above, namely, the latent heat of crystallization derived from the solidifying asthenosphere and convective and conductive heat sources from the asthenosphere, it predicts a significantly earlier thermal equilibration of the stretched lithosphere.

The effect of sediment accumulation on reducing lithospheric cooling and subsidence is illustrated in Figure 4. The sediment layer hinders cooling and mantle solidification at the base of the lithosphere. The final thickness reached by a cooling lithosphere covered by sediments is appreciably less than its initial unperturbed thickness and is less than the minimal thickness reached by a cooling lithosphere with no sediment cover. The total thermal subsidence is less for a basin with a sediment cover, depending on the thickness and conductivity of the cover, because the final equilibrium temperatures reached in the lithosphere are higher. The cumulative sediment thickness in Figure 4 is 13.3 km at 500 m.y. (no sediments were loaded after 250 m.y.) with an assumed thermal conductivity of 2.0 $Wm^{-1}K^{-1}$.

Another reason why thermal subsidence is reduced when sediments are present is that the sediment load itself displaces an equivalent volume of hot lithosphere (i.e., upwelled asthenosphere below the sedimentary basin) into the region below the base of the unperturbed lithosphere. The cooling and contraction of this material cannot be invoked as contributing to the thermal subsidence of the overlying sedimentary basin. This effectively advective cooling of the lithosphere, as sediments are loaded onto its surface, is the reason why the rate of thermal subsidence is actually greater for the sediment loading case during the first 100 m.y. or so in Figure 4. Also shown on Figure 4 (dotted line) is an exponentially decaying curve (with respect to time) such as those used by Bond and Kominz (1984), derived from McKenzie (1978), in modeling observed basin subsidence curves. The curve shown has an amplitude of 1.4 km and a decay constant of 62.8 m.y., based on the thermal parameters listed in Table 1 (cf. McKenzie, 1978). Although the presence of a "thermal blanketing effect," reducing the total thermal subsidence compared to McKenzie's (1978) predictions, has been noted previously (e.g., Beaumont et al., 1982; Nunn et al., 1984), the intrinsic effects of sedimentation in possibly controlling the lifespan of thermally subsiding basins has not been widely recognized (e.g., Stephenson et al., in press).

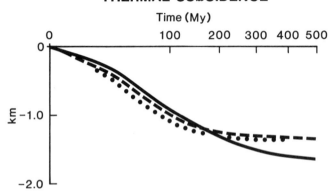

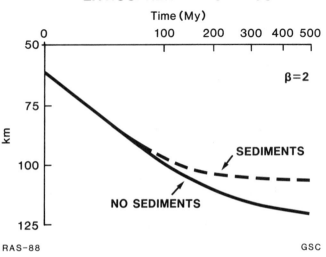

Figure 4. Thermal subsidence (top) and lithosphere thickness (bottom) as functions of $t^{1/2}$ with sedimentation (dashed curves) and with no sedimentation (solid curves), the former compared to the temporally decaying exponential curve $1.4 [\exp(-t/62.8)-1]$ (dotted); $\beta=2$; local   RAS-88   GSC

Spatial Intrabasinal Variations in Thermal Subsidence

Obviously, regional variations in the lithosphere thinning parameter, β, can produce regional variations in thermal subsidence within a sedimentary basin. The rate and amount of thermal subsidence vary according to β, although not necessarily as predicted by McKenzie's (1978) first-order model. The control exerted by the presence of a sedimentary layer, known in part as its thermal blanketing effect (e.g., Beaumont et al., 1982), on the coupled thermal state of the lithosphere-sedimentary basin system is such that the presence or absence of the sediment layer strongly affects the nature of the thermal subsidence (Fig. 4). This means that tectonic subsidence curves derived from backstripping techniques (e.g., Bond and Kominz, 1984) cannot be interpreted directly in terms of a water-filled-basin subsidence model. Figure 4 shows that for β=2 the maximum difference in thermal subsidence (for a thermal conductivity typical of a shale-dominated basin) is about 350 m. In terms of McKenzie's model this is equivalent to β 1.75 or 2.5, rather than β=2.

It follows that significant regional variations in sedimentation rates and thicknesses and of bulk thermal properties within the sediment layer also may produce discernible differences in tectonic subsidence curves, even where no major differences in lithospheric heating have occurred. Such a case might arise along the strike of a rifted continental margin that has most of its sediment load concentrated at major deltas. Figure 5 shows thermal subsidence for two basins filled with sediments with different thermal conductivities (solid lines). The thermal conductivity values chosen, 1.3 and 5.2 $Wm^{-1}K^{-1}$ are extreme lower and upper limits that would approximately correspond to those expected for basins filled with water-saturated shale and halite respectively. In both cases β=2 and local isostasy is in effect. The difference in thermal subsidence at 200 m.y. is more than 600 m. In terms of McKenzie's model this is equivalent to β 1.5 or 3, rather than B=2. Figure 5 also shows that nonnegligible thermal subsidence lasts longer for the high thermal conductivity case.

Figure 3 suggests that variations in asthenosphere partial melt beneath a rifted basin (e.g., Mutter et al., 1988) can effect discernible spatial differences in subsidence behavior. Such a case might occur if high-angle salients exist along the main rift trend or if a rifted margin is in part a "passive" margin and in part an "active" margin (cf. Keen, 1985). Because the basins in Figure 3 are filled only with water, the equilibrium thicknesses eventually attained by the lithosphere (not shown) does not depend on the degree of partial melt. The rate of thermal subsidence is affected by the degree of partial melt but the final amount of thermal subsidence is not. Thus the total thermal subsidence, for equivalent β, can be redistributed in time depending on the partial melt. As an example, Figure 5 shows the relative thermal uplift through time of a sedimentary basin overlying a 100% partial-melt zone compared to ambient 5% asthenospheric partial melt (dashed lines).

Complications in basin stratigraphy may arise from spatial variations in lithospheric stretching factors and asthenospheric state and/or sediment budgets and coupled thermal responses of the basin and lithosphere. These complications are amplified by the mechanical loading effects—or isostatic compensation—of the sedimentary strata. For example, it is well known that isostatic flexure can induce subsidence of cold, unthinned lithosphere flanking thermally subsiding basins.

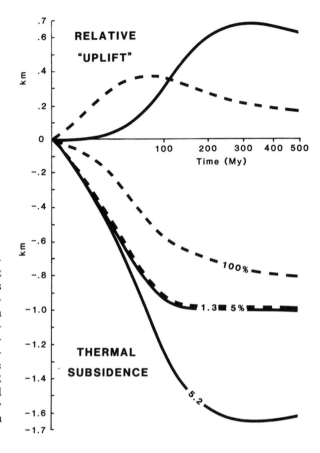

Figure 5. The influence of thermal conductivity of the sediment layer (solid lines with values shown) and the influence of degree partial melt (dashed lines with values shown) on thermal subsidence as functions of $t^{1/2}$. The relative "uplifts" of the low conductivity case (1.3 $Wm^{-1}K^{-1}$) versus the high conductivity case (5.2 $Wm^{-1}K^{-1}$) and of the high partial melt case (100%) versus the low partial melt case (5%) are shown at the top. $\beta=2$; local isostasy.

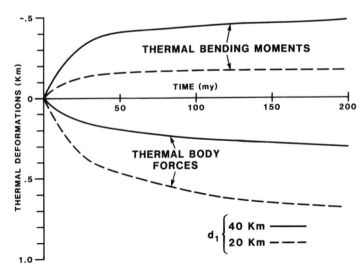

Figure 6. The influence of regional thermo-isostasy on deformations induced by thermal bending moments and thermal body forces at the center of a 100 km radius circular sedimentary basin on a lithosphere with an elastic layer d_1 of 40 km (solid curve) and 20 km (dashed curve). Sedimentation rate is 0.2 mm/yr; $\beta=2$.

Beaumont et al. (1982) assumed that the cooler part of the rheologically layered lithosphere is elastic, and that the mechanical loads of sediment and water were therefore compensated regionally. However, in their and other similar models, the isostatic response to the thermal loads—that is the thermal subsidence itself—was calculated assuming local isostasy. Not only is this inconsistent but it neglects thermal bending moments that are generated in a cooling, rheologically layered lithosphere. Bending moments arise because of the different rheological responses of the lithospheric layers to its overall cooling. Bills (1983) noted the possible role of thermal bending moments during vertical deformations of the lithosphere, and considered that these thermal flexural effects may significantly affect sedimentary basin formation.

Figure 6 illustrates the thermal effects of a cooling, thinned lithosphere beneath a circular basin for the present model. The elastic layer thickness of the lithosphere is taken to be either 20 km or 40 km. Sediment accumulation is assumed to proceed at a constant rate of 0.2 mm/yr until the basin is completely filled. The thermal body forces are generated predominantly by the cooling of the lower lithosphere and they induce subsidence. The thermal body-force subsidence is reduced appreciably if the elastic layer is thick (i.e., 40 km; solid lines), as the thermal load is more regionally compensated isostatically. By contrast, because of the combined effects of cooling of deeper and warming of shallow regions, the thermal bending moments generate a thermal uplift that increases in magnitude proportionately with elastic thickness. If the elastic layer thickness increases through time with cooling, it follows that the relative effects of the thermal bending moments will also increase through time. Figure 6 shows that the combined effects of the thermal bending moments and thermal body forces, for this particular circular basin, yield no net thermal subsidence if the elastic layer is 40 km thick. This clearly implies that the presence of thermal bending moments in a cooling lithosphere results in potentially significant complications in the regional isostatic compensation model of thermally subsiding sedimentary basins. In a two-dimensional model of a passive continental margin, for example, Stephenson et al., (in press) show that thermal bending forces, while attenuating thermal subsidence in areas flanking the rapidly cooling oceanic lithosphere (e.g., the continental rise), enhance subsidence beneath the continental shelf and may play a role in the subsidence of platformal basins cratonward of continental margins.

TEMPORAL INTRABASINAL VARIATIONS IN THERMAL SUBSIDENCE

Isostatic equilibrium, whether local or regional, usually has been treated as a balance of vertical forces—the weight of the superimposed load and a countervailing buoyancy force at the base of the crust or lithosphere. But isostatic equilibrium also balances the presence of horizontal stresses in the lithosphere such as illustrated by the vertical motions associated with thermal bending moments—and induced in-plane stresses—in the present model. It follows that changes of sufficient magnitude in the regional horizontal stress field also may generate discernible anomalies in sedimentary basin subsidence or uplift. This is the essence of the models of Cloetingh et al. (1985) and Cloetingh and Kooi (this volume) for a tectonic origin for third-order stratigraphic cycles.

Stress relaxation within zones of finite yield strength in the lithosphere is another tectonic mechanism that could produce sedimentary cyclicity within sedimentary basins. In the thermo-mechanical subsidence model considered here (Fig. 1), for example, the lower lithosphere layer, being of viscoelastic rheology, is assumed to possess no finite yield strength. The viscosity of this layer is considered to range with depth from an effective long-term value of $\sim 10^{23}$ Pa s to asthenospheric viscosities of $\sim 10^0$ Pa s (Nakiboglu and Lambeck, 1983). The implied relaxation-time constant of this layer is therefore less than about 10^{-1} m.y. (e.g., Ranalli, 1987). This means that for loads of greater duration than 10^{-1} m.y. all associated stresses in the viscoelastic layer are fully relaxed and the relaxation is assumed to take place by viscous flow. Even if this layer did possess a very small (Ranalli, 1987) yield strength, the model is not designed to resolve the resultant discontinuous, nonviscous nature of the relaxation. Since the relaxation-time constant is much smaller than both the thermal-conduction constant, about 10^2 m.y. for the parameters listed in Table 1, and the evolution period of sedimentary basins, also in the order of 10^2 m.y., the viscoelastic lower lithosphere model is fully appropriate when considering the long-term evolution of sedimentary basins.

The presence of a purely elastic upper lithosphere throughout the evolution of sedimentary basins implies a viscoelastic relaxation-time constant there much greater than 10^2 m.y. [In considering cases of local isostatic compensation of the sedimentary and thermal loads, the elastic layer is assumed to have effectively zero thickness.] The purely elastic layer, by definition, never fails. Sedimentary loads of several kilometers thickness, however, clearly are capable of generating flexural stresses within the elastic layer that should be sufficient to produce failure (e.g., Walcott, 1971)—whether by elastic-brittle failure near the surface or by elastic-plastic failure near the base of the "elastic" layer (e.g., Ranalli and Murphy, 1987). These kinds of failure could produce stratigraphic cyclicity within sedimentary basins with periodicities depending on the yield strengths inherent to the upper lithosphere and on the rate at which flexural stresses are accumulated—in this case, primarily a function of the sedimentation rate. In other words, it is possible that higher-order stratigraphic cycles in basins are themselves evidence for episodic failure within the upper lithosphere.

The model is assumed to consist of these uniformly viscoelastic and elastic layers so that it is analytically tractable. The stepwise change in effective model viscosity at the interface of the elastic and viscoelastic layers reflects in a crude way the expected temperature dependence of viscosity in the lithosphere. More likely (e.g., Courtney and Beaumont, 1983), the viscosity profile of the lithosphere is gradational. Thus, depending on the lithosphere geothermal gradient, a layer within the lithosphere will be characterized by viscoelastic relaxation times in the range 10^{-1} - 10^2 m.y., intermediate to those implied by the present model and of similar order of magnitude as the periodicities of several second- and higher-order stratigraphic cycles. The temperature-dependent viscoelastic model (Courtney and Beaumont, 1983) assumes no yield strength in the lithosphere at any depth, and does not resolve the nature of stress relaxation at times much less than the characteristic viscoelastic relaxation-time constants. Rather, it provides a model of stress relaxation at times much greater than a given viscoelastic relaxation-time constant. Again, relaxation at these "effective" viscosities, but within the presence of a yield strength, would be episodic or cyclic.

The ability at the present time to confidently estimate the rheology of the lithosphere, its effective viscosity and its strength as a function of depth and composition, for example, is highly imperfect (e.g., Ranalli, 1987). Cyclicity in sedimentary basins, which is observable and quantifiable, may be a far better clue toward understanding the nature of the lithosphere and of how the lithosphere subsides in response to tectonic forces and changes in tectonic forces associated with the evolution of sedimentary basins than the converse may be.

SUMMARY

For thermally subsiding sedimentary basins, at continental margins and other areas of thinned continental crust, vertical deformations of the lithosphere result from the purely mechanical effects of sedimentary loading as well as from changes in the ambient temperature field. The consequences of the latter are typically modeled as a purely exponentially decaying subsidence, itself an approximation to the transient solution for temperature in a conductively cooling uniform slab.

At the basin-wide scale, there are fundamental first-order thermal processes that imply a thermal subsidence function that is not purely exponential. These include the thermal effects of changes in asthenospheric partial melt at the base of the lithosphere accompanied by the introduction (or removal) of latent heat to (or from) the lithosphere; the implicit constraints afforded by having asymptotically thick lithosphere beneath oceans and continents; and the advective cooling combined with possible insulating effects of a thickening sedimentary sequence atop the cooling lithosphere. While these processes contribute to a thermal subsidence curve that is not purely exponential, the implied curve, for a given location within a sedimentary basin, remains functionally monotonic over the time scale of basin formation. Under certain circumstances the lifespan of thermal subsidence of extensional basins may be considerably greater than the conventionally adopted value of 200 m.y.

At the spatial intrabasinal scale, it is well recognized that a flexural (or regional isostatic) component in the purely mechanical effects of sedimentary loading can produce apparent differential subsidence or relative uplift. These effects are often predictable if they occur upon a spatially monotonic thermally subsiding base, but thermal processes (other than the regional variations in the degree of crustal thinning) also can produce intrabasinal differential subsidence or uplift. These thermally derived variations are amplified by the sediment loading mechanics and will, themselves, generate higher-order intrabasinal flexural variations. Thermal processes possibly important in this respect may include those arising from regional variations in the degree of partial melt in the subbasin asthenosphere, regional variations in the bulk thermal properties of basin sediment, and the consequences of the (usually ignored) regional isostatic compensation of the thermal body forces including the implicit accompanying effects of thermal bending moments. If these kinds of processes can be shown to be nonnegligible at the stratigraphic level then complex intrabasinal, three-dimensional stratigraphic patterns may emerge as a direct result of the basin subsidence driving mechanism.

At the temporal intrabasinal scale, it is speculated that there may be fundamental processes intrinsic to the basin-wide subsidence driving mechanism that might be relevant to the generation of stratigraphic cyclicity. These would be in addition to and interacting with external factors such as eustatic changes, changes in regional horizontal stresses, and changes in sediment supply. They may include episodic stress relaxation phenomena within the lithosphere (as in ductile/brittle failure at some lithospheric yield strength), induced both by stresses generated by sediment loading and by thermoelastic stresses.

ACKNOWLEDGMENTS

The author acknowledges the significant contributions made to the present work by his colleagues, S.M. Nakiboglu (King Saud University, Saudi Arabia) and M.A. Kelly (University of British Columbia). Thoughtful reviews of an earlier version of the manuscript by Jeffrey Nunn (Louisiana State University) and Dale Issler (Geological Survey of Canada, Calgary) were very helpful and are much appreciated.

REFERENCES CITED

Beaumont, C., Keen, C.E., and Boutilier, R., 1982, On the evolution of rifted continental margins: Comparison of models and observations for the Nova Scotian margin: Geophysical Journal of the Royal Astronomical Society, v. 70, p. 667-715.

Bills, B.G., 1983, Thermoelastic bending of the lithosphere: Implications for basin subsidence: Geophysical Journal of the Royal Astronomical Society, v. 75, p. 169-200.

Bond, G.C., and Kominz, M.A., 1984, Construction of tectonic subsidence curves for the early Paleozoic miogeocline, southern Canadian Rocky Mountains: Implications for subsidence mechanisms, age of breakup, and crustal thinning: Geological Society of America Bulletin, v. 95, p. 155-173.

Brown, K.C., and Beaumont, C., 1988, The effect of partial melting on syn-rift subsidence during pure extension of the lithosphere: Geological Association of Canada—Mineralogical Association of Canada—Canadian Society of Petroleum Geologists Program with Abstracts, v. 15, p. A14.

Cloetingh, S., and Kooi, H., 1989, Intraplate stresses: A new perspective on QDS and Vail's third order cycles, in Cross, T.A., ed., Quantitative Dynamic Stratigraphy: New Jersey, Prentice Hall, (this volume).

Cloetingh, S., McQueen, H., and Lambeck, L., 1985, On a tectonic mechanism for regional sea-level variations: Earth and Planetary Science Letters, v. 75, p. 157-166.

Cochran, J.R., 1983, Effects of finite rifting times on the development of sedimentary basins: Earth and Planetary Science Letters, v. 66, p. 289-302.

Courtney, R.C., and Beaumont, C., 1983, Thermally-activated creep and flexure of the oceanic lithosphere: Nature, v. 305, p. 201-204.

Hastaoglu, M.A., 1986, Numerical solution to moving boundary problems: Application to melting and solidification: International Journal of Heat and Mass Transfer, v. 29, p. 495-499.

Keen, C.E., 1985, The dynamics of rifting: Deformation of the lithosphere by active and passive driving forces: Geophysical Journal of the Royal Astronomical Society, v. 80, p. 95-120.

Kono, Y. and Amano, M., 1978, Thickening model of the continental lithosphere: Geophysical Journal of the Royal Astronomical Society, v. 54, p. 405-416.

Lambeck, K., and Nakiboglu, S.M., 1981, Seamount loading and stress in the ocean lithosphere 2: Viscoelastic and elastic-viscoelastic models: Journal of Geophysical Research, v. 86, p. 6961-6984.

McKenzie, D., 1978, Some remarks on the development of sedimentary basins: Earth and Planetary Science Letters, v. 40, p. 25-32.

Mutter, J.C., Buck, W.R., and Zehnder, C.M., 1988, Convective partial melting 1: A model for the formation of thick basaltic sequences during the initiation of spreading: Journal of Geophysical Research, v. 93, p. 1031-1048.

Nakiboglu, S.M., and Lambeck, K., 1983, A re-evaluation of the isostatic rebound of Lake Bonneville: Journal of Geophysical Research, v. 88, p. 10439-10447.

Nakiboglu, S.M., and Lambeck, K., 1985, Comments on thermal isostasy: Journal of Geodynamics, v. 2, p. 51-65.

Nunn, J.A., Scardina, A.D., and Pilger, R.H., Jr., 1984, Thermal evolution of the north-central Gulf Coast: Tectonics, v. 3, p. 723-740.

Oldenburg, D.W., 1975, A physical model for the creation of the lithosphere: Geophysical Journal of the Royal Astronomical Society, v. 43, p. 425-451.

Ozisik, M.N., 1980, Heat conduction: New York, John Wiley and Sons, 687 p.

Parker, E.C., Davis, P.M., Evans, J.R., Iyer, H.M., and Olsen, K.H., 1984, Upwarp of anomalous asthenosphere beneath the Rio Grande rift: Nature, v. 312, p. 354-356.

Ranalli, G., 1987, Rheology of the earth: Boston, Allen and Unwin, 366 p.

Ranalli, G., and Murphy, D.C., 1987, Rheological stratification of the lithosphere: Tectonophysics, v. 132, p. 281-295.

Solomon, S.C., 1973, Shear wave attenuation and melting beneath the Mid-Atlantic Ridge: Journal of Geophysical Research, v. 78, p. 6044-6059.

Stephenson, R.A., Embry, A.F., Nakiboglu, S.M., and Hastaoglu, M.A., 1987, Rift-initiated Permian to Early Cretaceous subsidence of the Sverdrup basin, in Beaumont, C., and Tankard, A.J., eds., Sedimentary basins and basin-forming mechanisms: Canadian Society of Petroleum Geologists Memoir 12, p. 213-231.

Stephenson, R.A., Nakiboglu, S.M., and Kelly, M.A., 1989, The effects of asthenosphere melting, regional thermo-isostasy, and sediment loading on the thermo-mechanical subsidence of sedimentary basins, in Price, R.A., ed., The origin and evolution of sedimentary basins and their energy and mineral resources: American Geophysical Union Geodynamics Series, (in press).

Walcott, R.I., 1972, Gravity, flexure, and the growth of sedimentary basins at a continental edge: Geological Society of America Bulletin, v. 83, p. 1845-1848.

8

Intraplate Stresses: A New Perspective on QDS and Vail's Third-Order Cycles

Sierd Cloetingh and Henk Kooi
Vening Meinesz Laboratory, University of Utrecht, The Netherlands
(Authors' present address: Department of Sedimentary Geology, Institute of Earth
Sciences, Free University, P.O. Box 7161, 1007 MC Amsterdam, The Netherlands)

Abstract

Fluctuation in lithospheric stresses is an important tectonic component of quantitative
dynamic basin stratigraphy, and provides a tectonic explanation for Vail's third-order cycles
in apparent sea levels. The gross onlap/offlap stratigraphic architecture of passive margin
basins can be simulated by models with changing horizontal stress fields. Modeling of the
U.S. Atlantic margin demonstrates that the inferred transience in the horizontal stress field
is qualitatively consistent with expectations based on what is known about plate kinematics
during the same time period. Out-of-phase intrabasinal cycles with, for example, relative
uplift at the flanks and increased subsidence at the basin center, such as the case for the Gulf
de Lions margin, also are predicted by the models. The large variations in estimates of
magnitudes of short-term changes in relative sea level between various basins around the
globe are in agreement with predictions of this tectonic model.

Introduction

During the last decade of substantial progress in geodynamic modeling, the role of
thermomechanical properties of the lithosphere has been emphasized in models of sedimentary basin evolution (e.g., Watts et al., 1982; Beaumont et al., 1982; Beaumont and Tankard,
1987). These models have assessed the contributions of a variety of lithospheric processes
to the vertical motions of lithosphere within sedimentary basins. These processes include
thermally induced contraction of the lithosphere amplified by the loading of sediments that
accumulate in these basins (Sleep, 1971), isostatic response to crustal thinning and stretching
(McKenzie, 1978), and flexural bending in response to vertical loading (Beaumont, 1978).

Simultaneously, major advances have been made in understanding the origins and
distributions of stress fields in plate interiors. Detailed analysis of earthquake focal mechanisms (Bergman, 1986), *in situ* stress measurements, and analysis of break-out orientations
obtained from wells (Bell and Gough, 1979; Zoback, 1985; Klein and Barr, 1986) have
demonstrated the existence of consistently oriented present-day stress patterns in the
lithosphere. Studies of paleostress fields in the lithosphere by analysis of microstructures

Quantitative Dynamic Stratigraphy (1989), T.A. Cross, ed., Prentice Hall, p. 127–148.

127

(Letouzey, 1986; Bergerat, 1987; Philip, 1987) have demonstrated temporal variations in the observed long-wavelength, spatially coherent stress patterns. This work has provided strong evidence for the occurrence of large-scale rotations in paleostress fields, and has shown (see Philip, 1987) that the state of stress can vary enough to produce quite different deformations on a relatively short time scale (approximately 5 m.y.). At the same time, numerical modeling (Richardson et al., 1979; Wortel and Cloetingh, 1981, 1983; Cloetingh and Wortel, 1985, 1986) has yielded better understanding of the causes of the observed variations in stress levels and stress orientations in lithospheric plates. These studies have demonstrated a causal relationship between the processes at plate boundaries and the deformation in the plate interiors (e.g., Johnson and Bally, 1986).

Although intraplate stresses play a crucial role during basin formation, their effect on the subsequent evolution of sedimentary basins largely has been ignored. The formation of sedimentary basins by lithospheric stretching, for example, requires tensional stress levels of the order of at least a few kbars (Cloetingh and Nieuwland, 1984; Houseman and England, 1986). Recent work by Cloetingh et al. (1985), Cloetingh (1986, 1988) and Karner (1986) has demonstrated that temporal fluctuations in intraplate stresses have important consequences for quantitative dynamic basin stratigraphy and may provide a tectonic explanation for short-term sea-level variations inferred from the stratigraphic record (Vail et al., 1977; Haq et al., 1987).

Simultaneously, we have explored (Cloetingh, 1986; Lambeck et al., 1987) using the stratigraphic record as a source of information on paleostress fields. Vail and co-workers have attributed cyclic variations in onlap/offlap stratigraphic patterns to glacially induced or other eustatic origins. This preference was based primarily on the inferred global synchroneity of sea-level variations, and the lack of a tectonic explanation for observed third-order cycles. Although other authors (e.g., Bally, 1982; Watts, 1982) argued for a tectonic cause of apparent sea-level variations, they were unable to identify a tectonic mechanism operating on a time scale appropriate to explain the observed short-term changes of sea level (Pitman and Golovchenko, 1983). A problem with the glacio-eustatic interpretation, however, may be the lack of significant Mesozoic and Cenozoic glaciation prior to the mid-Tertiary (Pitman and Golovchenko, 1983). In the absence of a glacio-eustatic control, possible tectonic, sediment yield or other climatic controls might be postulated. Plate dynamics and associated changes in stress levels in plate interiors offer a tectonic framework for observed short-term relative sea-level variations and are a parameter of concern to quantitative dynamic stratigraphy. Mechanisms for long-term changes in sea level (e.g., Heller and Angevine, 1985; Kominz, 1984) are beyond the scope of the present paper.

INTRAPLATE STRESS AND QUANTITATIVE DYNAMIC STRATIGRAPHY

Intraplate stresses modulate basin deflections caused by the primary driving mechanisms of basin subsidence as, for example, thermal contraction of the lithosphere amplified by sediment loading. Cloetingh et al. (1985) modeled a passive margin evolving through time in response to changing thermal regime and loading by sediments (Fig. 1a). They showed that vertical deflections of the lithosphere of up to a hundred meters may be induced by the action of horizontal stresses in the lithosphere with magnitudes up to a few kbars. They proposed that basement deflections induced by short-term changes in horizontal stress associated with changes in plate-tectonic regimes are capable of producing not only the magnitude but also the rate of Vail's short-term changes in apparent sea level. Figure 1b shows that the stress-induced subsidence/uplift perturbations change sign and magnitude with intrabasinal position, which provides a means to discriminate between this tectonic effect and eustatic contributions to the apparent sea level. The occurrence of out-of-phase intrabasinal cycles (see also Embry, 1988), for example, can be explained by this tectonic model. It is important to realize that the model relies on stress changes—from the perspective of the model a reduction in compression is equivalent to an increase in tension, and vice versa. An increase in compression, or equivalently a reduction of tension, causes relative uplift of the basin flank and increased subsidence at the basin center, whereas a reduction in compression, or equivalently an increase in tension, induces the opposite effect.

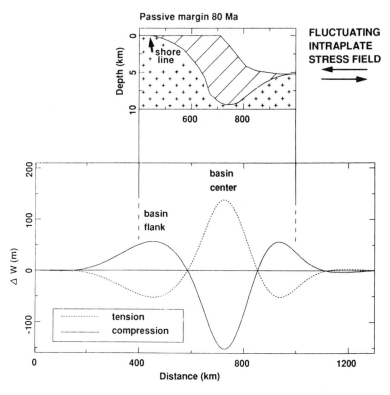

Figure 1. Flexural deflections at a sedimentary basin induced by changes in intraplate stress field. Top: an 80 Ma old passive margin initiated by stretching. The wedge of sediments flexurally loads an elastic plate. The thickness of the plate varies horizontally due to lateral changes in the temperature structure of the lithosphere. Bottom: Differential subsidence or uplift (meters) induced by a change to 1 kbar compression (solid line) and 1 kbar tension (dashed line).

To model passive margin processes, we considered an elastic lithosphere that cools with time after basin formation. Thus the response of the lithosphere to changes in intraplate stress field is time-dependent, not only because the sediment load increases with time, but also because of the changing mechanical properties of the lithosphere. Although sufficiently large to explain Vail's third-order cycles, the values for stress-induced basement deflections given in Figure 1b are conservative estimates, being based on an elastic model for the mechanical properties of the lithosphere. The incorporation of a more realistic (weaker) brittle-ductile rheology of the lithosphere based on extrapolation of rock-mechanics data (Goetze and Evans, 1979) significantly magnifies the values of the stress-induced differential motions within the basins.

Figure 2 illustrates the relative movement between sea level and the lithosphere at the flank of a flexural basin immediately landward of the principal sediment load as predicted by numerical calculations (Cloetingh et al., 1985) given for the elastic plate model. The synthetic stratigraphy at the basin edge is shown for three situations: long-term flexural widening of the basin with cooling (Watts, 1982) in the absence of an intraplate stress field (Fig. 2a); the same with a superimposed stress change to 500 bar compression at 50 Ma (Fig. 2b); and the same with a superimposed stress change to 500 bar tension at 50 Ma (Fig. 2c). As noted by Watts (1982), the thermally induced flexural widening of the basin (see Fig. 2a) provides an adequate explanation for long-term phases of coastal onlap. However, because it is a long-term change, it does not produce the punctuated character of sedimentary basin stratigraphy characterized by a succession of alternating rapid onlap and offlap phases. Inspection of Figures 2b and c demonstrates that the incorporation of intraplate stresses in geodynamic models of basin evolution can successfully produce sequence boundaries and the overall onlap/offlap characteristics associated with a punctuated stratigraphy.

Here we concentrate on the relationship between tectonics and stratigraphy of rifted basins, in particular the U.S. Atlantic margin, the North Sea and the Gulf de Lions. However, the effect of intraplate stress fields is equally important to other basins, such as foreland basins where lithosphere is flexed downward by sedimentary loads (Beaumont, 1981; Quinlan and Beaumont, 1984; Tankard, 1986). These authors interpreted unconformities in the Appalachian foreland basin as products of uplift of the peripheral bulge caused by viscoelastic relaxation of the lithosphere. However, intraplate tensional or compressional stresses, of which the latter is more natural in this tectonic setting, can amplify or reduce the height of the peripheral bulge by an equivalent amount and equally influence the stratigraphic record in foreland basins.

$$\longrightarrow$$

Figure 2. Synthetic stratigraphy for a 60 Ma old passive margin that was initiated by lithospheric stretching followed by thermal subsidence and flexural infilling of the resulting depression. Hachuring indicates the position of a sedimentary package bounded by isochrons of 50 Ma and 52 Ma after basin formation. (a) Continuous onlap associated with long-term cooling of the lithosphere in the absence of intraplate stress fields. (b) A transition to 500 bar in-plane compression at 50 Ma induces uplift of the peripheral bulge, narrowing of the basin and a phase of rapid offlap, which is followed by a long-term phase of gradual onlap due to thermal subsidence. (c) A transition to 500 bar in-plane tension at 50 Ma induces downwarp of the peripheral bulge, widening of the basin and a phase of rapid basement onlap.

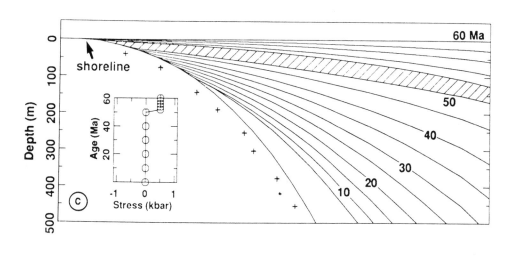

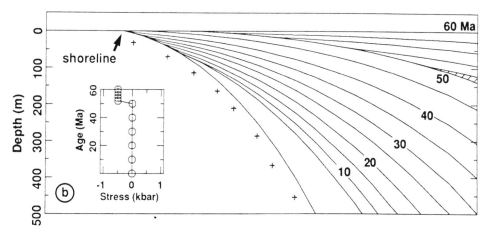

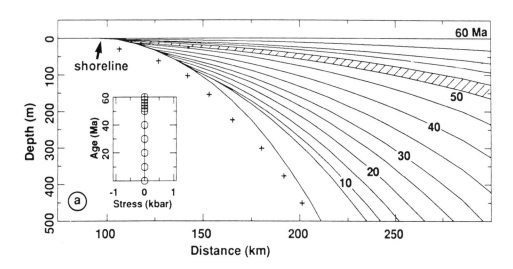

STRATIGRAPHIC MODELING OF THE U.S. ATLANTIC MARGIN

As demonstrated by Figure 2, the incorporation of intraplate stresses in elastic models of basin evolution predicts a succession of onlaps and offlaps characteristic of strata along the flanks of basins such as the U.S. Atlantic margin (Sleep and Snell, 1976; see Fig. 3). We selected the U.S. Atlantic margin for a numerical simulation of the observed stratigraphy for several reasons. The U.S. Atlantic margin stratigraphy has been extensively documented (e.g., Poag, 1985), and its evolution has been quantitatively modeled (Sleep and Snell, 1976; Watts and Thorne, 1984; Steckler et al., 1988). Sleep and Snell (1976) proposed a visco-elastic model of the lithosphere to account for the observed late-stage narrowing of the North Carolina margin. Watts and Thorne (1984) and Steckler et al. (1988) employed a two-layer stretching model adopting an elastic rheology of the lithosphere and zero intraplate stresses. They assumed global long-term and short-term sea-level fluctuations throughout the basin evolution. Our modeling approach resembles those by Watts and Thorne (1984) and Steckler et al. (1988) in that we adopt a stretching model for basin initiation, but differs by incorporating the effects of finite and multiple stretching phases and intraplate stresses. We use a finite-difference approach for the thermal calculations as discussed by (Verwer, 1977).

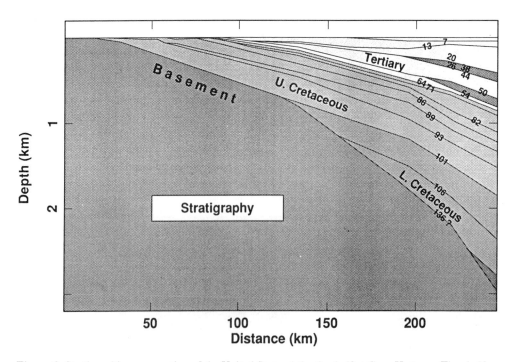

Figure 3. Stratigraphic cross section of the United States Atlantic shelf at Cape Hatteras. The shelf break is about 40 km from the right of the figure. Ages of stratigraphic unit boundaries are given in Ma (After Sleep and Snell, 1976).

Although of limited impact for the late-stage development of the U.S. Atlantic margin, the incorporation of finite stretching rates severely affects syn-rift and early post-rift subsidence and sedimentation (Jarvis and McKenzie, 1980; Cochran, 1983). There is general agreement that the initial rifting phase began in the Late Carnian (approximately 225 Ma), whereas sea-floor spreading began about 180 Ma (Manspeizer, 1985; Ziegler, 1988). Thus the initiation of subsidence is associated with a long period of rifting and stretching. Jurassic sediments were deposited only in the deeper part of the margin now located under the outer shelf. This may be explained by post-Jurassic widening of the basin due to a second stretching phase, or by subcrustal attenuation under the inner shelf part of the basin inhibiting its subsidence. That this thermal anomaly may have been very large is evident from the long duration (approximately 36 m.y.) of cooling after rifting. On the other hand, evidence for a period of extensional tectonics and Early Cretaceous northward propagation of the Atlantic rift (Ziegler, 1988) and results of subsidence analysis of the U.S. Atlantic margin (Greenlee et al., 1988) support the occurrence of multiple stretching phases. Therefore, we have adopted both subcrustal attenuation and a minor second stretching phase acting from 131-119 Ma. We also assume that as the basement subsides, the equilibrium profile of the margin will be maintained by sediments that fill the resulting depression to a constant water depth.

The stratigraphy modeled for an elastic plate—with the effective elastic thickness given by the depth to the $400°$ C isotherm—in the absence of intraplate stresses and ignoring eustatic changes is shown in Figure 4a. This figure demonstrates the well-known failure of conventional elastic models of basin evolution to predict basin narrowing with younger sediments restricted to the basin center. The observed narrowing of the basin during its late-stage evolution has been interpreted in previous modeling studies as either reflecting the response of the basin to a phase of visco-elastic relaxation (Sleep and Snell, 1976) or the response to a long-term eustatic fall (Watts and Thorne, 1984; Steckler et al., 1988). The total thickness of the Cenozoic sediments provides an independent constraint for the magnitude of the proposed long-term sea-level fall. From our modeling we obtain an upper estimate of approximately 100 m for the post-Late Cretaceous long-term lowering in sea level. We, therefore incorporated a long-term sea-level curve with a Late Cretaceous highstand of 100 m, a curve equivalent to the minimum curve of Kominz (1984). Inspection of the resulting stratigraphic model (Fig. 4b) demonstrates that, although the incorporation of long-term changes in sea level contributes to the Cenozoic narrowing of the margin, the long-term post-Late Cretaceous decline in sea level alone does not produce both the documented basin narrowing and the total thickness of sediments accumulated during this time. We propose that a large part of the observed nondepositional or erosional character of the shelf surface is caused by stress-induced uplift of the margin flank. Similarly, short-term changes in intraplate stress levels can produce the Early Eocene and Oligocene offlap phases. Figure 4c shows the best fit to the observed stratigraphy incorporating long-term sea-level changes after Kominz's (1984) minimum curve and a fluctuating intraplate stress field in the stratigraphic modeling. The observed stratigraphy can be simulated by relaxation of tensional intraplate stress fields during Mesozoic times and a post-Cretaceous transition to compressional stress, the level of which increases with time during the Tertiary.

Prior to rifting, eventually followed by continental break-up, tensional stresses increase. For example, rifting in the Southern Atlantic was not an instantaneous process, but occurred

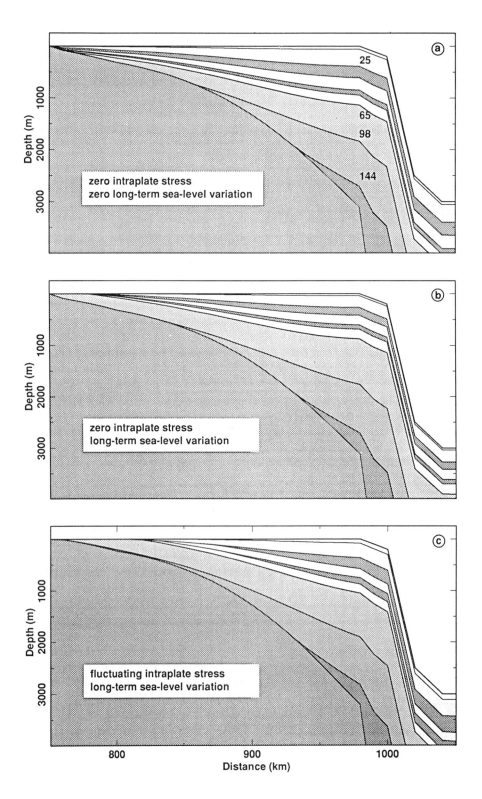

as discrete rifting phases with accompanying reduction of accumulated tensional stresses. This process might explain the enigmatic occurrence of high-frequency sea-level fluctuations during the Cretaceous that do not correlate with accelerations in plate spreading or increases in ridge lengths (Schlanger, 1986). Similarly, the correlation of short-term sea-level fluctuations along both sides of the Atlantic might be an expression of rifting-related accumulation and relaxation of tensional stresses. According to this view, increases of tensional stresses induce periods of apparent sea-level rise. These are followed by periods of sea-level lowering of shorter duration that are associated with the rapid relaxation of tensional stresses. Hence, in the period just prior to rifting, sea level should rise gradually. A major sea-level fall then occurs rapidly during the continental break-up phase. This predicted break-up unconformity commonly is observed in the stratigraphic record of passive margins, and a major lowering in sea level that coincides with the onset of the opening of the South Atlantic is shown in the Haq et al. (1987) curves. It is interesting to note that a break-up unconformity generally is not predicted by geodynamic models of sedimentary basins that do not incorporate intraplate stresses.

In summary, the incorporation of intraplate stresses in elastic models of basin evolution can, in principle, predict a succession of onlaps and offlaps such as observed along the flanks of the U.S. Atlantic margin and the Tertiary North Sea Basin (Kooi et al., 1988). This stratigraphic geometry can be interpreted as a natural consequence of short-term phases of mechanical widening and narrowing of basins by fluctuations in intraplate stress levels superimposed on the long-term broadening of the basin induced by cooling after its formation.

PALEOSTRESS FROM STRATIGRAPHIC MODELING: U.S. ATLANTIC MARGIN AND THE CENTRAL NORTH SEA

Figure 5 shows the paleostress field inferred from the stratigraphic model presented above. The stress levels used in elastic models for basin stratigraphy provide upper limits, as the incorporation of depth-dependent rheology (Goetze and Evans, 1979) in the models will lower the predicted stress levels. Similarly, the resolution of the paleostress curve is affected by the quality of the adopted sedimentation models and the availability of high-resolution paleobathymetric data. The long-term trend of the paleostress pattern is from overall tension during the Cretaceous to a stress regime of accumulating compression during the Tertiary. Superimposed on this long-term trend are more abrupt and shorter duration changes. Both the character of these changes and their timing are largely consistent with independent data on the kinematic evolution of the Central Atlantic (Klitgord and Schouten, 1986). Rifting in the Atlantic evolved from initiation of sea-floor spreading in the Gulf of Mexico–Central Atlantic–Ligurian Tethys at 175 Ma, and proceeded by a number of discrete steps (170, 150,

Figure 4. U.S. Atlantic margin stratigraphy modeled for elastic rheology of the lithosphere. (a) Modeled stratigraphy in the absence of intraplate stresses, assuming zero changes in long-term sea level. (b) Modeled stratigraphy in the absence of intraplate stresses, but adopting long-term changes in sea level after Kominz (1984). (c) Modeled stratigraphy showing the combined effect of long-term changes in sea level and a fluctuating intraplate stress field producing short-term sea-level fluctuations.

132, 119, 80, 67 Ma) to the start of spreading in the Northern Atlantic at 59 Ma. The paleostress curve inferred from the stratigraphic model suggests that rifting events during the period from about 180 to 140 Ma were associated with a major relaxation of tensional stresses, followed by a phase of renewed accumulation of tension. The paleostress curve for the Tertiary is of particular interest for a detailed comparison with the documented tectonic history of the Atlantic. The predicted phase of relaxation of tensional stresses and the transition to a more neutral stress regime around 50 Ma coincides with the termination of Thulean volcanism in the Northern Atlantic, the break-up of the Greenland-Rockall and Norwegian-Greenland sea (P.A. Ziegler, personal communication, 1988; see also Tucholke and Mountain, 1986), and the Eocene compressional phases in the Arctic Sverdrup Basin

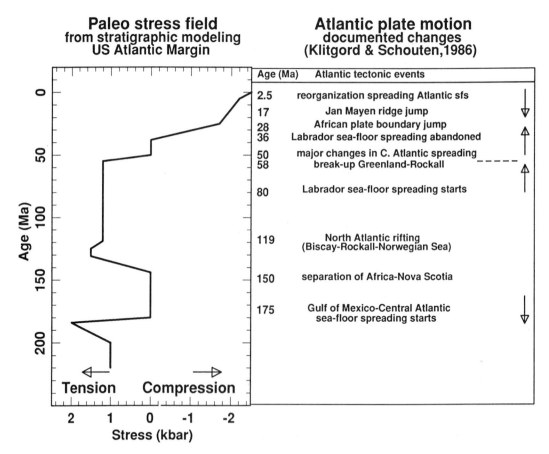

Figure 5. The paleo-stress curve inferred from the stratigraphic modeling (Fig. 4c) of the U.S. Atlantic margin. Tension is positive, compression is negative. Timing of kinematic events in the North/Central Atlantic is given at the right hand side (After Klitgord and Schouten, 1986).

(Embry, 1988). Similarly, the predicted transition to a more compressional stress regime coincides with the timing of the Caribbean orogeny, the Pyrenean orogeny and the cessation of spreading in the Labrador Sea (Klitgord and Schouten, 1986). As noted by Issler and Beaumont (1987), sea-floor spreading ended in the Labrador Sea simultaneously with widespread shelf shallowing, tectonism and coastal erosion, consistent with an increase in the level of compressional stresses. The change in the compressional stress level at the time of the mid-Oligocene regression (Fig. 5), coincides with a major reorganization in the Central Atlantic—the African plate boundary jump.

The paleostress curve derived from the apparent sea-level record of the North Sea area, assuming that the sea levels are controlled by the effects of intraplate stresses, shows a similar change from a long-term regime of overall tension during the Jurassic and Cretaceous to a regime of more compressional character during the Tertiary (Lambeck et al., 1987). These findings were confirmed recently by detailed stratigraphic modeling of the North Sea Central Graben, that incorporated the role of intraplate stresses (Kooi et al., 1988). The inferred paleostress curve (Fig. 6) is consistent with the observed transition from rift-wrench tectonics during the Mesozoic to compressional tectonics during the Tertiary in northwestern Europe (Ziegler, 1982). The paleostress curve appears to mirror the tectonic evolution of northwestern Europe in other respects as well: rifting episodes correspond to relaxation of tensional paleostresses, and Alpine orogenic phases correspond to episodes of increased compressional stress. From Late Eocene to Early Oligocene a stress regime of more tensional character is predicted concomitant with the timing of rifting in the European platform, an event that has inhibited to a large extent propagation of compressional stresses induced by Alpine collision into the North Sea area (Ziegler, 1982). As shown by Figure 6, this tensional phase and the predicted overall increase in the level of the post-Early Oligocene compression is consistent with paleostress data from the northwestern European platform (Letouzey, 1986; Bergerat, 1987).

These data demonstrate a large-scale rotation of the paleostress field in northwestern Europe from NE-SW oriented Late Oligocene/Early Miocene compression to the present NW-SE orientation of the largest compressive stress, a direction which is almost perpendicular to the strike of the modeled Central Graben Basins (Klein and Barr, 1986; Kooi et al., 1988). Similarly, the paleostress curve inferred from the stratigraphic modeling is compatible with observed correlations between the timing of unconformities in northwestern European platform basins and documented changes (Livermore and Smith, 1985; Savostin et al., 1986) in the kinematic evolution of the Tethys belt.

INTERBASINAL VARIATIONS IN MAGNITUDES OF RELATIVE SEA LEVEL

Although the Exxon curves (Vail et al., 1977; Haq et al., 1987) are based on data from basins throughout the world, they are heavily weighted toward the Northern Atlantic and North Sea areas. As noted by several authors (e.g., Miall, 1986; Hubbard, 1988; Hallam, 1988), the inferred global cycles may primarily reflect the seismic stratigraphic record of basins in a tectonic setting dominated by rifting events in the northern and central Atlantic. Summerhayes (1986) and Miall (1986) questioned the global character of the Exxon curves, pointing out that the synchroneity of the inferred sea-level changes may be widespread, but not necessarily global. Others (e.g., Kerr, 1984) regard the Exxon curves as tool applicable for

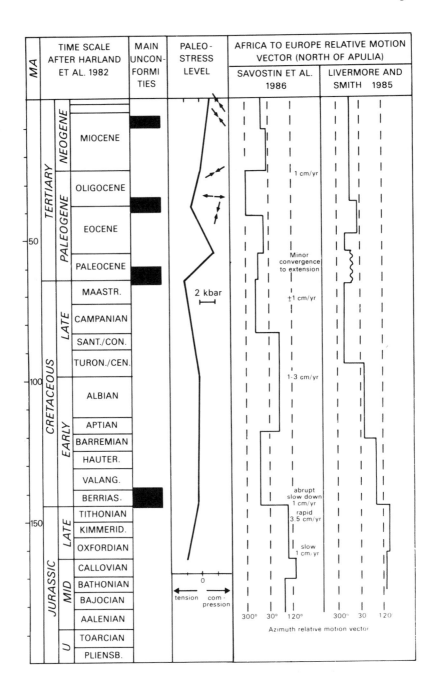

★   Anorogenic volcanism

↗   Principle compressive
         paleostress direction   } After Bergerat 1987
                                        and Letouzey 1986
← →   Principle tensional
         paleostress direction

worldwide correlation of unconformities. Similarly, there has been little agreement about the causes of short-term sea-level fluctuations. The assumption of global synchroneity has played a crucial role in arguments favoring a glacio-eustatic or a tectonic cause for short-term sea-level changes. Only fluctuations in sea level with magnitudes in excess of 50 m require stress changes of a magnitude to be related to major reorganizations at convergent plate boundaries, fragmentation of plates or collision processes. This observation explains the existence of the correlation noted by Bally (1982) in timing of plate reorganizations and rapid lowerings in sea level.

The regional character of intraplate stresses provides a basis for alternative interpretations of observed deviations from "global" sea-level cycles (e.g., Hubbard et al., 1985; Hallam, 1988; Embry, 1988). Whereas such deviations from a global pattern are a natural feature of our tectonic model (Cloetingh et al., 1985; Cloetingh, 1988), the occurrence of short-term deviations does not preclude the presence of global events elsewhere in the stratigraphic record. These are expected when major plate reorganizations and changes in intraplate stress fields occur simultaneously in more than one plate or when glacio-eustasy dominates. Major plate reorganizations occurred during the Mid-Oligocene (Engebretson et al., 1985) and the Early Cenozoic (Rona and Richardson, 1978; Schwann, 1985). The mid-Oligocene is a particularly tectonically active time in the northern and southern Atlantic, with the concomitant occurrence of a major Alpine folding phase (Ziegler, 1982) and uplift of the shelf along the Atlantic margins of Africa (Lehner and de Ruiter, 1977). Furthermore, differences in rheological structure of the lithosphere, which influence its response to applied intraplate stresses, might also explain differences in magnitudes of inferred sea levels such as observed between time-equivalent changes in the Tertiary North Sea region and the Gippsland Basin off southeastern Australia (Vail et al., 1977), and between the Jurassic North Sea and the Canadian Sverdrup Basin (Embry, 1988).

The decay of thermal subsidence with time after a heating or rifting event also has consequences for the magnitude of stress-induced sea-level changes. The position of coastal onlap reflects the position where the rate of subsidence equals the rate of sea-level fall. During application of stress the rate of subsidence is temporarily changed, and consequently the equilibrium point of the coastal onlap is shifted in position. The thermally induced rate of long-term subsidence decreases exponentially with age (Turcotte and Ahern, 1977). As noted by Thorne and Watts (1984), the production of offlapping stratigraphic geometries during late stages of passive margin evolution requires much lower rates of sea-level change than those needed to produce offlapping geometries during earlier stages of basin evolution. If these offlapping geometries are caused by fluctuations in intraplate stress levels, then the rate of changes of stress needed to create them also diminish with age during the flexural evolution of the basin. This is particularly relevant for assessing the relative contributions of tectonics and eustasy as a cause for Cenozoic unconformities. For example, Cenozoic unconformities developed at old passive margins in association with short-term basin narrowing could be produced by relatively mild changes in intraplate stress levels. Such late-stage narrowing of Phanerozoic platform basins and passive margins is frequently observed (Sleep and Snell, 1976), without clear evidence for active tectonism.

Figure 6. Synthetic paleo-stress curve as inferred from the stratigraphic modeling of the Central North Sea Basin (After Kooi et al., 1988). Also shown in this column are paleo-stress orientation data from Bergerat (1987). The columns on the right show Africa relative to Europe plate motion data from Savostin et al. (1986) and Livermore and Smith (1985).

Independent studies of the magnitude of the mid-Oligocene sea-level lowering indicate a value much smaller than previously thought. The magnitude of this sea-level fall, which is by far the largest shown in the Vail et al. (1977) and Haq et al. (1987) curves, is now estimated as between 50 m (Miller and Fairbanks, 1985; Watts and Thorne, 1984) and 100 m (Schlanger and Premoli-Silva, 1986). Hence, a significant part of the short-term sea-level record inferred from seismic stratigraphy might have a characteristic magnitude of a few tens of meters, which can be explained by relatively modest stress fluctuations. The superposition of a glacio-eustatic event and a major tectonic reorganization might explain the exceptional magnitude of the Oligocene sea-level lowering.

INTRABASINAL VARIATIONS IN THE DEVELOPMENT OF SEQUENCES

Discriminating regional tectonic events from eustatic signals in the stratigraphic record of individual basins is usually difficult, especially if biostratigraphic correlation is imprecise (Hallam, 1988). As noted previously, intraplate stresses cause opposite subsidence histories at the flanks and in the centers of basins. Because the sign and magnitude of the corresponding apparent sea-level change is a function of position within a basin, there is a means for testing within separate basins the effect of intraplate stresses and of distinguishing this mechanism from eustatic contributions. As an example, Figure 7 shows schematically the laterally varying expression of changes in intraplate stress levels on subsidence predicted by our modeling. For additional stratigraphic criteria that may discriminate stress-induced tectonic from eustatic controls on depositional cycles, see Embry (1988).

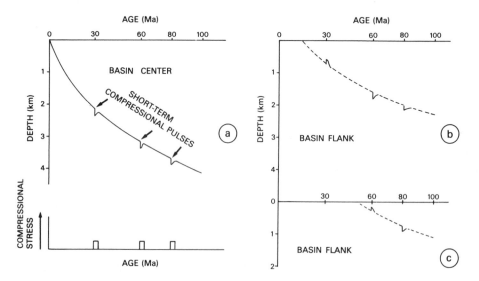

Figure 7. Effect of intraplate stresses on subsidence curves at three different positions a,b, and c in a rifted basin. (a) Effect of compression on subsidence predicted for a well in the basin center. (b) and (c) show the effect of compressional stress on subsidence for locations at the flanks of the basin, closer to the position of the flexural node. Note in these cases the different effects of compression at different time intervals, which are caused by widening of the basin during its long-term thermal evolution.

In contrast to the present-day tectonic setting of the U.S. Atlantic margin, which is affected primarily by far-field effects of ridge-push forces in the North-American plate (Richardson et al., 1979), the passive margins of the Mediterranean are located in an active tectonic setting dominated by the Africa-Eurasia collision. The Mediterranean margins, therefore, are natural laboratories for studying near-field effects on basin stratigraphy of intraplate stresses associated with collision, . Figure 8 is a stratigraphic cross section of the Gulf de Lions margin in the northwestern Mediterranean based on recent work at the Institut Francais du Petrole (Burrus et al., 1987). Also shown are subsidence-history curves for different positions along the margin. As noted by Burrus et al. (1987), the subsidence curves conform to predictions from thermal models of passive margin subsidence, except for the last 5 m.y. where they deviate markedly from the thermally predicted subsidence. Rapid excess subsidence of about 500 m occurred at the basin center, while uplift of a few hundred meters occurred at the shelf (Fig. 8). The thick offlap sequences and time-equivalent unconformities shown in Figure 8 correspond to the Messinian salinity crisis, a period marked by a drop in Mediterranean sea level and commonly attributed to its desiccation due to isolation of the Mediterranean from the major ocean basins (Bessis, 1986). In the stratigraphic modeling we have incorporated the Messinian sea-level drop and changes in paleobathymetry documented by Bessis (1986). The stratigraphy modeled assuming an elastic rheology of the lithosphere and incorporating a fluctuating intraplate stress level is displayed in Figure 9. The sign of the observed differential motions across the basin agrees with model predictions. Rapid vertical motions of the basin starting at 7 to 5 Ma coincide (Fig. 10) with the timing of a documented regional compressive phase (Burrus et al., 1987). Intraplate stresses are particularly effective for inducing large vertical differential motions at young passive margins, because of their lower flexural rigidity. This is consistent with and may explain the large-magnitude vertical motions of the lithosphere in the young Gulf de Lions basin. These findings also suggest that vertical motions of the lithosphere caused by late-stage compression during the post-rift phases of extensional basins can produce substantial errors in estimates of crustal extension derived from subsidence analysis with standard stretching models.

Hallam (1988) and Embry (1988) have shown that a significant number of Jurassic unconformities are confined to the flanks of the North Sea basins and the Sverdrup Basin, respectively. At the same time, the occurrence of a correlation between unconformities at the basin edge and the basin center (Wise and van Hinte, 1986) is not in conflict with the predictions of the tectonic model of Figure 1. According to this model uplift of the basin edge with exposure of the inner shelf of passive margins and steepening of the basin slope can be caused by intraplate compressional stresses or, equivalently, by relaxation of a tensional stress regime. As noted by Miller et al. (1987), the frequently observed correlation between unconformities on the shelf and in the deeper parts of continental margin basins might simply result from subaerial exposure of the shelves. These authors argued that "the material eroded from the exposed shelves could have increased sediment supply to the actually restricted submarine shelf, stimulating increased slope failure and submarine erosion." It seems that the essential factor controlling the intrabasinal correlation of unconformities is the ratio of surface gradient to differences in water depth across the basin. Hence, care should be taken in selectively interpreting the occurrence of intrabasinal correlations solely in terms of eustatic changes in sea level.

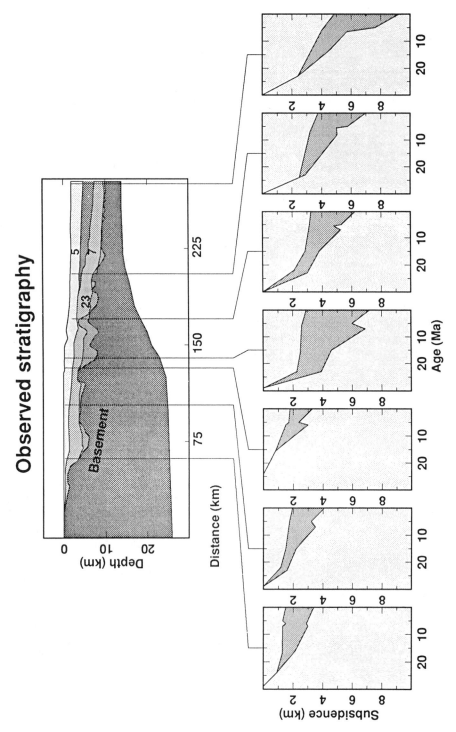

Figure 8. Documented stratigraphy Gulf de Lions passive margin (northwest Mediterranean). Lower part of the figure shows subsidence curves for different positions along stratigraphic cross section. Shading indicates unloading correction (After Burrus et al., 1987).

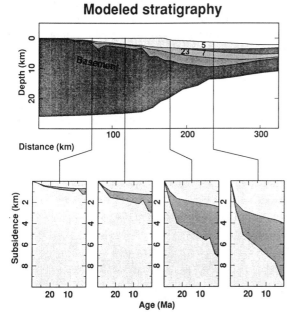

Figure 9. Modeled stratigraphy of the Gulf de Lions passive margin for an elastic rheology of the lithosphere, adopting a strong compressive phase starting at the 5 to 5 Ma time interval. Curves showing predicted subsidence are given in the lower part of the figure for positions along the modeled stratigraphic cross section.

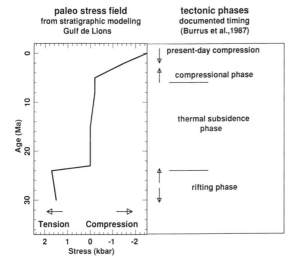

Figure 10. The paleostress field inferred from the stratigraphic modeling of the Gulf de Lions margin (Fig. 9) with timing of tectonic phases (after Burrus et al.,

The stratigraphic modeling of the U.S. east coast and Gulf de Lions passive margins described in this paper and similar modeling for the North Sea (Kooi et al., 1988; see also Lambeck et al., 1987), strongly suggest that tectonics might be the controlling factor underlying the apparent sea-level record, even during glacial periods.

CONCLUSIONS

Numerical modeling demonstrates that the incorporation of intraplate stresses in quantitative models of basin evolution can predict a succession of onlap and offlap patterns such as observed along the flanks of the U.S. Atlantic margin. Such a punctuated stratigraphy can be viewed as the natural consequence of short-term narrowing of basins due to moderate fluctuations in intraplate stress levels, superimposed on the long-term broadening of the basin due to thermal contraction. A paleostress field inferred from the U.S. Atlantic margin stratigraphy is characterized by a transition from overall tension during the Mesozoic to a more compressional regime during the Cenozoic. These findings are supported by a similar analysis of North Sea Basin stratigraphy, and strongly suggest that the short-term apparent sea-level record of the basins at both sides of the Northern and Central Atlantic reflects the tectonic evolution of the Atlantic and global tectonic effects. Differential subsidence across passive margins provides a criterion to discriminate eustatic from tectonic controls on sea-level fluctuations. Stress-induced subsidence and uplift explains the observed record of vertical motions in the Gulf de Lions basin.

ACKNOWLEDGMENTS

Thanks are due to Kurt Lambeck and Herb McQueen for various contributions. Julian Thorne, Ashton Embry, Rinus Wortel, Tom McGee and Peter Ziegler are thanked for useful discussions. Randell Stephenson, Tim Cross and Mike Steckler provided thoughtful reviews. Jean Burrus, Peter Ziegler, Ashton Embry and Anthony Hallam furnished preprints prior to publication.

REFERENCES CITED

Bally, A.W., 1982, Musings over sedimentary basin evolution: Philosophical Transactions of the Royal Society London, Series A, v. 305, p. 325-338.

Beaumont, C., 1978, The evolution of sedimentary basins on a viscoelastic lithosphere: Theory and examples: Geophysical Journal of the Royal Astronomical Society, v. 55, p. 471-498.

Beaumont, C., 1981, Foreland basins: Geophysical Journal of the Royal Astronomical Society, v. 65, p. 291-329.

Beaumont, C., Keen, C.E., and Boutilier, R., 1982, On the evolution of rifted continental margins: Comparison of models and observations for the Nova Scotia margin: Geophysical Journal of the Royal Astronomical Society, v. 70, p. 667-715.

Beaumont, C., and Tankard, A.J., 1987, eds., Sedimentary basins and basin-forming mechanisms: Canadian Society of Petroleum Geologists Memoir 12, 527 p.

Bell, J.S., and Gough, D.I., 1979, Northeast-southwest compressive stress in Alberta: Evidence from oil wells: Earth and Planetary Science Letters, v. 45, p. 475-482.

Bergerat, F., 1987, Stress fields in the European platform at the time of Africa-Eurasia collision: Tectonics, v. 6, p. 99-132.

Bergman, E.A., 1986, Intraplate earthquakes and the state of stress in oceanic lithosphere: Tectono-physics, v. 132, p. 1-35.

Bessis, F., 1986, Some remarks on the study of subsidence of sedimentary basins, application to the Gulf de Lions margin (western Mediterranean): Marine and Petroleum Geology, v. 3, p. 37-63.

Burrus, J., Bessis, F., and Doligez, B., 1987, Heat flow, subsidence and crustal structure of the Gulf of Lions (NW Mediterranean): A quantitative discussion of the classical passive margin model, *in* Beaumont, C., and Tankard, A.J., eds., Sedimentary basins and basin-forming mechanisms: Canadian Society of Petroleum Geologists Memoir 12, p. 1-15.

Cloetingh, S., 1986, Intraplate stresses: A new tectonic mechanism for fluctuations of relative sea level: Geology, v. 14, p. 617-621.

Cloetingh, S., 1988, Intraplate stresses: a new element in basin analysis, *in* Kleinspehn, K., and Paola, C., eds., New perspectives of basin analysis: New York, Springer-Verlag, p. 205-230.

Cloetingh, S., McQueen, H., and Lambeck, K., 1985, On a tectonic mechanism for regional sealevel variations: Earth and Planetary Science Letters, v. 75, p. 157-166.

Cloetingh, S., and Nieuwland, F., 1984, On the mechanics of lithospheric stretching and doming: A finite element analysis: Geologie en Mijnbouw, v. 63, p. 315-322.

Cloetingh, S., and Wortel, R., 1985, Regional stress field of the Indian plate: Geophysical Research Letters, v. 12, p. 77-80.

Cloetingh, S., and Wortel, R., 1986, Stress in the Indo-Australian plate: Tectonophysics, v. 132, p. 49-67.

Cochran, J.R., 1983, Effects of finite rifting times on the development of sedimentary basins: Earth and Planetary Science Letters, v. 66, p. 289-302.

Embry, A.F., 1988, A tectonic origin for third-order depositional sequences in extensional basins—Implications for basin modeling, *in* Cross, T.A., ed., Quantitative dynamic stratigraphy: New Jersey, Prentice-Hall, this volume.

Engebretson, D.C., Cox, A., and Gordon, R.G., 1985, Relative motions between oceanic and continental plates in the Pacific Basin: Geological Society of America Special Paper 206, 56 p.

Goetze, C., and Evans, B., 1979, Stress and temperature in the bending lithosphere as constrained by experimental rock mechanics: Geophysical Journal of the Royal Astronomical Society, v. 59, p. 463-478.

Greenlee, S.M., Schroeder, F.W., and Vail, P.R., 1988, Seismic stratigraphy and geohistory of Tertiary strata from the continental shelf off New Jersey: Calculation of eustatic fluctuations from stratigraphic data, *in* Sheridan, R.E., and Grow, J.A., eds., The Atlantic continental margin: U.S.: Boulder, Colorado, Geological Society of America, The Geology of North America, v. I-2, p. 399-416.

Hallam, A., 1988, A reevaluation of Jurassic eustasy in the light of new data and the revised Exxon curve, *in* Wilgus, C., ed., Sea level changes—An integrated approach: Society of Economic Paleontologists and Mineralogists Special Publication 42, in press.

Haq, B., Hardenbol, J., and Vail, P.R., 1987, Chronology of fluctuating sea level since the Triassic (250 million years to present): Science, v. 235, p. 1156-1167.

Heller, P.L., and Angevine, C.L., 1985, Sea level cycles during the growth of Atlantic type oceans: Earth and Planetary Science Letters, v. 75, p. 417-426.

Houseman, G., and England, P., 1986, A dynamical model of lithosphere extension and sedimentary basin formation: Journal of Geophysical Research, v. 91, p. 719-729.

Hubbard, R.J., 1988, Age and significance of sequence boundaries on Jurassic and Early Cretaceous rifted continental margins: American Association of Petroleum Geologists Bulletin, v. 72, p. 49-72.

Hubbard, R.J., Pape, J., and Roberts, D.G., 1985, Depositional sequence mapping to illustrate the evolution of a passive margin, *in* Berg, O.R., and Woolverton, D.G., eds., Seismic stratigraphy II—An integrated approach: American Association of Petroleum Geologists Memoir 39, p. 93-115.

Issler, D.R., and Beaumont, C., 1987, Thermal and subsidence history of the Labrador and West Greenland continental margin, *in* Beaumont, C., and Tankard, A.J., eds., Sedimentary basins and basin-forming mechanisms: Canadian Society of Petroleum Geologists Memoir 12, p. 45-69.

Jarvis, G.T., and McKenzie, D.P., 1980, Sedimentary basin formation with finite extension rates: Earth and Planetary Science Letters, v. 48, p. 42-52.

Johnson, B., and Bally, A.W., 1986, eds., Intraplate deformation: Characteristics, processes and causes: Tectonophysics, v. 132, p. 1-278.

Karner, G.D., 1986, Effects of lithospheric in-plane stress on sedimentary basin stratigraphy: Tectonics, v. 5, p. 573-588.

Kerr, A.R., 1984, Vail's sea-level curves aren't going away: Science, v. 226, p. 677-678.

Klein, R.J., and Barr, M.V., 1986, Regional state of stress in western Europe, *in* Stephensson, O., ed., Rock stress and rock stress measurements: Lulea, Centek Publishers, p. 33-44.

Klitgord, K.D., and Schouten, H., 1986, Plate kinematics of the Central Atlantic, *in* Vogt, P.R., and Tucholke, B.E., eds., The western North Atlantic region: Boulder, Colorado, Geological Society of America, The Geology of North America, v. M, p. 351-378.

Kominz, M.A., 1984, Oceanic ridge volumes and sea-level change—An error analysis, *in* Schlee, J.S., ed., Interregional unconformities and hydrocarbon accumulation: American Association of Petroleum Geologists Memoir 36, p. 109-126.

Kooi, H., Cloetingh, S., and Remmelts, G., 1988, Intraplate stresses and the stratigraphic evolution of the North Sea Central Graben: Geologie en Mijnbouw, in press.

Lambeck, K., Cloetingh, S., and McQueen, H., 1987, Intraplate stresses and apparent changes in sea level: The basins of north-western Europe, *in* Beaumont, C., and Tankard, A.J., eds., Sedimentary basins and basin-forming mechanisms: Canadian Society of Petroleum Geologists Memoir 12, p. 259-268.

Lehner, P., and De Ruiter, P.A.C., 1977, Structural history of Atlantic margin of Africa: American Association of Petroleum Geologists Bulletin, v. 61, p. 961-981.

Letouzey, J., 1986, Cenozoic paleo-stress pattern in the Alpine foreland and structural interpretation in a platform basin: Tectonophysics, v. 132, p. 215-231.

Livermore, R.A., and Smith, A.G., 1985, Some boundary conditions for the evolution of the Mediterranean region, *in* Stanley, D.J., and Wezel, F.C., eds., Geological Evolution of the Mediterranean Basin: New York, Springer Verlag, p. 83-98.

Manspeizer, W., 1985, Early Mesozoic history of the Atlantic passive margin, *in* Poag, C.W., ed., Geological evolution of the United States Atlantic margin: New York, Van Nostrand Reinhold, p. 1-23.

McKenzie, D.P., 1978, Some remarks on the development of sedimentary basins: Earth and Planetary Science Letters, v. 40, p. 25-32.

Miall, A.D., 1986, Eustatic sea level changes interpreted from seismic stratigraphy: A critique of the methodology with particular reference to the North Sea Jurassic record: American Association of Petroleum Geologists Bulletin, v. 70, p. 131-137.

Miller, K.G., and Fairbanks, R.G., 1985, Oligocene to Miocene carbon isotope cycles and abyssal circulation changes, *in* Sundquist, E.T., and Broecker, W.S., eds., The carbon cycle and atmospheric CO_2: Natural variations Archean to present: American Geophysical Union Geophysical Monograph 32, p. 469-486.

Miller, K.G., Fairbanks, R.G., and Mountain, G.S., 1987, Tertiary oxygen isotope synthesis, sea level history, and continental margin erosion: Paleoceanography, v. 2, p. 1-19.

Philip, H., 1987, Plio-quaternary evolution of the stress field in Mediterranean zones of subduction and collision: Annales Geophysicae, v. 5B, p. 301-320.

Pitman, W.C., III, and Golovchenko, X., 1983, The effect of sea level change on the shelf edge and slope of passive margins, *in* Stanley, D.J., and Moore, G.T., eds., The shelfbreak: Critical interface on continental margins: Society of Economic Paleontologists and Mineralogists Special Publication 33, p. 41-58.

Poag, C.W., 1985, ed., Geological Evolution of the United States Atlantic margin: New York, Van Nostrand Reinhold, 382 p.

Quinlan, G.M., and Beaumont, C., 1984, Appalachian thrusting, lithospheric flexure and the Paleozoic stratigraphy of the eastern interior of North America: Canadian Journal of Earth Sciences, v. 21, p. 973-996.

Richardson, R.M., Solomon, S.C., and Sleep, N.H., 1979, Tectonic stress in the plates: Reviews of Geophysics and Space Physics, v. 17, p. 981-1019.

Rona, P.A., and Richardson, E.S., 1976, Early Cenozoic plate reorganization: Earth and Planetary Science Letters, v. 40, p. 1-11.

Savostin, L.A., Sibuet, J.C., Zonenshain, L.P., Le Pichon, X., and Roulet, M.J., 1986, Kinematic evolution of the Tethys belt from the Atlantic Ocean to the Pamirs since the Triassic: Tectonophysics, v. 123, p. 1-35.

Schlanger, S.O., 1986, High-frequency sea-level fluctuations in Cretaceous time: An emerging geophysical problem, in Hsu, K.J., ed., Mesozoic and Cenozoic oceans: American Geophysical Union Geodynamic Series 15, p. 61-74.

Schlanger, S.O., and Premoli-Silva, I., 1986, Oligocene sealevel falls recorded in mid-Pacific atoll and archipelagic apron settings: Geology, v. 14, p. 392-395.

Schwan, W., 1985, The worldwide active middle/Late Eocene geodynamic episode with peaks at 45 and 37 My b.p. and implications and problems of orogeny and sea-floor spreading: Tectonophysics, v. 115, p. 197-234.

Sleep, N.H., 1971, Thermal effects of the formation of Atlantic continental margins by continental break up: Geophysical Journal of the Royal Astronomical Society, v. 24, p. 325-350.

Sleep, N.H., and Snell, N.S., 1976, Thermal contraction and flexure of mid-continent and Atlantic marginal basins: Geophysical Journal of the Royal Astronomical Society, v. 45, p. 125-154.

Steckler, M.S., Watts, A.B., and Thorne, J.R., 1988, Subsidence and basin modeling at the U.S. Atlantic passive margin, in Sheridan, R.E., and Grow, J.A., eds., The Atlantic continental margin: U.S.: Boulder, Colorado, Geological Society of America, The Geology of North America, v. I-2, p. 399-416.

Summerhayes, C.P., 1986, Sealevel curves based on seismic stratigraphy: Their chronostratigraphic significance: Palaeogeography, Palaeoclimatology, Palaeoecology, v. 57, p. 27-42.

Tankard, A.J., 1986, Depositional response to foreland deformation in the Carboniferous of eastern Kentucky: American Association of Petroleum Geologists Bulletin, v. 70, p. 853-868.

Thorne, J., and Watts, A.B., 1984, Seismic reflectors and unconformities at passive continental margins: Nature, v. 311, p. 365-368.

Tucholke, B.E., and Mountain, G.S., 1986, Tertiary paleoceanography of the western North Atlantic Ocean, in Vogt, P.R., and Tucholke, B.E., eds., The western North Atlantic region: Boulder, Colorado, Geological Society of America, The Geology of North America, v. M, p. 631-650.

Turcotte, D.L., and Ahern, J.L., 1977, On the thermal and subsidence history of sedimentary basins: Journal of Geophysical Research, v. 82, p. 3762-3766.

Vail, P.R., Mitchum Jr., R.M., and Thompson III, S., 1977, Global cycles of relative changes of sea level, in Payton, C.E., ed., Seismic stratigraphy—Applications to hydrocarbon exploration: American Association of Petroleum Geologists Memoir 26, p. 83-97.

Verwer, J.G., 1977, A class of stabilized three-step Runge-Kutta methods for the numerical integration of parabolic equations: Journal of Computational and Applied Mathematics, v. 3, p. 155-166.

Watts, A.B., 1982, Tectonic subsidence, flexure and global changes of sea level: Nature, v. 297, p. 469-474.

Watts, A.B., Karner, G.D., and Steckler, M.S., 1982, Lithospheric flexure and the evolution of sedimentary basins: Philosophical Transactions of the Royal Society of London, Series A, v. 305, p. 249-281.

Watts, A.B., and Thorne, J.R., 1984, Tectonics, global changes in sea level and their relationship to stratigraphical sequences at the U.S. Atlantic continental margin: Marine and Petroleum Geology, v. 1, p. 319-339.

Wise, S.W., and van Hinte, J.E., and others, 1986, Mesozoic-Cenozoic depositional environment revealed by Deep Sea Drilling Project Leg 93 drilling on the continental rise off the eastern Unites States, in Summerhayes, C.P., ed., North Atlantic palaeoceanography:: Geolological Society of London Special Publication, v. 21, p. 35-66.

Wortel, R., and Cloetingh, S., 1981, On the origin of the Cocos-Nazca spreading center: Geology, v. 9, p. 425-430.

Wortel, R., and Cloetingh, S., 1983, A mechanism for fragmentation of oceanic plates, in Watkins, J.S., and Drake, C.L., eds., Studies in continental margin geology: American Association of Petroleum Geologists Memoir 34, p. 793-801.

Ziegler, P.A., 1982, Geological atlas of Western and Central Europe, Shell Internationale Petroleum Maatschappij (The Hague): Amsterdam, Elsevier, 130 p.

Ziegler, P.A., 1988, Post-Hercynian plate reorganization in the Tethys and Arctic-North Atlantic, in Manspeizer, W., ed., Triassic-Jurassic rifting in North–America–Africa: Amsterdam, Elsevier, in press.

Zoback, M.D., 1985, Wellbore break-out and in-situ stress: Journal of Geophysical Research, v. 90, p. 5523-5530.

9

FROM GEODYNAMIC MODELS TO BASIN FILL—A STRATIGRAPHIC PERSPECTIVE

Teresa E. Jordan and Peter B. Flemings
Department of Geological Sciences & Institute for Study of the Continents, Snee Hall, Cornell University, Ithaca, New York 14853-1504 USA

ABSTRACT

Interpreting the conditions under which a series of strata accumulated is the central objective of traditional stratigraphic studies. The interpretations are based upon a large set of conceptual models. Our ability to interpret correctly the ancient conditions will be enhanced greatly when these conceptual models are translated into mathematical statements, linked together in logical series, and used to predict stratigraphies that result from choices of boundary and initial conditions. The natural system is so complex that our capacity to distinguish stratigraphic consequences of such basic natural occurrences as sea-level changes, tectonic events, and climatic changes depends upon experiments using quantitative dynamic models. Use of quantitative models also will enhance our understanding of the interactions that exist between various components of the Earth's surface environment, and between the surficial processes and solid-Earth phenomena.

To illustrate the direction and potential of such research, we present a model of the stratigraphic development of a nonmarine foreland basin. Tectonic subsidence of the basin is controlled by lithosphere rheology plus the geometry and timing of fault motions. Mass is redistributed at the surface of the Earth by a simple approximation of erosion and deposition, which responds to the constructional tectonic topography. Four types of variables exist in this simple model: flexural properties, structural history, erodibility of bedrock, and transport efficiency in the depositional regime. The outputs are cross sections illustrating topography in the mountain range and "facies" and time-lines within the basin. For experiments in which erodibility of the bedrock is considered the only variable, changing sediment yield controls the width, thickness, and facies characteristics of the basin. As erodibility increases, the basin widens and the peripheral bulge is subdued.

INTRODUCTION

A prime task for many geologists is the interpretation of the conditions under which sedimentary rocks accumulated. We have built and refined extensive sets of conceptual models which guide those interpretations (e.g., facies models, facies tract models). But in practice, geologists often are faced with nonunique interpretations; as far as we are able to

Quantitative Dynamic Stratigraphy (1989), T.A. Cross, ed., Prentice Hall, p. 149–163.

149

judge on a qualitative basis, many different original conditions and histories can create similar sedimentary lithologies and lithologic successions. Multiple working hypotheses are constructed, and often the best hypothesis cannot be determined. Even first-order features of the stratigraphic record are often difficult to interpret. For example, the existence of a major lithologic change in a conformable series of strata implies that a major change occurred at that location of the Earth's surface at a particular point in time. Why did it occur at that time? Which of the principal independent (but interacting) controls on strata was responsible: tectonics of the local basin; sediment yield; sea level or nonmarine base level; latitude/climate and ocean-basin configurations; or geologic age?

This situation could be improved upon if quantitative methods existed to explore the consequences of each of the alternative interpretations. If one could predict the stratigraphic section that would result from each hypothesis, the end products of alternative interpretations would probably differ. Those differences would provide new criteria for interpreting the rocks, perhaps requiring new and different observations.

This report examines opportunities for and state of the art in quantitative modeling of the stratigraphy of sedimentary basins. We consider "stratigraphy of a basin" to be the distribution through space and time of sedimentary rocks of varying lithologies. We restrict the discussion to the spatial and temporal scales inherent in the geodynamic mechanisms that control the subsidence of sedimentary basins: areas of the Earth that are 10s to 1000s of kilometers in length and width, and time scales of 10^5 to 10^8 years. We will consider the stratigraphic characteristics that are appropriate to a first-order description of the contents of a sedimentary basin. Traditionally, these are features that would appear on a map or cross section of an entire basin. As an example, a forward model of the development of paired foreland-"piggyback" basins is presented which illustrates the importance of sediment yield on the stratigraphy of a basin.

PREVIOUS WORK

The creation and fill of sedimentary basins have been modeled on two scales. Geodynamic models operate in terms of lithospheric mechanics at scales that are comparable to or that exceed the dimensions of the basin. Most synthetic stratigraphic models deal with sediment transport through and accumulation within an already-subsiding basin.

Geodynamic models describe basin subsidence as a response of the lithosphere to deformation or thermal changes. The emphasis has been on a quantity termed "tectonic subsidence," which is the subsidence that the Earth's surface would undergo were there no sediment (or water) to fill the newly created basin. In cases for which sediment has been assumed to fill the basin, the predicted geometry of strata has been compared to the geologic record. However, typically the sedimentary fill is considered to be a passive and minor component of basin evolution; it is frequently assumed that sediment fills the basin to a horizontal datum.

Synthetic stratigraphic models emphasize the depositional processes and products. Allen (1974, 1978), Leeder (1978) and Bridge and Leeder (1979) illustrate this approach to nonmarine systems, and Harbaugh and Bonham-Carter (1970) and Schwarzacher (1975) were pioneers in marine environments; Cross (this volume) includes other examples. Many such examples are geometrically- and stochastically-based forward models: given a certain sediment input and accommodation space as boundary conditions, the distribution and geometry of sediment can be predicted. Turcotte and Kenyon (1984) introduced this approach at the basin scale, considering stratal geometries that result from sea-level variations. Alternatively, geomorphological and sedimentological models that are based on the physics of sediment transport have been developed for stream grade adjustment in response to base-level changes (e.g., Snow and Slingerland, 1987), but the integrated consequence through time of such changes has not been considered quantitatively. Whereas many synthetic stratigraphic models embed one or more depositional systems in a larger subsiding region, in most cases the surface environment (in particular, sediment yield to the basin) is not modeled to evolve dynamically in response to independent or dependent controls.

From the point of view of the field geologist, neither type of existing quantitative stratigraphic model is very useful yet. First, they do not treat more than one or two of the independent controls, and thus are unable to help the geologist visualize the consequences of multiple working hypotheses. Second, models that begin with the geodynamics of a basin have not yielded predictions at levels which are testable against the wealth of existing stratigraphic observations. Although those models predict gross basin geometry as a function of time, they do not predict facies, nor lithologies, nor the temporal and spatial distribution of strata and unconformities. To achieve their potential utility, stratigraphic models must link the physics of surface processes (erosion and deposition) to the physics of tectonic processes. In nature, changes in one domain force responses in others; the same must occur in stratigraphic models.

Using quantitative dynamic models we will be able to approach complete and correct simulations of stratigraphic sequences by judiciously choosing initial and boundary conditions that are based on field data, and guided by results of previous models. Realistically, we may never reach the point of simulating stratigraphic sequences at the basin scale in pure forward models. However, the models will, at a minimum, enhance our ability to interpret field data, to grasp the physical processes that operated in the basin, and to understand the boundary and initial conditions of the natural basin system. Furthermore, the models will help us to evaluate which processes produced which recognizable responses, and how the independent processes interact. By revealing connections between processes and responses that are not intuitively obvious, the quantitative models will result in improved conceptual models. The simulation models should be viewed as experiments; each experiment guides observations, and coincidence or conflict between predictions and observations guide the next iteration of improvement in description of geological processes.

COMPLEX RESPONSES OF THE BASIN SYSTEM

The geometries and lithologies of strata are products of complex interactions of multiple tectonic and surficial processes, operating simultaneously and with feedback. Whereas lithospheric processes are the primary trigger to basin development, solid Earth behavior also is affected by processes occurring at the Earth's surface. Surficial processes largely have been ignored in geodynamic models, presumably because they are considered to be minor and to operate at different temporal and spatial scales than do geodynamic processes. Yet these surficial processes may have important controls on lithospheric-scale processes. If we can incorporate in our forward models complete descriptions of the independent variables, and test them against well-constrained basin histories, we can learn about possible complex relations between solid Earth and surficial processes. Below we describe some situations in which surficial processes may play an important role in the behavior of the crust. We emphasize foreland basins, but the possible interactions are equally important in other categories of sedimentary basins (e.g., Beaumont et al., 1982).

For a foreland thrust belt and basin pair, basin geometry is controlled by the progressive geometry of thrusts and folds, which controls the geometry of loads (Jordan, 1981; Beaumont, 1981); much remains to be learned about causal relations between erosion and deformation. Given present understanding of the mechanics of growth of a thrust-faulted wedge of deformed strata, Davis et al. (1983) and Dahlen and Suppe (1988) hypothesized that the internal deformation of a thrust belt is very sensitive to the slope of the rock/air interface (Fig. 1, a). Thus erosion, which depends on relief (Ahnert, 1970) caused by thrust deformation and on independent variables such as climate and bedrock lithology, is a first-order

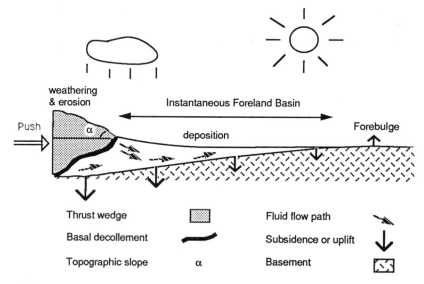

Figure 1. Cross-sectional cartoon of a foreland basin and the leading edge of a thrust belt.

control on thrust wedge deformation. Erosion not only determines the geometry and rate of sediment yield to the foreland basin, but it is also likely that it influences the timing and geometry of tectonic subsidence.

The thermal structure of a foreland basin and the crust beneath it places another control on width and erosion history of a basin; thermal structure is influenced by surficial processes in at least two ways. First, the accumulation of sediment in the basin disrupts the lithosphere's initial thermal structure, and it is the thermal structure that controls the elastic properties of the lithosphere (Kominz and Bond, 1986). Thus the timing and abundance of sediment yield, which are controlled by surficial processes as well as tectonically constructed topography, control the timing and magnitude of a thermal anomaly. Second, ground-water circulation is also a key factor in the thermal history of any basin, controlling the balance between conductive and convective heat transfer, which affects the creation and decay of a thermal anomaly. In a foreland basin, ground-water circulation is a major factor (Garven and Freeze, 1984; Beaumont et al., 1985) which is controlled by water-table configuration, basin dimensions, subsurface distribution of permeability, thermal structure of the lithosphere, and compaction history (Garven and Freeze, 1984). Although lithospheric activity controls the gross history and configuration of the basin, surface characteristics (climate, erosion rates, facies types) clearly influence ground-water circulation (Fig. 1). Added levels of complexity result because heat flow is influenced by compaction, diagenesis, and permeability of the foreland basin strata, which in turn are controlled by vertical and horizontal distribution of depositional textures, ground-water circulation and rock/fluid interactions.

At present, quantitative dynamic models of sedimentary basins are not sufficiently advanced to investigate these complex process-response interactions. As models evolve that integrate phenomena of different spatial and temporal scales (i.e., integrating regional-scale tectonic models with predictions of physical properties of rocks at successively finer scales), the opportunities to examine complex responses will improve. Models that will be capable of analyzing complex responses must be based upon accurate physical expressions of individual processes rather than on general relations that lump interacting processes together.

A MODEL OF THE STRATIGRAPHIC RECORD OF THRUST-RELATED BASINS

Until recently, geodynamic models of sedimentary basins have not included the dynamic processes of sediment yield, transport and deposition. Forward geodynamic models typically have treated sediment as a fluid, like water, that instantaneously fills a deflection in the lithosphere to a horizontal datum at sea level (e.g., Turcotte and Schubert, 1982). In other, pseudo-forward geodynamic models, observed basin stratigraphy is compared with model output and the model parameters are adjusted until the stratigraphy is simulated at an acceptable level. Although these models do not prescribe that the basin fill to a horizontal depositional surface at sea level, they nonetheless require that the fill be instantaneous (e.g., Beaumont, 1981; Jordan, 1981; Quinlan and Beaumont, 1984). Because these models were based upon and designed to mimic preserved stratigraphy, rather than to fill a lithospheric depression by dynamic processes of sediment transport and deposition, they have little stratigraphic predictive value.

The task of devising mathematical descriptions of all surficial processes involved in weathering, transport, and deposition of sediment and of linking them logically in a computer program might appear imposing. But the task may be simplified by recognizing that the geometry of lithospheric deformation imposes the initial constructional topography. The constructional topography constrains the range of erosion rates and depositional processes that may occur (see also Leeder and Gawthorpe, 1987). For example, passive margin basins contain the full spectrum of fluvial to abyssal marine depositional realms, whereas retroarc foreland basins may contain a more limited range of depositional systems. Therefore quantitative models of these two basin types can be tailored to the appropriate boundary conditions.

Model Framework

To illustrate the state of the art of quantitative dynamic stratigraphic models, and to describe some improvements to such models that we foresee are necessary, we use a nonmarine foreland basin model as an example. Related models are explored in more detail elsewhere (Flemings and Jordan, in review).

The intent of this model is to simulate the fill of a foreland basin in response to tectonically driven conditions. In their simplest case, foreland basins are created by flexure of the lithosphere in response to the load of a thickening foreland thrust belt. Constructional topography is produced by deformation in the thrust belt and regional isostatic compensation of this load. In this way, a mountain range–basin pair is created, whose topographic form is the sum of local strain and regional subsidence of the lithosphere (Fig. 1). The variables that affect the shape of the constructional topography are the thrust fault geometry (Fig. 2) and the flexural properties of the lithosphere, which are expressed in the model as the flexural rigidity of an elastic plate (Appendix 1).

A. initial condition of undeformed crust

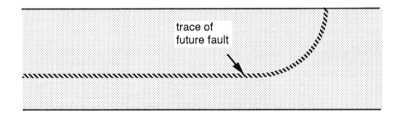

B. after deformation

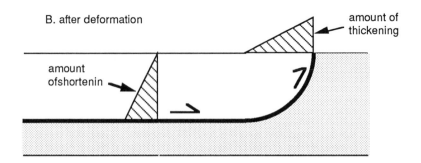

Figure 2. The structural model employed in experiments shown in Figure 4. A) Initial conditions of model. B) After an increment of shortening, a thickened zone is produced that acts as a load. In the model, the load is compensated by flexure and then acted on by erosion and deposition.

At each time step constructional topography is modified by surficial processes. Erosion is calculated as a function of the instantaneous slopes in the mountain belt, and deposition is calculated so that mass is conserved and the accommodation space is partially filled (Fig. 3). In nature, deformation in a thrust belt is a long-term process, and transfer of mass by erosion and deposition require similarly long periods. In the model, both tectonic processes and surface processes operate incrementally through a series of time steps (Appendix 1). Mass that is transferred by surficial processes is compensated by regional isostasy.

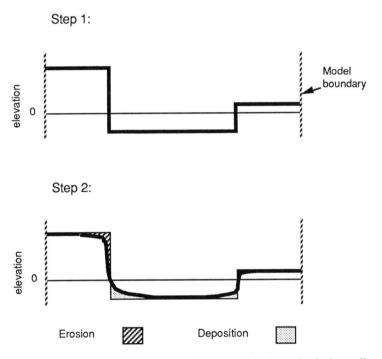

Figure 3. Elements of the surficial processes model. Topography is shown by the heavy line. Step 1 illustrates one time step of tectonic construction of topography. Step 2 illustrates the response by surface processes to the constructional topography.

As a first approximation, erosion and deposition are modeled by assuming that the rate of mass transport is proportional to slope. Mass conservation then requires that change in elevation at any point is proportional to the curvature of the slope. These relations describe diffusive processes and are treated mathematically with the diffusion equation (Appendix 1). The diffusion equation has a long history of utility in geomorphological studies that describe creep processes (Culling, 1965), fluvial transport (Begin et al., 1981), and deltaic processes (Kenyon and Turcotte, 1985).

It is possible that other analytical approaches might more accurately describe the complex spectrum of processes, such as chemical and mechanical weathering, hill-slope storage and transport, fluvial transport, and overbank deposition, that in concert transfer mass from mountain to basin (Anderson and Humphrey, 1988). Yet the net result of these processes is approximated reasonably well by the diffusion equation. In mountains, areas of rapid change in slope erode more rapidly than do graded surfaces; in basins, areas of rapid change in slope are filled by sediment most rapidly (Fig. 3). The net result is the smoothing of constructional topography. Flemings and Jordan (in review) discuss empirical studies that justify using the diffusion equation. We use two diffusion constants (which can be thought of as transport coefficients; Appendix 1), one for bedrock and soil in the mountain belt and one for sediment and recently deposited strata in the basin. Increasing the transport coefficient in the mountain belt increases the erosion rate and sediment yield. Increasing the transport coefficient in the basin increases the efficiency of sediment transport within the basin.

Results of the model are displayed in two ways (Fig. 4). First, the distribution of strata deposited during specific time steps is recorded; this is displayed as cross sections of the basin with time lines mapped through the strata. On these sections, the upper surface in the basin represents the instantaneous surface of deposition, and its slope is the instantaneous depositional gradient. The second portrayal of the data emphasizes those depositional slopes; these cross sections map the strata deposited on slopes within a given range of gradient. The depositional slopes mimic environmental characteristics and may be linked empirically to flow regimes and transport mechanisms.

The stratigraphic impact of varying either thrust history (geometry and time of motion), flexural rigidity, erosional production of sediment, or depositional regime in the basin can be tested with the model, if the explicit assumptions and simplifications are accepted. If a systematic or nonsystematic change in the depositional geometry or character through time in a foreland basin is noted, this forward model could be used to test hypotheses about the cause(s) of the change. Such hypotheses must be framed in terms of one or more variables of the model. For example, if it is observed that facies tracts shift toward the mountain belt through time, one might ask whether this is due to increased thrust load, to decreased flexural rigidity, to decreased sediment yield, or to decreased competence of the streams in the basin.

Results

The specific model of a thrusted mountain block and flanking sedimentary basins presented here is representative of basement uplifts in the foreland of various mountain systems (e.g., Rodgers, 1987), in particular the Rocky Mountain foreland of the western United States and the Sierras Pampeanas of Argentina. The structural model also applies to segments of some thin-skinned thrust systems. Compressional shortening is modeled as occurring on a listric reverse fault (Erslev, 1986; Jordan and Allmendinger, 1986) in which the width of the region

→

Figure 4. A synthetic stratigraphy of foreland (right) and piggyback (left) basins during 5 million years. The initial condition was a horizontal upper surface. The transport coefficient in the mountains (K_m) is varied in the three examples, while the thrust shortening rate (2 mm/yr), flexural rigidity (10^{30} dyne-cm), and transport coefficient in the basins (1000 m^2/yr) are held constant. The upper cross section of each pair shows strata deposited in 1 m.y. time intervals; the lower cross section shows depositional "facies" in those strata ("facies" A deposited on slopes 0.001; "facies" B deposited on slopes <0.001). The erosional topography of the mountain block (heavy line) is shown after 1, 3, and 5 m.y. of deformation. The dashed lines show the topographic axes of the basins.

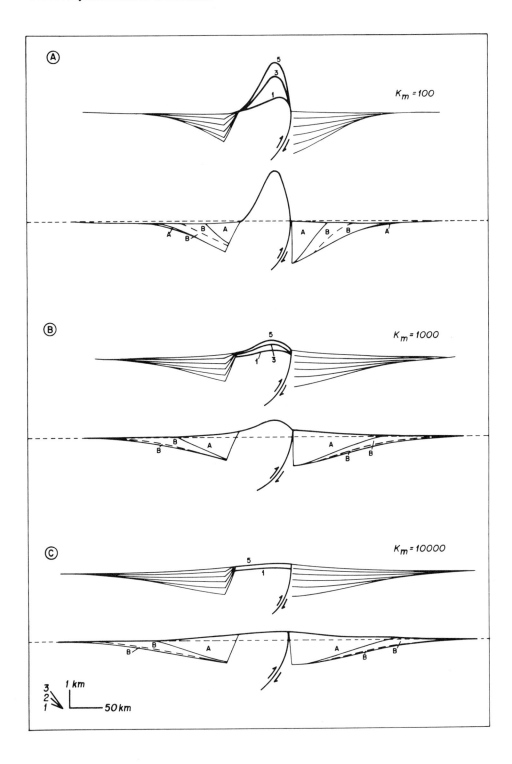

thickened by the rotation is equal to the depth to the basal décollement (Fig. 2). The rate of movement along the fault is 2 mm/yr, a rate appropriate to some ranges in the Sierras Pampeanas (Jordan and Allmendinger, 1986). The assumed flexural rigidity, 10^{30} dyne-cm, is a reasonable value for regions like the Rocky Mountain foreland of Wyoming and the Subandean foreland of Bolivia (Jordan, 1981; Lyon-Caen et al., 1985). Using a somewhat different structural model but similar erosional model, Karner (1986, 1987) modeled paired foreland and "perched" basins.

Figure 4 illustrates the evolution of a faulted mountain block and its basin during 5 million years. The case shown in Figure 4a illustrates the basic features. Two basins result from the structural configuration: the basin on the right side, adjacent to the faulted side of the mountain block, is a classic foreland basin. The basin on the left, which is not fault bounded, is a synclinal, "perched," or "piggyback" basin (Ori and Friend, 1984). Although the tectonic subsidence component is asymmetric, larger on the right than on the left side of the mountain (Fig. 3), sediment is shed from both sides of the mountain block. The mass of the sediment enhances subsidence on both sides, whereas subsidence beneath the mountain block is reduced.

To illustrate the potential insight that can be gained through such models into the dynamics of a basin, Figure 4 presents three different basin configurations and stratigraphies generated by changing the transport coefficient in the mountain block while the other variables are held constant. Flemings and Jordan (in review) explore the stratigraphic consequences of the other variables. By varying the transport coefficient in the mountain belt, the erosion rate of the thrust block is varied. Case 4a, with a low erosion rate, results in a thrust block with high elevation. Case 4c has such a high denudation rate that the mountain has little elevation above the neighboring basinal areas, and a large percentage of the thrusted mass is transferred to the basin producing a large volume of foreland basin strata. A geologic example of a low-relief thrusted mountain may occur in the Cretaceous foreland system of Wyoming. Warner and Royse (1987) noted that, in spite of rapid translation of the thick Absaroka thrust sheet, erosion was so rapid that little structural thickening occurred in the region of thrusting. Presumably the rapid erosion resulted in low paleorelief and produced a large sediment yield to the Absaroka-age foreland basin.

A consequence of the diffusion equation is that, as a thrust block of fixed wavelength is uplifted, the erosion rate is proportional to the elevation. As a result the thrust block exponentially approaches a steady-state topography in which uplift rate equals denudation rate (i.e., in each time step the total mass of material uplifted is removed by erosion). For a low rate of erosion (a small transport coefficient in the mountains), steady state is approached slowly (Fig. 4a), whereas, for a high erosion rate, steady state is achieved rapidly (the elevations at times 3, 4, and 5 Ma are barely distinguishable and could not be drafted at the scale of Figure 4c).

The volume of sediment deposited increases as the erosional transport coefficient increases (Fig. 4a,b,c). Because the transport coefficients in the basins are the same in each of the three runs, the surface slope in each of the basins is approximately equal. The basins must therefore broaden and, due to flexural subsidence under the sediment load, thicken. Without the guidance of the modeling, the increase in width of the basins could be misinterpreted easily as the result of increases in the flexural rigidity of the basement.

The depositional slope generated in the three cases is slightly different: steeper gradients develop in the basins with a greater sediment yield (Fig. 4). The "facies" cross sections illustrate strata deposited on surface slopes with gradients greater than 0.001 (A) and less than 0.001 (B) (Fig. 4). The model suggests that lower-gradient facies are deposited closer to the thrusted mountain front for cases of low sediment yield (Fig. 4a) than for cases of high sediment yield (Fig. 4c). In both the foreland and piggyback basins, a low sediment yield also results in a pronounced topographic axis in the depositional basin (marked by dashed line, Fig. 4a), with sediment on the distal side of the topographic axis supplied from the topographically high peripheral bulge. For a high sediment yield, sediment is so widely dispersed that topographic expression of the peripheral bulge is suppressed and the basin's topographic axis is at or near the depositional pinch-out on the distant flank of the basin (Figs. 4b,c).

The model cases (Figs. 4a,b,c) imply that a variation *through time* of erosion rate, and consequently sediment yield to the basin, would produce changes in the stratigraphy of a basin even if other factors remained constant. The Late Cretaceous and early Cenozoic Rocky Mountain foreland basins of Wyoming may represent a history of such changes. Uplift of ranges with fault geometries approximated by this model (Fig. 2) successively exposed bedrock possessing varying resistances to erosion: Mesozoic shale and sandstone, followed by Paleozoic carbonate and quartzite units, followed by Precambrian crystalline basement (Love, 1970). Flemings (1987) concluded that this variation in erosional resistance was a major contributor to the latest Cretaceous through Eocene stratigraphic history of the Wind River Basin. The synthetic stratigraphic model would simulate such a lithologic change if differing erosional transport coefficients were used to represent progressive stages of unroofing. Early stages of uplift (high coefficient) would result in topographies and basin geometries like those of Figure 4c, whereas late stages of uplift (low coefficient) would produce those shown in Figure 4a.

Future Work

In its present form, this forward model uses a combination of geometrical constraints, general approximations, and conservation laws. The model can be expanded by incorporating approximations of other aspects of the natural system. An additional goal is to improve the model by replacing the general approximations with more detailed or exact approximations, such as described below.

Slight increases in the predictive capacity can be achieved with minor changes. For instance, density differences between eroded and deposited material should be calculated in the mass balance, as well as progressive compaction and density changes during burial. Lithologic variations in the thrust belt might also be simulated and related to varying resistances to erosion. The structural model can be varied to simulate other mountain systems, such as a thrust wedge (Flemings and Jordan, in review).

More important increases in the predictive capacity depend upon incorporating approximations of other facets of the natural system. In the basin, sea level could be added to the model and simulation of marine deposition and transfer of material across the shoreline approximated appropriately. Thus, sea-level change could be treated as an independent variable. Thrust geometries and surficial processes need to be combined with descriptions of the thermal behavior of the lithosphere and with other rheological models of the lithosphere, in order to test the possible validity of those models. Whatever structural model is employed should be coupled to erosion such that the thrust belt structure responds dynamically to surface processes, in turn controlling the constructional topography (Dahlen and Suppe, 1988). Approximations of erosion other than diffusion should be tested to simulate the effects of climate and elevation.

In terms of making the model simulate anything more detailed than the predictions shown here, enormous opportunities exist. At this level, it will be important to make judicious decisions concerning whether to rely on gross approximations or to incorporate more exact descriptions of the physics and chemistry of the relevant processes. Statistical approximations of certain aspects of surficial processes would be appropriate. Erosion could be described more realistically, either based upon empirical generalizations or on the physics and chemistry of weathering, to predict sediment yield volumes as well as grain size distributions. Transport equations with appropriate bedload and suspended load behavior might be substituted for the diffusion equation (e.g., Paola, 1988). Increased detail will be a necessary step toward a goal of predicting permeability structure within a basin, which in turn will permit modeling of ground-water flow, convective heat transfer, and diagenetic reactions.

An added challenge is to adapt the model to three dimensions. In the case of foreland basins, a two-dimensional treatment can be justified temporarily because foreland thrust belts are commonly laterally extensive features. However, conserving mass within the cross section may not be appropriate where the foreland basin is superimposed on a more regional-scale drainage.

Conclusions

Stratigraphers will find that quantitative dynamic stratigraphic models have numerous applications because they provide a quantitative basis for interpretation of ancient strata. Opportunities to build upon geodynamic basin models, creating models with predictive capabilities, are excellent. Realistic mathematical descriptions of surficial processes need to be combined with the initial and boundary conditions determined by the geodynamic models. Quantitative dynamic stratigraphic models will illuminate the interactions between processes acting on Earth's surface and deformational and thermal phenomena that affect the lithosphere. The models must be tested by comparing model predictions against the stratigraphic records of well-studied basins.

The foreland basin model presented here illustrates the potential strengths of dynamic models that can be used to perform stratigraphic experiments. This model improves on previous foreland basin models by explicitly including surface transport of mass (erosion and deposition) in addition to subsurface transport (on thrust faults). The model allows us to investigate the stratigraphic consequence of several independent controls. In the specific experiment presented, as the erosional transport rate varies, the wavelength of the basin varies, despite the fact that the flexural rigidity of the Earth's lithosphere is kept constant.

Thus such variables as climate and source rock lithology can play important roles in controlling basin architecture.

The forward model presented here also illustrates the natural progression in quantitative dynamic stratigraphic modeling. A pre-existing conceptual model of foreland basin evolution has been mathematically expressed in terms of simple approximations of the physical processes. Future steps are needed that will replace the simple approximations with statistical approximations and more fundamental physical and chemical descriptions of the processes.

ACKNOWLEDGMENTS

We thank T. Hearn, D. Turcotte, T. Cahill and J. Abel for suggestions and guidance in modeling. Noye Johnson and Jim Beer inspired and cooperated with stratigraphic studies of foreland basins which led to the modeling. C. Lee Roark, T. Gubbels, G. de V. Klein, R. Anderson, and T. Cross critiqued a draft of the paper. G. Karner graciously compared progress on unpublished models. Tim Cross is thanked enthusiastically for organizing and leading a stimulating QDS workshop. We are grateful for the support of the Shell Oil Company Foundation and the donors to the Petroleum Research Fund of the American Chemical Society (18887-AC2). INSTOC contribution #94 .

APPENDIX 1

The two dimensional model iterates through 10,000 yr time intervals. The initial input is the crustal thickening created in a single time step along a rotational fault geometry (Fig. 2). Regional isostatic compensation of the input thickening generates a topography that is modified in each time step by progressive erosion and deposition (Fig. 3).

ISOSTASY   Sediment and thrust loads are compensated by an elastic lithosphere. The variation in deflection (w) as a function of position (x) for an elastic plate overlying an inviscid fluid due to a load (q(x)) is described by:

$$D \, d^4w/dx^4 + (\rho_m) \, gw = q(x) \tag{1}$$

where D is flexural rigidity, ρ_m is mantle density, and g is gravitational acceleration. The solution of (1) for a line load is given in Turcotte and Schubert (1982).

MASS TRANSPORT   Erosion and deposition are modeled as a diffusive process in two dimensions:

$$\partial h/\partial t = K \, \partial^2 h/\partial x^2 \tag{2}$$

where t is time, h is elevation and K is a transport coefficient with units of length^2/time. This equation is derived by combining the continuity equation

$$\partial h/\partial t = -\partial S/\partial x \tag{3}$$

with a sediment flux S that is proportional to slope

$$S = -K \, \partial h/\partial x. \tag{4}$$

Equation (2) expresses the law of conservation of mass and, neglecting density change and compaction, is appropriate for systems where mass transport is proportional to slope.

SOLUTION ALGORITHM DURING EACH ITERATION   The load due to crustal thickening is discretized into line loads and the deflection caused by each line load is calculated from the flexure equation (1); the isostatically compensated topography is the sum of these deflections. An implicit finite-difference approximation of the diffusion equation (Incropera and De Witt, 1985) is used to calculate the erosion and deposition that occur during the time step in response to the topography. The topography, depositional slope, and thickness of deposited or eroded material are recorded at each iteration.

REFERENCES CITED

Ahnert, F., 1970, Functional relationships between denudation, relief, and uplift in mid-latitude drainage basins: American Journal of Science, v. 268, p. 243-263.

Allen, J.R.L., 1974, Studies in fluviatile sedimentation: Implications of pedogenic carbonate units: Geological Journal, v. 9, p. 181-208.

Allen, J.R.L., 1978, An exploratory quantitative model for the architecture of avulsion-controlled alluvial suites: Sedimentary Geology, v. 21, p. 129-147.

Anderson, R.S., and Humphrey, N.F., 1989, Interaction of weathering and transport processes in the evolution of arid landscapes, in Cross, T.A., ed., Quantitative dynamic stratigraphy: New Jersey, Prentice-Hall (this volume).

Beaumont, C., 1981, Foreland basins: Geophysical Journal of the Royal Astronomical Society, v. 65, p. 291-329.

Beaumont, C., Boutilier, R., MacKenzie, A.S., and Rullkötter, J., 1985, Isomerization and aromatization of hydrocarbons and paleothermometry and burial history of Alberta foreland basin: American Association of Petroleum Geologists Bulletin, v. 69, p. 546-566.

Beaumont, C., Keen, C.E., and Boutilier, R., 1982, On the evolution of rifted continental margins: comparison of models and observations for the Nova Scotian margin: Geophysical Journal of the Royal Astronomical Society, v. 70, p. 667-715.

Begin, Z.B., Meyer, D.F., and Schumm, S.A., 1981, Development of longitudinal profiles of alluvial channels in response to base-level lowering: Earth Surface Processes and Landforms, v. 6, p. 49-68.

Bridge, J.S., and Leeder, M.R., 1979, A simulation model of alluvial stratigraphy: Sedimentology, v. 26, p. 617-644.

Culling, W.E.H., 1965, Theory of erosion on soil-covered slopes: Journal of Geology, v. 73, p. 230-254.

Davis, D., Suppe, J., and Dahlen, F.A., 1983, Mechanics of fold-and-thrust belts and accretionary wedges: Journal of Geophysical Research, v. 88, p. 1153-1172.

Dahlen, F.A., and Suppe, J., 1988, Mechanics, growth and erosion of mountain belts, in Clark, S. and Burchfiel, B.C., eds., Processes in continental lithospheric deformation: Geological Society of America Special Paper 218, p. 161-178.

Erslev, E.A., 1986, Basement balancing of Rocky Mountain foreland uplifts: Geology, v. 14, p. 259-262.

Flemings, P.B., 1987, The paleogeography of the Maastrichtian and Paleocene Wind River Basin: Interpreting Rocky Mountain Foreland deformation [M.A. Thesis]: Ithaca, New York, Cornell University, 175 p.

Flemings, P.B., and Jordan, T.E., 1987a, Sedimentary response in a foreland basin to thrusting: A forward modelling approach: Geological Society of America Abstracts with Programs, v. 19, p. 664.

Flemings, P.B., and Jordan, T.E., 1987b, Synthetic stratigraphy of foreland basins: EOS (American Geophysical Union Transactions), v. 68, p. 419.

Flemings, P.B., and Jordan, T.E., in review, A synthetic stratigraphic model of foreland basin development: Journal of Geophysical Research.

Garven, G., and Freeze, R.A., 1984, Theoretical analysis of the role of groundwater flow in the genesis of stratabound ore deposits. 2. quantitative results: American Journal of Science, v. 284, p. 1125-1174.

Harbaugh, J.W., and Bonham-Carter, G.F., 1970, Computer simulation in geology: New York, John Wiley & Sons, 575 p.

Incropera, F.G., and DeWitt, D.P., 1985, Introduction to heat transer: New York, John Wiley and Sons, 712 p.

Jordan, T.E., 1981, Thrust loads and foreland basin development, Cretaceous, western United States: American Association of Petroleum Geologists, v. 65, p. 2506-2520.

Jordan, T.E., and Allmendinger, R.W., 1986, The Sierras Pampeanas of Argentina: A modern analogue of Rocky Mountain foreland deformation: American Journal of Science, v. 286, p. 737-764.

Karner, G.D., 1986, On the relationship between foreland basin stratigraphy and thrust sheet migration and denudation: EOS (American Geophysical Union Transactions), v. 67, p. 1193.

Karner, G.D., 1987, The interplay between erosion and basin development: Yearbook, Lamont-Doherty Geological Observatory, Palisades, NY, p. 46-51.

Kenyon, P.M., and Turcotte, D.L., 1985, Morphology of a delta prograding by bulk sediment transport: Geological Society of America Bulletin, v. 96, p. 1457-1465.

Kominz, M.A., and Bond, G.C., 1986, Geophysical modelling of the thermal history of foreland basins: Nature, v. 320, p. 252-256.

Leeder, M.R., 1978, A quantitative stratigraphic model for alluvium, in Miall, A., ed., Fluvial sedimentology: Canadian Association of Petroleum Geologists Memoir 6, p. 587-596.

Leeder, M.R., and Gawthorpe, R.L., 1987, Sedimentary models for extensional tilt-block/half-graben basins, in Coward, M.P., Dewey, J.F., and Hancock, P.L., eds., Continental extensional tectonics: Geological Society of London Special Publication 28, p. 139-152.

Love, J.D., 1970, Cenozoic geology of the Granite Mountains area, central Wyoming: U.S. Geological Survey Professional Paper 495-C, 154 p.

Lyon-Caen, H., Molnar, P., and Suarez, G., 1985, Gravity anomalies and flexure of the Brazilian Shield beneath the Bolivian Andes: Earth and Planetary Science Letters, v. 75, p. 81-92.

Ori, G.G., and Friend, P.F., 1984, Sedimentary basins formed and carried piggyback on active thrust sheets: Geology, v. 12, p. 475-478.

Paola, C., 1988, Subsidence and gravel transport in alluvial basins, in Kleinspehn, K.L., and Paola, C., eds., New perspectives in basin analysis: New York, Springer-Verlag, p. 231-243.

Quinlan, G.M., and Beaumont, C., 1984, Appalachian thrusting, lithospheric flexure and the Paleozoic stratigraphy of the eastern interior of North America: Canadian Journal of Earth Sciences, v. 21, p. 973-996.

Rodgers, J., 1987, Chains of basement uplifts within cratons marginal to orogenic belts: American Journal of Science, v. 287, p. 661-692.

Schwarzacher, W., 1975, Sedimentation models and quantitative stratigraphy: Amsterdam, Elsevier, Developments in Sedimentology, v. 19, 382 p.

Snow, R.S., and Slingerland, R.L., 1987, Mathematical modeling of graded river profiles: Journal of Geology, v. 95, p. 15-33.

Turcotte, D.L., and Kenyon, P.M., 1984, Synthetic passive margin stratigraphy: American Association of Petroleum Geologists Bulletin, v. 71, p. 882-889.

Turcotte, D.L., and Schubert, G., 1982, Geodynamics, applications of continuum physics to geological problems: New York, John Wiley & Sons, 450 p.

Warner, M.A., and Royse, F., 1987, Thrust faulting and hydrocarbon generation: discussion: American Association of Petroleum Geologists Bulletin, v. 71, p. 882-889.

10

STRIPPER: An Interactive Backstripping, Decompaction and Geohistory Program

Raymond F. Gildner
Department of Geological Sciences and the Institute for the Study of the Continents
(INSTOC), Cornell University, Ithaca, NY 14853-1504 USA

Abstract

Backstripping and decompaction are two fundamental techniques of basin analysis. This paper presents an algorithm for the two processes based upon the work of Sclater and Christie (1980). Sclater and Christie's technique does not allow for exact solutions: Newtonian Iteration is used to rapidly converge on a numerical solution with a predetermined accuracy.

The code for a top-down, structured program in Microsoft Basic v. 3.0 for the Apple Macintosh is included which uses Macintosh's user-friendly interface. The portion of the program containing the algorithm that solves Sclater and Christie's equations is contained in the subroutine "Guts" for ease of translation into other languages.

Introduction

One goal of basin analysis is to reconstruct the subsidence history of a sedimentary basin. A technique was devised by Sclater and Christie (1980) for this, based upon principles of isostasy and using an empirically devised model for the compaction of sediments through time. It is not my purpose to defend the model, and the user should be familiar with the model and some caveats (see also Lerche, this volume).

From examination of empirical data from the North Sea, Sclater and Christie determined that the porosity of sediments decreased exponentially with depth as:

$$\phi = \phi_o\, e^{-cz} \tag{1}$$

where ϕ_o is the surface or initial porosity, c is the compaction coefficient (both ϕ_o and c are presumed constants for the material), and ϕ is the porosity at depth z. Using the density of sediment grains, and knowing the initial porosity, ϕ_o, it is then possible to calculate the mass on the stratigraphic column and thus the isostatic subsidence required to support this mass on the crust. Assuming a constant depth before and after deposition, any difference between the isostatic subsidence and the thickness of the stratigraphic column must represent the action of some process to increase the accommodation of the basin. The source of the accommodation could be tectonic subsidence, sea level variations or some other process. The identification of the source is the responsibility of the user. The Sclater and Cristie model solves for only the isostatic subsidence, and this simplification ignores the two- and three-dimensional effects of flexure.

The Sclater and Cristie model treats subsurface depth as the sole control on porosity. It should be recognized that it is not the depth *per se* that causes sediments to compact, but the pressure of the overburden. Such pressures would be determined by both the thickness of the overburden and its density. As the parameters for the compaction curves are not calibrated to pressure, but to depth, this criticism is largely compensated for by the empirical data. Since the mean density of most sedimentary rocks occur in a limited range (from about 2 to 3 g/cm³) this is as would be expected and is evidenced by the surprisingly good fit of the points from their data to the empirical curves. Sclater and Christie's paper should be consulted for details about the methods and data set used to construct their model.

ALGORITHMS

Sclater and Christie developed their model to the point of an irreducible equation. Their equation was converted to a differential equation and the numerical solution is sought iteratively using Newton's method. There are many methods for finding numerical solutions of differential equations which converge more rapidly than Newton's method (e.g., predictor-corrector and Runga-Kutta methods). However, in most cases, the accuracy of the solution these methods produce is unknown, or can only be determined after the fact. Newton's method was chosen for its simplicity and for the advantage of being able to require it to continue the iteration procedure until a desired accuracy has been achieved. In the listing provided in the appendix, the accuracy is rather arbitrarily set to one part in 10^{-5} m in the data statement in the subroutine "Guts."

In calculating isostatic subsidence, it is necessary to estimate the densities of the sediment, the supportive aesthenosphere, and the interstitial pore fluids. The density of the upper aesthenosphere is taken to be 3.33 g/cm³, the density of the pore fluids is taken to be 1.03 g/cm³, and both are written into the program. The density of the sediment refers to the density of the grains. For example, a quartz sandstone would use the density of quartz (2.65 g/cm³), that of a limestone would use the density of calcite (2.71 g/cm³). In theory, any density can be used. In practice, the program limits the user to somewhat more reasonable densities (between 0 and 5 g/cm³).

The estimated density of a layer is determined by use of the compaction information the program has determined through the iterative process. First, the mean porosity of the layer is determined as an integrated average. The mean density of the layer, ρ_{layer}, is then

$$\rho_{layer} = \rho_w \phi + \rho_g (1 - \phi) \tag{2}$$

where ρ_w, ρ_g are the densities of the pore water (1.03 g/cm³) and sediment grains, respectively, and ρ_{layer} is the mean density of the layer, determined as described above.

The calculations for the backstripping and decompaction are contained entirely in the subroutine "Guts." The remainder of the program, indeed the majority of the code, are "niceties" which use the Macintosh's Toolbox routines to make the program very user friendly. Two of these routines, the Button and the Edit Field, are used extensively and perhaps deserve explanation for those unfamiliar with the Macintosh operating system.

A button is a small figure on the screen, either a square or a circle (Stripper uses the square "check box" only). When the pointer is moved, using the mouse, into the interior of the button and the mouse's (physical) button is clicked, the computer knows to take a certain action. In most instances in Stripper, the box will now contain an "x," indicating that the action was chosen. Should the user decide that the action pertaining to the button was not wanted, a second point and clicking in the box will remove the "x" and the computer will not perform the indicated action.

An edit field is a device of the interface to facilitate the entry of information. It is a rectangle on the screen, often with associated text informing the user of the type or form of data expected. The mouse can be used to move between edit fields freely. However, before data can be entered, the field must be active, as will be indicated by the location of the blinking cursor. Any entry in an edit field can be changed before moving onto the next screen. The use of edit fields and buttons will become clear as one uses the program.

The program follows the recommendations of Working Group 6 of the QDS workshop in that it is organized into several modules. Those which occur in the listing before "Guts" are input modules, those which follow "Guts" are output and graphic modules.

INPUT

The user is presented with a number of choices in the program which are presented here in the order encountered in execution. The first screen presents the user who has not read this text with some background information, including reference to the original publication by Sclater and Christie and some information on the origin of the program.

The second screen prompts the user, via buttons, with the choice of either entering data via the keyboard or referring the program to a prepared data file. Although any word processor may be used to prepare a data file, the program has the option to save data entered by the keyboard into a file in the appropriate format. he program has some features which make the second process (prepared data file) easier and faster. Should the program access the prepared data file, there are no further input options: all information needed for the successful completion of the program are included in the file. The remainder of this section refers only to use of the option to input data from the keyboard.

The next screen prompts the user for information regarding the amount and form of the input data. First, the program requires the user to enter into an edit field the number of layers which are to be processed. This is done to encounter memory limitations as soon as possible. The programs declares all variables at the beginning, so if there is not enough memory to complete all the calculations necessary the program will "bomb" at the very beginning. Also on the third screen are buttons for the options for the form of the data. The data can be entered as depths to the base or thicknesses of the layers and can be entered in feet, meters or kilometers. The default form is thicknesses in kilometers.

The next screen is for the input of information for each layer. In the background, a second screen will print the information you have just entered, as the computer understood it. In the foreground, the screen prompts the user for each layer. At the top is the number of the layer to be entered. On the left are four edit fields. The uppermost is for the thickness of (or depth to) the layer. It will treat the data as necessary for its type, as you have instructed it in the previous screen. The lower three edit fields allow the input of the density, initial porosity, ϕ_0, and compaction coefficient, c. Default values (from the original paper by Sclater and Christie, 1980) for limestone, sandstone and shale can be entered by clicking the appropriate box on the right. The last button, labelled "Other," allows the user to enter unique values. One of these buttons must be activated before the computer will accept the data.

After the data for the last layer have been entered, the program will proceed with the decompaction and backstripping. The program calculates each stage, from the present backward in time, one layer at a time, from the surface to the base of the section. It will mark its progress on the screen by printing the stage which it has begun to calculate.

OUTPUT

Once all calculations are complete, the program will present the user with a list of output options. The user chooses which option is desired via buttons, then chooses the appropriate button at the the bottom of the screen. The output options are:

1. View the output table on the screen. The program will list on the screen the depths to the base of each layer, the isostatic subsidence, and the unexplained subsidence ("nonisostatic subsidence") for every stage in the decompaction, from the youngest to the oldest.
2. Print the output table. The programs sends the above data to the printer.
3. View the section sketches on the screen. The program creates a simple plot of the depth to the base of each section in kilometers (y-axis) through all the stages (x-axis). Since there are no times attributed to each layer, the x-axis divided into equal increments (Fig. 1).
4. Incorporate times and water depths into sketch, thus producing a geohistory diagram. The program asks for the input of the times and water depths associated with the base of each layer, then plots the depth to the base of each section in kilometers (y-axis) through all the associated times (x-axis). See Figure 1 for an example.

    Plots may be printed after they have been inspected on the screen. The user can print the screen display using a key sequence from the keyboard: simultaneously press Command-Shift-4. This information is displayed on the screen before the data are plotted. Alternatively, the program prints the picture on the Clipboard, allowing the user to paste the image into the Scrapbook or directly into a word processing or graphics program.
5. Print the input data.
6. Write input data to file. The program writes the input data of a text document in the form required for the "Data from file" input option. The first line is the number of layers in the data set; the remaining lines are the depth, in kilometers, the density, initial porosity and the compaction coefficient for each layer (each value separated by a comma).

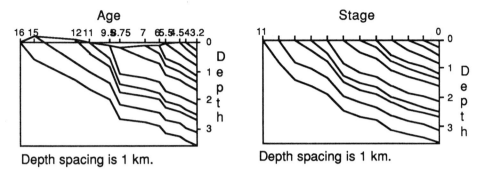

Figure 1. Sample output from the Geohistory(left) and the Sketch (right) options for output.

PROGRAM SPEED AND ACCURACY

The program performs all calculations in single precision (8 significant digits). The results from the program were compared with results of similar calculations using several different algorithms performed on a VAX 11/750 mainframe computer. It was found that the calculations agreed to all significant digits. The program has been used, in its various early versions, in both the classroom and in research. In all cases, it has performed without error.

The speed of the program in Microsoft Basic (binary version) v. 3.0 is quite fast. However, the speed increase with a compiled version is quite impressive and compilation is recommended to the user. The program listing (Appendix) will compile without further modification using Microsoft Basic Compiler v. 1.0.

ACKNOWLEDGMENTS

Dr. Teresa Jordan's Basin Analysis course was the initial impetus behind the progam. Spafford Ackerly provided very helpful ideas in the earliest stages, and Tim Cross expressed the interest which was the final impetus to work out the remaining bugs.

REFERENCE CITED

Sclater, J.G., and Christie, P.A.F., 1980, Continental stretching: An explanation of the post-mid-Cretaceous subsidence of the central North Sea basin: Journal of Geophysical Research, v. 85, p. 3711-3739.

APPENDIX: PROGRAM LISTING

[Words in all CAPITAL letters are reserved words in Microsoft Basic v. 3.0. Single words followed by a colon (e.g., Stripper:) are subroutine titles.]

```
Stripper:
'This  program  backstrips  a  given  section  as  per  Sclater  &  Christie,  1980,
JGR  85:  3711-3739.
'R.F.  Gildner,  1987/1988
outtext$  =  "No  Output"
```

```
nlayers = 0
DEF  FNf(a,b,c)  =   EXP(-c(a)*z(b,c))

WINDOW  CLOSE  1
GOSUB   Screen1
GOSUB   DataIn
GOSUB   Guts
GOSUB   OutOpts
FOR  i=1 TO 4
       WINDOW CLOSE i
NEXT  i
GOSUB   Finished

Screen1:
     GOSUB wind1
       PRINT "This program calculates the subsidence history for a column as
per the"
       PRINT "        method of Sclater and Christie [1980, JGR 85:3711-3739]."
       PRINT "       If you are not familiar with that work, I recommend you read"
       PRINT "       it before proceeding."
       PRINT "If you are familiar with the paper, the program should be
largely"
       PRINT "       self explanatory.    "
       PRINT "Default values for initial porosity, density and compaction
coefficient"
       PRINT "        for sandstone, limestone and shale are from said paper."
       PRINT :PRINT
       PRINT "         This program is written in MS Basic(b) v3.0, and com-
piled"
       PRINT "                              using MS Basic Compiler v1.0."
       PRINT "         1987, RFG.  Freeware:  copy at will.  <RETURN> to con-
tinue."
       INPUT "",a
       WINDOW CLOSE 1
       RETURN

DataIn:
     GOSUB Wind2
       BUTTON 1, 1, "New data", (10,30)-(140,70),2
       BUTTON 2, 1, "Data from file", (160,30)-(290,70),2
       WHILE DIALOG(0)<>1
     WEND
       ButtonID = DIALOG(1)
       WINDOW CLOSE 2
       ON ButtonID GOSUB NewIn, ReadIn
       IF LEN(a$)=0 AND ButtonID=2 GOTO DataIn
       BUTTON CLOSE 0
       RETURN
ReadIn:
        a$=FILES$(1,"TEXT")
       IF LEN(a$)=0 THEN RETURN
       OPEN a$ FOR INPUT AS #1
       INPUT#1, nlayer
       n = nlayer+1
       DIM z(n,n), c(n), f0(n), den(n), age(n)
       DIM fbar(n,n), tect(n), lith$(n),cage(n)
       dimmed=1
       n = n-1
       z(1,1) = 0
       GOSUB wind1
       FOR layer=1 TO n
               INPUT #1, z(1,layer+1), c(layer),f0(layer),den(layer)
               PRINT z(1,layer+1),  c(layer),f0(layer),den(layer)
```

```
        NEXT layer
        CLOSE #1
        RETURN
NewIn:
    GOSUB Wind3:
        CALL MOVETO(100,33):PRINT "Number of Layers"
        EDIT FIELD 1," ",(220,20)-(260,35)
        BUTTON 1, 2, "Thickness",(100,70) - (200,100),2
        BUTTON 2, 1, "Depth", (300,70) - (400,100),2
        BUTTON 3, 1, "Feet",(100,120)-(200,150),2
        BUTTON 4, 1, "Meters",(200,120)-(300,150),2
        BUTTON 5, 2, "Kilometers",(300,120)-(400,150),2
        BUTTON 6, 1, "OK",(50,170)-(150,200)
        BUTTON 7, 1, "Quit",(300,170)-(400,200)
      k1=1
       choice1$ = "Thickness"
      choice2$ = "km"
      buttonidle1:
           WHILE DIALOG(0)<>1
         WEND
           nlayers = VAL(EDIT$(1))
           ButtonID = DIALOG(1)
             ON ButtonID GOSUB Thick, Deep, Feet, Meter, Km,Accept,Finished
         RETURN
    Accept:
      BUTTON CLOSE 0
      IF nlayers<2 THEN
         BEEP
          GOTO NewIn
      END IF
      n = nlayers +1
       DIM z(n,n), c(n), f0(n), den(n), age(n)
        DIM fbar(n,n), tect(n), lith$(n),cage(n)
      dimmed=1
      n = n-1
      z(1,1) = 0
      WINDOW CLOSE 3
      GOSUB wind1
       PRINT "c = compaction coefficient,   f0 = initial porosity"
      GOSUB Wind4
      FOR layer =1 TO n
             BUTTON 1,1,"Limestone",(250,55)-(385,73),2
             BUTTON 2,1,"Sandstone",(250,78)-(385,97),2
             BUTTON 3,1,"Shale",(250,102)-(385,121),2
             BUTTON 4, 1,"Other",(250,126)-(385,145),2
             BUTTON 5,1,"OK",(205,155)-(395,175)
             BUTTON 6,1,"Quit",(5,155)-(195,175)
           GOSUB MakeBox
            EDIT FIELD 2,"",(30,98)-(85,113)
            EDIT FIELD 3,"",(150,98)-(195,113)
            EDIT FIELD 4,"",(100,130)-(145,145)
            EDIT FIELD 1,"",(100,65)-(195,80)
          chose=0
buttonidle2:
             WHILE DIALOG(0)<>1:WEND
             ButtonID = DIALOG(1)
            t$ = EDIT$(1)
              ON ButtonID GOSUB LS,SS,Sh,NewRx
             IF ButtonID=5 THEN
                  IF chose=0 THEN GOTO buttonidle2
             ELSEIF ButtonID=6 THEN
                GOTO Finished
             ELSE
```

```
                    GOTO buttonidle2
            END IF
              GOSUB BadChecks
            CLS
                 PRINT USING "##.###";z(1,layer+1):PRINT c(layer),  f0(layer),
den(layer)
              WINDOW OUTPUT 4
        NEXT layer
        WINDOW CLOSE 4
        RETURN
Thick:
        choice1$ = "Thickness"
        BUTTON 1,2
        BUTTON 2,1
         GOTO buttonidle1
Deep:
        choice1$ = "Depth"
        BUTTON 1,1
        BUTTON 2,2
         GOTO buttonidle1
Feet:
        choice2$ = "Feet"
        BUTTON 3,2
        BUTTON 4,1
        BUTTON 5,1
        k1 = .0003048
         GOTO buttonidle1
Meter:
        choice2$ = "Meters"
        BUTTON 3,1
        BUTTON 4,2
        BUTTON 5,1
        k1 = .001
         GOTO buttonidle1
Km:
        choice2$ = "Km"
        BUTTON 3,1
        BUTTON 4,1
        BUTTON 5,2
        k1 = 1
         GOTO buttonidle1
MakeBox:
        LINE (0,50)-(400,50)
         LINE (200,50)-(200,150)
         LINE (0,150)-(400,150)
        CALL TEXTFONT(7)
        CALL TEXTSIZE(36)
        CALL MOVETO(140,35)
        label$ = "Layer" + STR$(layer)
        PRINT label$
        CALL TEXTFONT(0)
        CALL TEXTSIZE(12)
        CALL MOVETO(10,70)
        PRINT choice1$
        PRINT "    in ";choice2$
        CALL MOVETO(15,110)
          PRINT "c";PTAB(95);"density"
          PRINT:PRINT PTAB(80);"f0"
        RETURN
LS:
        chose=1
        BUTTON 1,2
        BUTTON 2,1
```

```
      BUTTON 3,1
      BUTTON 4,1
        EDIT  FIELD  2,"0.71",(30,98)-(85,113)
        EDIT  FIELD  3,"2.71",(150,98)-(195,113)
        EDIT  FIELD  4,"0.7",(100,130)-(145,145)
        EDIT FIELD  1,t$,(100,65)-(195,80)
      lith$(layer) = "Ls"
      c(layer) = .71
      f0(layer) = .7
      den(layer) = 2.71
      RETURN
SS:
      chose=1
      BUTTON 1,1
      BUTTON 2,2
      BUTTON 3,1
      BUTTON 4,1
        EDIT  FIELD  2,"0.27",(30,98)-(85,113)
        EDIT  FIELD  3,"2.65",(150,98)-(195,113)
        EDIT  FIELD  4,"0.49",(100,130)-(145,145)
        EDIT  FIELD  1,t$,(100,65)-(195,80)
      lith$(layer) = "SS"
      c(layer) = .27
      f0(layer) = .49
      den(layer) = 2.65
      RETURN
Sh:
      chose=1
      BUTTON 1,1
      BUTTON 2,1
      BUTTON 3,2
      BUTTON 4,1
        EDIT  FIELD  2,"0.51",(30,98)-(85,113)
        EDIT  FIELD  3,"2.72",(150,98)-(195,113)
        EDIT  FIELD  4,"0.63",(100,130)-(145,145)
        EDIT  FIELD  1,t$,(100,65)-(195,80)
      lith$(layer) = "Sh"
      c(layer) = .51
      f0(layer) = .63
      den(layer) = 2.72
      RETURN
NewRx:
      chose=1
      BUTTON 1,1
      BUTTON 2,1
      BUTTON 3,1
      BUTTON 4,2
      lith$(layer) = "Other"
      c(layer) = VAL(EDIT$(2))
      den(layer) = VAL(EDIT$(3))
      f0(layer) = VAL(EDIT$(4))
        EDIT  FIELD  1,t$,(100,65)-(195,80)
      RETURN
BadChecks:
        IF  c(layer)<=0  OR  c(layer)>=1  THEN
              bad$ = "c"
              GOTO baddata
            ELSEIF  den(layer)<=0  OR  den(layer)>=5  THEN
              bad$ = "density"
              GOTO baddata
            ELSEIF  f0(layer)<=0  OR  f0(layer)>=1  THEN
              bad$ = "f0"
              GOTO baddata
```

```
      END IF
       IF  choice1$="Thickness"  THEN
                  z(1,layer+1) = z(1,layer) + k1*(VAL(EDIT$(1)))
           ELSE
                  z(1,layer+1) = k1*(VAL(EDIT$(1)))
      END IF
        IF  z(1,layer+1)<=z(1,layer)  THEN
          bad$ = "depth"
          GOSUB baddata
      END IF
      WINDOW OUTPUT 1
      RETURN
baddata:
      WINDOW 1
       PRINT TAB(20);"data out of range: " + bad$
       WINDOW 4
       GOSUB MakeBox
       GOTO buttonidle2

Guts:
      WINDOW 1
      FOR stage = 2 TO n
          PRINT stage
          FOR layer = 1 TO n
              z(stage,layer) = 0
          NEXT layer
          FOR layer = stage TO n
              K = f0(layer)/c(layer)
                 D1 = z(stage,layer) + z(stage-1,layer+1) - z(stage-1,layer)
                  D2 = - K*(FNf(layer,stage-1,layer) - FNf(layer,stage-1,layer+1)
    FNf(layer,stage,layer))
              D = D1 + D2
                 R = z(stage,layer+1) + K*FNf(layer,stage,layer+1) - D
                 dR = 1 - f0(layer)* FNf(layer,stage,layer+1)
                ZLAST = z(stage,layer+1) - R/dR
                 WHILE ABS(z(stage,layer+1) - ZLAST) >= .00001
                    z(stage,layer+1) = ZLAST
                     R = z(stage,layer+1) + K*FNf(layer,stage,layer+1) - D
                    dR = 1 - f0(layer)* FNf(layer,stage,layer+1)
                    ZLAST = z(stage,layer+1) - R/dR
              WEND
          NEXT layer
      NEXT stage
      FOR stage=1 TO n
          FOR layer=1 TO n
              K = f0(layer)/c(layer)
                 X = FNf(layer,stage,layer) - FNf(layer,stage,layer+1)
                Y = z(stage,layer+1) -   z(stage,layer)
                IF Y > 0 THEN fbar(stage,layer) = K*X/Y
          NEXT layer
      NEXT stage
      FOR stage = 1 TO n
          FOR layer = stage TO n
              rho = (fbar(stage,layer)*(1.03) + (1-
fbar(stage,layer))*den(layer))   *   (z(stage,layer+1)-z(stage,layer))
z(stage,   n+1)
              sum = sum + rho
          NEXT layer
              tect(stage) = z(stage,n+1)*(3.33 - sum)/(3.33)
          sum = 0
      NEXT stage
      RETURN
```

```
OutOpts:
      ButtonID = 0
      GOSUB wind1
      FOR i=1 TO 6
          ans(i)=0
      NEXT i
        BUTTON  1,1,"View  the  output  table  on  screen",(10,10)-(480,40),2
        BUTTON  2,1,"Print  the  output  table",(10,45)-(480,75),2
        BUTTON  3,1,"View  the  section  sketches  on  screen",(10,80)-(480,110),2
        BUTTON  4,1,"Incorporate  times  or  water  depths  into  sketch  (Geohistory
diagram)",(10,115)-(480,145),2
        BUTTON  5,1,"Print  the  input  data",(10,150)-(480,180),2
        BUTTON  6,1,"Write  input  data  to  file",(10,185)-(480,205),2
        BUTTON  7,1,"OK",(20,210)-(235,240)
        BUTTON  8,1,outtext$,(255,210)-(470,240)
      outtext$="Quit"
ButtonIdleOut:
      WHILE DIALOG(0)<>1
      WEND
      ButtonID = DIALOG(1)
      IF ButtonID<=6 THEN
          GOSUB Switch
          GOTO ButtonIdleOut
      ELSEIF ButtonID=8 THEN
          WINDOW CLOSE 4
          RETURN
      ELSE
          GOSUB Exec
      END IF
      GOTO OutOpts
      RETURN
Switch:
      i = ButtonID
      IF BUTTON(i)=1 THEN
          ans(i)=1
          BUTTON i,2
      ELSE
          ans(i)=0
          BUTTON i,1
      END IF
      RETURN
Exec:
      WINDOW CLOSE 1
      IF ans(2)=1 OR ans(5)=1 THEN
          LPRINT DATE$,"Stripper",TIME$
          LPRINT
      END IF
      IF ans(6)=1 THEN GOSUB FileOut
      IF ans(5)=1 THEN GOSUB DataPrint
      IF ans(1)=1 THEN GOSUB OutScreen
      IF ans(2)=1 THEN GOSUB OutPrint
      IF ans(3)=1 THEN
          GOSUB Instruct
          GOSUB SketchScreen
      END IF
      IF ans(4)=1 THEN
          GOSUB Instruct
          GOSUB SketchGeoHist
      END IF
      test=0
      FOR i=1 TO 6
          test=test+ans(i)
      NEXT i
```

```
              IF test=0 GOTO Finished
        RETURN
FileOut:
        a$=FILES$(0)
        IF LEN(a$)=0 THEN RETURN
        OPEN a$ FOR OUTPUT AS #2
        WRITE#2, n
        FOR layer=1 TO n
                WRITE#2, z(1,layer+1), c(layer),f0(layer),den(layer)
        NEXT layer
        CLOSE #2
        WINDOW CLOSE 2
        RETURN
DataPrint:
        GOSUB wind1
        CALL TEXTFONT(0)
        CLS
        PRINT "For some unkown reason, the Imagewriter needs to be reset if"
        PRINT "    you have printed out any of the figures using the"
        PRINT "       <";CHR$(17);"-Shift-4> method.  Please turn it on and off
now."
        PRINT "Then type <return>."
        CALL TEXTFONT(1)
        INPUT "",a
        WIDTH LPRINT ,8
        LPRINT TAB(41);"-c*z"
        LPRINT   "INPUT";TAB(37);"f0*e"
        LPRINT   "","Lith-"," Depth to","","","density"
        LPRINT "Layer","ology"," base (km)","   c","   f0","g/cc"
        FOR layer=1 TO n
            LPRINT
layer,lith$(layer),z(1,layer+1),"",c(layer),f0(layer),den(layer)
        NEXT layer
        LPRINT
        WINDOW CLOSE 1
        RETURN
OutScreen:
        GOSUB wind1
        FOR stage=1 TO n
            PRINT :PRINT"stage "stage,
                PRINT TAB(5);"Non-isostatic subsidence="tect(stage),"Airy
isostacy="z(stage,n+1)-tect(stage)
            PRINT TAB(12); "layer","depth","thickness"
            FOR layer= stage TO n
                IF layer=stage THEN
                        thickness=z(stage,layer+1)
                ELSE
                        thickness=z(stage,layer+1) - z(stage,layer)
                END IF
                    PRINT USING "#############";INT(layer);
                PRINT SPC(4);
                    PRINT USING "###########.####";z(stage,layer+1),thickness
            NEXT layer
            INPUT "";a$
        NEXT stage
        WINDOW CLOSE 1
        RETURN
OutPrint:
        LPRINT "OUTPUT"
        FOR stage=1 TO n
            LPRINT"stage "stage,
                LPRINT TAB(15);"Non-isostatic subsidence=";
                LPRINT USING "##.####";tect(stage);
```

```
            LPRINT " km","Airy isostacy=";
                LPRINT USING "##.####";z(stage,n+1)-tect(stage);
         LPRINT " km"
            LPRINT TAB(11);"layer"," depth (km)","thickness (km)"
         FOR layer= stage TO n
             IF layer=stage THEN
                    thickness=z(stage,layer+1)
             ELSE
                    thickness=z(stage,layer+1) - z(stage,layer)
             END IF
                LPRINT USING "#############";INT(layer);
             LPRINT SPC(15);
                LPRINT USING "###.####";z(stage,layer+1);
             LPRINT SPC(6);
                LPRINT USING "###.###";thickness
         NEXT layer
         LPRINT
      NEXT stage
      RETURN
SketchScreen:
      CLS
       OPEN "Clip:Picture" FOR OUTPUT AS #1
       PICTURE ON
        LINE (50,50)-(350,200),,b
        CALL MOVETO(190,20)
       PRINT "Stage"
        CALL TEXTSIZE(9)
        CALL MOVETO(345,40)
       PRINT "0"
        CALL MOVETO(35,40)
       PRINT n
        yinc = 150/z(1,n+1)
       xinc = 300/n
       FOR stage =1 TO n
            xl = 350 - INT((stage)*xinc)
            LINE (xl,50)-(xl,45)
          xr = xl + xinc
          FOR layer = 2 TO n+1
                yr = 50 + INT(z(stage,layer)*yinc)
                yl = 50 + INT(z(stage+1,layer)*yinc)
                LINE (xl,yl)-(xr,yr)
          NEXT layer
       NEXT stage
        a = 10^(INT ( LOG( z(1,n+1) ) / LOG(10) ) )
        yinc = INT(a*yinc)
       bb = 50
       i = 0
       WHILE bb<200
            LINE (350,bb)-(355,bb)
            CALL MOVETO(357,bb+4)
          PRINT i*a
          i = i + 1
          bb = bb + yinc
       WEND
        CALL TEXTSIZE(12)
        CALL MOVETO(50,220)
       PRINT "Depth spacing is ";a;" km."
        CALL MOVETO(385,95)
       PRINT "D"
        CALL MOVETO(385,115)
       PRINT "e"
        CALL MOVETO(385,135)
       PRINT "p"
```

```
        CALL  MOVETO(385,155)
        PRINT "t"
        CALL  MOVETO(385,175)
        PRINT "h"
        CALL  HIDECURSOR
        PICTURE OFF
         image$=PICTURE$
         PRINT#1,image$
        CLOSE #1
         PICTURE (5,5)
         INPUT "",a
         CALL SHOWCURSOR
         INPUT "",a
         WINDOW CLOSE 1
         RETURN
SketchGeoHist:
        INPUT "What is the age of the top of the first layer";reage
        age(1)=reage
        agelast = age(1)
        FOR stage = 2 TO n+1
AgeIn:
            PRINT   "What is the age for the base of layer ";stage-1
            IF stage = 2 THEN PRINT "This is the youngest layer."
            IF stage = n+1 THEN PRINT "This is the oldest layer."
            INPUT age(stage)
            IF age(stage)<=agelast THEN
                PRINT "You just violated the Law of Superposition."
                PRINT "What would Nicolas Steno say?"
                PRINT "Try again.  You must enter an age greater than ";age-
last;"."
                GOTO AgeIn
            ELSE
                agelast=age(stage)
            END IF
        NEXT stage
        CLS
          INPUT "Do you wish to include initial water depths? IN km!!! (y/n)?
", a$
         a$=UCASE$(a$)
        IF a$<>"Y" THEN GOTO Draw
        FOR stage = 1 TO n+1
             PRINT  "What is the water depth at time";stage
            INPUT wd
             FOR layer=stage TO n+1
                  z(stage,layer)=z(stage,layer) + wd
            NEXT layer
        NEXT stage
Draw:
        CLS
         OPEN "Clip:Picture" FOR OUTPUT AS #1
        PICTURE ON
          LINE  (50,50)-(350,200),,b
         CALL MOVETO(190,20)
         PRINT "Age"
         yinc = 150/z(1,n+1)
          xinc = 300/(age(n+1)-age(1))
        FOR stage = 1 TO n+1
             cage(stage) = age(stage) - age(1)
        NEXT stage
        CALL TEXTSIZE(9)
        FOR stage =1 TO n
             xr = 350 - INT(cage(stage)*xinc)
             xl = 350 - INT(cage(stage+1)*xinc)
```

```
            LINE (xr,50)-(xr,45)
            CALL MOVETO(xr-5,40)
           PRINT age(stage)
           FOR layer = stage+1 TO n+1
                yr = 50 + INT(z(stage,layer)*yinc)
                yl = 50 + INT(z(stage+1,layer)*yinc)
               LINE (xl,yl)-(xr,yr)
          NEXT layer
      NEXT stage
      FOR stage=1 TO n
                yl = 50 + INT(z(stage+1,stage+1)*yinc)
                xl = 350 - INT(cage(stage+1)*xinc)
                yr = 50 + INT(z(stage,stage)*yinc)
                xr = 350 - INT(cage(stage)*xinc)
               LINE (xr,yr)-(xl,yl)
      NEXT stage
      a = 10^(  INT ( LOG(z(1,n+1)) / LOG(10) )  )
      CALL TEXTSIZE(9)
      yinc = INT(a*yinc)
     bb = 50
     i = 0
      WHILE bb<=200
          LINE (350,bb)-(355,bb)
          CALL MOVETO(357,bb+4)
        PRINT i*a
        i = i + 1
        bb = bb + yinc
     WEND
      CALL TEXTSIZE(12)
      CALL MOVETO(50,220)
      PRINT "Depth spacing is ";a;" km."
      CALL MOVETO(385,95)
     PRINT "D"
      CALL MOVETO(385,115)
     PRINT "e"
      CALL MOVETO(385,135)
     PRINT "p"
      CALL MOVETO(385,155)
     PRINT "t"
      CALL MOVETO(385,175)
     PRINT "h"
      LINE (50,50)-(50,45)
      CALL TEXTSIZE(9)
      CALL MOVETO(45,40)
      PRINT age(n+1)
      CALL TEXTSIZE(12)
      CALL HIDECURSOR
      PICTURE OFF
       image$=PICTURE$
       PRINT#1,image$
      CLOSE #1
      PICTURE (5,5)
      INPUT "",a
      CALL SHOWCURSOR
      INPUT "",a
      WINDOW CLOSE 1
      RETURN
Instruct:
      GOSUB wind1
      CALL TEXTFONT(0)
      PRINT "In order to print this now:"
      PRINT "     0.)  Make sure the Imagewriter is on!"
      PRINT "     1.)  Type <Return> to begin"
```

```
      PRINT "        2.)  Type <";CHR$(17);"-Shift-4> after the image has ap-
peared."
     PRINT "                      The image will print.
     PRINT "        3.)  Type <Return>."
     PRINT "You can also save the output to the clipboard and paste it into"
     PRINT "     your drawing program, or directly into a document.  To do
so:"
     PRINT "        4.) After typing <Return>, the cursor will reappear.  Access
the
     PRINT "          Scrapbook under the apple menu."
     PRINT "        5.)  Select Paste under the Edit menu"
     PRINT "        6.)  Press <Return> to continue."
     CALL TEXTFONT(1)
     INPUT "",a
     RETURN
Finished:
     FOR i=1 TO 5
          WINDOW CLOSE i
     NEXT i
       WINDOW  5,,(10,30)-(150,150),3
     BEEP:BEEP
      CALL TEXTFONT(7)
      CALL TEXTSIZE(48)
      CALL MOVETO (5,45)
      PRINT "Done."
       BUTTON 1, 1, "Do it again", (15,65)-(140,85),2
       BUTTON 2, 1, "Quit", (15,90)-(135,110),2
       WHILE DIALOG(0)<>1
     WEND
      IF DIALOG(1)=1 THEN
          IF dimmed=1 THEN
                    ERASE z,c,f0,den,age,fbar,tect,lith$,cage
               dimmed=0
          END IF
          BUTTON CLOSE 0
          WINDOW CLOSE 5
          GOTO Stripper
     ELSE
          BUTTON CLOSE 0
          WINDOW CLOSE 5
     END IF
     END
     RETURN
wind1:
      WINDOW  1,,(3,22)-(509,290),3
      RETURN
Wind2:
      WINDOW  2,  "",(53,40)-(353,140),3
      RETURN
Wind3:
      WINDOW  3,"",(3,40)-(490,253),3
      RETURN
Wind4:
      WINDOW  4,,  (50,100)-(450,280),3
      RETURN
```

11

Maturity Modeling of Sedimentary Basins: Approaches, Limitations and Prediction of Future Developments

Douglas W. Waples
Department of Geology and Geological Engineering, Colorado School of Mines, Golden, CO 80401, and Geochemical Consultant, 5400 East Sixth Ave., Denver, CO 80220 USA

Abstract

Application of maturity modeling to predict organic maturity and hydrocarbon generation began about twenty years ago, but has become common within the exploration community only in the last few years. Lopatin-type models have been more popular than kinetic models, but the latter, which are inherently superior, will be preferred in the future if ways can be found to calibrate them with measured maturity data.

Although current applications usually utilize geothermal gradients derived from logging temperatures to create subsurface temperature profiles, use of measured or estimated basement heat flows and thermal conductivities of rocks is likely to become increasingly popular in coming years. More research is required to establish norms for both heat flows and conductivities.

Debate has begun on the issue of how to model maturation in faulted sections, but more research is needed to resolve the geological questions. Application of maturity models to the cracking of hydrocarbons has not yet achieved wide popularity, but may do so in the future.

Corrections for compaction effects have not been routine in the past, and are probably of relatively minor significance compared to other uncertainties in maturity modeling. However, compaction will become an increasingly important consideration as maturity models are integrated with geodynamic and backstripping models designed to reconstruct structure, facies, faulting and other geologic phenomena. These models will be two or three dimensional, and will be run on geological/geophysical workstations currently under development. It is therefore likely that maturity modeling gradually will lose its separate identity over the next few years.

Introduction

Although research on the use of maturity modeling to determine coal rank (Karweil, 1956; Lopatin, 1971) and the progress of oil generation (Tissot, 1969) began long ago, petroleum explorationists were slow in adopting this technology. Until the recent popularization of Lopatin's method to describe oil generation (Waples, 1980), few exploration geologists had even heard of maturity modeling. Today, however, modeling of hydrocarbon generation has become common within oil companies of all sizes and the basic technology of maturity modeling is fairly well established. Numerous software programs, both proprietary and commercial, permit geologists themselves to perform maturity calculations on personal computers, minicomputers, or mainframes. Some of these programs are quite easy to use.

More sophisticated variations of the basic maturity-modeling package are now being developed; many of these applications will be unfamiliar even to those geologists who have experience with maturity modeling. Examples of less-common applications are: modeling the cracking of oil to gas; use of heat flow and rock conductivities instead of geothermal gradients; distinction between slow and fast generation; consideration of the effects of igneous activity; and correction for compaction effects. These options are likely to be used more commonly in the future.

Most maturity modeling to date has been applied to individual wells. A few groups have developed two- or even three-dimensional models, in which several wells or control points can be modeled simultaneously. These maturity models eventually will be incorporated into comprehensive geodynamic and backstripping models designed to reveal many facets of the geologic history of an area.

Present Status of Maturity Modeling

Lopatin-type versus Kinetic Models

Two fundamentally different types of maturity models are in use today: those based on Lopatin's (1971) method, and the kinetic models based on Tissot's (1969) work, especially as elaborated by Tissot and Espitalie (1975). Lopatin's method has been much more popular among exploration geologists for several reasons: it was the first model to become accessible to large numbers of exploration geologists through Waples' (1980) commentary in the AAPG Bulletin; it requires minimal computational power and thus lends itself to hand calculations (an important consideration in the early 1980s); data input and output are in familiar forms to geologists; and published calibrations between calculated maturities (called TTI values) and measured maturity levels are available. As a result, TTI values have become the *lingua franca* for most published maturity modeling studies, including some which actually used kinetic models to perform the calculations.

Kinetic models, in contrast, have not yet been accepted as widely by the geologic community, although they have been embraced enthusiastically by researchers as a topic worthy of extensive investigation. Reasons for their slow acceptance by exploration geologists include publication of most of the early literature in French, the necessity for computers to carry out calculations, their greater degree of mathematical complexity, lack of popularization of kinetic models in literature read by exploration geologists, lack of attention to making data input and output comfortable for geologists, and failure of kinetic models to provide a convenient calibration between measured and calculated maturities.

Lopatin-type and kinetic models have been reviewed and compared recently by Wood (1988). There is little question that kinetic models are intrinsically superior to Lopatin's model, which severely simplifies the very complex kinetics of hydrocarbon generation (Tissot et al., 1987). Furthermore, the kinetic models allow one to consider an infinite number of mixtures of kerogen types, whereas Lopatin-type models use only a single type. Many companies that have dedicated research efforts to maturity modeling have focused on developing their own proprietary kinetic models, but little information on these proprietary models is available in the literature. The kinetic parameters published by Tissot and Espitalie (1975) are now mainly of historical interest; more recent work by that group is largely proprietary (but see Ungerer et al., 1986, for some information). The publications of the group at Lawrence Livermore Laboratory (e.g., Sweeney et al., 1986; Burnham et al., 1987) probably represent some of the best data currently available to the public.

One weakness of Lopatin's method is its demonstrable inaccuracy when cooking times are very short or very long, or where cooking temperatures are unusually high or low (e.g., Katz et al., 1982; Schmoker, 1986; Wood, 1988). Many of the criticisms that have been leveled at Lopatin's method, however, seem exaggerated in importance. Given the many other poorly constrained assumptions that must be made in any maturity modeling project (such as estimation of paleotemperatures), the inherent weaknesses in Lopatin's method are not always the most serious limitation.

The availability of a direct calibration between TTI and vitrinite reflectance is an extremely valuable feature of Lopatin's method. By comparing measured and calculated maturities one can determine whether the postulated geologic/geothermal input data are acceptable, and if not, how one might begin to correct them. For example, adjustments can be made in the amount of erosional removal, present-day temperature profile, paleotemperature profiles, or any combination of these.

Kinetic models, in contrast, focus on the progress of hydrocarbon generation rather than on changes in vitrinite reflectance. Because there are no universally accepted standards for measuring hydrocarbon generation directly, kinetic models cannot be used to check the accuracy of the geologic and geothermal input data.

The numerous kinetic models currently used by oil companies and consulting groups actually differ very little among themselves. All apparently give satisfactory results, at least for the data sets used in developing and testing the models. Furthermore, since the uncertainty in input data (e.g., erosional removal or paleotemperatures) is often much greater than the intrinsic differences among the various kinetic models, it seems reasonable to conclude that all the available kinetic models will adequately reproduce real-world situations. It is questionable whether further research toward developing better kinetic models will actually improve the quality of modeling. Instead, efforts should be expended in developing better ways of determining present and past subsurface temperatures.

Subsurface Temperature Profiles

Until recently virtually all maturity models utilized geothermal gradients to specify subsurface temperatures. In the typical case, an average surface temperature (estimated or measured) and a generalized geothermal gradient obtained from one or more logging runs are used to construct the present-day subsurface temperature profile. Then, the same surface temperature usually is extrapolated back into the past, although it can be varied in response to evidence or hunches that paleoclimates were different from the modern one. Little attention has been paid in the published literature to the effects of glaciation or permafrost, or to whether the "surface temperature" should properly refer to the temperature at the sea surface or sea floor at the time of deposition.

The use of logging temperatures to determine true subsurface temperatures is fraught with perils. Most logging temperatures must be corrected for the effects of incomplete equilibration of the mud within which the thermometer is suspended. Typical corrections are on the order of 15°C, but there is no guarantee of their accuracy. Nonsystematic errors of several degrees in the corrected temperatures are likely.

A second problem with logging temperatures is that there are almost never enough measurements to establish a detailed depth-temperature profile. In most cases, where only a single measurement near total depth is available, one has few options except to assume a constant gradient from the surface. In the rare cases where logs are run at several depths during drilling, it may be possible to establish "dogleg" gradients, but coverage of the section is invariably incomplete. Many workers shy away from inferring dogleg gradients, even when the appropriate data are available, preferring instead to draw a best-fit straight line through the various corrected temperatures.

Thus with available data and logging practices it is almost impossible to determine a realistic present-day subsurface temperature profile. Even more severe is the difficulty of estimating paleogeothermal gradients. For many years maturity models utilized these inadequate estimates because no other data were available. Now, however, an alternative may be emerging: use of basement heat flow and thermal conductivity of the rocks in the section.

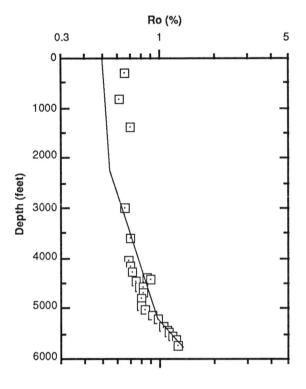

Figure 1. Vitrinite-reflectance values and a line representing calculated maturities (TTI values expressed as R_o equivalents) for a well in which the upper part of the section is of much higher thermal conductivity than the lower part. In this case maturity modeling using constant basement-heat-flow values through time and variable rock conductivities through the section are able to reproduce the nonlinearity of the measured maturity profile.

The major advantage of using the heat-flow/conductivity method is that the complex subsurface temperature profiles created in this way are more realistic than the simple ones derived from geothermal gradients. In those rare cases where equilibrated temperature profiles have been obtained over the entire well, temperatures predicted by the heat-flow/ conductivity method agree well with the measured values, even when the thermal profiles are complex. Finally, when rocks of strongly contrasting conductivities are juxtaposed, both the temperature and vitrinite-reflectance profiles are nonlinear. These phenomena also can be reproduced in maturity models using the heat-flow/conductivity method (Fig. 1).

The biggest hindrances to broader application of the heat-flow/conductivity method are the lack of data on contemporary basement heat flows and the difficulties in estimating paleoheat flows. Furthermore, such a model should also include convective heat transfer, most of which is nonvertical. Third, decay of radioactive elements provides an additional, internal source of heat that is independent of basement heat flow. Fourth, more data are needed on general rock conductivities; measurement of conductivities must become routine. Attempts have already been made to address many of these concerns (Doligez et al., 1986; Gosnold and Fischer, 1986; Luheshi and Jackson, 1986; Goblet et al., 1986; Hermanrud, 1986; Ryback, 1986; Palciauskas, 1986). Finally, the concept of reconstructing heat flow in the past could be an illusion if heat flow in the earth is actually a chaotic process, as it can be in convective media (Gleick, 1987).

Thus far there have been few published examples of maturity modeling using heat flows and rock conductivities. The current prevalence of the more convenient but much less realistic geothermal-gradient method, in my view, renders moot any serious debate about the advantages of kinetic models over the Lopatin type.

Decompaction

The earliest maturity models, including the work calibrating TTI values to vitrinite reflectance (Lopatin, 1971; Waples, 1980), ignored compaction effects. Reasons included the computational difficulties in correcting for compaction and the belief that any error in neglecting compaction was minor and systematic. In such studies contemporary stratal thicknesses were assumed equal to the original depositional thicknesses.

The net effect of considering compaction effects is to bury the sediments more rapidly during their earlier (shallower and cooler) history and more slowly during their later (deeper and hotter) history (Fig. 2). The result of this change in burial trajectory is to increase the calculated TTI values by about 30% (Fig. 3). If compaction effects are considered, the TTI-R_o calibration of Waples (1980) should not be used. Dykstra's (1987) scale (Fig. 3) is the only published alternative. Failure to recalibrate the TTI-R_o scale can in some cases lead to errors in the timing of maturation (Fig. 2).

Consideration of compaction effects is still an inexact science. One or more generalized compaction equations (e.g., Falvey and Middleton, 1981) must be selected to describe the change in thickness of a layer as a function of depth, and a compaction constant for each equation must be obtained. The model must describe the compaction behavior of various rock types, including mixed lithologies. Philosophies and methods of dealing with phenomena such as leaching, cementation, overpressuring (undercompaction), hydrocarbon expulsion, and possible decompaction during erosional periods also must be developed. It is likely that some or all of these considerations have been addressed in proprietary models, but published data are generally lacking.

Many people believe that correction for compaction in maturity models is optional rather than absolutely necessary. Because compaction curves for many sections tend to be rather similar, the error in neglecting compaction may be fairly constant from example to example. Furthermore, the lack of well-constrained compaction equations and the inaccuracy of such equations in the real world of lithologic heterogeneity, overpressuring, and cementation makes it very difficult to predict compaction precisely.

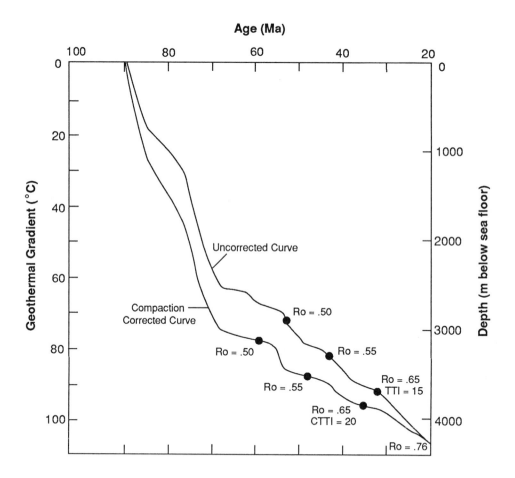

Figure 2. Comparisons of burial trajectories and timing of maturation for a rock using burial trajectories ignoring compaction (upper) and decompacting the present rock layer (lower). Reprinted from Dykstra (1987) by permission of American Association of Petroleum Geologists.

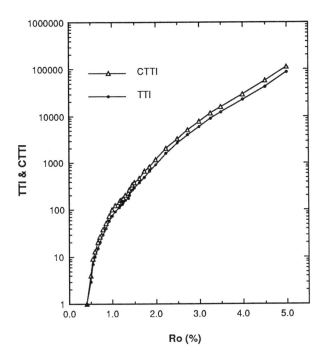

Figure 3. Comparison of TTI values ignoring compaction effects (TTI) and those calculated by decompacting the sedimentary column (CTTI) with vitrinite-reflectance values (R_o). For a given R_o value CTTI values are always higher than TTI values because the burial trajectory brings the rocks to high temperatures earlier when compaction is considered (see Fig. 2). However, the practical difference is slight. Data taken from Dykstra (1987).

Faulting

In recent years some of the problems of maturity modeling in thrust terrains have been debated (e.g., Angevine and Turcotte, 1983; Furlong and Edman, 1984; Warner et al., 1987). Much of the discussion centers upon whether or not thrusting is slow enough to permit complete or nearly complete thermal equilibrium at all times between the two thrust sheets. If thermal equilibrium is not maintained, hot rocks at the base of the hanging wall will be emplaced directly above the cooler strata of the footwall. The resulting abnormal thermal profile will cause more intensive heating (and anomalously high maturity) at the top of the footwall, as well as retardation of maturity in the hanging wall because it is insulated from basement heat flow.

Although there are no strong signs that this question will be resolved soon, many workers are interested in modeling maturity in faulted sections. At the moment, however, technical limitations also play a role: the maturity-modeling computer programs currently available generally are not able to handle faulting.

Data Output

Although data output traditionally has been different for the kinetic and the Lopatin-type models, these differences are now beginning to disappear. Lopatin-type models traditionally have shown burial-history curves, usually with isotherms superimposed (Fig. 4). Prior to the advent of computer calculations, such diagrams had to be constructed in order to carry out the hand calculations. It also has been traditional to show the development of maturity on the burial-history curves, usually in the form of oil- and gas-generation windows (Fig. 4). These diagrams give a clear picture of the data used in constructing the models, as well as the level of maturity now and at any time in the past for any horizon. It is particularly easy to discuss timing of generation using such burial-history diagrams.

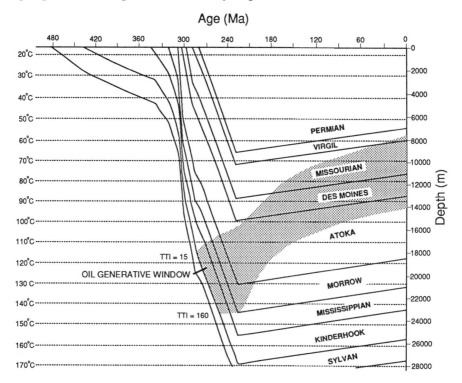

Figure 4. Family of burial-history curves superimposed on isotherms. Shaded area represents the oil-generation window defined by TTI values equivalent to 0.65% R_o (top) and 1.3% R_o (base). From Waples (1985). Reprinted by permission of I.H.R.D.C.

By contrast, output of kinetic models was originally in the form of a plot of maturation (expressed as f, the fractional realization of the kerogen's original source potential) as a function of time for individual selected strata (Fig. 5). Such a plot also displays timing of hydrocarbon generation, although in a different format than that employed by the Lopatin-type models. Until recently, burial-history curves and isotherms were seldom shown for kinetic models because the computer could perform the maturity calculations without drawing those diagrams. Without complementary burial-history curves, however, it is more difficult for geologists to visualize the geologic, thermal, and maturity history of the strata of interest. The failure of kinetic models to fully utilize burial-history curves may partly explain the reluctance of the exploration community to embrace kinetic models.

Since van Hinte (1978) published the details of geohistory modeling (taking sea level as the datum rather than the sea floor, as is done for burial-history curves), geohistory diagrams have been used occasionally in maturity modeling. In some cases (e.g., Fig. 6), the inclusion of information on water depths can be closely linked to changes in thermal regimes. In such

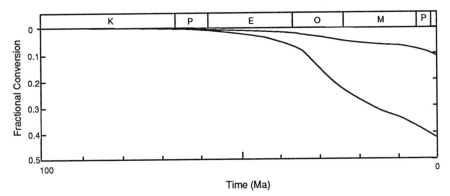

Figure 5. Fractional conversion (*f*) of kerogen's hydrocarbon-source potential to hydrocarbons as a function of time for two rock layers. Fractional conversion was calculated using the kinetic model of Tissot and Espitalie (1975). The slope of each curve is proportional to the rate of hydrocarbon generation. The top of the oil-generation window is usually taken as $f = 0.1$.

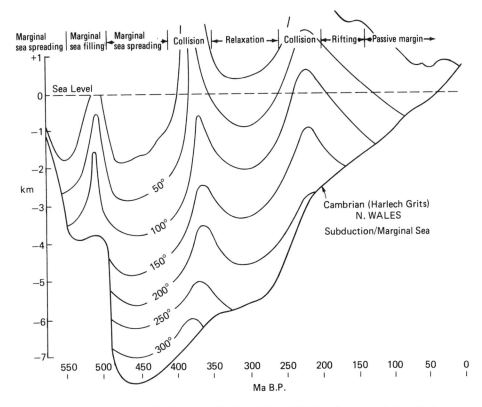

Figure 6. Geohistory diagram for Cambrian Harlech Grits of Wales, showing relationship between subsurface temperature profile and tectonic history. Sea level is taken as the datum, rather than the sea floor as for burial-history curves. From Siever and Hager (1981). Reprinted by permission of the Canadian Society of Petroleum Geologists.

cases the use of geohistory curves in maturity modeling is fully justified. In most cases, however, no close link exists between water depth and thermal history. In such situations the use of geohistory curves for maturity modeling is likely to confuse rather than clarify. The primary use of geohistory curves is for study of paleoenvironments and their relationships to lithofacies, organic facies, and basin evolution.

ANTICIPATED FUTURE DEVELOPMENTS

One consequence of the increasing computerization of maturity modeling is that kinetic models will almost certainly displace Lopatin-type models. At the present time, however, Lopatin-type models still enjoy two important advantages: tradition and the published TTI-R_o calibration that allows modelers to check the accuracy of their geologic and geothermal reconstructions. In order for kinetic models to replace the Lopatin-type models completely, they will have to offer the user a convenient way to compare measured and calculated maturities.

How this comparison will be accomplished is still uncertain. Because kinetic models are not intrinsically coupled to vitrinite reflectance, any such calibration would be artificial and nonuniversal. Perhaps the solution will be to calibrate kinetic models to transformation ratios (production indices), which are readily available from Rock-Eval pyrolysis; after all, production indices are supposed to measure hydrocarbon generation. Sweeney et al. (1986) were successful in preliminary attempts at developing such a calibration. Staining and expulsion, however, frequently make production indices difficult to interpret.

Alternatively, both Lopatin-type and kinetic models may be useful for different purposes. The geologic and geothermal history first could be verified by fitting TTI to R_o values, and then kinetic models could be used to determine the details of hydrocarbon generation. This approach would take advantage of the unique features of each type of model: Lopatin-type models' calibration to vitrinite reflectance and their consequent indifference to variations in kerogen type, and the kinetic models' ability to characterize subtle variations in organic facies and to replicate hydrocarbon-generation kinetics more precisely.

As kinetic models become more popular it is likely that some of the features of Lopatin-type modeling will be adopted. One obvious advantage of the Lopatin-type models is their emphasis on burial-history curves and displays of the evolution of the hydrocarbon-generation window (Fig. 4). Adaptation of these displays to kinetic models, an easy task, will be important to explorationists.

Although it is not yet clear which kinetic model will become most popular in the future, it is almost certain that numerous models will continue to coexist. Because of the inherent similarities of the various models, geologists will cease to care which particular one was used. The most likely candidates for wide acceptance are the Lawrence Livermore model (e.g., Sweeney et al., 1986) and the IFP-Beicip model, a refinement of the Tissot-Espitalie (1975) model, especially if the IFP's kinetic parameters are published or if sales of Beicip's computer program boom.

It is almost certain that consideration of compaction will become increasingly popular in future modeling, but not because compaction is of great importance in maturity calculations themselves. Instead, the emphasis on compaction effects will derive from a desire to incorporate maturity modeling into two- and three-dimensional backstripping packages designed to describe the geologic evolution of sedimentary basins. In such situations considerations of compaction may be much more important in analyzing fluid flow and migration pathways, structural development, and depositional facies, than in maturity calculations themselves. Much work will go into developing more reliable compaction equations, including consideration of "anomalies" like undercompaction, hydrocarbon expulsion, cementation, and leaching.

One of the most important questions for the future of maturity modeling is whether the heat-flow/conductivity method will displace the geothermal-gradient method for estimating subsurface temperatures. One problem—development of a good data base on rock conductivities—should not be difficult to solve. A more serious problem will be to obtain a large data base of modern basement heat-flow values in order to establish norms for the various types of basins and nonbasinal settings in the world today. A third problem will be to extrapolate these data to the past. It is likely that our knowledge of basement heat flows in both the present and the past will remain poor for at least a few years. Finally, the effects on nonconductive heat flow (such as igneous activity, hydrothermal activity, and water flow in aquifers) must be included. Although these problems will not be simple to solve, I hope that their existence does not prevent people from attempting to use the heat-flow/conductivity method in the interim.

There are two problems in future application of maturity modeling to faulted (especially thrust faulted) sections. The easy problem—to develop a computer model capable of handling multiply faulted sections—should be solved in 1989. The second, much more serious, problem is to estimate the magnitude of the thermal perturbation introduced into both the hanging wall and the footwall as the result of thrusting. As the references cited earlier show, the debate is far from ended. It is not clear that conclusions eventually drawn from a well-studied area like the Western Overthrust Belt in the U.S. will be applicable to other thrusted regimes, in which the rate of thrusting, thickness of thrust sheets, lithologies (and hence rock conductivities), and degree of hydrothermal activity could be quite different.

Recent research on hydrocarbon expulsion from source rocks suggests that expulsion is most efficient when generation is intense and rapid (Cooles et al., 1986). Determining rates of maturation as well as levels of maturity is therefore an important aspect of maturity modeling. Although recording of generation rates has been an integral part of kinetic models since their inception, the concept is new to Lopatin-type models. Distinguishing between periods of active generation and periods of slow generation or stagnation (Fig. 5) will become a routine part of all modeling, and much more attention will be paid to generation rates in interpreting the results of modeling.

Although application of maturity modeling to cracking of oil to gas, or of wet gas to dry gas, has always been possible in principle (Tissot and Espitalie, 1975; Waples, 1980), in practice this aspect has been lamentably neglected. Establishment of kinetic parameters for cracking has not received the research attention that oil generation has, and there are few published applications in which preservation limits have been examined by modeling (e.g., Waples, 1988). Future research undoubtedly will refine our relatively crude understanding of the kinetics of cracking processes. This increased knowledge, coupled with increased interest in establishing economic basement with computer printouts rather than the drill bit, will lead to more applications of maturity modeling to predict hydrocarbon-preservation limits.

It seems certain that much of the maturity modeling in the future will be conducted as a component part of larger, computerized, two- or three-dimensional basin models. Such models will incorporate backstripping, stratigraphic architecture, lateral and vertical facies relationships, structural development, faulting and fluid movement. The most important features of these models, for the purposes of maturity modeling, will be their ability to provide a framework for considering conductive and convective heat transfer, and relating compaction to fluid movement.

The input data for such maturity modeling will therefore have to be compatible with data required for other geological and geophysical purposes. Moreover, these comprehensive basin models will undoubtedly encourage entry of data directly from digitized logs, seismic sections, geologic cross sections, or other data sources. Maturity modeling will thus be reduced, as is proper, to describing one aspect of the total geological picture.

The computing power required for comprehensive basin models will doubtless exceed the capabilities of stand-alone personal computers. Since this type of modeling will evolve as an adjunct to digitizing, mapping, and interpretation programs and concepts already being developed for geological and geophysical workstations, the future of maturity modeling will probably reside in those same workstations.

REFERENCES CITED

Angevine, C.L., and Turcotte, D.L., 1983, Oil generation in overthrust belts: American Association of Petroleum Geologists Bulletin, v. 67, p. 235-241.

Burnham, A.K., Braun, R.L., Gregg, H.R., and Samoun, A.M., 1987, Comparison of methods for measuring kerogen pyrolysis rates and fitting kinetic parameters: American Chemical Society Preprint, 27 p.

Cooles, G.P., Mackenzie, A.S., Quigley, T.M., 1986, Calculation of petroleum masses generated and expelled from source rocks, in Leythaeuser, D., and Rullkotter, J., eds., Advances in organic geochemistry 1985: Pergamon, p. 235-245.

Doligez, B., Bessis, F., Burrus, J., Ungerer, P., and Chenet, P.Y., 1986, Integrated numerical simulation of the sedimentation, heat transfer, hydrocarbon formation, and fluid migration in a sedimentary basin: The THEMIS model, in Burrus, J., ed., Thermal models in sedimentary basins: Paris, Editions Technip, p. 173-198.

Dykstra, J., 1987, Compaction correction for burial history curves: Application of Lopatin's method for source rock determination: Geobyte, v. 2, no. 4, p. 16-23.

Falvey, D.A., and Middleton, M.F., 1981, Passive continental margins: Evidence for pre-breakup deep crustal metamorphic subsidence mechanisms: Oceanologica Acta, v. 4, p. 103-114.

Furlong, K.P., and Edman, J.D., 1984, Graphical approach to determination of hydrocarbon maturation in overthrust terrains: American Association of Petroleum Geologists Bulletin, v. 68, p. 1818-1824.

Gleick, J., 1987, Chaos: Making a new science: New York, Viking, 352 p.

Goblet, P., Ledoux, E., and de Marsily, G., 1986, Possibilities of abnormal flow in sedimentary basin: Some examples, in Burrus, J., ed., Thermal models in sedimentary basins: Paris, Editions Technip, p. 235-246.

Gosnold, W.F., and Fischer, D.W., 1986, Heat flow studies in sedimentary basins, in Burrus, J., ed., Thermal models in sedimentary basins: Paris, Editions Technip, p. 199-218.

Hermanrud, C., 1986, On the importance to the petroleum generation of heating effects from compaction derived water: An example from the northern North Sea, in Burrus, J., ed., Thermal models in sedimentary basins: Paris, Editions Technip, p. 247-270.

Hood, A., Gutjahr, C.C.M, and Heacock, R.L., 1975, Organic metamorphism and the generation of petroleum: American Association of Petroleum Geologists Bulletin, v. 59, p. 986-996.

Karweil, J., 1956, The metamorphism of coals from the perspective of physical chemistry: Zeitschrift der Deutschen Geologischen Gesellschaft, v. 107, p. 132-139 (in German).

Katz, B.J., Liro, L.M., Lacey, J.E., White, H.W., and Waples, D.W., 1982, Time and temperature in petroleum formation: Application of Lopatin's method to petroleum exploration: Discussion and Reply: American Association of Petroleum Geologists Bulletin, v. 66, p. 1150-1152.

Lopatin, N.V., 1971, Time and temperature as factors in coalification: Izvestiya Akademiya Nauk SSSR, Seriya Geologicheskaya, no. 3, p. 95-106 (in Russian).

Luheshi, M.N., and Jackson, D., 1986, Conductive and convective heat transfer in sedimentary basin, *in* Burrus, J., ed., Thermal models in sedimentary basins: Paris, Editions Technip, p. 219-134.

Palciauskas, V.V., 1986, Models for thermal conductivity and permeability in normally compacting basins, *in* Burrus, J., ed., Thermal models in sedimentary basins: Paris, Editions Technip, p. 323-338.

Ryback, L., 1986, Amount and significance of radioactive heat sources in sediments, *in* Burrus, J., ed., Thermal models in sedimentary basins: Paris, Editions Technip, p. 311-322.

Schmoker, J.W., 1986, Oil generation in the Anadarko Basin, Oklahoma and Texas: Modeling using Lopatin's method: Oklahoma Geological Survey Special Publication 86-3, 40 p.

Siever, R., and Hager, J.L., 1981, Paleogeography, tectonics, and thermal history of some Atlantic margin sediments, *in* Kerr, J.W., and Fergusson, A.J., eds., Geology of the North Atlantic Borderlands: Canadian Society of Petroleum Geologists Memoir 7, p. 95-117.

Sweeney, J.J., Burnham, A.K., Braun, R.L., 1986, A model of hydrocarbon production in the Uinta Basin, *in* Burrus, J., ed., Thermal models in sedimentary basins: Paris, Editions Technip, p. 547-561.

Tissot, B., 1969, First data on the mechanism and kinetics of the formation of petroleum in sediments: Revue de l'Institut Francais du Petrole, v. 24, p. 470-501 (in French).

Tissot, B., and Espitalie, J., 1975, Thermal evolution of organic material in sediments: Applications of a mathematical simulation: Revue de l'Institut Francais du Petrole, v. 30, p. 743-777 (in French).

Tissot, B.P., Pelet, R., and Ungerer, P., 1987, Thermal history of sedimentary basins, maturation indices, and kinetics of oil and gas generation: American Association of Petroleum Geologists Bulletin, v. 71, p. 1445-1466.

Ungerer, P., Espitalie, J., Margnif, F., and Durand, B., 1986, Use of kinetic models of organic matter evolution for the reconstruction of paleotemperatures. Application to the case of the Gironville Well (France), *in* Burrus, J., ed., Thermal models in sedimentary basins: Paris, Editions Technip, p. 531-546.

van Hinte, J.E., 1978, Geohistory analysis—Application of micropaleontology in exploration geology: American Association of Petroleum Geologists Bulletin, v. 62, p. 201-222.

Waples, D.W., 1980, Time and temperature in petroleum formation: Application of Lopatin's method to petroleum exploration: American Association of Petroleum Geologists Bulletin, v. 64, p. 916-926.

Waples, D.W., 1985, Geochemistry in Petroleum Exploration: Boston, International Human Resources Development Corp., 232 p.

Waples, D.W., 1988, Novel application of maturity modeling for hydrocarbon prospect evaluation: Houston, Proceedings, Twelfth World Petroleum Congress, v. 2, p. 3-8.

Warner, M.A., Royse, F., Edman, J.D., and Furlong, K.P., 1987, Thrust faulting and hydrocarbon generation: Discussion and reply: American Association of Petroleum Geologist Bulletin, v. 71, p. 882-896.

Wood, D.A., 1988, Relationship between thermal maturity indices calculated using Arrhenius equation and Lopatin method: American Association of Petroleum Geologists Bulletin, v. 72, p. 115-134.

12

Modeling Fluid/Rock Interactions in Sedimentary Basins

Wendy J. Harrison
Department of Geology and Geological Engineering, Colorado School of Mines, Golden,
CO 80401 USA

Abstract

Predicting fluid/rock interactions in sedimentary basins is a critical component in understanding a diverse range of problems including porosity/permeability evolution, diagenesis of sediments, the origin of certain ore deposits, and the origin, migration and attenuation of groundwater contaminants. There are three commonly used methods of predicting these interactions: empirical porosity decline curves, facies/burial history diagenetic models, and theoretical fluid/rock interaction models. Each has a different scale of relevance and effectiveness.

Empirical porosity decline curves integrate the combined effects of mechanical compaction and mineral precipitation/dissolution reactions on sediment porosity. When the time-temperature index (TTI) for thermal history is utilized as opposed to burial depth, porosity trends can be compared among basins of variable burial and thermal histories. This approach apparently provides data of adequate quality for problems such as basin paleohydrologic analysis. Diagenetic models interpreted from petrographic observations are generally only predictive at a reservoir scale. When undertaken thoroughly, facies *versus* burial controls can be unravelled and used successfully to develop predictive porosity models. Diagenetic modifications at unconformities can be predicted from stratigraphic models alone. Theoretical reaction-path models provide the most detailed predictions of fluid/rock interactions. These models need calibration with petrographic and geochemical observations to refine predictions.

Two case studies are presented to illustrate the degree to which fluid/rock interactions can be modeled successfully: a diagenetic model for dolomite formation in the Gippsland basin, Australia, and a predictive model for the origin and attenuation of leachates from processed oil shale. At present, predictive models of fluid/rock interactions in sedimentary basins are not limited by numerical code availability, rather by demonstrated applications. Successful predictions will need to integrate stratigraphic and facies geometries, and paleohydrologic and burial histories, with theoretical models of chemical reactions. Model calibration using observations will be necessary until sufficient expertise is developed to make accurate forward models of porosity and permeability evolution.

INTRODUCTION

A sediment at deposition consists of an assemblage of mineral grains, rock fragments, clay and organic matrix, biologic/skeletal material, and in the case of many carbonate and evaporitic rocks, a mixture of chemically precipitated phases. Generally, this depositional assemblage is in chemical disequilibrium. A combination of mechanical and chemical diagenetic processes cause the progressive equilibration of the initial assemblage as time, temperature and depth increase during burial in a sedimentary basin. Ultimately, a metamorphic rock is formed which, although possibly not completely equilibrated, has undergone a significant redistribution of components in response to external pressure and temperature conditions (Rumble and Spear, 1983; Ferry, 1984; Rumble et al., 1986). During burial, sedimentary rocks with depositional porosities of 40% and more (Pryor 1973; Schmoker and Halley, 1982; Baldwin and Butler, 1985) are compacted and cemented resulting in a reduction of porosity and permeability. By the time the transition to metamorphic conditions occurs, only fracturing makes a significant contribution to porosity and permeability. To predict porosity/permeability decline in the subsurface—of interest to exploration geologists and production engineers in the petroleum industry, and to groundwater hydrologists concerned with contaminant migration and the effective utilization of groundwater resources—we must be able to predict the mechanical and chemical reactions that occur in a sediment after deposition. In this paper, I review the extent to which we can predict fluid/rock interactions today and the needs we have in this field to improve our capabilities.

FLUID/ROCK INTERACTIONS IN SEDIMENTARY BASINS

Burial diagenesis in sedimentary rocks occurs through a combination of mechanical and chemical processes, documented from diverse petrographic and geochemical studies of sedimentary rocks. Mechanical compaction is most effective in early, shallow burial when, under ideal circumstances, uncemented sands with perfectly spherical grains can reach an equilibrium intergranular porosity of approximately 26% (hexagonal close packing; Housenecht, 1987). Sedimentary rocks containing ductile grains such as mica and clay matrix can lose almost all their porosity and permeability through mechanical compaction alone.

Porosity loss beyond that induced mechanically involves chemical compaction (pressure solution) and cementation. Pressure solution may be an efficient diagenetic process under lithostatic pressures and high geothermal gradients (Nagtegaal, 1978; Housenecht, 1984, 1987) and is thus restricted to deeper and/or hotter parts of sedimentary basins. Intergranular porosity lost by pressure solution is irreversible (Housenecht, 1987). Cementation can occur at any time during burial history and dissolution of grains and authigenic cements makes this form of porosity loss reversible.

Although many complex mineral reactions have been reported, the volumetrically important ones in siliciclastic sequences are those that introduce quartz and calcite cements and that dissolve feldspar, calcite and rock fragments to produce secondary, leached grain porosity (Blatt, 1979; Land et al., 1987). The source for most cement and the mechanism for removing dissolved constituents is probably fluid moving through the subsurface. Consequently, understanding the paleohydrologic evolution of a sedimentary basin is a key variable in diagenetic predictions (e.g., Siever, 1986).

Diagenetic reactions do not occur randomly. In compacting sedimentary basins there is a temperature- and depth-related distribution of diagenetic products that varies with the starting composition of the sediment. In the case of siliciclastic rocks, this starting composition can be estimated fairly accurately from stratigraphic and facies models. For example, in arkosic sediments a typical early alteration assemblage of *kaolinite+smectite+calcite* is replaced during burial first by a *quartz+kaolinite+illite+calcite* assemblage then, at greater depths, by an *albite+2M mica/illite+chlorite+ankerite* assemblage (Boles, 1978, 1982; Hoffman and Hower, 1979; Ramsayer and Boles, 1986; Land et al., 1987). In volcaniclastic rocks the succession is *cristobalite+smectite+clinoptilolite*, followed by *quartz+analcime+illite+calcite* and then by *albite+laumontite+illite+calcite* (Hoffman and Hower, 1979; Surdam and Boles, 1979; Crossey et al., 1984; Mathieson, 1984). These final assemblages are essentially those of zeolite facies metamorphism (Winkler, 1967).

The concept of diagenetic facies can be developed by analogy with metamorphic rocks. The definition of metamorphic facies utilizes pressure and temperature as the controlling variables, although fluid pressure is sometimes significant (Ferry, 1984). However, diagenetic facies need to be defined in terms of a hydrologic variable which may be more important than temperature, pressure and time. This is because sediments are more permeable than metamorphic rocks and thus more open to fluid movement. Chemical equilibrium is reached more rapidly in permeable sedimentary rocks than in less permeable strata (Boles and Ramsayer, 1987; Whitney and Northrop, 1987).

We can divide fluid/rock interactions into two groups for the purposes of this review: those that occur naturally as part of the burial process, and those that are induced by man's activity. The single most critical variable in predicting diagenetic reactions is the volume and chemistry of fluid that passes through the rock; thus it is convenient to classify all fluid/rock reactions on the basis of Galloway's (1984) classification of hydrologic regimes in a sedimentary basin.

Reactions in the Meteoric Hydrologic Regime

The depth of penetration of fresh water is often 2 km and greater in many sedimentary basins (Galloway, 1984; Meisler et al., 1985; Harrison and Bethke, 1988) and thus chemical reactions induced by activities such as disposal of brines by well injection (Kreitler, in press), tailings-pond seepage, contaminant migration from industrial land farms or coal mines, and the potential seepage from underground storage of nuclear waste (Wolery, 1980) all occur in this hydrologic regime. It is reasonable to assume that groundwaters are of relatively low salinities (unless in close proximity to evaporite or other salt deposits) and that flow rates in aquifers are typically centimeters to meters per year.

Diagenetic reactions include the leaching of quartz and chert, alteration of K-feldspar and plagioclase to kaolinite, and the dissolution of high-Mg calcite and aragonite (Fig. 1). The formation of dolomite in both carbonate and siliciclastic rocks occurs where nonsaline meteoric waters mix with more saline formation waters at the interface between meteoric and compactional regimes (Rosen and Holden, 1985; Ward and Halley, 1985; Harrison et al., 1987; Yin, 1988). Smectites are stable under diagenetic conditions (Jones, 1986) and in manmade chemical environments such as steam flooding of reservoirs for enhanced oil recovery (Hutcheon, 1984).

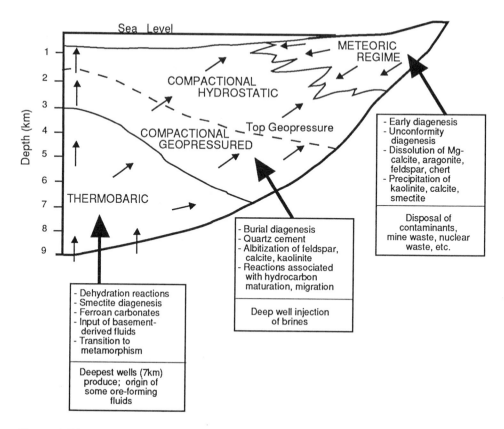

Figure 1. Diagenetic and environmentally significant fluid/rock interactions within the principal hydrologic regimes in an actively filling sedimentary basin.

Mineral solubilities are typically low at the pressures and temperatures found in the diagenetic regime (Blatt, 1979), and large volumes of fluid moving through subsurface strata are therefore implicated in accounting for observed precipitation and dissolution reactions (Bjorlykke, 1979; Land et al., 1987). Within this near-surface meteoric hydrologic regime, flow rates of meters per year (Galloway, 1984) may be high enough to account for leaching of K-feldspar and formation of kaolinite sometimes hundreds of meters from the leached grains (Bjorlykke, 1979, 1984; Land et al., 1987). Mobilization of silica has also been attributed to rapidly moving meteoric waters (Blatt, 1979).

Reactions in the Compactional Regime

In an actively subsiding and filling basin, pore waters are driven out of the basin due to the weight of the accumulating overlying sediments (Bethke, 1985). Flow is principally strata-parallel in permeable aquifers, moving from deeper to shallower parts of the basin. In shallow compacting sediments, flow may be vertical and upwards, across homogeneous lithologies (Harrison and Bethke, 1988). In this hydrologic regime, many of the classic reactions of burial diagenesis occur, such as albitization of feldspar, silica cementation, and calcite precipitation. The generation of secondary porosity through leaching of calcite and feldspar by thermally-generated CO_2 and organic acids is assumed to occur in this regime.

Unlike the meteoric regime, where waters move rapidly and have low salinities, pore waters in the compactional regime are typically more saline, may have significant organic-acid conecentrations (Morton and Land, 1987), and move at rates of a few centimeters to fractions of a millimeter per year (Galloway, 1984; Bethke, 1985; Harrison and Bethke, 1988). Additionally, there is a finite volume of such water available leading to the concept of a water volume problem in reconciling observed diagenetic patterns, particularly volumes of silica cement (Bjorlykke, 1979; Blatt, 1979; Boles and Franks, 1979) and albitization of feldspar (Boles, 1982; Morton and Land, 1987).

Some workers have explored the possibility of enhancing alumino-silicate and carbonate solubilities to circumvent the water volume problem. Schmidt and MacDonald (1979) proposed that porosity in several major oil fields on the North Slope and North Sea was secondary, produced by dissolution of calcite cement, which was enhanced by high pCO_2 produced during thermal maturation of hydrocarbons. Franks and Forrester (1984) calculated that a maximum of 10% secondary porosity in a sandstone could be produced by calcite dissolution due to thermally-generated CO_2 in Gulf Coast Cenozoic sediments. Surdam et al. (1984) proposed that complexing of Al^{3+} by oxalic and acetic acids, also associated with hydrocarbon generation, could enhance plagioclase solubility and account for kaolinite precipitation distant from feldspar dissolution sites. Solubility of quartz may also be increased by formation of organic complexes (Bennett et al., 1988).

These two mechanisms for enhancing solubilities do not seem entirely satisfactory, however. Lundegard and Land (1986) argued that CO_2 volumes from thermal decarboxylation are insufficient to account for the amounts of feldspar dissolution observed in Gulf Coast sediments. Measured organic-acid concentrations in formation waters (Carothers and Kharaka, 1978; Kharaka et al., 1986; Fisher, 1987; MacGowan and Surdam, 1988) are lower than those reported in Surdam's experiments where increased aluminum solubility was demonstrated. Acetate is the dominant dissolved organic complex in formation waters, whereas experimental studies show that salicylate and oxalate complexes are more stable than acetate complexes of aluminum and silicon (Surdam et al., 1984; Bennett et al., 1988). Kharaka et al. (1986) and Drez and Harrison (1987) used available thermodynamic data to show that aluminum-acetate and aluminum-oxalate complexes are not particularly effective in enhancing plagioclase solubility. Equilibrium constants favor formation of calcium and magnesium oxalates but aluminum oxalate was found to contribute less than 1% to total dissolved aluminum. Nevertheless, there is an apparent link between thermal decomposition of organic matter, hydrocarbon migration and some kinds of diagenetic modifications (Boles and Ramsayer, 1987; MacGowan and Surdam, 1988; Shock, 1988), and secondary porosity generated in this manner can be locally volumetrically significant (Boles, 1984).

As the basin matures and ceases filling and subsiding, fluids in the compactional regime become stagnant and are eventually replaced as water of meteoric origin infiltrates to greater depths (Galloway, 1984; Kreitler, in press). Pore fluid pressures in the compactional regime can be in excess of hydrostatic, or geopressured (Bethke, 1986a). Release of pore pressure at the transition zone to hydrostatic pressures causes diagenetic reactions such as calcite and silica cementation (Bruton and Helgeson, 1983). Release of pore pressure in producing hydrocarbon wells can cause significant well-bore scale precipitation and formation damage (Riese et al., 1987).

Diagenetic Reactions in the Thermobaric Regime

In the deepest parts of sedimentary basins, thermal decomposition of hydrous minerals releases additional pore water into the compacted sediments (Galloway, 1984). Additional water may be contributed from the basement, possibly metamorphic waters or ancient pore waters from basement complexes (e.g., Morton and Land, 1987). The diagenetic alteration of smectite to illite (Perry and Hower, 1972) as well as the formation of deep basinal brines implicated in the origin of Mississippi Valley-type ore deposits and certain uranium and copper deposits (e.g., Galloway, 1984; Sverjensky, 1984, 1987) may be typical reactions in this regime. Minerals forming well-bore scale in hydrocarbon production wells as deep as 6.7 km (22,000') such as those in Mobile Bay, Alabama, (McBride et al., 1987) may precipitate from fluids that originate in this hydrologic regime. Little is known about the temporal evolution of the thermobaric regime in sedimentary basins. If temperature is a key control on mineral dehydration reactions, then thermobaric fluids may occur at fairly shallow depths and may be dominant in basins with elevated heat flow.

PREDICTING FLUID/ROCK INTERACTIONS IN SEDIMENTARY BASINS

The previous discussions indicate that with some knowledge of the initial composition of a sediment and of its burial history and paleohydrologic evolution it may be possible to predict diagenetic modifications with some degree of success (e.g., Siever, 1986). To an extent this is true. There are three common approaches to predicting diagenetic reactions: empirical porosity decline curves, empirical diagenesis-facies/burial history correlations, and theoretical fluid/rock interaction models. Each approach has a different scale of application and different assumptions and, of course, a variable degree of success.

Porosity Decline Curves

At the broadest scale, porosity decline curves, which show average porosity as a function of burial depth, represent the sum of compactional and chemical diagenetic effects for a basin. Several of such curves are shown in Figure 2, one for carbonates, two for sandstones and an envelope of values for shales; the loss of porosity appears to approximate an exponential function. These curves can be compiled fairly rapidly from core-plug porosity measurements or even electrical wire-line logs for a basin of interest. Fuchtbauer (1974) has extended this approach by constructing porosity decline curves on the basis of sandstone grain size, showing that porosity declines more rapidly in fine-grained than in coarser-grained sandstones. The principal limitation of such an approach, even if the data are of an appropriate scale, is that comparisons of porosity trends in one basin with those in another are inappropriate because, for a given present-day depth, a multitude of burial histories are possible. Progression to a predictive porosity model is difficult (Schmoker and Gautier, 1988).

A recent improvement to this approach includes the use of the Time-Temperature Index, or TTI (Waples, this volume) instead of burial depth to predict porosity. These decline curves are illustrated in Figure 3 from the work of Schmoker and Gautier (1988). Use of TTI eliminates the porosity scatter created by variable thermal histories when rocks of similar

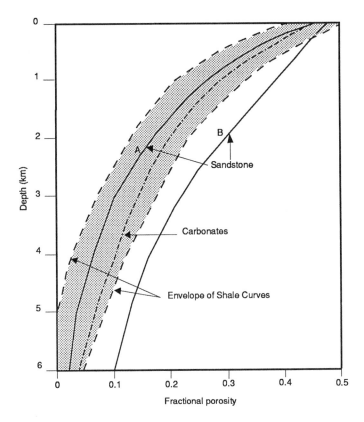

Figure 2. Porosity decline curves for sandstones (A, from Bethke, 1986a; B, from Schmoker and Gautier, 1988), carbonates (Schmoker and Halley, 1982), and shales (Butler and Baldwin, 1985).

present-day burial depths are compared. Thermal maturity models are used frequently in basin analysis (Waples, this volume) and therefore may be used to predict porosity trends.

The limitation to this approach is principally one of scale. Whereas time/depth-averaged porosity trends may be adequate for basin-scale analyses, they may be inadequate for reservoir- or field-scale predictions. For example, the porosity of the deep Norphlet Formation (Gulf Coast Jurassic) at almost 7 km (22,000') averages 7-8%, both on the basis of TTI-porosity decline curves (Fig. 3) and detailed point-counting of petrographic thin sections (McBride et al., 1987). The measured range in porosity is from 0 to 22%, and is determined by the distribution of several major cements including quartz, anhydrite, halite and illite. The average value yields little information about either this range or the relative importance and timing of compaction (16% of porosity loss), pressure solution (3% of porosity loss), cementation (8.5% of porosity loss), and secondary porosity development. Although the average value is adequate for regional paleohydrologic analyses and for hydrocarbon maturation and migration models (Bethke, 1985; Garven, in press), it is inadequate for reserves predictions and field development. In such applications economic operations may depend on an assessment of reservoir porosity to within a few percent, and a porosity cut-off value that is much higher than the average value.

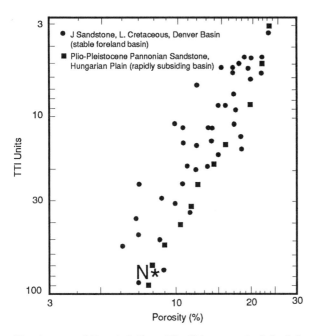

Figure 3. Porosity versus TTI trends for sandstones from two basins having different thermal histories (from Schmoker and Gautier, 1988). The Norphlet Formation (Jurassic) in Alabama at almost 7 km depth is marked by *N. TTI for the Norphlet is from the Smackover Formation, the inferred source for the Norphlet gas (McBride et al., 1987). A thermal history somewhat similar to that of the Pannonian sandstone is assumed giving a predicted porosity for the Norphlet sandstones of about 8%.

Facies- and Burial-Specific Diagenetic Models

At a smaller scale than regional basin-wide porosity trends, detailed petrographic and geochemical studies provide evidence for the diversity of fluid/rock interactions that occur in a sedimentary basin. When these studies are comprehensive, sequences of diagenetic events can be unravelled and can be used to infer past paleohydrologic conditions (McBride et al., 1987). The two principal controls on the fluid/rock interactions that actually occur are original mineralogical composition (e.g., arkose, volcaniclastic, quartzose sandstone) and the paleohydrologic history of the strata. Paleohydrologic history integrates the effects of burial and thermal histories with the stratal architecture and facies geometries within which the sediment is positioned. Under ideal circumstances, diagenetic studies need to be integrated with facies and stratigraphic studies to be most effective in predicting fluid/rock interactions. However, the usual situation is that the two are studied separately, frequently by different investigators and results are rarely integrated!

Exceptions exist and have shown that it is possible to unravel the relative importance of facies controls versus burial history controls in some instances. The following examples illustrate a situation where facies-controlled diagenesis is evident, using the work of Stonecipher et al. (1984), and a situation where diagenetic reactions associated with unconformities are relatively easy to predict.

FACIES VERSUS BURIAL HISTORY CONTROLS, FRONTIER FORMATION   Stonecipher et al. (1984) have shown that original, facies-specific distributions of clay particles in sandstones from the Frontier Formation, Wyoming, have controlled the pattern of fluid flow during subsequent burial and resulted in a facies-specific cementation pattern. In upper parts of offshore sand ridges in the Frontier Formation, well-sorted, matrix-free, quartzose sandstones have lost almost all their porosity and permeability due to quartz cementation. Sandstones and siltstones lower in the sand ridges were not cemented because an original depositional clay matrix inhibited fluid flow and cementation. These lower parts of the sand ridges now have better porosity and permeability than the upper parts (Fig. 4a), although the lower, matrix-rich lithologies compacted more than the matrix-free tops of the sand ridges. The initially clean, porous and permeable sandstones focused fluid flow during burial and thus were more susceptible to cementation by silica precipitated from the moving pore fluids. In this example, an accurate facies model would allow porosity and permeability distributions to be predicted after basic diagenetic trends had been evaluated petrographically.

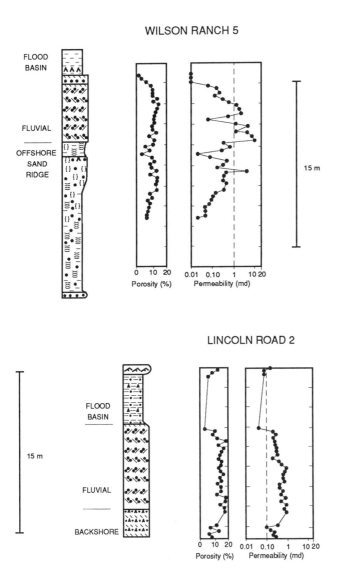

Figure 4. Porosities and permeabilities from the Frontier Formation, Wyoming (modified from Stonecipher et al., 1984). A. (top) Wilson Ranch well. The offshore sand ridge shows a trend of increasing porosity and permeability towards the top of the sand body due to decreasing matrix content. Preferential cementation of the cleanest sands at the top of the ridge has caused a reversal in this trend. Average porosity in the overlying fluvial sandstone is 13%, and average permeability is 10 millidarcies. B. (bottom) Lincoln Road well. Average porosity in this fluvial sandstone is similar to that in the fluvial sandstone in Fig. 4a, 15%, but the average permability is about ten times lower, 1 millidarcy, due to kaolinite formation. This point-bar sandstone is in a more proximal location to an inferred source of meteoric water.

Although others have reported a similar correlation between matrix clay and lack of cementation in sandstones (Pettijohn et al., 1972; Melvin and Knight, 1984), this correlation is not universal. In a second example from Stonecipher et al. (1984), quite different cementation histories occur in a fluvial point bar sandstone at two locations despite identical depositional conditions. In this example, the stratal geometries in which the sandstones are deposited are different, resulting in different fluid flow patterns during burial and thus different fluid/rock interaction patterns. Figures 4a and b show porosity/permeability data for these two fluvial sandstones. Average porosity is essentially identical in both locations (13% and 15% in the Wilson Ranch and Lincoln Road wells, respectively), whereas average permeabilities are 10 millidarcies in the Wilson Ranch well and only 1 millidarcy in the Lincoln Road well. The principal cause of permeability loss is the formation of authigenic clays, particularly kaolinite.

Kaolinite formed as a result of infiltration of meteoric water. The paleorecharge area was closer to the Lincoln Road well and thus this well has more kaolinite. The Wilson Ranch well was probably too far downdip to be affected by infiltration. The fluvial sandstones in the Lincoln Road well are also much thicker than those in Wilson Ranch. Focusing of fluid flow in the thinner sand in the Lincoln Road well created a more open system and faster flowing pore fluids than in the Lincoln Road well, kinetically restricting cementation (Stonecipher et al., 1984). These arguments lead to two conclusions: (1) in a single formation, facies-controlled depositional compositions may cause a specific diagenetic pattern; and (2) burial history and stratal geometries can impart a variable diagenetic pattern in two sandstones of originally identical facies.

In both these examples, petrographic and geochemical data are necessary to calibrate a diagenetic model before predictions can be made from facies distributions and stratigraphic models. In the next example, sufficient knowledge of typically occurring fluid/rock interactions has been acquired that regional scale porosity modifications can be predicted from a stratigraphic model alone.

UNCONFORMITY-RELATED DIAGENESIS   The infiltration of fresh, meteoric water into strata that are subaerially exposed at some stage in their burial history by uplift and erosion is probably the easiest form of diagenetic modification to predict. Chemical reactions associated with the infiltration of weakly acidic water, produced by the interaction of rainwater with atmospheric CO_2 and soil-zone bacterial CO_2, are well known. These reactions include dissolution of feldspars and quartz and precipitation of kaolinite in siliciclastic rocks, and the dissolution of aragonite and high-Mg calcite and the formation of low-Mg calcite (e.g., Bjorlykke, 1984; Land, 1986) in carbonate rocks. Although many of these dissolution reactions will produce leached-grain secondary porosity, the precipitation of kaolinite in pores and pore throats can decrease permeability.

The reservoir quality of several major hydrocarbon-producing fields is governed by porosity and permeability enhancement created by meteoric water incursion at unconformities. Leaching of chert and precipitation of kaolinite in the Prudhoe Bay Field (Melvin and Knight, 1984) may have been caused by infiltration of meteoric water at the Lower Cretaceous unconformity (Fig. 5). Meteoric infiltration into several North Sea reservoirs (Bjorlykke and Brensdal, 1986) probably occurred at the Kimmeridgian unconformity. A number of unconformities produced during the Jurassic and Lower Cretaceous were responsible for allowing fresh-water leaching of calcite in the Gulf Coast Smackover Formation and, downdip of the leached calcite, the formation of dolomite from mixing of saline and fresh pore waters (Grabowski et al., 1987).

Using techniques of seismic stratigraphy (Vail et al., 1977), unconformities can be identified from seismic data and at a regional scale, risk can be assigned to possible porosity enhancement. A more accurate assessment can be made if lithologic distributions are known and some boundary conditions can be placed on paleotopography and paleoclimate which will both control the extent of meteoric influx. In addition to assessing diagenetic modifications, risk of hydrocarbon biodegradation and water washing can be assessed.

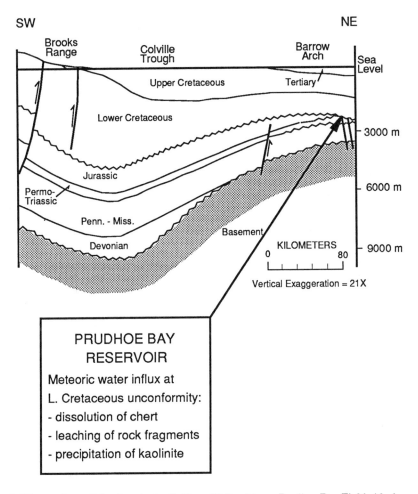

Figure 5. Diagenetic modifications in the Sadlerochit Sandstone, Prudhoe Bay Field, Alaska, caused by infiltration of meteoric water at the Lower Cretaceous unconformity (modified from Melvin and Knight, 1984).

Theoretical Models of Fluid/Rock Interactions

Many of the mass transport problems summarized above can be constrained through reaction-path modeling in which multiple chemical equilibria among minerals, aqueous fluids, and gases are computed under a variety of physical conditions. This represents the most detailed predictive approach and its development will be crucial for accurate models at the scale of reservoir production problems, contaminant transport predictions and diagenetic heterogeneities that may redefine depositionally-determined aquifer and aquitard arrangements.

REACTION-PATH MODELS   Reaction-path modeling is an established technique for evaluating processes involved in the formation of a variety of hydrothermal ore deposits (Helgeson, 1969, 1979; Brimhall, 1980; Reed, 1983; Sverjensky, 1984, 1987; Janecky and Seyfried, 1984; Bowers et al., 1985; Gitlin, 1985). The power of a reaction-path approach is that boundary conditions can be set for geochemically feasible diagenetic pathways. Critical variables can be identified readily and a theoretical framework can be developed for interpretation of observations. There is always iterative interaction between theoretical models and petrographic/geochemical data.

Reaction-path modeling has been minimally used in sedimentary diagenesis despite a clear need for quantitative analysis of the multiple mineral-fluid reactions involved (Plummer et al., 1983) for two principal reasons. First, an equilibrium thermodynamic approach may be perceived to be incorrect for sediment diagenesis which typically occurs at low pressures and temperatures. This perception stems from frequent petrographic observations of disequilibrium mineral assemblages and documented, kinetically-controlled reactions such as the diagenesis of smectite (Bethke and Altaner, 1986) and the precipitation of dolomite (Hardie, 1987). Second, the importance of mass transport as a diagenetic process has only recently been documented through the collection of basin-wide diagenetic observations (Land et al., 1987) and through the construction of basin paleohydrologic models that constrain fluid movement pathways (Garven and Freeze, 1984a; Bethke, 1985).

Although sedimentary mineral assemblages are often in apparent disequilibrium, the concepts of local equilibrium and irreversible thermodynamics (Helgeson, 1979) are still appropriate. Successful calculations are made by considering only a portion of the sedimentary rock, essentially decoupling unreactive parts of the sediment, for a defined problem. This approach is analogous to the hydrologists' use of *representative elementary volume* (Freeze and Cherry, 1979). Recent reaction-path codes (*SOLVEQ*, Reed, 1983; *EQ3NR*; Wolery; 1983; *GT*, Bethke, 1986b) have advanced options over earlier versions (*PATHI* , Helgeson, 1969; *EQ3/6*, Wolery, 1979), and calculations can now be made for systems in which all, or just selected parts, of the calculations are rate-limited. Examples include the redox disequilibrium option and the availability of several different rate laws for precipitation and dissolution of minerals (Wolery et al., 1984; Delany et al., 1986). Such options allow, for example, the construction of a single reaction-path model that honors the known kinetic data for smectite diagenesis, the bacterially-mediated reduction of SO_4^{-2}, mineral compositions that are solid solutions, and the equilibrium solubilities of quartz and K-feldspar.

MODEL OPTIONS   All reaction-path models calculate irreversible reactions with reference to 1000 g of aqueous fluid. A mineral assemblage is allowed to react with this fluid, or the fluid can be examined alone under a given set of external variables. In simple reaction-path models, a fluid composition can be heated or cooled in a closed system and changes in mineral saturations and speciation of dissolved components can be examined. The reaction path is one of variable pressure or temperature. More complex reaction paths react a mineral assemblage with a fluid. In the most simple of this model type, a mineral or rock is incrementally titrated into the fluid. This is equivalent to the engineering model of a continuously-stirred reactor. The reaction path is thus one of decreasing water to rock ratio. At each step, previously precipitated minerals are allowed to redissolve (or back react) if the solution becomes appropriately undersaturated.

A more complex model, sometimes termed "flow-through," prohibits the back reaction of the precipitated minerals; it is thus a model of fluid flowing through an aquifer where minerals—once precipitated—are left behind as the evolving fluid flows ahead. Additional variants on these models include mixing one water with another and flushing one water through an existing rock/pore water system. In all these cases, the reaction path is one of variable rock to water ratio, or in the case of a water mixing model, the ratio of one fluid to another fluid. It is possible to superimpose temperature and pressure gradients on these model calculations in which case the reaction path involves multiple variables and results are correspondingly more difficult to interpret. Table 1 shows some different diagenetic scenarios represented by various path options.

Several reaction-path models have additional features that may be relevant to diagenetic and near-surface chemical problems. For example, it is possible to model nonequilibrium behavior in several ways. The rate at which reactants are titrated into the water can be adjusted, simulating kinetic controls. In this way a mineral resistant to reaction can be modeled by reducing the rate at which it is added to the fluid while the reverse technique would be useful for an unstable mineral, such as a high-temperature polymorph. The formation of minerals can be suppressed, allowing the precipitation of minerals, known to be restricted by kinetic limitations, to be inhibited in an equilibrium model. It also allows, for example, a water to be saturated with respect to amorphous silica without precipitating quartz.

The $EQ3/6$ code has three different rate laws which may be chosen to model kinetically-controlled reactions (e.g., Delany, 1985). Redox disequilibrium, commonly reported in groundwaters (Lindburg and Runnels, 1984), can also be accommodated in calculations. Solid-solution models are available for some minerals and more recent releases of $EQ3/6$ and GT allow mineral equilibria in high-salinity solutions to be modeled with the Pitzer equations or with the Harvie-Moller-Weare formulation (Harvie et al., 1984) substituting for the normal Debye-Huckel activity extrapolations. Most of these options are probably not needed in modeling diagenetic reactions in sedimentary basins, however they may be crucial for modeling shorter term groundwater problems such as contamination, acid mine drainage, and pollutant attenuation through ion exchange and adsorption.

Table 1 Possible reaction path models for fluid/rock interactions in sedimentary basins.

Chemical System	Model	Reaction Path	Interpretation
A. Fluid			
	(i) closed system equilibrium	pressure &/or temperature	Species distribution; saturation calculations
	(ii) open system, flow through	pressure &/or temperature	Evolution of ore-forming solutions; bore-hole scale precipitation
B. Fluid + Rock			
	(i) closed system equilibrium	fluid/rock ratio & P/T gradient	Burial diagenesis of isolated sedimentary rock
	(ii) as (i) with kinetic constraints	as B(i) & time	Interpretation of laboratory experiments
	(iii) open system, flow-through	as B(i)	Leachate generation; formation of hydrothermal ores & alteration haloes
	(iv) open system, flush, with kinetics	as B(i) & time	Experimental column leaching studies; leachate generation
C. Fluid A + Fluid B			
	(i) closed system equilibrium	fluid A:B ratio	Mixing of groundwater with contaminant; coastal aquifer mixing
D. Fluid A + rock, mixed with Fluid B			
	(i) open system, flush, with/without kinetics	fluid A:B ratio	Mixing zone diagenesis; unconformity-related diagenesis; steam, CO_2 or alkali flood for enhanced oil recovery

Chemical Systems B-D can include pressure and/or temperature gradients as additional reaction-path variables.

MODEL LIMITATIONS   The two principal limitations on the model results are how well the chosen reaction path matches actual history and how well the available thermodynamic and kinetic data are known. In the first case, trial and error, as well as accurate burial histories and petrographic and geochemical data are used iteratively to obtain the best match between theoretical predictions and observations. In the second case, the user must check the completeness of the supporting data base for the problem being solved. This is particularly important for trace elements and aqueous metal complexes for which thermodynamic data are often very incomplete. Little is to be gained from blaming failure of theoretical models to predict reality on inadequate thermodynamic data because, in most cases, reaction-path models have a common source of data, much of it dating to the 1930s (e.g., Helgeson et al., 1978)! Thermodynamic data and lack of adequate data are the common thread that binds the different reaction-path models together. Geochemists actively involved in the collection of basic thermodynamic quantities are indeed appreciated by the modeling community.

Some rules of thumb in reaction-path modeling and the interpretation of results are: (1) the most simple model that can adequately account for the observations should be constructed, (2) an equilibrium closed-system model should always be calculated as a reference point, (3) rate constants for kinetically-controlled reactions are much less well known that the equilibrium thermodynamic data and thus should be used with caution, and (4) it is important to have an exact problem to solve. This last rule of thumb, sometimes known as problem-posing, is probably the most likely cause of model failure or success.

PROBLEM POSING   This step represents the conceptual link between a perceived geochemical problem and its representation in terms of the variables in a reaction-path model. Most of the recent reaction-path models have supporting thermodynamic data bases that include hundreds of minerals, aqueous complexes and species and gas phases. Scanning the data base may reveal a diversity of minerals such as troilite, grossular, disordered dolomite, beidellite, and tremolite. Clearly many of these phases are not relevant to a given problem. However, the diversity of the data bases gives reaction-path models their predictive ability, and sometimes gives the user apparent problems in reconciling predictions with observations. It is thus important to clearly perceive a sequence of events that must either be replicated in a model or be tested to find the path and reactions most compatible with observation.

For example, if we wished to model the diagenetic reactions in a marine sandstone, the problem could be posed as the closed-system equilibrium reactions between a mineral assemblage typical of a slightly impure sand (quartz, feldspar, rock fragments and some smectite) and seawater in a thermal gradient representing the temperatures reached during burial. The assumption is that there is no external source of either water or dissolved material during burial, possibly a realistic scenario for an isolated point-bar sand. The calculated mineral assemblage, shown in Figure 6, is typical of greenschist-grade metamorphism and is clearly an inadequate diagenetic model in most instances. If we consider the burial history of the sand more carefully however, realizing that the pore-water composition has almost certainly evolved during burial and may have been completely flushed and replaced by water of meteoric origin or water charged by organic acids (Galloway, 1984) it is clear that a closed-system model is too simple. The correctly posed problem needs to account for changing pore-water chemistries, contributions from hydrocarbon generation, and an open system in which water can move through the hypothetical reservoir sandstone. Under these conditions, in a thermal gradient, the sequence of predicted minerals is more in agreement with diagenetic observations (Fig. 6), and it is unnecessary to resort to "inadequate thermodynamic data" to explain the failure of the first model.

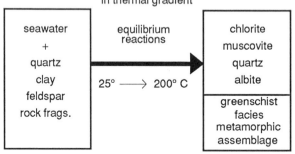

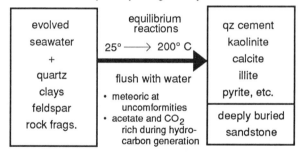

Figure 6. Two alternate reaction path models for the burial diagenetic reactions in a marine sandstone. The predicted mineral assemblage in model 2 is more commonly observed.

INTEGRATION OF FLUID FLOW AND CHEMICAL MODELS   Reaction-path models are independent of scale and time. Water volumes and chemistries can be constrained in these calculations by integrating the results of paleohydrologic models (e.g., Bethke, 1985; Garven and Freeze, 1984a, b). These models allow calculations of the geological development of fluid pressures and directions and velocities of fluid movement due to the processes of compaction, convection and topographic drive (Bethke, in press). Distributions of past hydrologic regimes, fluid pressures and velocities can be successfully integrated into the problem-posing stage of fluid/rock interaction models (e.g., Harrison et al., 1987).

MODEL OUTPUT   Calculated results for the mixing of a pore water in equilibrium with a typical feldspathic sandstone with a water of meteoric origin are shown in Figure 7 (Bethke et al., 1988). In this model we wish to evaluate the effect of meteoric water flushing an arkosic sandstone previously buried but now exposed at the surface. Such a model might be expected to predict the kinds of diagenetic alterations found beneath an unconformity in the geological record or in the meteoric regime in a compacting basin. The reaction-path variable in this model is salinity. The origin of the calculation is a 0.5m NaCl brine in equilibrium with quartz, feldspar, calcite and pyrite. As meteoric water replaces the saline water a series of reactions occur that include the replacement of calcite by dolomite, the dissolution of first albite and then K-feldspar, and the precipitation of kaolinite and a smectite (Na-nontronite) as the iron is released to solution during pyrite dissolution and reincorporated in a clay. Many of these predicted reactions are reported in sandstones beneath unconformities. It is also possible to calculate the volume changes associated with each stage of mineral alteration (Fig. 7C) and using a reference point, calculate the net effect on sandstone porosity, which could then be used to infer reservoir quality or modifications to hydraulic conductivity.

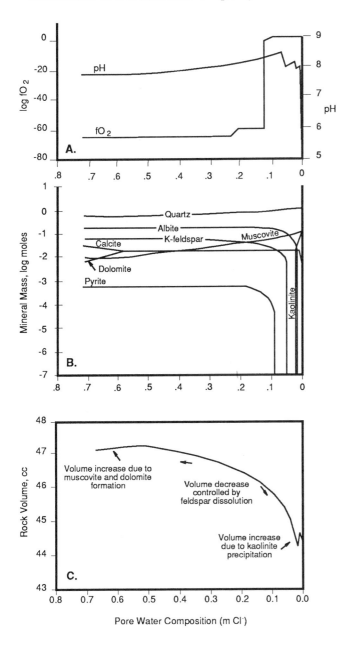

Figure 7. Diagenetic reactions induced in a sandstone by flushing it with meteoric water and replacing an initally saline (1m NaCl) porewater. 7A shows changes in pore-water pH and oxygen fugacity, 7B shows mineral reactions, and 7C shows volume changes accompanying mineral precipitation and dissolution reactions (modified from Bethke et al. 1988).

CASE STUDIES IN MODELING FLUID/ROCK INTERACTIONS

Two case studies are presented to illustrate how predictive models of fluid/rock interactions can be developed. The first involves the development of a diagenetic model for the Gippsland Basin, offshore southeast Australia, which can be integrated into hydrocarbon exploration strategies. The second involves a model for leachate generation from processed oil shales which was developed to address environmental concerns arising from the intended commercial development of oil shales as a hydrocarbon resource in the late 1970s and early 1980s.

Diagenesis in the Latrobe Group, Gippsland Basin

The Gippsland Basin, southeast Australia (Fig. 8) has been the major offshore hydrocarbon resource in Australia since the 1960s (Threlfall et al., 1976). The basin produces from the Latrobe Group which is a fluvial to shallow marine sequence of Late Cretaceous to early Eocene age overlying the Strzelecki Group which fills a rift valley of Jurassic to Lower Cretaceous age (Fig. 9). The source for hydrocarbons has been inferred to be a basinward facies of the Latrobe Group (Bodard et al., 1984) as well as coal-bearing strata within the producing field (Yin, 1988). Landward, nonmarine facies of the Latrobe include coal measures that are commercially developed (Thompson, 1980).

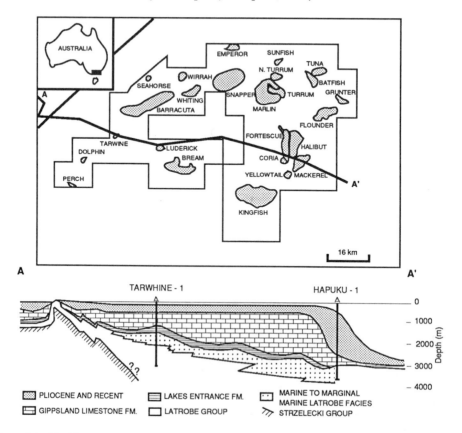

Figure 8 A. (Top) Locations of Gippsland basin oil and gas fields. **B.** (Bottom) Cross section along line A-A' (after Kuttan et al., 1986).

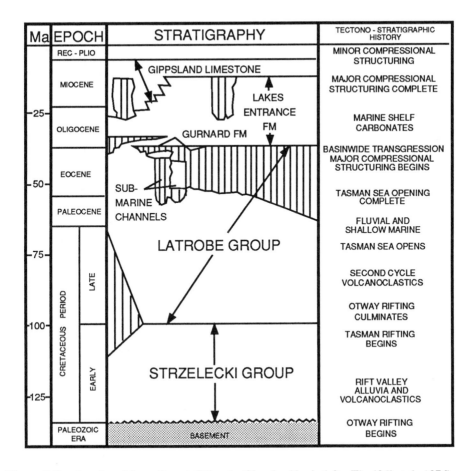

Figure 9. Stratigraphy of the sediments filling the Gippsland basin (after Threlfall et al., 1976).

Reservoir-quality sandstones are found in both fluvial and shallow marine facies. Compositionally, the sandstones are quartz-rich with K-feldspar, detrital kaolinite and illite, and lithic fragments (Table 2). The principal controls on reservoir quality are matrix content in the marine sandstones and dolomite cement in the fluvial sandstones (Fig. 10). K-feldspar has been leached and authigenic kaolinite precipitated, and quartz cement averages a few percent. Limpid, poikilotopic dolomite cement occurs as nodules and continuous horizons in wells where it can amount to as much as 40% of the rock volume, completely destroying porosity and permeability. It formed relatively early in the depositional history of the Latrobe fluvial sediments and shows evidence for growth and minor dissolution under cathodoluminescent light. Reservoir porosities and permeabilities can exceed 25% (Kuttan et al., 1986) and 2 darcies in uncemented fluvial sandstones.

Table 2 Sandstone compositional analysis—Latrobe Group.

| | Fluvial Channel | | Marine | |
	Range (%)	Average (%)	Range (%)	Average (%)
Grains				
quartz	44-73	60	30-76	57
k-feldspar	0-17	5	0-17	6
lithic fragments	0-27	6	0-19	6
Matrix	0-9	4	0-37	10
Cements				
carbonate	0-40	10	0-18	1
quartz	0-10	2	0-9	3
clays	0-7	1	0-8	1
k-feldspar	0-1	trace	0-1	trace
Visible Porosity				
intergranular	0-25	10	0-33	14
leached grain	0-8	2	0-7	2
Intergranular Volume	19-43	28	20-45	29

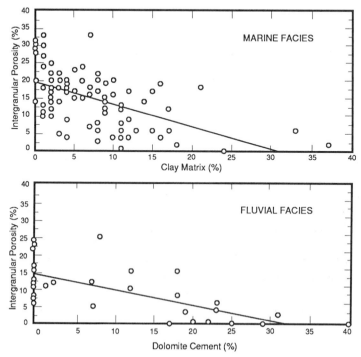

Figure 10. Intergranular porosity varies with clay matrix in marine sandstones and with dolomite cement in fluvial sandstones.

DIAGENETIC MODELS   Relatively few papers have been published on the diagenesis of Latrobe Group sediments. Bodard et al. (1984) suggested that kaolinite, dolomite and quartz are produced by burial diagenesis. Thermogenic CO_2 created acidic waters which caused K-feldspar leaching and dolomite dissolution. All porosity was interpreted to be secondary, and significant quantities of carbonate were therefore assumed to have been leached.

Yin (1988) proposed that the dolomite cement was an early phase, whereas the dissolution of K-feldspar and the formation of kaolinite were related to the effects of organic acids produced during hydrocarbon maturation. This hypothesis is currently awaiting evaluation by this author while the thermodynamic data base of a reaction-path model is updated to support an adequate number of organic acids and metal complexes.

In this study, dolomite is proposed to have formed early in the burial history, and kaolinite, quartz and K-feldspar leaching can all be produced as result of freshwater leaching and mixing in aquifer system that exists in the Latrobe Group sandstones.

PALEOHYDROLOGY AND GEOCHEMISTRY   The paleohydrologic evolution of the Gippsland Basin is critical in understanding the diagenetic history. An onshore outcrop area of the Latrobe Group recharges an aquifer that extends over 60 km offshore and to greater than 2 km subsea (Thompson, 1980). Salinities range from 500-4,000 ppm. The freshwater wedge overlies normal compactional pore waters whose salinities vary from 10,000 to 40,000 ppm (Kuttan et al., 1986). The aquifer is presumed to discharge offshore along faults (Fig. 11). Although the exact history of the freshwater lens is unknown, the structure onshore that creates the recharge zone is at least Oligocene in age (Thompson, 1980), implying that the existing hydrologic regimes may be representative of those present for much of the basin's history. During lowstands of sea level, increased topographic drive may have caused meteoric waters to infiltrate more basinward strata, creating the mixed salinity zone beneath the present day fresh water (e.g., Meisler et al., 1985).

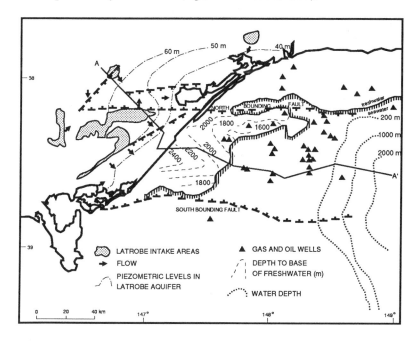

Figure 11 A. Hydrology of the Latrobe aquifer (after Thompson, 1980).

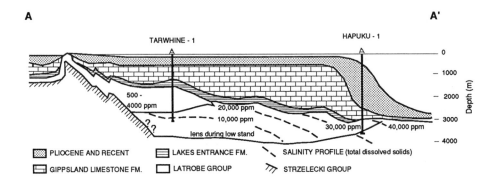

Figure 11 B. Salinity distribution along cross-section A-A' (after Kuttan et al., 1986).

Measurements of stable carbon and oxygen isotopes in the dolomite show considerable scatter (Fig. 12). No significant depth trends are seen in either isotope, suggesting that burial diagenetic reactions and dolomite produced from thermally-generated carbon are not significant. The best interpretation of the data is that the dolomite formed in a brackish water environment at temperatures <60°C, assuming a rainwater composition of about -9% δO^{18}, consistent with the paleolatitude of the Gippsland Basin during Tertiary times. Clearly the range of isotopic carbon compositions dictates that at least some of the carbon in the dolomite must have been contributed by thermal decarboxylation.

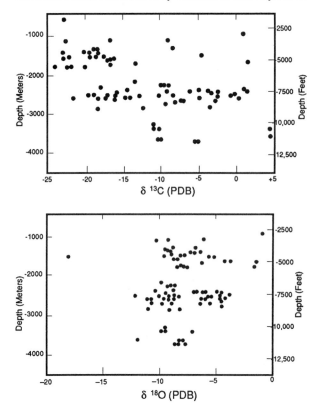

Figure 12 A. (Top) Carbon isotopic composition of dolomite cement (Analyses by D.L. Shettel). **B.** (Bottom) Oxygen isotopic composition of dolomite cement (Analyses by D.L. Shettel).

REACTION-PATH CALCULATIONS   The *EQ3/6* reaction-path model was used to evaluate whether mixing fresh water and saline pore waters could account for the formation of dolomite and for the other diagenetic reactions in the Latrobe Group. End-member water chemistries shown in Table 3 were used. Temperatures were assumed to be 25°-60°C, as suggested by interpretations of isotopic data.

Table 3 Water analyses used for mixing calculations

	Freshwater (Barracouta A-3)	Saline Water (Kingfish A-19)
Ca^{2+}	32 mg/l	220 mg/l
Mg^{2+}	9	1000
Na^+	2943	11,000
SO_4^{2-}	1461	900
HCO_3^-	1135	198
Cl^-	2953	19,000
TDS	8570 mg/l	32,342 mg/l
pH	7.0	5.6
Mg/Ca (mol)	0.24	7.58

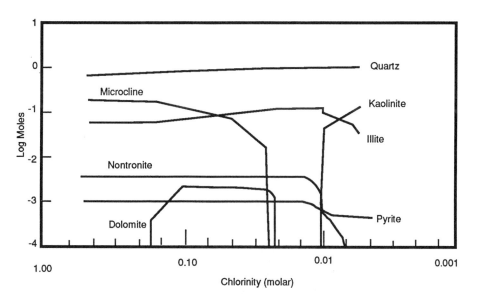

Figure 13. Predicted mineral reactions in the Latrobe Group sandstones as a result of mixing pore-waters from the freshwater lens with those from the saline zone. Dolomite precipitates from a pore water of intermediate composition.

Model results are shown in Figure 13. The stability field for dolomite scans an intermediate salinity range and forms when the mixed salinity water has adequate Mg^{+2} and HCO_3^- concentrations to reach saturation with respect to dolomite. At low salinities dolomite formation is limited by low Mg availability, whereas at high salinities it is limited by HCO_3^- availability. Other reactions that occur as a result of fluid mixing are the dissolution of K-feldspar, and the formation of kaolinite and illite, all petrographically observed reactions. Observations sometimes include evidence for dolomite and quartz leaching which can also be accounted for by this model. If waters become more saline, for example during a rise in sea level when saline water will replace freshwater, dissolution of both dolomite and quartz are predicted.

APPLICATIONS TO EXPLORATION AND FIELD DEVELOPMENT   A model of water mixing adequately explains the formation of dolomite in the Latrobe Group. Although mixing zones are typically rather narrow (Freeze and Cherry, 1979), numerous changes in sea level during Tertiary times (Haq et al., 1986) created a broad zone of intermediate salinity pore waters (c.f., Atlantic coastal plain, Meisler et al., 1985) and dolomite formed over a laterally extensive area. Dolomite has been found in the Hapuku-1 well, close to the limit of the known field (Bodard et al., 1984) and well within the saline water zone at the present day. A lowstand of 200 m below present-day sea level during the Oligocene (Haq et al., 1986) would have been adequate to cause fresh water to reach this location (Fig. 11b)

This model reconciles an observation that dolomite is found in sedimentary sequences underlying major marine transgressive sequences (Partridge, 1976). Rises in sea level would result in sediments with more dense pore waters overlying sediments with fresh waters in pore spaces. This unstable configuration would result in rapid mixing of the two water types. Although formation of dolomite is not restricted to such locations it is clear from these chemical calculations why the correlation exists. Drilling to greater depths (economic basement is about 3 km) will not decrease the risk of dolomite cementation unless the stratal geometries change significantly.

A combination of facies and stratigraphic models for this basin which include sequence stratigraphic analysis to identify major transgressive surfaces can significantly improve reservoir-quality predictions both for field development above economic basement and for exploration deeper in the basin. This model has integrated stratigraphy, paleohydrology, geochemistry and petrography in accounting for dolomite cementation; because the diagenetic processes involved were identified, a regionally predictive model can be developed.

Leachate Generation from Processed Oil Shales

In the late 1970s and early 1980s, commercial development of the Green River Formation oil shales in Colorado and Utah was anticipated in the face of high prices for imported oil. A key issue in the extraction process development involved the problem of solid-waste disposal (Ferraro and Nazaryk, 1982). After retorting (oil extraction by thermal processes), the shale has changed significantly both in physical and chemical properties. To maximize oil recovery, the shale is crushed; during thermal retorting mineralogical changes further breakdown the rock fragments, creating a spent material that has a coarse, powder-like texture, a high surface area and a compacted porosity of >40%.

Table 4 Mineralogical changes during retort processes. Data summarized from Cole et al. (1978) and Williamson et al. (1980).

Initial Mineralogy	Surface retort (e.g., TOSCO II) 450°-500°C	Modified *in situ* (e.g., PARAHO) 700°-1000°C
quartz	unchanged	reacted
calcite	unchanged	decarbonated
dolomite	unchanged	decarbonated
clays	dehydrated	reacted
zeolites	dehydrated	dehydrated
pyrite	pyrrhotite and Fe-oxides	Fe-oxides
		New Minerals Formed
		diopside
		forsterite
		monticellite
		ackermanite
		CaO, MgO

Chemical changes depend on the thermal process used and are dominated by whether the carbonate minerals that make up over 60% of the oil shale break down (Table 4). When temperatures exceed 600-700°C, decarbonation reactions occur and the waste shale may contain CaO, MgO, and high-temperature silicate minerals such as ackeramnite, monticellite and diopside.

Plans for disposal of this high volume, high surface area and chemically reactive material involved storage in canyons that cut below the Green River Formation into the underlying Uinta Formation sandstones. Although climatic conditions in western Colorado are arid, runoff from rainstorms collects in the canyons and can potentially interact with the spent shale. Although economic conditions have postponed development of commercial oil shale processing indefinitely, the approach used in this study is amenable to other contaminant problems.

REACTION-PATH CALCULATIONS   Reaction-path models were used to evaluate two problems: 1) the controls on chemical variations in leachate compositions and the range of expected leachate chemistries; and 2) the potential for contaminant attenuation by the underlying Uinta sandstones if leachates drained from the spent shale.

To address the first question, two end-member flow conditions were investigated (Fig. 14). The evolution of the leachate was tracked as it percolated through a spent shale pile, reacting with increased amounts of solid material, and the evolution of the leachate was tracked as progressively more pore volumes of fresh groundwater infiltrated the pile. From these calculations, two key controls on leachate chemistry were identified: the oxidation state of the spent shale, and the temperature at which the shale was retorted. Under partially saturated conditions when oxygen is available in the shale, high concentrations of dissolved sulfate and thiosulfate are produced from oxidation of pyrite (Table 5). High-temperature processing creates a leachate with very alkaline pH, composed solely of Na^+ and SO_4^{-2}. Low-

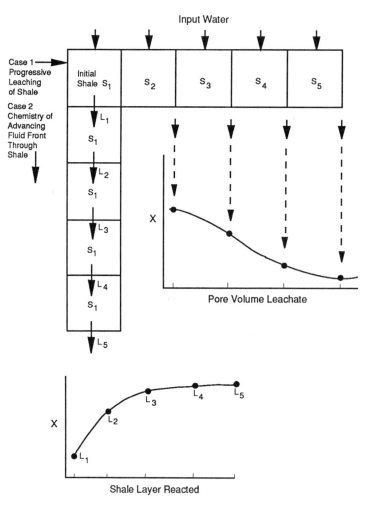

Figure 14. Conceptual model for reaction path calculations of retorted oil shale leachates. X represents a compositional variable in the leachate. S1-S5 are progressively changing shale compositions and L1-L5 are progressively changing leachate compositions.

temperature processing creates a more neutral leachate with significant bicarbonate content (Figs. 15 and 16). In both extraction methods, an early mineral precipitate in the spent shale pile is gypsum, which in reality may act as a cementing agent, effectively sealing the pile and limiting further groundwater infiltration!

Calculations showed that the predicted fate of leachate, once it escapes from the canyons and infiltrates the underlying Uinta sandstones, depends on the availability of smectitic clays in that rock (Fig. 17). These clays were calculated to attenuate sulfate and reduce pH of the leachate through the precipitation of gypsum.

Table 5 Comparison of model and experimental leachates. Paraho Shale data from Garland et al. (1979) and Killkelly et al. (1981).

	Experimental Paraho Shale*	Calculated using EQ3/6 Low T process	Calculated using EQ3/6 High T process
pH	11.1	7.3	12.8
Eh (v)	-0.1	-0.19	-0.6
S^{2-}	no data	0	3400
SO_4^{2-}	22,000 mg/l	14,025 mg/l	27,487 mg/l
$S_2O_3^{2-}$	2120	no data	no data
Na^+	10,400	17,634	16,646
K^+	1230	41.5	555
Ca^{2+}	530	5.7	0.4
Mg^{2+}	140	0.26	0
HCO_3^-	no data	28,494	2.5
CO_3^{2-}	2145	108	2541

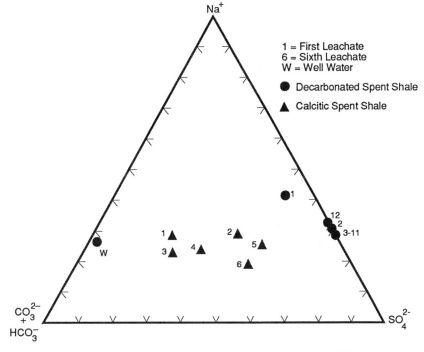

Figure 15. Major ion compositions of model leachates, pore volumes 1 to 6. Well water composition from Weeks et al. (1974). Leachates from decarbonated shales contain no ($HCO_3^- + CO_3^{2-}$). All model leachates are a potential source of groundwater sulphate.

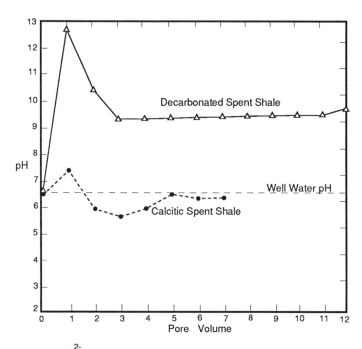

Figure 16. Evolution of leachate pH produced from low- and high-temperature processed oil shales.

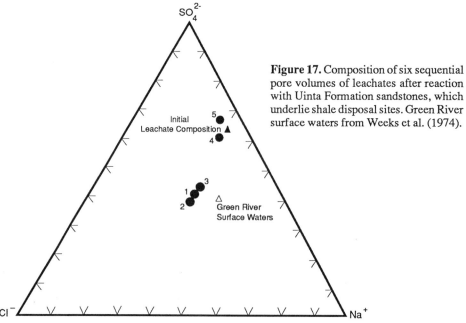

Figure 17. Composition of six sequential pore volumes of leachates after reaction with Uinta Formation sandstones, which underlie shale disposal sites. Green River surface waters from Weeks et al. (1974).

MODEL UINTA FORMATION COMPOSITIONS
1. 0 MOLES MONTMORILLONITE, 1 MOLE (QZ + PLAG)
2. 0.1 MOLES MONTMORILLONITE, 0.9 MOLE (QZ + PLAG)
3. 0.3 MOLES MONTMORILLONITE, 0.7 MOLE (QZ + PLAG)
4. 0.6 MOLES MONTMORILLONITE, 0.4 MOLE (QZ + PLAG)
5. 1.0 MOLES MONTMORILLONITE, 0.0 MOLE (QZ + PLAG)
   QZ : PLAG = 1:1
   QZ = QUARTZ : PLAG = PLAGIOCLASE

MODEL CALIBRATION AND APPLICATION   Calibration of calculated chemistries shows agreement between experimental leaching studies and theoretical predictions, illustrated by the data for fluoride in leachates (Fig. 18). Such calibration gives confidence that the model assumptions are realistic and that the models can be used predictively. When model results are compared with pore fluids extracted from a spent shale pile produced by operation of an experimental pilot processing plant, the theoretical compositions bracket the observed leachate chemistries (Fig. 19). During operation of the pilot plant it is probable that a variety of temperature conditions were used, thus creating a spent shale and leachate of intermediate composition.

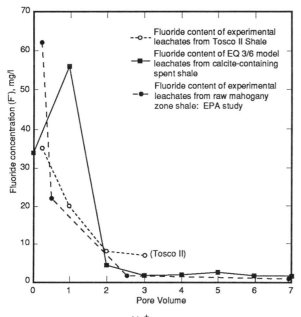

Figure 18. Comparison of experimental (McWhorter, 1980) and field leachate fluoride concentrations (Stollenwerk and Runnells, 1981) with model predictions made in this study.

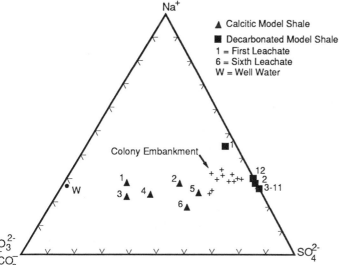

Figure 19. Comparision of leachate compositions found in an experimental oil shale disposal site (Battelle, 1981) with predicted leachate compositions made in this study. Pore fluids from the Colony Embankment are a chemical mixture of the two model leachate types, suggesting partial decarbonation of the retorted shale, presumably due to experimentation with pilot plant operations.

STATUS AND FUTURE DIRECTIONS

Our ability to predict fluid/rock interactions in sedimentary basins varies with the level of detail required. Empirical porosity-decline curves are remarkably successful at predicting subsurface trends over thousands of meters and provide data of relevance for basin-scale models, for example, paleohydrologic reconstructions (Bethke, 1985; Harrison and Bethke, 1988). Careful, detailed petrographic and geochemical studies can allow construction of empirical models at a reservoir scale, but usually such models are not sufficiently general to be applied regionally or in other basin settings. The prediction of unconformity-associated porosity development is an exception. More studies at this scale in which stratigraphic and facies models are integrated with diagenetic models are needed to develop the experience we need to make predictive generalizations.

Theoretical models of fluid/rock interactions can be used successfully to predict diagenesis (Harrison et al., 1987) and other chemical reactions associated with man's interaction with the environment (Miller et al., 1977; Wolery, 1980; Riese et al., 1987). Limitations to this approach at present include the need to calibrate and extend the calculations using field or analytical data as the forward models are not entirely reliable. Another limitation is the quality of supporting thermodynamic and kinetic data. This can only be resolved by diverse applications to a variety of problems through which unreliable data can be identified and eventually replaced. The integration of the interactions between organic and inorganic species in aqueous fluids will greatly improve the completeness of our models of diagenetic processes and contaminant transport.

Stratigraphic and facies models provide critical input and must play a much greater role if our success in predicting fluid/rock interactions is to improve. Figure 20 shows schematically how stratigraphy, facies geometries and burial history interface with models of fluid flow and chemical reactions. In some areas, the interfaces already exist, particularly in the construction of paleohydrologic models from burial history and stratigraphic data (e.g., Bethke, 1985). There is also an emerging use of chemical models, constrained by facies-controlled lithologic distributions and paleohydrologic analysis (e.g., Harrison et al., 1987).

The interface of chemical models with two-dimensional models of fluid flow has been started in some engineering applications related to groundwater hydrology (e.g., SUTRA, a USGS model that allows simulations of chemical transport). However, these combined flow and chemical-reaction models have limited application in geologic problems because the fluid-flow equations do not account for changes to the sedimentary system over geologic time and the chemical reactions are oversimplified to reduce computational requirements. More complex combined flow-transport models are being developed (e.g., Lichtner et al., 1986; Moore, 1987; Ortleva, 1987). These models will probably not be available for routine use for three to five years. Results from calculations involve geochemical and hydrologic variables and are complex to interpret. It is realistic to expect that unless the results of separate fluid-flow and chemical models are first fully understood, and the limitations of separate models clearly prevent interpretation of the problem, that the results of combined flow-transport models will be inadequately and inaccurately interpreted. Significant progress in the area of predicting fluid/rock interactions can be made by applying individual codes. In the last three years major advances in both types of models have occurred so that code availability is no longer the limiting step (Garven, 1984a,b; Delany et al., 1986; Bethke, 1985, 1986b; Bethke, in press). Clever applications to diverse problems will dictate the need for future model development.

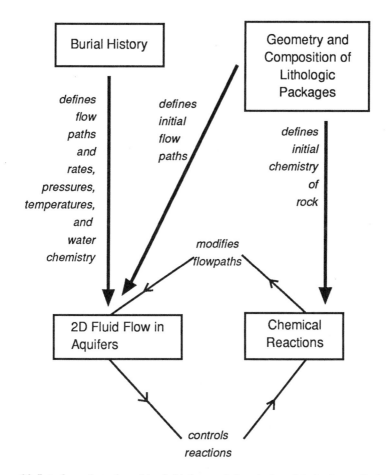

Figure 20. Interface of stratigraphic, fluid-flow and chemical models in the prediction of fluid/rock interactions in sedimentary basins.

ACKNOWLEDGMENTS

Parts of this paper represent work performed while the author was employed by Exxon Production Research Company, Houston, Texas. Many of the ideas presented in this paper were developed through conversations with A.E. Bence, D.R. Pevear, S.T. Paxton, L.L. Summa, and P.H. Monaghan at Exxon Production Research Company. The concept of diagenetic facies in sedimentary basins grew from discussions with A.E. Bence and D.R. Pevear. Gippsland Basin diagenetic models were a collaborative effort of this author, R.H. McCallister, Exxon Company, U.S.A. and D.L. Shettel, Mifflin and Associates. Studies of oil shale leachate chemistry were initiated by this author and P.H. Monaghan (retired). The author thanks EPR for permission to publish this research and for many years of support in the area of fluid/rock interaction modeling. The author's experience with the reaction-path models EQ3/6 and GT has grown from collaborative research with C.M. Bethke, University of Illinois. Careful reviews of this paper by A.E. Bence, T.A. Cross, and D.R. Pevear are appreciated.

REFERENCES CITED

Baldwin, B., and Butler, C.O., 1985, Compaction curves: American Association of Petroleum Geologists Bulletin, v. 69, p. 622-626.

Battelle, 1981, Evaluation of moisture and chemical status of Colony retorted shale embankment. Report prepared for Exxon Company, USA, Denver.

Bennett, P.C., Meeker, M.E., Seigel, D.I., and Hassett, J.P., 1988, The dissolution of quartz in dilute aqueous solutions of organic acids at 25°C: Geochimica et Cosmochimica Acta, v. 52, p. 1521-1531.

Bethke, C.M., 1985, A numerical model of compaction-driven groundwater flow and heat transfer and its application to the paleohydrology of sedimentary basins: Journal of Geophysical Research, v. 90, p. 6817-6828.

Bethke, C.M., 1986a, Inverse hydrologic analysis of the distribution and origin of Gulf Coast-type geopressured zones: Journal of Geophysical Research, v. 91, p. 6535-6545.

Bethke, C.M., 1986b, Reduced basis method for calculating heterogeneous geochemical equilibria and multicomponent reaction paths: Geochemical Modelling Workshop (convened by Lawrence Livermore National Laboratory), Fallen Leaf Lake, September 14-17, 1986.

Bethke, C.M., 1989, Modelling subsurface flow in sedimentary basins: Geologische Rundschau (in press).

Bethke, C.M., and Altaner, S.P., 1986, Layer by layer mechanism of smectite illitization and application to a new rate law: Clays and Clay Minerals, v. 34, p. 136-145.

Bethke, C.M., Harrison, W.J., Upson, C., and Altaner, S.P., 1988, Supercomputer analysis of sedimentary basins: Science, v. 239, p. 233-237.

Bjorlykke, K., 1979, Cementation of sandstones: Journal of Sedimentary Petrology, v. 49, p. 1358-1359.

Bjorlykke, K., 1984, Formation of secondary porosity: How important is it?, in McDonald, D.A., and Surdam, R.C., eds., Clastic diagenesis: American Association of Petroleum Geologists Memoir 37, p. 277-289.

Bjorlykke, K., and Brensdal, A., 1986, Diagenesis of the Brent sandstone in the Statfjord Field, North Sea, in Gautier, D.L., ed., Roles of organic matter in sediment diagenesis: Society of Economic Paleontologists and Mineralogists Special Publication 38, p. 158-167.

Blatt, H., 1979, Diagenetic processes in sandstones, in Scholle, P.A., and Schlager, P.R., eds., Aspects of diagenesis: Society of Economic Paleontologists and Mineralogists Special Publication 26, p. 141-157.

Bodard, J., Wall, V., and Cass, R.A., 1984, Diagenesis and the evolution of Gippsland basin reservoirs: Australian Petroleum Exploration Association Journal, v. 24 p. 314-335.

Boles, J.R., 1978, Active ankerite cementation in the subsurface Eocene of southwest Texas: Contributions to Mineralogy and Petrology, v. 68, p. 13-22.

Boles, J.R., 1982, Active albitization of plagioclase, Gulf Coast Tertiary: American Journal of Science, v. 282, p. 165-180.

Boles, J.R., 1984, Secondary porosity reactions in the Stevens sandstone, San Joaquin Valley, California, in McDonald, D.A., and Surdam, R.C., eds., Clastic diagenesis: American Association of Petroleum Geologists Memoir 37, p. 217-224.

Boles, J.R., and Franks, S.G., 1979, Clay diagenesis in Wilcox sandstones of southwest Texas: Implications of smectite diagenesis on sandstone cementation: Journal of Sedimentary Petrology, v. 49, p. 55-70.

Boles, J.R., and Ramsayer, K., 1987, Diagenetic carbonate in Miocene sandstone reservoir, San Joaquin basin, California: American Association of Petroleum Geologists Bulletin, v. 71, p. 1475-1487.

Bowers, T.S., and Taylor, H.P., 1985, An integrated chemical and stable isotope model of the origin of midocean ridge hot spring systems: Journal of Geophysical Research, v. 90, p. 12,583-12,606.

Bowers, T.S., Von Damm, K.L., and Edmond, J.M., 1985, Chemical evolution of mid-ocean ridge hot springs: Geochimica et Cosmochimica Acta, v. 49, p. 2239-2253.

Brimhall, G.H., Jr., 1980, Deep hypogene oxidation of porphyry copper potassium silicate protore at Butte, Montana: A theoretical evaluation of the copper re-mobilization hypothesis: Economic Geology, v. 75, p. 384-409.

Bruton, C.J., and Helgeson, H.C., 1983, Calculation of the chemical and thermodynamic consequences of differences between fluid and geostatic pressure in hydrothermal systems: American Journal of Science, v. 283-A, p. 540-588.

Carothers, W.W., and Kharaka, Y.K., 1978, Aliphatic acid anions in oil-field waters—Implications for origin of natural gas: American Association of Petroleum Geologists Bulletin, v. 62, p. 2441-2453.

Cole, R.D., Liu, J.M., Smith, G.V., Huckley, C.C., and Saporoschenko, M., 1978, Iron partitioning in oil shale of the Green River Formation, Colorado: A preliminary Mossbauer study: Fuel, v. 57, p. 727-732.

Crossey, L.J., Frost, B.R., and Surdam, R.C., 1984, Secondary porosity in laumontite-bearing sandstones, in McDonald, D.A., and Surdam, R.C., eds., Clastic diagenesis: American Association of Petroleum Geologists Memoir 37, p. 225-239.

Delany, J.M., 1985, Reaction of Topopah Spring tuff with J-13 water: A geochemical modeling approach using the EQ3/6 reaction path code: Lawrence Livermore National Laboratory Report URCL-53631, 46 p.

Delany, J.M., Puigdomenech, I., and Wolery, T.J., 1986, Precipitation kinetics option for the EQ6 geochemical reaction path code: Lawrence Livermore National Laboratory Report UCRL-53642, 31 p.

Drez, P.E., and Harrison, W.J., 1987, Do organic acids play a role in diagenesis.?: American Association of Petroleum Geologists Research Conference, "Prediction of reservoir quality through chemical modelling," Park City, Utah.

Ferraro, P., and Nazaryk, P., 1982, Assessment of the cumulative environmental impacts of energy development in northwestern Colorado: A status report: Proceedings of the 15th Oil Shale Symposium, Colorado School of Mines, p. 494-504.

Ferry, J.M., 1984, A biotite isograd in south-central Maine, U.S.A.: Mineral reactions, fluid transfer, heat transfer: Journal of Petrology, v. 25, p. 871-893.

Fisher, J.B., 1987, Distribution and occurrence of aliphatic acid anions in deep subsurface waters: Geochimica et Cosmochimica Acta, v. 51, p. 2459-2468.

Franks, S.G., and Forester, R.W., 1984, Relationships among secondary porosity, pore-fluid chemistry and carbon dioxide, Texas Gulf Coast, in McDonald, D.A., and Surdam, R.C., eds., Clastic diagenesis: American Association of Petroleum Geologists Memoir 37, p. 63-81.

Freeze, R.A., and Cherry, J.A., 1979, Groundwater: New Jersey, Prentice-Hall, 604 p.

Füchtbauer, H., 1974, Sediments and sedimentary rocks—1 (second edition, English translation): New York, Hafner Publishing, 494 p.

Galloway, W.E., 1984, Hydrogeologic regimes of sandstone diagenesis, in McDonald, D.A., and Surdam, R.C., eds., Clastic diagenesis: American Association of Petroleum Geologists Memoir 37, p. 3-15.

Garland, T.P., Wildung, R.E., and Harbert, H.P., 1979, Influence of irrigation and weathering reactions on the composition of percolates from retorted oil shales in field lysimeters: Proceedings of the 12th Oil Shale Symposium, Colorado School of Mines, p. 52-57.

Garven, G., and Freeze, R.A., 1984a, Theoretical analysis of the role of groundwater flow in the genesis of stratabound ore deposits. 1. Mathematical and numerical model: American Journal of Science, v. 284, p. 1085-1124.

Garven, G., and Freeze, R.A., 1984b, Theoretical analysis of the role of groundwater flow in the genesis of stratabound ore deposits. 2. Quantitative results: American Journal of Science, v. 284, p. 1125-1174.

Garven, G., 1989, A hydrogeologic model for the formation of the giant oil sands deposits of western Canada sedimentary basin: American Journal of Science (in press).

Gitlin, E., 1985, Sulfide remobilization during low temperature alteration of sea-floor basalt: Geochimica et Cosmochimica Acta, v. 49, p. 1567-1579.

Grabowski, G.J., Jr., Williams, S.C., Kick, R.M., Harrison, W.J., McFarlan, E., Jr., Reeckman, S.A., and Kauffman, J., 1987, Aquifer model for early diagenesis and porosity prediction, Smackover Formation (Upper Jurassic), northern Gulf basin: Abstracts for the Society of Economic Paleontologists and Mineralogists Mid-Year Meeting, Austin, Texas.

Haq, B.U., Hardenbol, J., and Vail, P.R., 1986, Chronology of fluctuating sea levels since the Triassic (250 million years ago to present): Science, v. 285, p. 1156-1161.

Hardie, L.A., 1987, Dolomitization: A critical view of some current ideas: Journal of Sedimentary Petrology, v. 57, p. 166-183.

Harrison, W.J., and Bethke, C.M., 1988, Paleohydrologic analysis of geopressure development and infiltration of meteoric waters in the Gulf basin: EOS (American Geophysical Union Transactions), v. 69, p. 360.

Harrison, W.J., McCallister, R.H., and Shettel, D.L., 1987, A diagenetic model for dolomite formation in Latrobe Group sandstones, Gippsland basin, Australia: American Association of Petroleum Geologists Research Conference, "Prediction of reservoir quality through chemical modelling," Park City, Utah.

Harvie, C.E., Moller, N., and Weare, J.H., 1984, The prediction of mineral solubilities in natural waters: The $K-Na-Mg-Ca-H-Cl-SO_4-H-HCO_3-CO_3-H_2O$ system to high ionic strength at 25°C: Geochimica et Cosmochimica Acta, v. 48, p. 723-751.

Helgeson, H.G., 1969, Thermodynamics of hydrothermal systems at elevated temperatures and pressures: American Journal of Science, v. 267, p. 729-804.

Helgeson, H.G., 1979, Mass transfer among minerals and hydrothermal solutions, in Barnes, H.L., ed., Geochemistry of hydrothermal ore deposits (second edition): New York, Holt, Rinehart and Winston, p. 568-610.

Helgeson, H.G., Delany, J.M., Nesbitt, H.W., and Bird, D.K., 1978, Summary and critique of the thermodynamic properties of the rock-forming minerals: American Journal of Science, v. 278, p. 1-229.

Hoffman, J., and Hower, J., 1979, Clay mineral assemblages as low grade metamorphic geothermometers: Application to the thrust faulted disturbed belt of Montana, U.S.A., in Gautier, D.L., ed., Roles of organic matter in sediment diagenesis: Society of Economic Paleontologists and Mineralogists Special Publication 38, p. 55-81.

Housenecht, D.A., 1984, Influence of grain size and temperature on intergranular pressure solution, quartz cementation and porosity in a quartzose sandstone: Journal of Sedimentary Petrology, v. 54, p. 248-361.

Housenecht, D.A., 1987, Assessing the relative importance of compactional processes and cementation to the reduction of porosity in sandstones: American Association of Petroleum Geologists Bulletin, v. 71, p. 633-642.

Hutcheon, I., 1984, A review of artificial diagenesis during thermally enhanced recovery, in McDonald, D.A., and Surdam, R.C., eds., Clastic diagenesis: American Association of Petroleum Geologists Memoir 37, p. 413-430.

Janecky, D.R., and Seyfried, W.E., Jr., 1984, Formation of massive sulfide deposits on oceanic ridge crests: Incremental reaction models of mixing between hydrothermal solutions and seawater: Geochimica et Cosmochimica Acta, v. 48, p. 2723-2738.

Jones, B.F., 1986, Clay mineral diagenesis in lacustrine sediments, in Mumpton, F.A., ed., Studies in diagenesis: U.S. Geological Survey Bulletin 1578, p. 291-300.

Kharaka, Y.K., Law, L.M., Carothers, W.W., and Goerlitz, D.G., 1986, Role of organic species dissolved in formation waters from sedimentary basins in mineral diagenesis, *in* Gautier, D.L., ed., Roles of organic matter in sediment diagenesis: Society of Economic Paleontologists and Mineralogists Special Publication 38, p. 111-123.

Killkelly, M.K., Harbert, H.P., III, and Berg, W.A., 1981, Field studies on Paraho retorted oil shale lysimeters: Leachate, vegetation, moisture, salinity and runoff, 1977-1980: Colorado State University, National Technical Information Service, PB 81-234742.

Kreitler, C.W., 1989, Hydrology of sedimentary basins: Journal of Hydrology (in press).

Kuttan, K., Kulla, J.B., and Neumann, R.G., 1986, Freshwater influx in the Gippsland basin: Impact on formation evaluation, hydrocarbon volumes and hydrocarbon migration: Australian Petroleum Exploration Association Journal, v. 26, p. 242-249.

Land, L.S., 1986, Limestone diagenesis—Some geochemical considerations, *in* Mumpton, F.A., ed., Studies in diagenesis: U.S. Geological Survey Bulletin 1578, p. 129-138.

Land, L.S., Milliken, K.L., and McBride, E.F., 1987, Diagenetic evolution of Cenozoic sandstones, Gulf of Mexico sedimentary basin: Sedimentary Geology, v. 50, p. 195-225.

Lichtner, P.C., Oelkers, E.H., and Helgeson, H.C., 1986, Interdiffusion with multiprecipitation/dissolution reactions: Transient model and the steady-state limit: Geochimica et Cosmochimica Acta, v. 50, p. 1951-1967.

Lindburg, R.D., and Runnels, D.D., 1984, Groundwater redox reactions: An analysis of equilibrium state applied to Eh measurements and geochemical modelling: Science, v. 225, p. 925-927.

Lundegard, P.D., and Land, L.S., 1986, Carbon dioxide and organic acids: Their role in porosity enhancement and cementation, Paleogene of the Texas Gulf Coast, *in* Gautier, D.L., ed., Roles of organic matter in sediment diagenesis: Society of Economic Paleontologists and Mineralogists Special Publication 38, p. 129-147.

MacGowan, D.B., and Surdam, R.C., 1988, Difunctional carboxylic acids anions in oil field waters: Organic Geochemistry, v. 12, p. 245-259.

Mathieson, M., 1984, Diagenesis of Plio-Pleistocene nonmarine sandstones, Cagayan basin, Philippines: Early development of secondary porosity in volcanic sandstones, *in* McDonald, D.A., and Surdam, R.C., eds., Clastic diagenesis: American Association of Petroleum Geologists Memoir 37, p. 177-195.

McBride, E.F., Land, L.S., and Mack, L.E., 1987, Diagenesis of eolian and fluvial feldspathic sandstones, Norphlet Formation (Upper Jurassic), Rankin County, Mississippi, and Mobile County, Alabama: American Association of Petroleum Geologists Bulletin, v. 71, p. 1019-1034.

McWhorter, D.B., 1980, Reconnaissance study of leachate quality from raw mined oil shale-laboratory columns: National Technical Information Service, PB 81-129017.

Meisler, H., Leahy, P.P., and Knobel, L.L., 1985, Effect of eustatic sea-level changes on saltwater-freshwater relations in the northern Atlantic coastal plain: U.S. Geological Survey Water Supply Paper 2255, 28 p.

Melvin, J., and Knight, A.S., 1984, Lithofacies, diagenesis and porosity of the Ivishak Formation, Prudhoe Bay area, Alaska, *in* McDonald, D.A., and Surdam, R.C., eds., Clastic diagenesis: American Association of Petroleum Geologists Memoir 37, p. 177-195.

Miller, D.G., Piwinski, A.J., and Yamauchi, R., 1977, The use of geochemical-equilibrium computer calculations to estimate precipitation from geothermal brines: Lawrence Livermore National Laboratory Report URCL-52197, 34 p.

Moore, C.H., 1987. A computer simulation using REACTRAN, of arkose diagenesis in seawater at 100°C: Temporal evolution of the system: American Association of Petroleum Geologists Research Conference, "Prediction of reservoir quality through chemical modelling," Park City, Utah.

Morton, R.A., and Land, L.S., 1987, Regional variations in formation water chemistry, Frio Formation (Oligocene), Texas Gulf Coast: American Association of Petroleum Geologists Bulletin, v. 71, p. 191-206.

Nagtegaal, P.J.C., 1978, Sandstone-framework instability as a function of burial diagenesis: Journal of the Geological Society of London, v. 135, p. 101-105.

Partridge, A.D., 1976, The geological expression of eustacy in the early Tertiary of the Gippsland basin: Australian Petroleum Exploration Association Journal, v. 16, p. 73-80.

Perry, E., and Hower, J., 1972, Burial diagenesis in Gulf Coast pelitic sediments: Clays and Clay Minerals, v. 8, p. 165-177.

Pettijohn, F.J., Potter, P.E., and Siever, R., 1972, Sand and sandstone: New York, Springer-Verlag, 618 p.

Plummer, L.N., Parkhurst, D.L., and Thorstenson, D.C., 1983, Development of reaction models for groundwater systems: Geochimica et Cosmochimica Acta, v. 47, p. 665-687.

Pryor, W.A., 1973, Permeability-porosity patterns and variations in some Holocene sand bodies: American Association of Petroleum Geologists Bulletin, v. 57, p. 162-189.

Ramsayer, K., and Boles, J.R., 1986, Mixed-layer illite/smectite minerals in Tertiary sandstones and shales, San Joaquin basin, California: Clays and Clay Minerals, v. 34, p. 115-124.

Reed, M.H., 1983, Calculation of multicomponent chemical equilibria and reaction processes in systems involving minerals, gases and an aqueous phase: Geochimica et Cosmochimica Acta, v. 46, p. 513-528.

Riese, W.C., Riese, A.C., and Reed, M.H., 1987, Geochemical modelling applied to the analysis of formation damage with an example from the Gulf of Mexico: American Association of Petroleum Geologists Research Conference, "Prediction of reservoir quality through chemical modelling," Park City, Utah.

Rosen, M.R., and Holdren, G.R., Jr., 1985, Origin of dolomite in Chesapeake Group (Miocene) siliciclastic sediments: An alternative model to burial dolomitization: Journal of Sedimentary Petrology, v. 56, p. 788-798.

Rumble, D., and Spear, F.S., 1983, Oxygen isotope equilibration and permeability enhancement during regional metamorphism: Journal of the Geological Society of London, v. 140, p 175-207.

Rumble, D., Ferry, J.M., and Hoering, T.C., 1986, Oxygen isotope geochemistry of hydrothermally altered synmetamorphic granite rocks from south-central Maine, USA: Contributions to Mineralogy and Petrology, v. 93, p. 420-428.

Schmidt, G.A., and McDonald, D.A., 1979, The role of secondary porosity in the course of sandstone diagenesis, in Scholle, P.A., and Schlager, P.R., eds., Aspects of diagenesis: Society of Economic Paleontologists and Mineralogists Special Publication 26, p. 175-207.

Schmoker, J.W., and Halley, R.B., 1982, Carbonate porosity versus depth: A predictable relationship for south Florida: American Association of Petroleum Geologists Bulletin, v. 66, p. 2561-2570.

Schmoker, J.W., and Gautier, D.L., 1988, Sandstone porosity as a function of thermal maturity: Geology, v. 16, p. 1007-1010.

Shock, E.L., 1988, Organic acid metastability in sedimentary basins: Geology, v. 16, p. 886-890.

Siever, R., 1986, Burial diagenesis of sandstones, in Mumpton, F.A., ed., Studies in diagenesis: U.S. Geological Survey Bulletin 1578, p. 237-248.

Stollenwerk, K.G., and Runnells, D.O., 1981, Composition of leachate from surface retorted and unretorted Colorado oil shale: Environmental Science and Technology, v. 15, p. 1340-1346.

Stonecipher, S.A., Winn, R.D., Jr., and Bishop, M.C., 1984, Diagenesis of the Frontier Formation, Moxa arch: A function of sandstone geometry, texture and composition and fluid flow, in McDonald, D.A., and Surdam, R.C., eds., Clastic diagenesis: American Association of Petroleum Geologists Memoir 37, p. 289-316.

Surdam, R.C., and Boles, J.R., 1979, Diagenesis of volcanic sandstones, *in* Scholle, P.A., and Schlager, P.R., eds., Aspects of diagenesis: Society of Economic Paleontologists and Mineralogists Special Publication 26, p. 227-243.

Surdam, R.C., Bosese, S.W., and Crossey, L.J., 1984, The chemistry of secondary porosity, *in* McDonald, D.A., and Surdam, R.C., eds., Clastic diagenesis: American Association of Petroleum Geologists Memoir 37, p. 37-62.

Sverjensky, D.A., 1984, Oil field brines as ore-forming solutions: Economic Geology, v. 79, p. 23-37.

Sverjensky, D.A., 1987, The role of migrating oil field brines in the formation of sediment-hosted Cu deposits: Economic Geology, v. 82, p. 1130-1141.

Thompson, B.R., 1980, The Gippsland sedimentary basin. A study of the onshore area [PhD. thesis]: Melbourne, Australia, University of Melbourne, 254 p.

Threlfall, W.E., Brown, B.R., and Griffith, B.R., 1976, Gippsland basin offshore, *in* Leslie, R.B., Evans, E.J., and Knight, C.L., eds., Economic Geology of Australia and Papau New Guinea, 3. Petroleum: Australasian Institute of Mining and Metallogy Monograph 7, p. 41-67.

Vail, P.R., Mitchum, R.M., Todd, R.G., Widmier, J.M., Thompson, S., III, Sangree, J.B., Bubb, J.N., and Hatlelid, W.G., 1977, Seismic stratigraphy and global changes in sea level, *in* Payton, C.E., ed., Seismic stratigraphy—Applications to hydrocarbon exploration: American Association of Petroleum Geologists Memoir 26, p. 49-212.

Ward, W.C., and Halley, R.B., 1985, Dolomitization in a mixing zone of near-seawater composition, Late Pleistocene, northwestern Yucatan peninsula: Journal of Sedimentary Petrology, v. 55, p. 407-420.

Weeks, J.B., Leavesly, G.H., Welder, F.A., and Saulnier, G.J., Jr., 1974, Simulated effects of oil shale development on the hydrology of Piceance basin, Colorado: U.S. Geological Survey Professional Paper 980, 83 p.

Whitney, G., and Northrup, H.R., 1987, Diagenesis and fluid flow in the San Juan basin, New Mexico—Regional zonation in the mineralogy and stable isotope composition of clay minerals in sandstone: American Journal of Science, v. 287, p. 353-382.

Williamson, D.L., Melchior, D.C., and Wildeman, T.R., 1980, Changes in iron minerals during oil shale retorting. Proceedings of the 13th Oil Shale Symposium, Colorado School of Mines, p. 337-349.

Winkler, H.G.F., 1967, Petrogenesis of metamorphic rocks: New York, Springer-Verlag, 231 p.

Wolery, T.J., 1979, Calculation of chemical equilibria between aqueous solutions and minerals: The EQ3/6 software package: Lawrence Livermore National Laboratory Report UCRL-52658, 79 p.

Wolery, T.J., 1980, Chemical modelling of geologic disposal of nuclear waste: Progress report and a perspective: Lawrence Livermore National Laboratory Report UCRL-52748, 66 p.

Wolery, T.J., 1983, EQ3NR. A computer program for geochemical aqueous speciation-solubility calculations: User's guide and documentation: Lawrence Livermore National Laboratory Report UCRL-53414, 191 p.

Wolery, T.J., Isherwood, D.J., Jackson, K.J., Delany, J.M., and Puigdomenech, I., 1984, EQ3/6 status and applications: Lawrence Livermore National Laboratory Report URCL-91884, 12 p.

Yin, P., 1988, Generation and accumulation of hydrocarbons in the Gippsland basin, S.E. Australia [PhD. thesis]: Laramie, Wyoming, University of Wyoming, 249 p.

13

GLOBAL CYCLOSTRATIGRAPHY—A MODEL

Martin A. Perlmutter and Martin D. Matthews
Texaco E&P Technology Division, P.O. Box 770070, Houston, Texas 77215-0070 USA

ABSTRACT

Cyclostratigraphy is the study of cyclic depositional patterns produced by climatic and tectonic processes. This paper describes a global scale, process driven, forward cyclostratigraphic model that integrates Milankovitch induced, short-term climatic changes with long-term tectonic basin evolution. Lake levels and eustasy are considered as functions of both these variables.

This model uses both conceptual first principles and an ordinal numerical scale to evaluate the processes of sediment weathering, erosion and transport within different climates and to estimate sediment yield to a basin. Redistribution of sediment within a basin is determined using conceptual and volumetric constraints.

Principal controls on a terrestrial depositional system are assumed to be: basic global climatic pattern for a specific time interval; climatic change with time; petrology of the provenance area; regional topography and bathymetry; and change in uplift and subsidence rates with time. Milankovitch oscillations cause climatic patterns to vary between predictable end members. This permits the construction of maps of cyclostratigraphic belts indicating those regions of the earth's surface affected by the same range of climates. Superposition of the predicted climatic sequence on the tectonic framework of a basin permit inference of stratigraphic relationships and simulations of seismic responses. The model is designed for clastic systems and is illustrated for a half-graben basin.

INTRODUCTION

Cyclostratigraphy is defined as the study of cyclic depositional patterns in the geologic record produced by the interaction of climate and tectonic processes. This paper outlines the global-scale forward model of clastic cyclostratigraphy utilized at Texaco over the last four years. Our cyclostratigraphic model integrates Milankovitch induced, short-term climatic changes with the long-term, tectonic evolution of a basin to examine the potential stratigraphic consequences. Eustasy is considered a function of these climatic and tectonic variables.

This cyclostratigraphic model was developed using a combination of conceptual, qualitative and quantitative approaches. The fundamentals of cyclostratigraphy are based on the first principles of weathering, erosion, and sediment transport and deposition. Evaluation of the relative weathering (including percentage of coarse sediment), erosion and transport in different climates was accomplished using an ordinal numerical scale because there is a lack of reliable quantitative information on these variables as a function of climate. Studies of these basin processes in contemporary settings are inadequate for several reasons: (1) they represent a limited range of climates; (2) their results generally ignore climate or combine the effects of climate with those of elevation and/or relief; (3) they are concerned only with suspended and dissolved load; (4) flood conditions are generally ignored; and (5) potential imbalances between sediment production and yield are overlooked.

Currently, the cyclostratigraphic model distributes sediment delivered to a basin into depositional environments using a combination of conceptual and volumetric constraints. In the future, sediment delivery to and partitioning within a basin will be modeled by a computer simulation of the physical processes of erosion, transport and deposition, combined with faulting, subsidence, compaction, and changes of sea or lake level. This paper discusses the fundamental assumptions of the model, the tectonic framework of a basin as a control on long-term deposition and stratigraphy, Hadley circulation and changes in Hadley circulation as a result of Milankovitch oscillations, the construction of cyclostratigraphic belt maps, integration of climate and tectonics in a rift setting to predict stratigraphic sequences through a Milankovitch cycle, and the simulation of synthetic seismic sections from predicted stratigraphy.

Background

The stratigraphic records of both continental and marine basins are frequently interpreted as cyclic. Orders or classes of cycles have been identified, ranging from those forming over millions of years to those forming over tens of thousands of years (Fig. 1). Changes in sea level, tectonics or climate, or a combination of these, have been proposed as the cause of these cycles.

Sea-level change was perhaps the first factor recognized as having the capacity to produce repetitive stratigraphic sequences. As early as 1874, Newberry proposed the rise and fall of sea level as the cause of a fining-upward sequence overlain by a coarsening-upward sequence. Wanless and Shepherd (1936) proposed glacio-eustasy, and associated climatic changes, as the cause of Paleozoic cyclothems. Numerous investigators still regard eustasy as the dominant control on sedimentary processes and sedimentation patterns (e.g., Vail et al., 1977; Posamentier and Vail, 1989).

Models attributing tectonics as the primary driving force of depositional cycles also have been popular. Williams (1891) concluded that sedimentary cycles are most likely caused by "a diminishing angle of slope . . . under uniform conditions of precipitation" (i.e., cycles of uplift and erosion under constant climate). The more well known clastic sedimentary models, including those of Twenhofel (1932), Krynine (1942), and Weller (1956) emphasize a similar interplay of tectonics, provenance and erosion. Recent modifications include Bott and Johnson (1967), Veizer and Jansen (1985) and Blair (1986).

Although the effects of climate on clastic deposition are well documented and discussed in treatises on sedimentation (e.g., Gilbert, 1895; Penck, 1914; Twenhofel, 1932; Krynine, 1942; Garner, 1959; Van Houten, 1964; Dott, 1964; Crook, 1967; Young et al., 1975; Fairbridge, 1976; Olsen, 1980; Le Tourneau, 1985), they are commonly neglected when interpreting clastic stratigraphic sequences (studies of Pleistocene strata generally being an

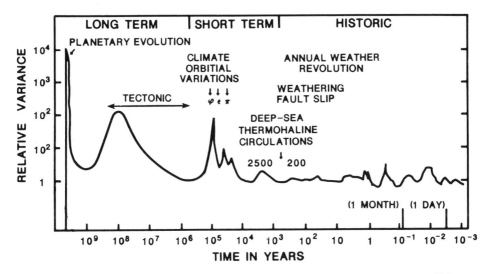

Figure 1. Estimated temporal spectrum of the controls on sedimentation (after Berger, 1980).

exception). Climatic changes generally are considered to cause only minor overprints to a tectonically controlled stratigraphic system rather than a critical factor that determines the system. In contrast, climatic changes caused by Milankovitch oscillations are now increasingly accepted as the cause for repetitive carbonate sequences (e.g., Van Houten, 1964; Barron et al., 1985; Herbert and Fischer, 1986; Dean and Gardner,1986; Goodwin et al., 1986; and Cotillon, 1987).

MODEL ASSUMPTIONS

The fundamentals of cyclostratigraphy are based on sedimentologic first principles, well known and much discussed by many previous authors. Our cyclostratigraphic model adds a unifying framework in which these principles can be placed. Within this framework the relative importance of various geologic factors may be evaluated in time and space, and basin-wide depositional styles and facies evolution may be estimated or predicted, thus linking sedimentology and stratigraphy.

Depositional cycles can be modeled by the time integration of the sedimentary processes producing sediment and transporting it to a basin, the depositional conditions within the basin and the preservation potential of various depositional subenvironments (Fig. 2). Cyclostratigraphy assumes that the principal controls on a depositional system are: (1) basic global climatic pattern for a specific time interval; (2) climatic change with time; (3) petrology of the provenance area; (4) regional topography and/or bathymetry; and (5) rates of uplift and subsidence.

The geographic position of a basin during a particular time interval dictates the climatic influences on basin stratigraphy. Regular and predictable oscillations of the earth's orbit and

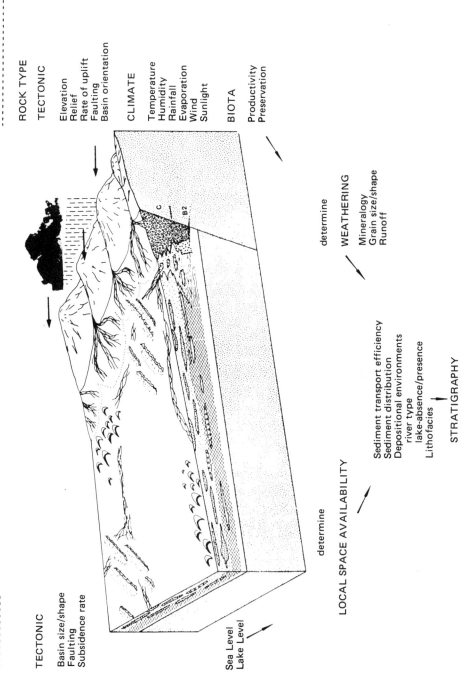

PRODUCTION/TRANSPORTATION

ROCK TYPE

TECTONIC

Elevation
Relief
Rate of uplift
Faulting
Basin orientation

CLIMATE

Temperature
Humidity
Rainfall
Evaporation
Wind
Sunlight

BIOTA

Productivity
Preservation

determine

WEATHERING

Mineralogy
Grain size/shape
Runoff

STRATIGRAPHY

Sediment transport efficiency
Sediment distribution
Depositional environments
river type
lake-absence/presence
Lithofacies

determine

LOCAL SPACE AVAILABILITY

Sea Level
Lake Level

TECTONIC

Basin size/shape
Faulting
Subsidence rate

ACCOMMODATION

Figure 2. Factors affecting the production, transport and deposition of sediment in a basin.

rotation, with periods of approximately 20,000, 40,000, 100,000 and 400,000 yr (Mi-lankovitch cycles), cause variations in climatic patterns between predictable end members. The regular change in climate directly affects weathering processes and rates, runoff, and potential sediment transport. These variables, interacting with provenance and topography, control grain-size distribution, mineralogy and sediment yield. The regular variations of these parameters, impressed upon a tectonic framework of uplift and subsidence, control sediment distribution within a basin. The stratigraphic record can then be predicted based on sedimentologic responses to a predictable set of climatic changes and tectonic basin evolution.

Several assumptions were made in constructing the model to simplify a complex series of real-world relationships among processes and responses.

(1)   Milankovitch induced climatic changes are a fact.

(2)   Solar radiation has varied within the same limits since (at least) the Cambrian, with orbital cycles occurring at essentially the same amplitudes and frequencies. The earth's rotation rate has decreased at a constant slow rate through the same time period.

(3)   Milankovitch oscillations are simulated by a nested set of cosine functions. The range of climates experienced by a particular region during the course of a Milankovitch cycle is associated with the particular phase of the curve. For convenience, we divide a cycle into 60° intervals, each representing a specific climate (Fig. 3). The derivative of a cosine function indicates that the rate of climatic change is at a minimum for the two end-member climates (maximum—phase A, and minimum—phase C), and is relatively rapid in the transitional climates (B1/B2 and D1/D2). Thus, the end-member

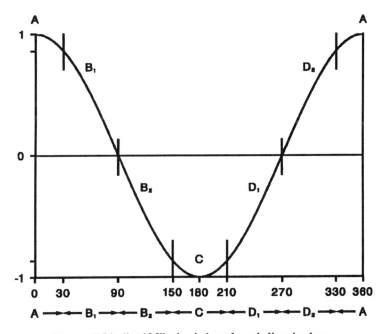

Figure 3. Idealized Milankovitch cycle and climatic phase.

weathering and transport conditions tend to approach or be in equilibrium, whereas the transitional climates are more dynamic, undergoing continual change and exhibiting a "memory" of preceding conditions.

(4)   The dynamic interaction of the atmosphere, oceans and landmasses have not changed significantly since the Cambrian.

(5)   The effect of climate on rock type is the primary agent in producing the textural properties of sediments. It is generally assumed in the literature that this property is primarily determined by relief, slope and transport distance. However, Garner (1959) clearly demonstrated that grain size is primarily a function of climate.

(6)   The effects of short-term, individual fault-slip events within a basin of deposition cause lateral shifts in the position of pre-existing depositional environments. This changes local sediment characteristics without changing the overall sedimentary properties. In addition, each step is aperiodic and unidirectional within the overall subsidence regime. Therefore, a sedimentary cycle produced by fault slip will be aperiodic and asymmetric because of the rapid nature of faulting and the more gradual return due to the processes of basin filling. Short-term, individual fault-slip events within the sediment source area do not cause a significant change in the characteristics of the sediments delivered to the basin because a single event cannot produce a significant change in climate or transport.

(7)   Long-term tectonic evolution of a basin is capable of creating new depositional environments. The balance between subsidence and sedimentation rates determines whether internal or external drainage is developed. Creating new environments, therefore, is a function of the growth history of a basin and the sum of many individual fault-slip events. A tectonically induced depositional cycle is produced as the sediment source areas are raised and subsequently eroded. These types of cycles are likely to be long term and asymmetric. A minimum change in elevation or relief of 500 m is estimated to be necessary before there is a significant change in climate and, therefore, sedimentary characteristics. This number may increase to greater than 3000 m based on an evaluation of the change in relative humidity and temperature with altitude (see Hay et al., 1982, Fig. 7). Assuming a maximum denudation rate of 90 cm/ kyr (estimate for a tropical/very humid environment; Leopold et al., 1964), the minimum time possible to remove 500 m would be 550,000 yr, not counting isostatic compensation. Uplift rates range from 5 cm/kyr on rift margins (Seidler and Jacoby, 1981) to 800 cm/kyr in orogenic belts (Schumm, 1963). Thus, the time required for an uplift of 500 m ranges between about 10 m.y. and 60 kyr. The time necessary to cause a sedimentary cycle as a result of uplift producing a new climate and erosion allowing it to evolve back to its original condition is, therefore, at least 610,000 yr, and can be as high as 10.5 m.y. or longer, ignoring isostasy.

(8)   Sediment production is determined for a granitic source terrain.

(9)   Sediment production is highest in temperate/humid and subhumid climates. Production decreases toward polar climates because of the decreasing role of thermal shock, toward arid climates because of the scarcity of water, and toward tropical/very humid climates because of the increased role of chemical weathering and solution. Sediment production associated with glaciers is added directly to the value for the base climate.

(10)   The ability to transport sediment into a basin is directly related to runoff, vegetative

cover and seasonality of precipitation. Sediment transport rates are greatest for tropical/dry, mediterranean and monsoonal climates (Schumm, 1977; Wilson, 1973). Transport values are corrected for glaciation and glacial and melting when appropriate.

(11)   The coarse fraction (sand size or larger) of the total sediment production is estimated by the balance between physical and chemical weathering. It is expected to be minimal under tropical/very humid conditions, where chemical weathering is dominant and coarse material is derived from the stable quartz component. Under these conditions, the coarse fraction is limited to about 10% of the mass of granitic parent material (Williams et al., 1982). As chemical weathering decreases and physical weathering increases, the proportion of coarse material increases to a value of 70% for temperate/arid conditions. Glaciers are expected to add an equal volume of coarse and fine material and so do not effect the value for the base climate.

(12)   The model does not deal directly with the historic time-scale events of avulsion, migration of depositional environments or individual floods. These are considered functions of short-term climatic patterns and long-term tectonic trends.

(13)   The effects of compaction on accommodation space are assumed relatively constant and overwhelmed by other effects.

TECTONIC FRAMEWORK

Modeling clastic sedimentary cycles necessitates the identification and integration of both tectonic and climatic controls. The effects of each control on basin stratigraphy must be observable, understood and/or predictable if models are to be effective and accurate descriptors of the real world. Sedimentary textures, depositional patterns and stratigraphic thicknesses reflect the interaction of tectonically and climatically induced variations. There are several significant differences between climatically and tectonically controlled driving forces, including time scales of operation, periodicity, environmental changes and textural changes, that can be used to discern the cause of depositional cycles.

Long-term changes in depositional regimes are dominated by basin position with respect to global climate belts, tectonic style, stage of basin development and provenance area. Factors such as basin subsidence (accommodation space), margin uplift (drainage area, maximum potential sedimentation rates, potential deliverable sediment volume, local climate) and continental drift (regional climate) are incorporated in the model. Together, these factors form a depositional framework that evolves relatively slowly over time.

Different types of basins (e.g., rift, wrench, forearc, sag) produce depositional regimes characteristic of each stage of their development. A half-graben was chosen for demonstration of the model because it can contain a wide range of depositional environments within a relatively narrow zone. Two end-member tectonic conditions are considered: high margins—subsidence rate greater than sedimentation rate; and low margins—subsidence rate less than sedimentation rate.

High uplift rates of rift margins are typical of the early to middle stages of rifting (Seidler and Jacoby, 1981). Basin subsidence rates are greater than sediment supply rates at these stages because rift-margin drainage areas are commonly underdeveloped. This tectonic regime commonly produces interior drainage and starved basins, and lakes are formed if runoff is sufficient.

In later stages of rifting, tectonic uplift of margins and subsidence of the basin slows or ceases, and rift shoulders are extensively dissected and eroded. Under these conditions, subsidence rates are less than sediment supply rates. Rift-margin drainage areas tend to be large, producing and transporting more sediment than can be accommodated in the basin. With sufficient runoff, an axial-river system develops and sediment is carried out of the basin, selectively removing fine-grained material from the system.

SHORT-TERM CLIMATIC CHANGES IN ATMOSPHERIC CIRCULATION

Milankovitch oscillations occur with periods of about 20,000, 40,000, 100,000 and 400,000 yr (Milankovitch, 1941). The different frequencies cause amplitudes to constructively and destructively interfere, creating beats that enhance or mute climatic effects (Lockwood, 1980). The climatic progression caused by Milankovitch oscillations directly controls the sediment input to, and distribution within, a basin. The first step in recognizing how the sediment delivery system to a basin can produce cycles as a result of climatic variability is to understand how regional climate can change over the duration of a Milankovitch cycle. To do this, it is important that climate be understood in the context of the global atmospheric circulation pattern.

Basic Hadley Circulation and Climate

The earth's atmospheric circulation, from equator to pole, is typically simplified into three cells, Hadley, Ferrel and Polar, composing Hadley Circulation (Hanwell, 1980). Temperature and humidity are related to the circulation of these atmospheric cells (Fig. 4). Temperatures generally decrease from the equator to the pole, and for convenience we have directly related temperature to each cell: Hadley—tropical; Ferrel—temperate; Polar—polar. Humidity, however, is a manifestation of the position within an individual cell. Humid conditions generally occur near upwelling arms of cells, and humidity decreases to a minimum (arid conditions) near downwelling arms.

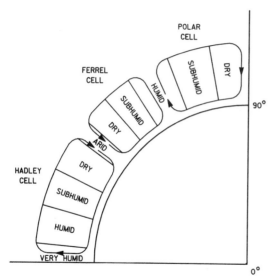

Figure 4. Simplified view of Hadley circulation cells showing estimated latitudinal distribution of relative humidity.

Modifications to Basic Hadley Circulation and Climate

The global distribution of climates is a function of Hadley circulation and therefore generally zonal in nature (Fig. 5). However, this ideal climatic zonation is complicated by monsoonal circulation, anticyclonic flow around mid-latitude highs, proximity to oceans, basin elevation, and high mountains or deep valleys (Figs. 5–8). Effects differ on east and west sides of continents, and vary by the season.

The monsoonal shift of the Intertropical Convergence Zone (ITCZ), caused by the geographic configuration of continents and oceans in mid-latitudes and the differential heating of land and sea, disrupts the basic zonal pattern in the tropics (Figs. 5–7). Monsoonal circulation creates an envelope in which environments are modified from the climates expected purely from Hadley circulation, producing alternating wet and dry seasons. Monsoons also "buckle" climatic belts outside the envelope, shifting environments away from their expected position. The absolute effect of a monsoon is dependent on the atmospheric pressure difference between land and sea, wind direction and wind fetch over land or water, and can be enhanced by the position of mountain ranges (Fein and Stevens, 1987).

In mid-latitude regions, the zonal nature of the downwelling arm of the Hadley and Ferrel cells can be disrupted due to the differential heating of land and sea. As a result, wind circulation around mid-latitude high-pressure cells (anticyclonic flow) can also play a large role in transferring heat and moisture onshore or offshore (Figs. 5 and 7). On east sides of continents, winds tend to blow onshore in equatorward areas, eliminating potential deserts by increasing precipitation (east coast effect). On west sides of continents, winds tend to blow onshore in poleward latitudes, making the environment more equable (west coast effect). The magnitude of the effect is dependent on wind direction, coastline orientation and oceanic currents.

Ocean currents transfer warm equatorial waters toward the poles and cold polar waters toward the equator (Figs. 5 and 7). Warm, surface boundary currents provide heat and moisture onshore to lower mid-latitude eastern coasts and upper mid-latitude western coasts. Cold, surface boundary currents can cool and provide moisture to mid-latitude western coasts. Upwelling of cold, deep currents cool western coasts and may provide moisture, depending on the prevailing wind direction.

Orographic effects on regional or local climates are dependent on the temperature and humidity of the air mass at sea level, the total change in altitude, the lapse rate and the prevailing wind direction (Fig. 5). As air rises, it cools adiabatically and increases in relative humidity. As air descends, it warms adiabatically and decreases in relative humidity. An orographic effect will occur when prevailing winds force air over mountain ranges, tending to produce moister conditions on the windward side and drier conditions on the leeward side.

An Example of Hadley Circulation During a Milankovitch Cycle

The relative sizes and positions of the three circulation cells vary seasonally as a function of differences in winter/summer insolation (Lockwood, 1980; and Frakes, 1979). Variations in solar radiation on the scale of Milankovitch oscillations are expected to shift the basic Hadley circulation pattern in much the same manner (Glennie, 1984). The generalized positions of

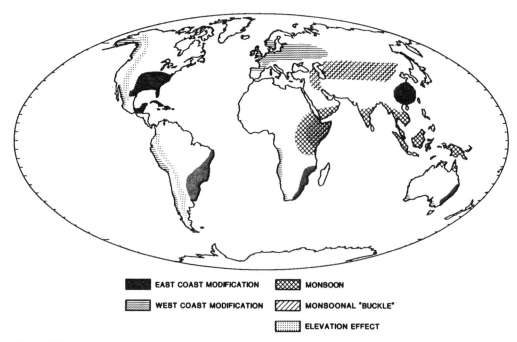

Figure 6. Regions of the present earth's surface which deviate from a simple zonal climatic distribution.

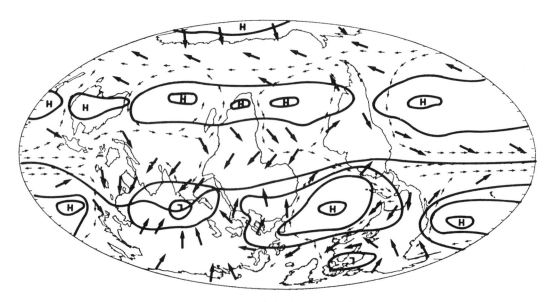

Figure 7. Global atmospheric circulation patterns and ocean currents for the present-day northern hemisphere summer. Large arrows indicate wind patterns. Small arrows indicate ocean currents.

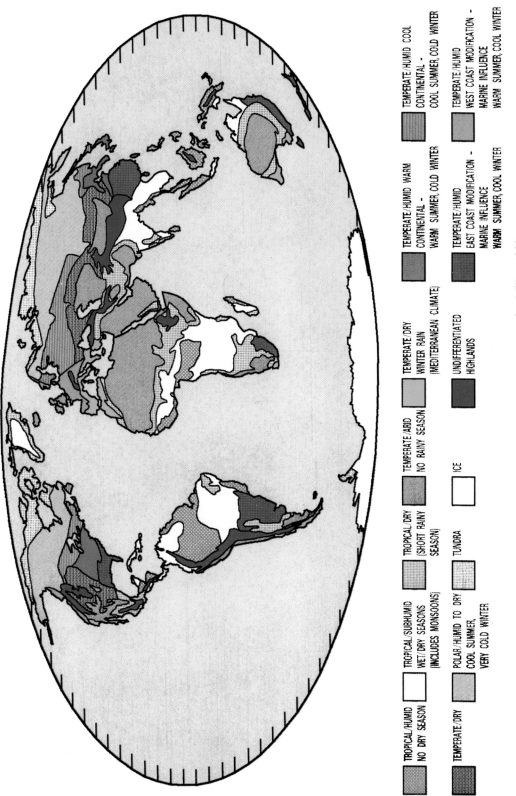

Figure 5. Distribution of present climatic environments (after Miller, 1966).

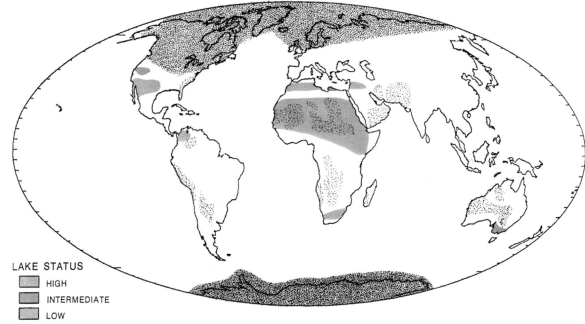

Figure 10. Distribution and levels of lakes during the last climatic minimum, 20,000 to 21,000 yrBP (after Street and Grove, 1979). The light stippled pattern represents locations of dune fields and the dark stippled pattern represents locations of glaciers (after Sarnthein and Diester-Haass 1977).

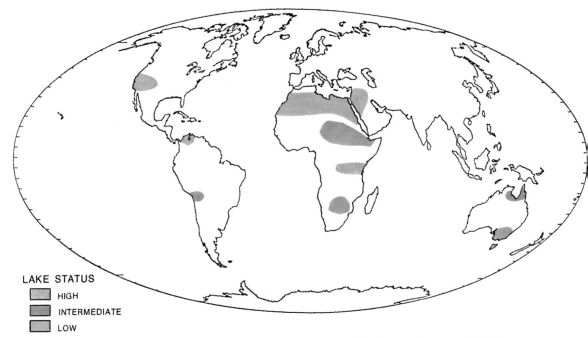

Figure 11. Distribution and levels of lakes during the last climatic warming event, 11,000 to 12,000 yrBP (after Street and Grove, 1979).

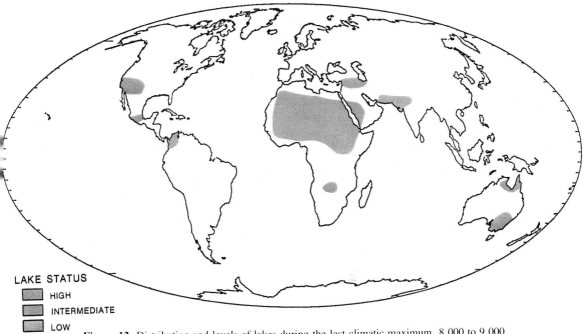

Figure 12. Distribution and levels of lakes during the last climatic maximum, 8,000 to 9,000 yrBP (after Street and Grove, 1979).

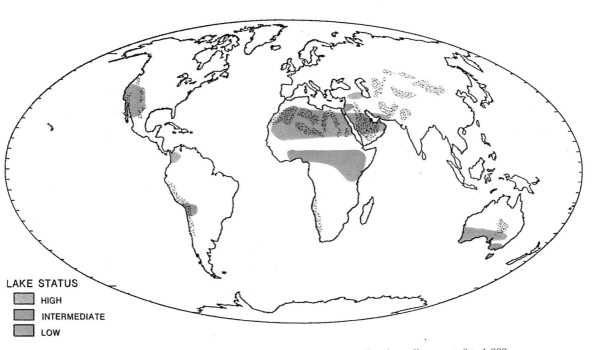

Figure 13. Distribution and levels of lakes during the present climatic cooling event, 0 to 1,000 yrBP (after Street and Grove, 1979). The light stippled pattern represents locations of dune fields (after Sarnthein and Diester-Haass, 1977).

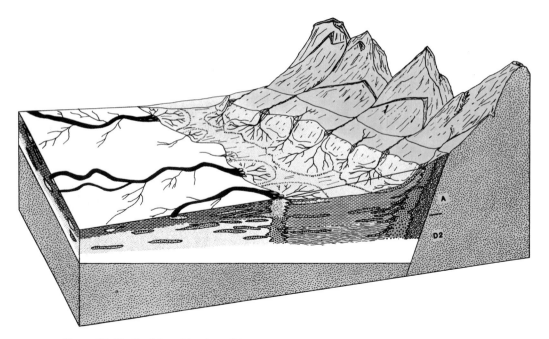

Figure 17. Idealized depositional conditions and stratigraphic relationships for a rapidly subsiding basin with elevated margins: belt #3, phase A (climatic maximum). Conditions are Tropical/Humid.

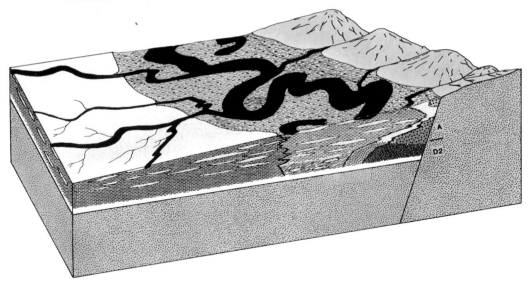

Figure 18. Idealized depositional conditions and stratigraphic relationships for a slowly subsiding basin with low margins: belt #3, phase A (climatic maximum). Conditions are Tropical/Humid.

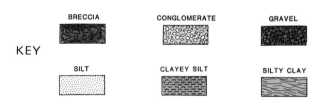

KEY

BRECCIA

CONGLOMERATE

GRAVEL

SILT

CLAYEY SILT

SILTY CLAY

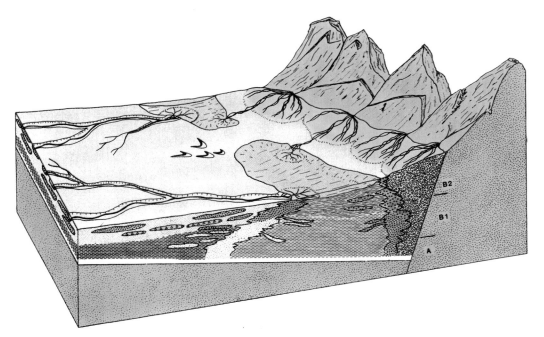

Figure 19. Idealized depositional conditions and stratigraphic relationships for a rapidly subsiding basin with elevated margins: belt #3, phases B1 and B2 (cooling). Conditions are Tropical/Subhumid in phase B1 and Tropical/Dry in phase B2.

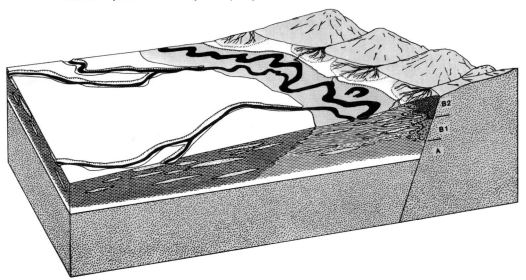

Figure 20. Idealized depositional conditions and stratigraphic relationships for a slowly subsiding basis with low margins: belt #3, phases B1 and B2 (cooling). Conditions are Tropical/Subhumid in phase B1 and Tropical/Dry in phase B2.

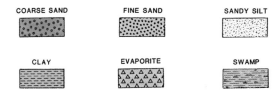

COARSE SAND   FINE SAND   SANDY SILT

CLAY   EVAPORITE   SWAMP

Figure 21. Idealized depositional conditions and stratigraphic relationships for a rapidly subsiding basin with elevated margins: belt #3, phase C (climatic minimum). Conditions are Temperate/Arid.

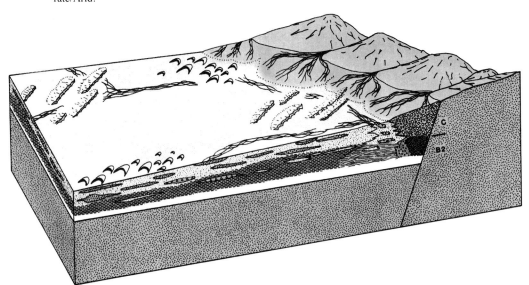

Figure 22. Idealized depositional conditions and stratigraphic relationships for a slowly subsiding basin with low margins: belt #3, phase C (climatic minimum). Conditions are Temperate/Arid.

KEY

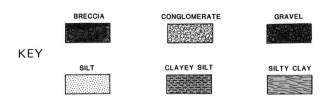

Figure 23. Idealized depositional conditions and stratigraphic relationships for a rapidly subsiding basin with elevated margins: belt #3, phases D1 and D2 (warming). Conditions are Tropical/Dry in phase D1 and Tropical/Subhumin in phase D2.

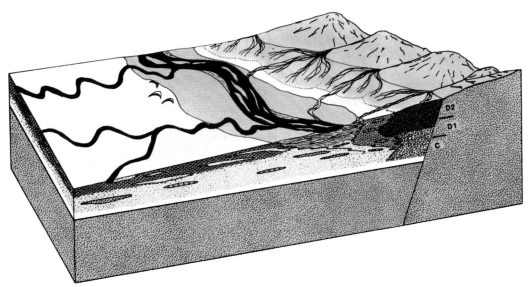

Figure 24. Idealized depositional conditions and stratigraphic relationships for a slowly subsiding basin with low margins: belt #3, phases D1 and D2 (warming). Conditions are Tropical/Dry in phase D1 and Tropical/Subhumin in phase D2.

COARSE SAND   FINE SAND   SANDY SILT

CLAY   EVAPORITE   SWAMP

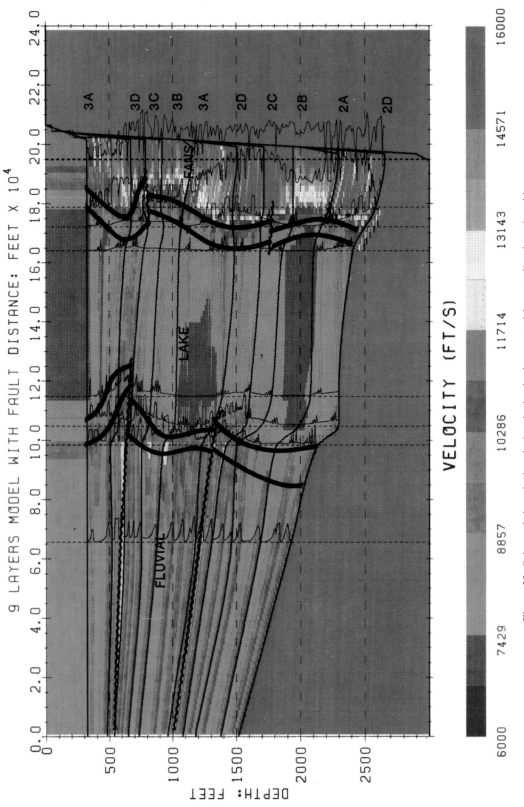

Figure 28. Color-coded synthetic velocity plot by depth generated from predicted stratigraphic columns. Higher velocities (red) represent coarser sediments. Lower velocities (blue) represent finer sediments.

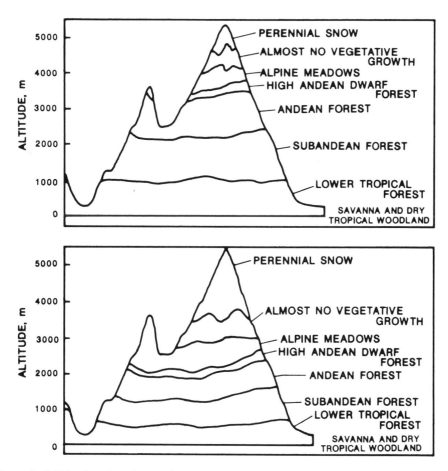

Figure 8. Shift in elevation of vegetative zones between the present and last climatic minimum (after van der Hammen, 1974). (a) Approximate elevations of present vegetation belts in the Eastern Cordillera, Columbia. (b) Approximate elevations of vegetative zones during the last climatic minimum. Note a downward shift of about 1200 m compared with the present.

Hadley circulation cells, as the earth proceeds through a Milankovitch cycle, are shown in Figure 9. This figure indicates the expected changes between two end-member conditions, the climatic minimum (Fig. 9a) and the climatic maximum (Fig. 9c). At the climatic minimum the Polar cell expands and the Hadley cell contracts, whereas at the climatic maximum the Polar cell contracts and the Hadley cell expands. Thus, as these temperature controlled cells migrate across the earth's surface, relative humidity belts migrate with them. Computer simulations of global climate are just beginning to be used to effectively evaluate Milankovitch induced depositional changes (e.g., Glancy et al., 1986).

To provide an understanding of the potential magnitude of these shifts, data from the end of the Wisconsin glacial period to the Holocene are used as examples. Data availability and

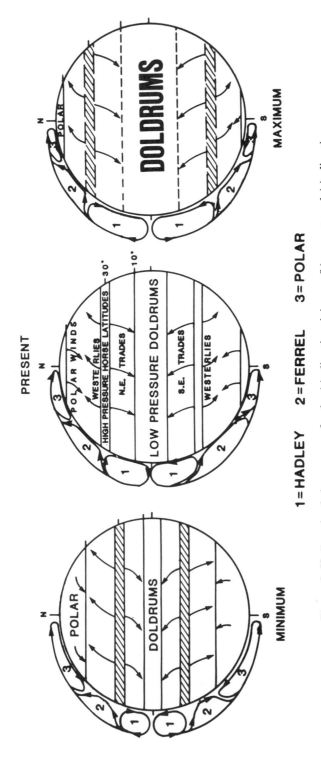

Figure 9. Hadley circulation patterns for the (a) climatic minimum, (b) present, and (c) climatic maximum (after Glennie, 1984).

244

extreme climatic conditions between glacial and interglacial periods make changes in circulation patterns more interpretable. Shifts in circulation were estimated using maps of the global distribution of lakes and lake levels (Street and Grove, 1979) and the positions of aeolian dune fields (Sarnthein and Diester-Haass, 1977). These maps are shown in Figures 10–13, with dune fields superimposed on Figures 10 and 13.

CLIMATIC MINIMUM: 20,000–21,000 YRBP (FIG. 10)   The efficiency of the (equatorial) upwelling arm of the Hadley cell, or ITCZ, is reduced because of the depressed insolation rate (Fairbridge, 1986). Lower summer temperatures over Asia, Africa and the equatorial Atlantic and Indian Oceans reduce the latitudinal shift and efficiency of monsoonal conditions, reducing moisture transport across Africa and producing the low lake levels in African equatorial regions.

Adiabatic warming in the downwelling arms of the Hadley and Ferrel cells causes high evaporation rates, with most deserts forming in this belt. Observations and interpretations of positions of low lake levels and active dune fields indicate that this zone existed as close to the equator as 15° to 20° latitude at this time. Adiabatic cooling in the upwelling arms of the Ferrel and Polar cells causes a secondary belt of relatively high precipitation. At the climatic minimum, the position of high and intermediate lake levels indicate this zone was displaced toward the equator to around 40° latitude.

WARMING TRANSITION: 11,000–12,000 YRBP (FIG. 11)   As the earth warmed, Hadley and monsoonal circulation became more efficient, carrying more moisture and shifting away from the equator. This resulted in the deepening of lakes in equatorial areas and southern Saharan Africa, to about 20° north. Concurrently, the downwelling arm of the Hadley/Ferrel cells moved poleward through the area between 25° and 35° latitude, displacing the position of low lake levels toward the poles as the desert zone migrated through this belt. Lakes fed from alpine sources, such as those in southwest North America, were persistent because of high-altitude glaciers and snow fields. The upwelling arm of the Ferrel/Polar cells moved poleward as the Polar cell shrank.

CLIMATIC MAXIMUM: 8,000–9,000 YRBP (FIG. 12)   The Hadley cell reached its maximum size and efficiency, while the Polar cell correspondingly shrank to its minimum size. The ITCZ directly affected regions as far as 10° from the equator. Monsoonal circulation expanded to its most poleward position, forming lakes and rivers in previously arid areas. Monsoons affected regions up to 35° north latitude in Asia. The downwelling arm of the Hadley/Ferrel cells migrated poleward to 35° to 40° latitude (except in monsoonal areas where it was displaced or modified), drying out lakes in the North American southwest and the northern part of the Arabian Peninsula. The upwelling arm of the Ferrel/Polar cells is estimated to have been located near 70° latitude.

COOLING TRANSITION: 0–1,000 YRBP (FIG. 13)   Lower atmospheric and oceanic temperatures began to limit moisture transfer from the ocean to land. The ITCZ became less efficient, with monsoonal circulation patterns contracting back toward the equator. The position of low lakes levels migrated toward the equator, returning to the poleward fringes of the region occupied by high lake levels during the climatic maximum. Deserts became reestablished between 30° and 20° latitude as the downwelling arm of the Hadley/Ferrel cell boundary moved through the region. Lakes between 60° and 35° deepened as the Ferrel cell migrated equatorward and brought a renewed increase in precipitation to this area. Note that the lake level distribution at this time is very similar to that at 11,000 to 12,000 yrBP.

CLIMATIC CHANGE AND CYCLOSTRATIGRAPHIC BELTS

Cyclostratigraphic Belts

The idealized variation of temperature and relative humidity with respect to latitude, as a function of the distribution of Hadley Circulation at the climatic maximum (phase A) and minimum (phase C), is depicted in Figure 14. These climatic changes have been utilized to delineate ten latitudinal cyclostratigraphic belts for the present land/sea distribution shown in Figure 15.

A cyclostratigraphic belt is defined as a region of the earth's surface that undergoes similar climatic changes during constructive interference of the four dominant Milankovitch oscillations. Thus, it represents an estimate of the end-member climates at a particular location and, for any arbitrary time interval, the range of climates should be contained within these end members. Belt boundaries and their end-member climates during maximum and minimum conditions are indicated on the right-hand side of Figure 14. Note that while temperature always decreases from the climatic maximum to minimum, this is not always the case for humidity. For example, the area between 15° and 20° latitude (belt #3) is tropical/humid during the climatic maximum and temperate/arid during the climatic minimum, whereas the area between 35° and 40° (belt #5) is temperate/arid during the climatic maximum and temperate/humid during the climatic minimum. Assuming fluvial sediment input is a function of humidity (i.e., runoff), cyclostratigraphic analysis indicates that the timing of sediment input to continental margins with respect to sea level varies with latitude.

Modifications to Cyclostratigraphic Belts

The idealized climatic sequence represented by a cyclostratigraphic belt occasionally must be modified to account for the effects of monsoonal circulation, mid-latitude anticyclonic wind flow, ocean currents and elevation. These effects can shift, intensify or diminish during a Milankovitch cycle (Fig. 16), and they can be quite pronounced between the climatic maximum and the climatic minimum.

Insolation variations during Milankovitch cycles cause significant changes in the position, strength and efficiency of a monsoon (Fairbridge, 1986). Monsoonal shift of the ITCZ and potential moisture transfer from the ocean are maximized during the climatic maximum because of enhanced heating of continents at mid-latitudes and higher sea-surface temperatures. The effectiveness of summer monsoonal circulation is significantly reduced during the climatic minimum, when insolation is lower and low-pressure systems over land masses are weaker. However, strong winter cooling can strengthen continental high-pressure cells and may produce monsoonal effects in the opposite hemisphere. The estimated

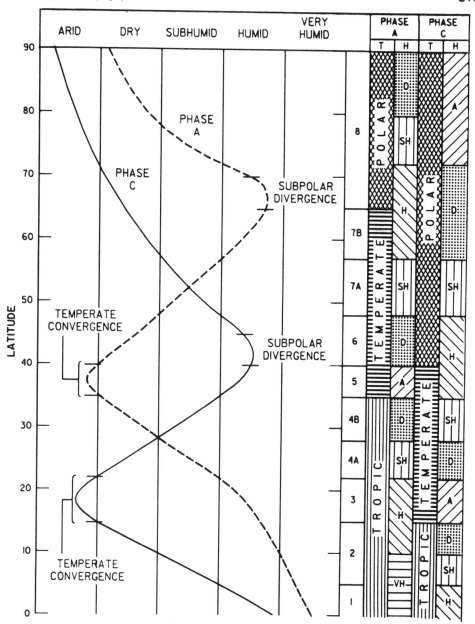

Figure 14. Climatic change and cyclostratigraphic belts. The horizontal scale is estimated relative humidity and the vertical scale is latitude. The solid line represents the estimated climatic zonation by latitude during the climatic maximum. The dashed line represents the estimated climatic zonation by latitude for the climatic minimum. Latitudinal ranges for maximum and minimum conditions of temperature and humidity are shown to the right of the diagram. Cyclostratigraphic belts 1 to 8 represent zones that exhibit distinct climatic/environmental end members during a complete Milankovitch cycle. For example, belt #1 varies from tropical/very humid to tropical/humid and belt #3 varies from tropical/humid to temperate/arid.

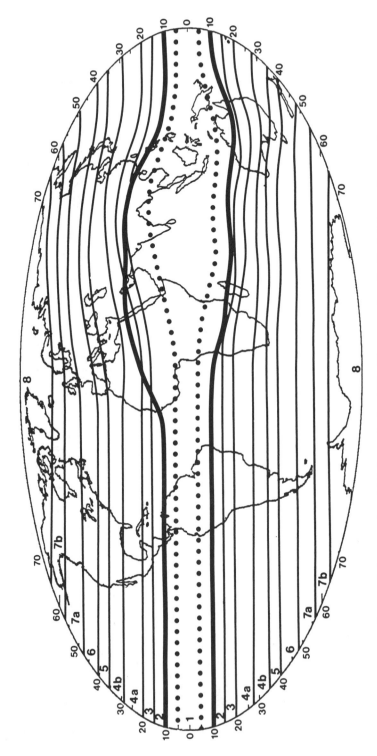

Figure 15. Cyclostratigraphic belt map for 0 to 3 Ma. Climates are described by belt for maximum and minimum conditions. Dotted line indicates position of the ITCZ at the climatic minimum and the solid line indicates its position at the climatic maximum.

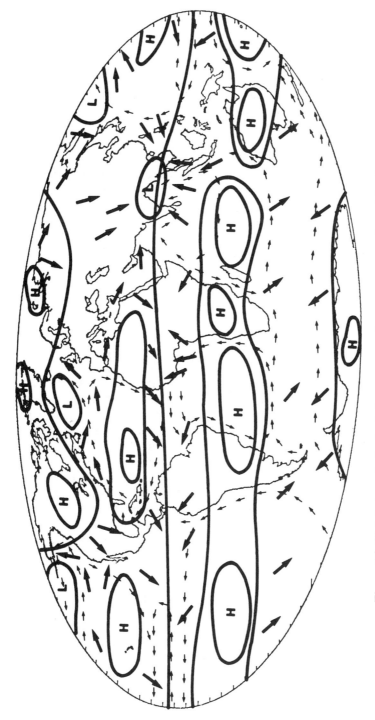

Figure 16. Global atmospheric circulation patterns and ocean currents estimated for the northern hemisphere summer for last climatic minimum. Large arrows indicate wind patterns. Small arrows indicate ocean currents.

positions of the ITCZ for climatic maximum and minimum are shown in Figure 15.

At the climatic maximum, anticyclonic flow around ocean-centered, mid-latitude high-pressure cells shifts poleward. The previously described regions of east and west coast effects also shift poleward and are expected to be at their maximum strength and areal extent. By contrast, at the climatic minimum anticyclonic flow and the associated regions of east and west coast effects shift toward the equator.

The alternation of maximum and minimum climatic conditions should, by analogy with the atmosphere, cause changes in oceanic current systems. Ocean boundary currents should become warmer and shift poleward during the climatic maximum. Upwelling zones, produced by a combination of wind direction and coastline orientation, also shift poleward. At the climatic minimum, boundary currents should be cooler and shift equatorward, and upwelling zones should also shift equatorward. Manabe (1969) and Weyl (1970) have discussed the rationale behind these assumptions.

The effects of elevation on environment vary significantly between maximum and minimum climatic conditions because of a change in the lapse rate (van der Hammen, 1974; Schubert and Medina, 1982). At the climatic maximum, the lapse rate is relatively low. This reduced temperature gradient expands and shifts environments to higher elevations (Fig. 8). At the climatic minimum, lapse rates are relatively high, and environments contract and shift toward sea level.

Constructing Cyclostratigraphic Belt Maps for the Past

Creation of cyclostratigraphic belt maps for the geologic past with different continental distributions, sea levels and orogenic belts is accomplished by deriving the seasonal atmospheric circulation patterns for conditions of climatic maximum and minimum, similar to the methods employed by Parrish and Curtis (1982). This information is simplified into idealized Hadley circulation. Cyclostratigraphic belts are then positioned to reflect regions of the earth's surface that should be influenced by the same climatic range. Transitional climates are interpolated from maximum and minimum conditions. Appropriate modifications to each belt are then applied to the climatic maximum, minimum and transitions. Atmospheric circulation patterns and climatic variations represented by cyclostratigraphic belts are spatially constrained by the geographic position of specific climatic indicators including evaporites, red beds, phosphorites, coals, mineralogic assemblages, sediment grain properties and depositional sequences.

PREDICTED STRATIGRAPHY FOR A HALF-GRABEN

Stratigraphy of a basin is controlled by temporal variabilities in sediment input and sediment distribution processes within the basin. Sediment input is controlled by climatic change over the course of a Milankovitch cycle, topography and provenance. Changes in the balance between rates of sediment input and basin subsidence determine variations in the general distribution of depositional environments. This cyclostratigraphic model integrates this evolution and migration of depositional environments over time, resulting in regular, predictable, lateral and vertical stratigraphic variations, or cycles.

Climatically Produced Stratigraphy

Theoretically, the time integral of the progression of environments through a climate cycle should yield a symmetric signature. However, the resulting thickness of the sedimentary record is not commonly symmetric over a climate cycle because of two effects: (1) depositional rates vary over a cycle; and (2) the change in energy conditions associated with a particular climatic succession has the potential to produce erosional surfaces, redistributing sediment and destroying part of the depositional record.

The expected depositional and stratigraphic changes over the course of a Milankovitch cycle are illustrated for a half-graben basin located in cyclostratigraphic belt #3 in an area unaffected by monsoons. This region exhibits the widest climatic variability and therefore displays the most dynamic stratigraphic changes.

General Distribution of Sediments within a Half-Graben

For the purposes of this illustration, two end-member tectonic conditions are modeled for a rift: high margins with subsidence rate greater than sedimentation rate, and low margins with subsidence rate less than sedimentation rate. Provenance areas on both sides of the graben are considered to occur at approximately the same range of elevations, allowing both margins to produce sediments of similar type, texture and mineralogy in similar volumes per unit area. The general morphology of the half-graben and associated drainage basins follow those described by Rosendahl et al. (1986). Sediments delivered to the basin on the hinge side generally are finer and better sorted than those on the fault side. The provenance area on the hinge side is farther from the basin depocenter and has a gentler average slope than the fault side. Longer transport distance coupled with lower slope increase the effects of abrasion but decrease transport efficiency, biasing the sediments delivered to the basin toward the fine-grained end of the spectrum.

The volume of sediments delivered to a rift basin from the hinge side is expected to be greater than the volume of sediments transported from the fault side. Although higher average slopes on the fault side are intuitively expected to cause higher erosion rates and therefore a relatively larger volume of sediments, larger drainage areas supplying the hinge side more than compensate for lower slopes. This imbalance of sediment supply, coupled with more rapid subsidence rate on the fault side, cause the depositional divide of the basin to be located faultward of the geometric center of the basin.

Predicted Stratigraphic Changes Through a Milankovitch Cycle

A climatic cycle has been divided into six phases: climatic maximum (phase A), climatic minimum (phase C), cooling transition (phases B1 and B2), and warming transition (phases D1 and D2). The expected changes in the depositional regime are displayed in Figures 17 to 24.

PHASE A: TROPICAL/HUMID; CLIMATIC MAXIMUM (FIGS. 17 AND 18)   Weathering is dominated by biochemical processes. End products are predominantly clays, silts, and dissolved species with some stable sand-sized material. Coarse material is estimated at about 15% of the total.

Runoff is relatively high, resulting in an energetic, efficient sediment transport system in areas of high relief. Areas of low slope may be vegetated and transport systems sluggish. High runoff results in the formation of either a large lake or a wide, meandering axial river system, depending on the tectonic conditions.

PHASE B1: TROPICAL/SUBHUMID; INITIAL COOLING (FIGS. 19 AND 20)   Weathering processes shift from biochemical to a combination of biochemical and physical. End products become coarser and unstable minerals, such as feldspar, are not as completely altered to clays. Coarse material is estimated to increase to about 30% of the total. Runoff decreases but the increase in production of sediment in the provenance area maintains the volume of sediment delivered to the basin. The size of lakes and axial rivers decreases. The position of the fault-side limit of the lake/axial-river facies begins to move away from the fault due to the increase in coarse-grained sediment delivered to alluvial fans. Local erosion may occur near the lake margin. Hinge-side rivers become underfit.

PHASE B2: TROPICAL/DRY; LATE COOLING (FIGS. 19 AND 20)   Weathering products are produced dominantly by physical processes, and end products become coarser. Unstable minerals are only slightly weathered to clay. Coarse material is estimated to increase to about 40% of the total. Runoff continues to decrease, reducing sediment transport potential to the basin. However, decreasing runoff reduces plant cover, increasing the availability of sediment for transport. The effect on sediment yield of this nonlinear relationship is complex. Although there may be a tendency for yield to increase slightly as climate changes from subhumid to dry (Schumm, 1977), we have assumed that the overall yield decreases as it comes into balance with the decreased production rates of a drier climate. Late in this phase sediment storage in the provenance area and on basin margins may occur. The decrease in runoff results in a further decrease in the size of lake/axial-river systems. The position of the fault-side limit of the lake/axial-river facies moves away from the fault due to the expansion of alluvial fans. Hinge-side rivers continue to shrink and may become braided and dunes may develop locally.

PHASE C: TEMPERATE/ARID; CLIMATIC MINIMUM (FIGS. 21 & 22)   Weathering is entirely dominated by physical processes. End products are the coarsest expected for the entire climatic cycle and coarse material is estimated to be about 70% of the total. Unstable minerals are common. Runoff is at a minimum and transport efficiency is low. Rivers are ephemeral, with flash floods depositing most sediment on alluvial fans. Most sediment is stored in provenance areas or on basin margins. Lakes containing a sufficient volume of water to persist through the climatic minimum are likely to become saline. Lakes with less water may become playas or evaporate completely. Winds redistribute sediment into dune fields.

PHASE D1: TROPICAL/DRY; EARLY WARMING (FIGS. 23 AND 24)   Weathering processes shift from physical to a combination of physical and biochemical. End products become finer. Unstable minerals partially alter to clay. Coarse material is estimated to decrease to about 40% of the total. Runoff increases, resulting in the initiation of a "flushing" phase in which sediment previously stored in the provenance area and on basin margins is flushed into the basin. This process continues until an equilibrium between sediment production and erosion/transportation is established. Runoff increases, increasing the size of the lake/axial-river system. Increase in transport efficiency within the basin also may produce local erosional events resulting in unconformities. A climatically induced sediment flush can create an unusual juxtaposition of simultaneous transgressional and regressional lake/axial-river margins. In lacustrine regimes, for example, a hinge-side transgression occurs because the dominant effect is the increase in water volume. On the fault-side, however, local regressions can occur if local deposition on alluvial fans from nearby provenance areas is sufficient to push the margin away from the fault over previously existing lake facies. Margins of axial-river systems may undergo similar changes.

PHASE D2: TROPICAL/SUBHUMID; LATE WARMING (FIGS. 23 & 24)   Weathering becomes more dominated by biochemical processes. End products become finer and unstable minerals become more completely altered to clays. Coarse material is estimated to decrease to about 30% of the total. Runoff continues to increase, expanding the sizes of the lake/axial-river systems, and extends the flush of sediment into the early part of this phase. Later in the phase, when all the available stored sediment has been transported and the rivers are in equilibrium with the sediment producing capacity of the climatic conditions of this phase, the deposition rate and the volume of coarse sediment are reduced. Hinge-side lake/axial-river margins may experience a transgression while fault-side margins initially experience a regression. Later in the phase, both margins migrate back toward the fault in a "normal" transgression.

SEISMIC SIMULATION OF STRATIGRAPHY

Predicted stratigraphy can be used to generate synthetic seismic and borehole geophysical profiles. These profiles can then be compared with real-world data for testing and refining models and for aiding in the recognition of seismic facies. As an example, the ideal stratigraphic facies relationships predicted by cyclostratigraphic modeling have been composited for a half-graben basin with interior drainage that has drifted from climatic belt #2 to belt #3 (Fig. 25). For simplicity, the transition between belt 2–phase D2 and belt 3–phase A is assumed to occur instantaneously, and one complete Milankovitch cycle is illustrated for each belt. Synthetic stratigraphic columns at eight locations across this rift were developed using (1) predicted depositional environments and their relative locations, (2) estimated thickness of each stratigraphic unit deposited during each phase of the climatic cycle, and (3) predicted erosional episodes.

   Synthetic velocity log profiles were created from the predicted stratigraphic columns (Fig. 26). Log responses were modeled using ideal grain-size distributions for each depositional environment and a simplifying assumption that grain size and velocity were directly related. Log shapes were estimated from Klein (1985). Impedance contrasts generated from the synthetic velocity logs were convolved with a zero-phase Ricker wavelet to create a synthetic seismic section in time (Fig. 27).

   A color coded, synthetic velocity plot by depth (Fig. 28) was also generated from the predicted stratigraphic columns. Several depositional features are recognizable on this display. Subtle velocity contrasts near the center of the lake represent sediments with low

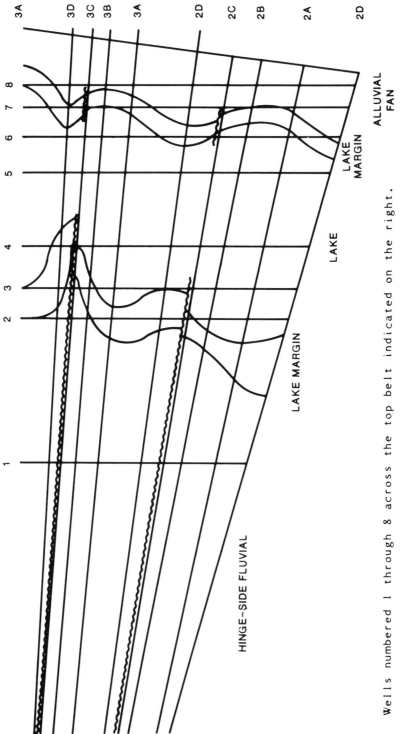

Wells numbered 1 through 8 across the top belt indicated on the right.

Figure 25. Synthetic stratigraphic section for a half-graben located in belt #2 (lower sequence) and belt #3 (upper sequence). For convenience, the migration of the basin from belt #2 to belt #3 is assumed to occur instantaneously between belt #2, phase D2 and belt #3, phase A.

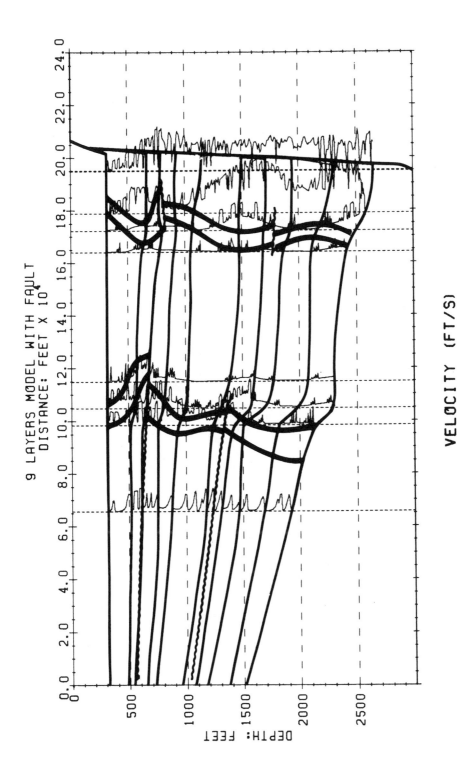

Figure 26. Synthetic stratigraphic section for a half-graben located in belt #2 (lower sequence) and belt #3 (upper sequence) exhibiting velocity profiles. Predicted basin stratigraphy has been used to generate eight artificial stratigraphic columns. Velocity profiles were simulated from these columns assuming ideal grain-size distributions and a simplifying assumption that velocity is directly related to grain size.

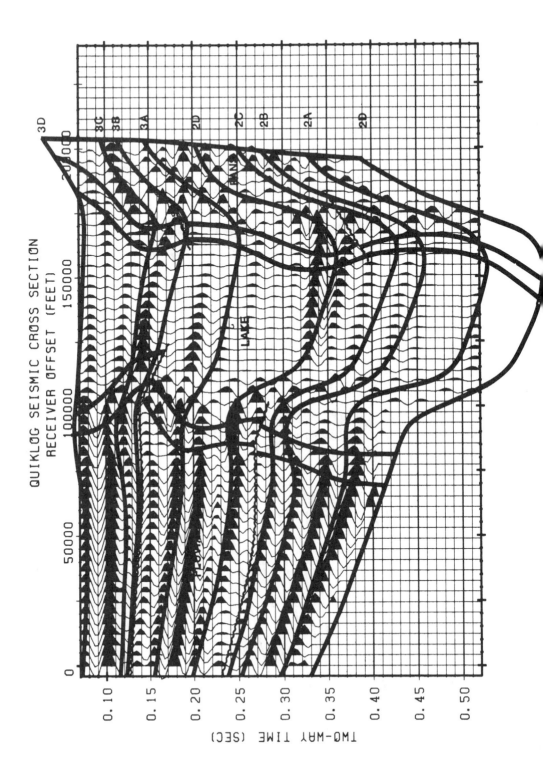

velocities (related to high clay content) deposited near the climatic maximum and slightly higher velocities (related to increased silt content) near the climatic minimum. High velocity contrasts in the lacustrine environment represent an increase in the frequency of turbidity flows. These occur in the flush phases of climate belts #2 and #3. The hinge-side fluvial sequence shows two episodes of higher velocities. These are associated with the increase in sand content of fluvial sediment during the climatic minimum and the following flush phase. Facies changes on the fault side of the basin are more difficult to recognize because higher slopes compress the distribution of depositional environments and reduce sorting.

MODEL TESTING AND FUTURE PLANS

This cyclostratigraphy model has been tested in a variety of rift basins and one foreland basin ranging from Triassic to Pleistocene. It has predicted successfully the variety of depositional environments and the general stratigraphic patterns in these basins. However, additional rigorous tests are needed to compare actual with predicted stratigraphic changes and event synchroneity across a basin. The ability to carry out such tests may be limited until the development of independent means for obtaining a time resolution on the order of 10,000 yr or even 100,000 yr in terrestrial basins.

First principles used to generate the stratigraphic responses also need to be refined, particularly for use in quantitative process-response models. These refinements include: (1) better estimates of the "equilibrium" distribution of grain sizes and mineralogic maturity produced for a constant climate and a given suite of bedrock lithologies; (2) better understanding of the duration for the weathering profile produced by a particular climate and suite of rocks to reach equilibrium and leave a characteristic signature; (3) better ability to quantitatively predict the distribution of sediments and environments in a basin as functions of tectonics and climatic change; and (4) a better understanding of how successive climates modify the stratigraphic record of previous climates.

We are extending the model into other depositional regimes, including marine systems. We also are developing the methodology to integrate the predicted climatic and stratigraphic relationships with well and seismic data in order to constrain correlations of subsurface lithologies and act as input to fluid flow models for hydrocarbon migration, waste (including radioactive) management, and water supply.

ACKNOWLEDGMENTS

We wish to acknowledge the following individuals for providing feedback, incentive, aid and encouragement: R. Anderson, R.E. Byrd, J.B. Carnes, T. Cross, K. Dempster, S. Edwards, D. Fulton, M. Gonyou, T. Jordan, F. H. Nicolai, F. A. Seamans, T. Urwongse and M. Walsh, as well as numerous members of Texaco's exploration and research staff for stimulating discussions. We also wish to thank Texaco Inc. for permission to publish.

REFERENCES CITED

Barron, E.J., Arthur, M.A., and Kauffman, E.G., 1985, Cretaceous rhythmic bedding sequences: A plausible link between orbital variations and climate: Earth and Planetary Science Letters, v. 72, p. 327-340.

Berger, A., 1980, The Milankovitch astronomical theory of paleoclimates: A modern review: Vistas in Astronomy, v. 24, p. 103-122.

Blair, T.C., 1986, Tectonic and hydrologic controls on cyclic alluvial fan, fluvial and lacustrine rift basin sedimentation, Jurassic-lowermost Cretaceous Todos Santos Formation, Chiapas, Mexico: Journal of Sedimentary Petrology, v. 57, p. 845-862.

Bott, M.H.P., and Johnson, G.A.L., 1967, The controlling mechanism of Carboniferous cyclic sedimentation: Quarterly Journal of the Geological Society of London, v. 122, p. 421-441.

Cotillon, P., 1987, Bed-scale cyclicity of pelagic Cretaceous successions as a result of world-wide control: Marine Geology, v. 78, p.109-123.

Crook, K.A.W., 1967, Tectonics, climate and sedimentation: 7th International Sedimentological Congress Proceedings, v. 1, p. 1-3.

Dean, W.E., and Gardner, J.V., 1986, Milankovitch cycles in Neogene deep-sea sediment: Paleoceanography, v. 1, p. 539-553.

Dott, R.H., 1964, Superimposed rhythmic stratigraphic patterns in mobile belts: Kansas Geological Survey Bulletin 169, v. 1, p. 69-86.

Fairbridge, R.W., 1986, Monsoons and paleomonsoons: Episodes, v. 9, p. 143-149.

Fairbridge, R.W., 1976, Effects of Holocene climatic change on some tropical geomorphic processes: Quaternary Research, v. 6, p. 529-556.

Fein, J.S., and Stevens, P.L., 1987, Monsoons: New York, Wiley Interscience, 632 p.

Frakes, L.A., 1979, Climates throughout geologic time: New York, Elsevier, 310 p.

Garner, H.F., 1959, Stratigraphic-sedimentary significance of contemporary climate and relief in four regions of the Andes Mountains: Geological Society of America Bulletin, v. 70, p. 1327-1368.

Gilbert, G.K., 1895, Sedimentary measurement of Cretaceous time: Journal of Geology, v. 3, p. 121-127.

Glancy, T.J., Barron, E.J., and Arthur, M.A., 1986, An initial study of the sensitivity of modeled Cretaceous climate to cycles insolation forcing: Paleoceanography, v. 1, p. 523-537.

Glennie, K.W., 1984, Early Permian-Rotliegend, in Glennie, K.W., ed., Introduction to the petroleum geology of the North Sea: Boston, Blackwell Scientific Publications, p. 41-60.

Goodwin, P.W., Anderson, E.J., Goodman, W.M, and Saraka, L.J., 1986, Punctuated aggradational cycles: Implications for stratigraphic analysis: Paleoceanography, v. 1, p. 417-429.

Hanwell, J., 1980, Atmospheric processes: Boston, George Allen and Unwin, 97 p.

Hay, W.W., Behensky, J.F., Barron, E.J., and Sloan, J.L., 1982, Late Triassic-Liassic paleoclimatology of the proto-central North Atlantic Rift System: Paleogeography, Paleoclimatology, Paleoecology, v. 40, p. 13-30.

Herbert, T.D., and Fischer, A.G., 1986. Milankovitch climatic origin of mid-Cretaceous black shale rhythms in central Italy: Nature, v. 321, p. 739-743.

Klein, G.deV., 1985, Sandstone depositional models for exploration for fossil fuels: Boston, International Human Resources Development Corporation, 209 p.

Krynine, P.D., 1942, Differential sedimentation and its products during one complete geosynclinal cycle: 1st Congreso Panamericano de Ingeniería de Minas y Geologiá, v. 2 Geologiá, pt. 1, p. 537-561.

Leopold, L.B., Wolman, M.G., and Miller, J.P., 1964, Fluvial processes in geomorphology: San Francisco, Freeman and Co., 522 p.

Le Tourneau, P.M., 1985, Alluvial fan development in the Lower Jurassic Portland Formation, central Connecticut—Implications for tectonics and climate, *in* Robinson, G.P., and Froelich, A.J., eds., Proceedings of the Second U.S. Geological Survey Workshop on the Mesozoic Basins of the Eastern U.S.: U.S. Geological Survey Circular 946, p. 17-26.

Lockwood, J.G., 1980, Milankovitch theory and ice ages: Progress in Physical Geography, v. 4, p. 79-87.

Manabe, S., 1969, Climate and the ocean circulation. Parts I and II: Monthly Weather Review, v. 97, p. 739-805.

Milankovitch, M., 1941, Kanon der Erdbestranlung and seine Anwendung auf das Eiszeitenproblem (Canon of insolation and the ice age problem): Transactions of Royal Serbian Academy, Belgrade (Yugoslavia) Special Publications, v. 132, Section of Mathematical and Natural Sciences, v. 33, 674 p. (English translation by the Israeli program for scientific translation. Published by U. S. Department of Commerce and the National Science Foundation, Washington, DC, NTIS Report SFCSI-COMM(TT-67-514010-1-2), 1969, 507 p.)

Miller, A., 1966, Meteorology: Columbus, Ohio, C. E. Merrill Publishing Co., 128 p.

Newberry, J.S., 1874, Circles of deposition in American sedimentary rocks: American Association for the Advancement of Science Proceedings, v. 22, p. 185-196.

Olsen, P.E., 1980, Fossil great lakes of the Newark Supergroup in New Jersey, *in* Manspeizer, W., ed., Field studies of New Jersey geology and guide to field trips: 52nd Annual Meeting of the New York State Geological Association: Newark, New Jersey, Rutgers University, p. 352-98.

Parrish, J.T., and Curtis, R.L., 1982, Atmospheric circulation, upwelling and organic-rich rocks in the Mesozoic and Cenozoic eras: Paleogeography, Paleoclimatology, Paleoecology, v. 40, p. 31-66.

Penck, A., 1914, The shifting of climatic belts: The Scottish Geographical Magazine, v. 30, p. 281-293.

Pickard, G.L., 1975, Descriptive physical oceanography: New York, Pergamon Press, 214 p.

Posamentier, H.W., and Vail , P.R., 1989, Eustatic controls on clastic deposition. II. Sequence and systems tract models, *in* Wilgus, C.K., et al., eds., Society of Economic Paleontologists and Mineralogists Special Publication 42 (in press).

Rosendahl, B.R., Reynolds, D.J., Lorber, P.M., Burgess, C.F., McGill, J., Scott D., Lambiase, J.J., and Derksen, S.J., 1986, Structural expressions of rifting: Lessons from Lake Tanganyika, Africa, *in* Frostick, L.E. et al., eds., Sedimentation in African rifts: Geological Society of London Special Publication 25, p. 29-43.

Sarnthein, S. and Diester-Haass, L., 1977, Eolian-sand turbidites: Journal of Sedimentary Petrology, v. 47, p. 868-890.

Schubert, C., and Medina, E., 1982, Evidence of Quaternary glaciation in the Dominican Republic: Paleogeography, Paleoclimatology, Paleoecology, v. 39, p. 281-294.

Schumm, S.A., 1963, The disparity between present rates of denudation and orogeny: U.S. Geological Survey Profesisonal Paper 454-H, 13 p.

Schumm, S.A., 1977, The fluvial system: New York, John Wiley and Sons, 338 p.

Seidler, E., and Jacoby, W.R., 1981, Parameterized rift development and upper mantle anomalies: Tectonophysics, v. 73, p. 53-68.

Street, F.A., and Grove, A.T., 1979, Global maps of lake level fluctuations since 30,000 yr B.P.:Quaternary Research, v. 12, p. 83-118.

Twenhofel, W.H., editor, 1932, Treatise on sedimentation: Baltimore, Maryland, Williams and Wilkens Co., 926 p.

Vail, P.R., Mitchum, R.M., and Thompson, S., III, 1977, Seismic stratigraphy and global changes of sea level, Part 3: Relative changes in sea level from coastal onlap, *in* Payton, C.E., ed., Seismic stratigraphy—Applications to hydrocarbon exploration: American Association of Petroleum Geologists Memoir 26, p. 63-81.

van der Hammen, T., 1974, The Pleistocene changes of vegetation in tropical South America: Journal of Biogeography, v. 1, p. 3-26.

Van Houten, F.B., 1964, Cyclic lacustrine sedimentation, Upper Triassic Lockatong Formation, central New Jersey and adjacent Pennsylvania: Kansas State Geological Survey Bulletin 169, v. 2, p. 497-531.

Veizer, J., and Jansen, S.L., 1985, Basement and sedimentary recycling—2: Time dimensions to global tectonics: Journal of Geology, v. 93, p. 625-43.

Wanless, H.R., and Shepherd, F.P., 1936, Sea level and climatic changes related to late Paleozoic cycles: Geological Society of America Bulletin, v. 47, p. 1177-1206.

Weller, J.M., 1956, Argument for diastrophic control of late Paleozoic cyclothems: American Association of Petroleum Geologists Bulletin, v. 40, p. 17-50.

Weyl, P.K., 1970, Oceanography: New York, John Wiley and Sons, 535 p.

Williams, H., Turner, F.J., and Gilbert, C.M., 1982, Petrography and introduction to the study of rocks in thin sections: San Francisco, W.H. Freeman, 626 p.

Williams, J.L., 1891, On cycles of sedimentation: American Geologist, v. 8, p. 315-324.

Wilson, L., 1973, Variations in mean annual sediment yield as a function of mean annual precipitation: American Journal of Science, v. 278, p. 335-349.

Wing, S.L., 1982, Relation of paleovegetation to geometry and cyclicity of some fluvial carbonaceous deposits: Journal of Sedimentary Petrology, v. 54, p. 52-66.

Young S.W., Basu, A., Mack, G., Darnell, N., and Suttner, L.J., 1975, Use of size-composition trends in Holocene soil and fluvial sand for paleoclimatic interpretation: Proceedings of the Ninth International Sedimentological Congress, v. 1, p. 201-206.

14

Mass-Balanced Paleogeographic Maps: Background and Input Requirements

William W. Hay [1], Christopher N. Wold [2], and Christopher A. Shaw [3]

[1] Museum, Cooperative Institute for Research in Environmental Sciences, and Department of Geology, Campus Box 449, University of Colorado, Boulder Colorado 80309 USA

[2] Institut für Allgemeine und Angewandte Geophysik, Ludwig-Maximillians Universität, Theresienstrasse 41/4, D-8000 München 2, Federal Republic of Germany

[3] U.S. Geological Survey, Branch of Sedimentary Processes, Denver Federal Center, Box 25046, MS 939, Lakewood Colorado 80225 USA

Abstract

Mass-balanced paleogeographic maps are quantitative reconstructions of the earth's surface at times in the past. They are based on the principle of mass balance: tectonic, erosion and sedimentation processes acting on the reconstructed surface over a given interval of time cause the mass of sediment eroded to be equal to the mass of sediment deposited. Input files consist of: 1) definition of the unit areas to form a grid within a region that has acted as a closed source-sink system with respect to detrital sediment; 2) present average elevation of each unit area; 3) representative lithostratigraphic columns for each unit area keyed to a chronostratigraphic framework; 4) a regional mass-age distribution of the sediment; 5) an estimate of the mass of sediment that was deposited during each time interval; 6) age of ocean crust and age assignments for the contracting material beneath thinned passive continental margins; 7) a regional sea-level curve; and 8) specification of an appropriate time step. These data are then manipulated by the program through process constants to produce mass-balanced paleogeographic maps (Shaw and Hay, this volume). The mass-age distributions have a high information content and provide a summary of the geologic history of the region.

Introduction

It is apparent that topography has an important direct influence on atmospheric circulation (Gill, 1982), and is a critical input for numerical general circulation models, or GCMs (Barron and Washington, 1984; Barron, 1985). It also has become evident that topographic relief strongly influences local depositional patterns both directly and indirectly through its influence on climate (Hay et al., 1982). Topographic relief ultimately controls drainage, and may cause the runoff from large land areas to enter the ocean in sensitive areas where fresh water input may cause stratification or changes in deep water formation (Hay, 1983). Predictions of wind-driven coastal upwelling in the past are highly dependent on assumptions about relief of the land areas (Barron, 1985; Parrish and Barron, 1986). Physical oceano-graphic studies demonstrate the importance of onshore topography (Roden, 1961; Gill, 1982) and offshore paleobathymetry (Garvine, 1973; Peffley and O'Brien, 1976; Blanton et al., 1981; Gill, 1982; McClain et al., 1986) in localizing coastal upwelling. These studies also suggest that more detailed paleotopographic and paleobathymetric reconstructions might be useful in hydrocarbon exploration.

Reconstructions of ocean floor paleobathymetry are straightforward, and have utilized the age-elevation relationship for ocean crust (Menard, 1969; Sclater et al., 1971; Parsons and Sclater, 1977). Reconstructions of the paleobathymetry of continental margins overlying thinned continental or transitional crust utilize the similarity of the subsidence there to that of ocean crust, as suggested by Sleep (1971), but requires special treatment (Steckler and Watts, 1978). Other reconstructions of paleotopography have been either wholly qualitative or have been based on general principles of plate tectonics (Kinsman, 1975; Southam and Hay, 1981; Hay, 1984; Hay et al., 1987), but these hypothetical reconstructions were unconstrained by geologic data.

Only recently has it become feasible to use mass balance to provide constraints on paleotopographic and paleobathymetric reconstructions. As a result of extensive marine seismic reflection profiling and study of materials recovered by the scientific ocean drilling programs, documentation of sedimentary budgets on regional and global scales are now adequate to consider many regions as closed systems with respect to detrital sediment and the earth as a closed system with respect to all sediment. This means that it now is possible to apply principles of mass balance to the interpretation of geologic history. Mass balance provides much more rigorous constraints on the reconstruction of paleogeography than the qualitative methods which have been employed until now. To distinguish this new generation of interpretations of earth history, we have coined the term "mass-balanced paleogeographic maps."

The principle on which the maps are produced is simple. In the absence of intervening tectonic events, the present topography is modified from older topography through erosion and the denudation rate is proportional to elevation. Therefore, by replacing sediment from depositional sites onto the source area, we reconstruct the topography. In practice, the procedure is more complex because the sites of deposition and potential source areas change with time.

The reduction of topographic relief and elevation through erosion and the accumulation of the eroded sediment in basins often is considered a diffusive process, making reconstruc-tion of paleogeography impossible. However, as we began to develop the techniques for producing mass-balanced paleogeographic maps we discovered that there is much more memory in the system than is apparent initially. On the short term erosion may be a random process, but on geologic time scales it is controlled rigorously by broad topographic uplifts (Hay, 1983), local relief, rock type and climate (Holland, 1978). Climate controls the rate of weathering of soil parent materials (Lasaga et al., 1985). It also determines the volume and temporal distribution of runoff that transports the detrital sediment to the site of deposition and the dissolved load to temporary storage in the oceanic reservoir. Mass-balanced paleogeographic reconstructions involve analysis of the spatial distribution of sediment mass; if some of this sediment mass occurs in parts of the general source region, the possible sites from which detrital sediment was derived are restricted. Thus mass-balanced paleogeographic maps yield more than shoreline position, paleobathymetric and paleotopographic information. They also specify overburden depths in the past, suggest specific source areas, and provide a wealth of detailed information testable by other geologic observations.

PREVIOUS WORK

Paleogeographic maps are attempts to display the surface of the earth as it was in the geologic past. Until the 1960s knowledge of geology of the earth was essentially restricted to the land areas; 70% of the globe was unknown. Nevertheless, Termier and Termier (1952) produced a suite of global maps that indicated deep oceanic areas, shallow seas, coastal plains, low hills and high mountains. Their interpretations were based on lithologic and paleontologic information. Vinogradov (1967, 1968a, 1968b, 1969), in the Atlases of Lithological-Paleogeographical Maps of the USSR, developed qualitative paleogeographic maps that show a variety of depth and elevation zones and include information on the probable geomorphologic configuration of the land and subsea surfaces. Similar paleogeographic maps have been prepared for the Rocky Mountain region (Mallory, 1972), Europe (Ziegler, 1982), and China (Wang, 1985). However, it is evident from cursory examination of these maps that mass is not conserved, that is, the amounts of sediments shown as deposited during a specific time interval would require unrealistic erosion in the source areas shown.

The advance to quantified reconstruction of paleobathymetry occurred with the discovery that after the formation of ocean lithosphere at sea-floor spreading centers, heat loss by conduction results in cooling and contraction of the lithosphere. The contraction with age has the form of an exponential decay, and thus the depth of oceanic crust can be predicted from knowledge of its age (Menard, 1969; Sclater et al., 1971).

For the first 80 m.y. after its creation, oceanic lithosphere subsides in proportion to the square root of time (Davis and Lister, 1974); after 70 m.y., the plate subsides exponentially to arrive at a constant depth of 6400 m, 130 m.y. after its creation (Parsons and Sclater, 1977).

This two-stage relation between depth and age can be explained by the formation of a thermal boundary layer (Sclater et al., 1980). The thermal boundary layer model requires that the oceanic lithosphere attain a constant thickness after a given time, and that this lithosphere remain in equilibrium with the asthenosphere with a constant temperature along the lithosphere/asthenosphere boundary.

The initial subsidence rate is proportional to the square root of the age of the lithosphere, because it reflects contraction and cooling as the plate loses heat to the seawater and moves away from the ridge. At about 80 m.y. after its formation, the oceanic lithosphere has attained a constant thickness of 125 km. The cooling of the lithosphere from above is no longer the major factor causing the plate to subside, and it is buoyed up by the hot asthenosphere, with a temperature at the lithosphere/asthenosphere boundary of about $1330°$ C. Further subsidence of the plate is explained by a slower cooling rate within the lithosphere, and by sediment loading.

Equation (1) describes the subsidence of oceanic lithosphere younger than 70 m.y., and is proportional to the square root of time (t) where t is in Ma, and d(t) is the depth in meters at a particular time t. The first term on the right side of Equation (1) is the depth (in m, relative to the present-day sea level) at which ocean ridges are formed (Sclater et al., 1971). The

second term on the right side of Equation (1) is t multiplied by a proportionality constant
determined by the fit to data, from Parsons and Sclater (1977).

$$d\left(t\right) = 2500 + 350\left(t^{0.5}\right) \tag{1}$$

Equation (2) describes the subsidence of ocean floor older than 70 m.y. The first term on
the right side of Equation (2) is the maximum depth reached by oceanic lithosphere (6400 m),
before being subducted. The 3200, in the second term on the right side of Equation (2), is the
intercept from the log (elevation) versus age in terms of the plate model for t = 0, from Parsons
and Sclater (1977). The exponential term was determined by the fit to data.

$$d\left(t\right) = 6400 - 3200\left(e^{-t/62.8}\right) \tag{2}$$

Using these formulae, a number of paleobathymetric reconstructions for specific ocean
basins have subsequently been published (Sclater, Abbott and Thiede, 1977; Sclater,
Hellinger and Tapscott, 1977; Sclater, 1985; Barron and Harrison, 1980). Unfortunately,
these equations are not applicable to all ocean floor, i.e., they do not apply to marginal seas
and areas of residual depth anomalies. Residual depth anomalies are substantial areas of the
ocean floor which lie more than 400 m above or below the depth predicted by the depth-age
Equations (1) and (2). The figure of ±400 m, for estimating residual depth anomalies is chosen
because it represents approximately 10 percent of the total subsidence of a mid-ocean ridge
from 2500 m at a spreading center, to 6400 m when at thermal equilibrium on very old
lithosphere (Sclater et al., 1985).

Many residual depth anomalies are related to hot spots. Heestand and Crough (1981)
examined the effect of hot spot swells on an empirical age-depth relationship by sorting age-
depth data as well as distance from the nearest hot spot track. Their study was on the Atlantic
ocean, and they computed a relationship of:

$$d\left(t\right) = 2900 + 300\left(t^{0.5}\right) \tag{3}$$

for ocean floor unaffected by hot spot uplift. 2900 m is the average depth of formation of mid-
ocean ridges in the Atlantic, and $300t^{0.5}$ is the proportionality constant times the square root
of time. Heestand and Crough (1981) arrived at this equation by fit to their data from the
Atlantic. They believed it was applicable to ocean floor of all ages, unaffected by hot spot
swells in the Atlantic. However, because thicknesses of oceanic and continental crust are
utilized in our computations of paleobathymetry (Shaw and Hay, this volume), we use the
equations of Parsons and Sclater (1977).

From studies of passive continental margins it became evident that the subsidence
required to accommodate sediment accumulation could not be explained solely as the result
of sediment loading, as had been suggested by Dietz (1963) and Watts and Ryan (1976). Sleep
(1971) recognized that the Atlantic continental (passive) margins follow a subsidence curve
similar to that of oceanic lithosphere. Using this idea Steckler and Watts (1978) were able to
remove the effects of sediment loading and then make quantitative hindcasts of the
paleobathymetry/subsidence/burial history of passive continental margins. Van Hinte (1978)
developed a method for constructing graphic age-depth plots, called geohistory diagrams,
which incorporate information on sea-level change, subsidence, and burial history. These
graphic representations of geologic history at a point on the earth's surface provided much
of the background necessary for the development of mass-balanced paleogeographic maps.

Recently researchers have begun to develop models to produce quantitative reconstructions of basins and the hydrocarbon maturation and concentration processes that operate within them. Numerical solutions have evolved from the conceptual models that applied geological processes and rates to a set of initial conditions, producing basin reconstructions constrained by the geologic record (Tissot and Welte, 1978; Waples, 1980; Welte and Yukler, 1981; Ungerer et al., 1984; Schmoker, 1986; Bethke et al., 1988; King, in press). Most of these models are two dimensional, but a few are three dimensional. However, virtually all quantitative models to date have treated basins as open systems. These models do not take into account the source area geometry which controls the amount of sediment delivered to the basin and the transport path the sediment travels to its final site of deposition, nor do they estimate the overburden that existed at particular times in the past.

Until the advent of reflection seismic studies and of the ocean drilling programs and offshore exploration, knowledge of the geology of the continental margins and ocean basins was so incomplete that it was impossible to treat land source areas and marine sediment repositories as closed systems. Research during the past two decades has added enormously to the fund of knowledge on the offshore of some regions, and the information from a few of these has recently been synthesized. The atlases produced for the proposed Ocean Margin Drilling Program provide a data base suitable for constructing regional paleogeographic maps based on mass balance. We have used five of these atlases as the basis for data compilations for the Gulf of Mexico (Buffler et al., 1984) and the Atlantic margin of the U.S. (Ewing and Rabinowitz, 1983; Bryan and Heirtzler, 1984; Shor and Uchupi, 1984a, 1984b). For the onshore geology we used the data compiled by Cook and Bally (1975) in their stratigraphic atlas of North and Central America, supplemented with information on the Atlantic margin and western Atlantic Basins from Sheridan and Grow (1988) and Vogt and Tucholke (1986).

To produce mass-balanced paleogeographic maps, we use the interrelationships between sedimentary rocks, topography and erosion processes observed today. Except for information on densities of sedimentary materials and pore fluids, on compaction of sediments with burial, and a sea-level curve for the region, no other geologic information is required for the reconstructions. The present topography, existing sediments and sedimentary rocks, and erosion process rates are considered better known than the vertical tectonic history of the region, so the former are prescribed in the model and uplift or subsidence is a free variable predicted by the model.

The shoreline is imposed on the reconstructed relief by the sea-level curve, and is independent of any information about marine or nonmarine accumulation of sediment. Thus a wide variety of other geologic information is available to test the reconstructions: marine versus nonmarine sediments, paleontologic interpretation of elevation or water depth, sedimentologic indications of surface slope, elevation or depth, paleoclimatic interpretations derived from fossils, evidence from burial diagenesis of sediments, and so forth. If the reconstructions are obviously incorrect, these other kinds of information may be useful in changing process variables, but they never enter directly into the reconstruction process. The maps hindcast many features and conditions suitable for testing by wholly independent information. Because the maps are quantitative expressions of manipulations of stratigraphic, present topographic, and erosional data, they and the assumptions involved in their production must be rigorously defined.

DEFINITION

A mass-balanced paleogeographic map is a quantitative reconstruction of a portion of the surface of the earth for a specific moment in the past. The map is based on conservation of mass so that tectonic, erosion, sedimentation, and isostatic processes acting on the surface over a given interval of time would erode a mass of sediment or rock equal to the mass of sediment deposited.

ASSUMPTIONS

The major assumptions that underlie the construction of mass-balanced paleogeographic maps are:
1) Sediment accumulating at any given time is derived mostly from the erosion of pre-existing older sediment or sedimentary rock with a relatively small increment from the weathering of igneous and metamorphic rocks (Garrels and Mackenzie, 1971; Holland, 1978; Veizer and Jansen, 1979, 1985). 2) The proportion of the sediment mass lost through subduction, metamorphism, and melting during any given time interval is small in relation to the total mass of sediment within the region being investigated (see Veizer and Jansen, 1985). 3) The earth's surface within the region being considered is in isostatic equilibrium. 4) The relief of the earth's surface is controlled by internal processes related to plate tectonics, except for erosion, sea-level changes, and isostatic adjustments. General uplift and subsidence are a result of heating or cooling of the lithosphere or of lateral advection of lithospheric material having greater or lesser density.

GENERATING INPUT FILES

The primary input files define the geographic boundaries of the unit areas, the average elevations, lithostratigraphy and chronostratigraphy in the unit areas, and the age of ocean crust or age of ocean crust adjacent to the margin in oceanic areas and passive margins, respectively. From these primary data files the mass-age distribution for sediment is compiled, and from this we calculate an estimate of the sedimentary masses that were eroded and deposited in the past. A regional sea-level curve must be specified, but at this stage cannot be derived from the primary data files for the reasons given by Burton et al. (1987). A time step appropriate for the reconstructions is selected, based on the mass-age distribution for the sediment and the sea-level curve.

Definition of Unit Areas

The mass-balanced paleogeographic maps are generated from a data grid. In order for the mass balance to be valid, the data grid must be a closed system with respect to detrital sediment, including all of the sources and sinks. The size of the unit areas chosen for the grid determines the horizontal spatial resolution of reconstructions. The grid could be based on squares having sides of equal length, but if maps are to be produced in different projections this will result in a large file of records of the latitude and longitude positions of all of the corners. In most instances it is more practical to define the unit areas in terms of a latitude and longitude grid . Equal latitude and longitude increments define "grid squares" which are square only at the equator and have varying proportions and areas away from the equator. The disadvantages of the complexity of "grid squares" of varying size are overwhelmed by the computational advantages and by the ready availability of software for many different map projections. In low and mid-latitudes the convergence of the meridians will not cause serious problems in the spacing of data, but at high latitudes this could become a problem, introducing an artificial "grain" into the maps.

Compilation of Present-Day Average Elevations

The reconstructions shown by Shaw and Hay (this volume) use a 1° square average elevation data file produced by the U.S. Central Intelligence Agency; this file has a vertical resolution of 100 m. More detailed compilations, such as ETOPO 5 that contains global average elevations at a resolution of 5 minutes of latitude and longitude, are available from the NOAA National Geophysical Solar/Terrestrial Data Center in Boulder, CO.

Compilation of Representative Lithostratigraphic Columns

The data base for the reconstructions is a set of "average" or "representative" lithostratigraphic columns keyed to a chronostratigraphic framework. Each column (or known absence of one) is representative of one unit area ("grid square") in the defined region of the reconstruction. Each column should extend to igneous or metamorphic basement. The age of the metamorphism must predate the times of interest in the reconstruction.

Each lithostratigraphic column should contain the following information for each recognizable unit: name, thickness, lithology, $CaCO_3$ content of sediments (if known), C_{org} content of sediments (if known), porosity (if known), age, and indication of whether the sediment was deposited in marine or nonmarine environments.

In order of preference, the potential data sources for the lithostratigraphic columns are: 1) columns constructed from a complete set of lithologic and isopach maps, synthesized by averaging real data; 2) drill holes logged to basement located near the center of the unit area with well-established stratigraphy; 3) composite columns constructed from surficial measured sections extending across the unit area; 4) columns from interpreted seismic sections; and 5) columns interpolated from adjacent areas. Many columns will of necessity be compiled from more than one kind of data source. In the offshore regions depicted in the atlases of the Ocean Margin Drilling Program, the stratigraphic information is usually presented in terms of two-way travel time between seismic stratigraphic horizons recognizable over wide areas. The two-way travel time data must be converted into stratigraphic thicknesses in meters using the results of seismic refraction studies. In the offshore Gulf of Mexico and Atlantic margin there are only a few wells for calibration of the seismic stratigraphic interpretations (COST wells and recently released industry wells), and the stratigraphic subdivisions are at best of epoch or stage length. More detailed stratigraphic subdivisions have been provided from industry sources.

Compiling the Mass-Age Distributions of Total and Detrital Sediments

The sediment consists of a detrital component and a precipitated component. The detrital component was transported as discrete particles from the source to the site of deposition. The detrital component defines the regional closed system. The precipitated component was transported in solution by rivers, groundwater and glaciers to an interior basin or to the sea where it entered the global oceanic reservoir. After a period of residence in the sea, it was precipitated by organisms or directly from seawater or interstitial waters. For the precipitated component that resided in the ocean, the closed system is global. Once deposited, both the detrital and precipitated sediments are subject to subsequent erosion and are involved in sedimentary cycling.

Both the total mass and the mass of the detrital fraction of sediment is compiled for each stratigraphic unit and age increment for each column. These are multiplied by the absolute areas of the grid squares represented by each column, and the masses for the entire region are obtained by summation. The geologic age increments, which are of unequal length, are then normalized to equal time intervals having durations equivalent to the time-steps used in the reconstructions. The results are mass-age distributions of the total sediment and of the detrital component of the sediment in the region to be reconstructed, expressed in equal time increments.

The mass-age distributions are spatially integrated expressions of the geologic history of the region. The total sediment mass-age distribution indicates the overall rates of sediment cycling (erosion and redeposition) and hence offers clues about the rates of specific processes that have affected the region. Hay (1981) and Hay and Behensky (1981) noted that the uplift and subsidence history of passive margins is indicated by volume-age or mass-age compilations. It also may be that mass-age distributions contain better values for long-term processes rates than the geologically instantaneous measurements on rivers and other transport agents made in their modern perturbed condition. Because the rates of erosion of detrital sediment are related to elevation (Garrels and Mackenzie, 1971; Hay and Southam, 1977; Hay et al., 1987), the mass-age distribution of detrital sediment can suggest, for example, uplift history, changes in the size of drainage basins, and changes in the nature of the source rocks being weathered.

Estimating the Amounts of Sediment Deposited in the Past

Reconstructing the amounts of sediment that were deposited in the past is crucial to the reconstruction of mass-balanced paleogeographic maps. This is because reconstructed sediment masses are put back on the potential source areas to reconstruct paleotopography.

The total existing global sedimentary mass is about 2700×10^{21}g, and the global rate of loss of sediment to subduction and metamorphism has been estimated to be 1×10^{21}g/m.y. (Veizer and Jansen, 1985; Hay et al., in press). Less than 0.05% of the total sedimentary mass is lost every million years. This is an order of magnitude less than the rate of internal recycling within the global sedimentary mass, which is in the order of 0.4% of the total sedimentary mass every million years (Veizer and Jansen, 1985).

Sediment cycling studies (Garrels and Mackenzie, 1971; Moore and Heath, 1977; Veizer and Jansen, 1979, 1985) suggest that youngest sediments are in the greatest jeopardy of being recycled and that older sediments become progressively more secure. Because each of the existing sediment masses in the mass-age distribution has been affected by erosion subsequent to initial deposition, the mass existing today is less than the mass that existed immediately after deposition. Garrels and Mackenzie (1971) suggested that the shape of the global distribution is as though there were two exponential decay processes operating on the sedimentary mass, a fast decay process representing the removal of young unconsolidated sediment by erosion, and a slow decay that represents the gradual destruction of older lithified sediments. Based on the assumptions that: 1) relatively young sediment is recycled more rapidly than older, more secure sediment; and 2) that deviations from the exponential decay curve fitted to a present mass-age distribution represent variations in the ancient sediment fluxes, we have applied the inverse of a decay curve of the form $M = Ae^{\lambda t}$ to reconstruct the sediment mass which was deposited at every time step (see Fig. 1). M is the mass of sediment which was deposited during an interval centered on time t in the past, A is the mass of sediment deposited during an interval of equivalent length prior to the present, and λ is the

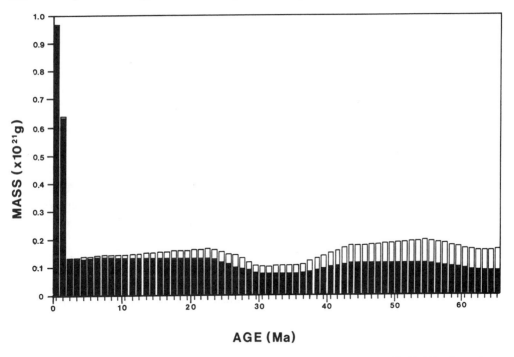

AGE (Ma)

Figure 1. Present day mass-age distribution of Cenozoic sediment in the northwest Gulf of Mexico and its drainage basin shown in black. The mass-age distribution is divided into 1 m.y. intervals; because the original information is in terms of series of the Tertiary, there appear to be long intervals of constant mass. Data presently available do not permit determination of differences within intervals of epoch length. In order to avoid sudden jumps in the reconstructions across epoch boundaries, the present-day mass-age distribution data have been smoothed using a seven point rolling average for the interval between 3 and 62 Ma. Shown as white increments to the bars is the estimate of the masses of sediment originally deposited but subsequently eroded and recycled to become younger sediment. The formula used to reconstruct the masses of sediment that existed in the past is based on the entire present-day mass-age distribution of Phanerozoic sediment in the northwest Gulf of Mexico and its drainage basin.

decay constant determined from the mass-age distribution of all Phanerozoic sediment in the Gulf of Mexico and its drainage area, and t is the mid-point of the time step expressed in million years. For a discussion of this method, see Hay et al. (in press).

The reconstructed masses of detrital sediment that existed in the past are a quantitative, spatially averaged abstract of the geologic history of the region. If the size of the drainage area remained constant, the reconstructed mass of detrital sediment is directly proportional to the average elevation of the drainage basin and vice versa. Major changes in the amounts of sediment deposited in the past can be interpreted as indicative of either rapid uplift or changes in the size of the drainage basin. Reconstructed mass-age distributions often show a characteristic slow decay of the amount of sediment deposited after an orogenic event as the mountains are gradually worn away; if there is no independent evidence for subsequent uplift other than the isostatic response to unloading, we refer to this as "simple topographic decay." Figure 1 suggests that in the region draining to the Gulf of Mexico, the Laramide Orogeny in the early Paleogene was followed by a quiescent interval in which simple topographic decay occurred; this was subsequently followed by an episode of uplift in the Oligocene with subsequent simple topographic decay; this was then followed by very rapid uplift and/or expansion of the drainage basin at the end of the Pliocene. Projecting into the future, it is apparent that it will be hundreds of millions of years before the High Plains will be eroded to near sea level. Thus, even with the very crude chronostratigraphic framework available from seismic data in the Gulf of Mexico, the outline of the geologic history of the region draining to the Gulf of Mexico is contained in the reconstruction of masses of detrital sediment deposited in the past.

Age of Ocean Crust and Thinned Continental Margin

The age of the ocean crust is required for the paleobathymetric calculation for each grid square on the ocean floor. For the Gulf of Mexico reconstructions (Shaw and Hay, this volume) the subsidence related to contraction of aging ocean crust was neglected because 1) spreading occurred in a short time interval about 160 Ma so that the subsidence over the past 65 m.y. has been relatively slight, and 2) there are conflicting ideas concerning the location and age of ocean crust in the Gulf of Mexico. For the Atlantic margin we used the age of ocean crust shown on the maps of Emery and Uchupi (1984). For the age of contracting mantle material beneath thinned continental margin crust we assigned an age equal to that of the adjacent oceanic crust.

Selection of a Sea-Level Curve

Our reconstructions are calculated relative to sea level at times in the past. Sea level can be held constant, which although unrealistic, is useful in testing the sensitivity of the model to changes in other parameters. Alternatively, a regional sea-level curve can be entered as a data file. If we use the eustatic sea-level curve of Haq et al. (1987), most of North America is flooded in the Paleogene, for which time they indicate a sea level as much as 225 m higher than today. We have found that the shoreline positions of Rainwater (1967) for the Gulf Coast can be duplicated in our reconstructions by reducing the amplitude of the Haq et al. sea-level curve by a factor of three to four, and accordingly use a reduced amplitude version of the Haq et al. curve for most model runs.

According to studies of basin sediment geometry by Burton et al. (1987), interpretations of sea-level fluctuations based on seismic cross-sections may be exaggerated. This is attributed to the effects of sediment compaction and tectonic movements that may distort eustatic changes by as much as a factor of five; this may account for the apparent overestimation of eustatic change by Haq et al. (1987).

One of the sea-level files should be the average position of sea level during each time step. But for reconstructing a shoreline position, it is better to use a file that gives the instantaneous position of sea level at the end of each time step.

Selection of the Appropriate Time Step

In order to understand the geologic development of a region, time steps used for the reconstructions must be short enough to resolve the major events in the geologic history of the area. Because the computation time for each time step is equal, there is an advantage in choosing time steps of relatively long duration, but short enough to resolve geologic changes. However, long time steps will produce maps which are averages over long intervals and may not resemble any paleogeography which actually existed. In areas we have studied, the northwestern Gulf of Mexico and its drainage basin and the Atlantic margin of the U.S., time steps no longer than 1 m.y. are required to follow geologic changes during the late Cenozoic, and these, of course, do not resolve changes within the Pleistocene.

The time steps are somewhat independent of the temporal resolution of the litho- and biostratigraphic data base in that they may be shorter than the stratigraphic resolution can justify. But their apparently unrealistically short length is required in order for the computations to bridge large sediment-flux changes resulting from rapid uplift, sea-level change, or reorganization of drainage basins. Clearly, the resolution of events in the geologic history of an area will depend on the resolution available in the data base, but time steps finer than the chronostratigraphic resolution inherent in the data base may be required to achieve sediment mass balance. For the studies of the Gulf of Mexico region presented in Shaw and Hay (this volume) we used only the stratigraphic subdivisions given by Buffler et al. (1984) and Cook and Bally (1975).

Conclusions

Mass-balanced paleogeographic maps are quantitative reconstructions of the geomorphic configuration of the earth's surface at moments in the geologic past. They are based on principles of sedimentary mass balance and isostasy, and they predict uplift or subsidence of continental source regions as a variable output. The primary input files define the unit areas, their average elevations, and the lithostratigraphy and chronostratigraphy in the unit areas. In ocean basins and passive margins, the age of oceanic crust and age of the contracting lithosphere beneath thinned continental margins are entered. Finally, a regional sea-level curve also is required. From these primary data files the mass-age distribution for sediment is compiled, and from this we calculate an estimate of the sedimentary masses that were eroded and deposited in the past. The length of the time steps for the reconstructions is determined from the mass-age distributions and the sea-level curve.

Spatial resolution of the model is determined by the size of the unit areas for which elevation and geologic information are available, whereas temporal resolution depends on the chronostratigraphic resolution of the geologic data. The models are constructed without the use of geologic information on the position of the shoreline, so that this as well as a variety of other geologic data can be used as a test of the accuracy of the reconstruction.

One fundamental attribute of our reconstruction method is the conversion of the traditional geologic time scale, divided into intervals of unequal length (e.g., periods, epochs and ages), into a numerical geologic time scale, divided into intervals of equal length. Although the numerical time scale often has been used in studies of the changing shape and paleobathymetry of ocean basins, it has not been applied extensively to reconstructions of continental areas. The use of the numerical time scale is essential for relating different process rates, and is required for quantitative mass-balanced paleogeographic reconstructions.

ACKNOWLEDGMENTS

The authors have benefited from conversations with a number of colleagues, particularly John R. Southam, Michael J. Rosol and Christopher G. A. Harrison. We would like to thank T. Cross, C. Paola and M. Perlmutter for their helpful reviews. This work has been supported by grants OCE 8409369 and OCE 8716408 from the National Science Foundation and 19274-AC2 from the American Chemical Society's Petroleum Research Fund, by the U.S. Geological Survey, and by gifts from Texaco, Inc.

REFERENCES CITED

Barron, E.J., 1985, Numerical climate modeling, a frontier in petroleum source rock prediction: Results based on Cretaceous simulations: American Association of Petroleum Geologists Bulletin, v. 69, p. 448-459.

Barron, E.J., and Harrison, C.G.A., 1980, An analysis of past plate motions: The South Atlantic and Indian Oceans, in Davies, P.A., and Runcorn, S.K., eds., Mechanisms of continental drift and plate tectonics: London, Academic Press, p. 89-109.

Barron, E.J., and Washington, W.M., 1984, The role of geographic variables in explaining paleoclimates: Results from Cretaceous climate model sensitivity studies: Journal of Geophysical Research, v. 89, p. 1267-1279.

Bethke, C.M., Harrison, W.J., Upson, C., and Altaner, S.P., 1988, Supercomputer analysis of sedimentary basins: Science, v. 239, p. 261-267.

Blanton, J.O., Atkinson, L.P., Pietrafesa, L.J., and Lee, T.N., 1981, The intrusion of Gulf Stream water across the continental shelf due to topographically-induced upwelling: Deep-Sea Research, v. 28A, p. 393-405.

Bryan, G.M., and Heirtzler, J.R., editors, 1984, Eastern North American continental margin and adjacent ocean floor, 28° to 36° N and 70° to 82° W: Woods Hole, Massachusetts, Ocean Margin Drilling Program, Regional Atlas Series, Atlas 5, Marine Science International, 41 + a-h p.

Buffler, R.T., Locker, S.D., Bryant, W.R., Hall, S.A., and Pilger, R.H., Jr., 1984, Ocean margin drilling program: Gulf of Mexico atlas: Woods Hole, Massachusetts, Ocean Margin Drilling Program Regional Atlas Series, Atlas 6, Marine Science International, 36 p.

Burton, R., Kendall, C.G.St.C., and Lerche, I., 1987, Out of our depth: On the impossibility of fathoming eustasy from the stratigraphic record: Earth Science Reviews, v. 24, p. 237-277.

Cook, T.D., and Bally, A.W., 1975, Stratigraphic atlas of North and Central America: Princeton, New Jersey, Princeton University Press, 272 p.

Davis, E.E., and Lister, C.R.B., 1974, Fundamentals of ridge crest topography: Earth and Planetary Science Letters, v. 21, p. 405-413.

Dietz, R.A., 1963, Collapsing continental rises: An actualistic concept of geosynclines and mountain building: Journal of Geology, v. 71, p. 314-333.

Emery, K.O., and Uchupi, E., 1984, The geology of the Atlantic Ocean: New York, Springer Verlag, xix + 1050 p.

Ewing, J.I., and Rabinowitz, P.D., editors, 1983, Eastern North American continental margin and adjacent ocean floor, 34° to 41° N and 68° to 78° W: Woods Hole, Massachusetts, Ocean Margin Drilling Program, Regional Atlas Series, Atlas 4, Marine Science International, 32 + a-h p.

Garrels, R.M., and Mackenzie, F.T., 1971, Evolution of Sedimentary Rocks: New York, W.H. Norton & Co., 397 p.

Garvine, R.W., 1973, The effect of bathymetry on coastal upwelling of homogeneous water: Journal of Physical Oceanography, v. 3, p. 47-56.

Gill, A.E., 1982, Atmosphere-ocean dynamics: New York, Academic Press, xv + 662 p.

Haq, B.U., Hardenbol, J., and Vail, P. R., 1987, Chronology of fluctuating sea levels since the Triassic (250 million years age to present): Science, v. 235, p. 1156-1166.

Hay, W.W., 1981, Sedimentological and geochemical trends resulting from the breakup of Pangaea: Oceanologica Acta, v. 4 Supplement, p. 135-147.

Hay, W.W., 1983, Significance of runoff to paleoceanographic conditions during the Mesozoic and clues to locate sites of ancient river inputs: Proceedings of the Joint Oceanographic Assembly, 1982, Department of Fisheries and Oceans, Ottawa, p. 9-17.

Hay, W.W., 1984, The breakup of Pangaea: Climatic, erosional, and sedimentological effects: Proceedings, 27th International Geological Congress, Moscow, v. 6 (Geology of Ocean Basins), p. 15-38.

Hay, W.W., Barron, E.J., Behensky, J.F., Jr., and Sloan, J.L., II, 1982, Triassic-Liassic paleoclimatology and sedimentation in proto-Atlantic rifts: Paleogeography, Paleoclimatology, Paleoecology, v. 40, p. 13-30.

Hay, W.W., and Behensky, J.F., Jr., The northern Gulf of Mexico as an anomalous passive margin: Gulf Coast Association of Geological Societies Transactions, v. 31, p. 309-313.

Hay, W.W., Rosol, M.J., Sloan, J.L.,II, and Jory, D.E., 1987, Plate tectonic control of global patterns of detrital and carbonate sedimentation: in Doyle, L.J., and Roberts, H.H., eds., Carbonate-clastic transitions: Amsterdam, Elsevier, Developments in Sedimentology, v. 42, p. 1-34.

Hay, W.W., and Southam, J.R., 1977, Modulation of marine sedimentation by the continental shelves, in Anderson, N.R., and Malahoff, A., eds., The fate of fossil fuel CO_2 in the oceans: New York, Plenum Press, Marine Science Series, v. 6, p. 569-604.

Hay, W.W., Sloan, J.L.,II, and Wold, C.N., 1988, The mass-age distribution of sediments on the ocean floor and the global rate of loss of sediment: Journal of Geophysical Research (in press).

Heestand, R.L., and Crough, S.T., 1981, The effects of hot spots on the oceanic age-depth relation: Journal of Geophysical Research, v. 86, p. 6107-6114.

Holland, H.D., 1978, The chemistry of the atmosphere and oceans: New York, Wiley-Interscience, ix + 351 p.

King, J.D., 1988, Burial history refinement by direct application of geochemical reaction kinetics: Computers in Geoscience, 16 p. (in press).

Kinsman, D.J.J., 1975, Rift valley basins and sedimentary history of trailing continental margins: in Fischer, A.G., and Judson, S., eds., Petroleum and global tectonics: Princeton, New Jersey, Princeton University Press, p. 83-126.

Lasaga, A.C., Berner, R.A., and Garrels, R.M., 1985, An improved geochemical model of atmospheric CO$_2$ fluctuations over the past 100 million years, *in* Sundquist, E.T., and Broecker, W.S., The Carbon Cycle and Atmospheric CO$_2$: Natural Variations Archaean to Present: American Geophysical Union, Washington DC, Geophysical Monograph 32, p. 397-411.

Mallory, W.W., editor, 1972, Geologic atlas of the Rocky Mountain region: Denver, Rocky Mountain Association of Geologists, 331 p.

McClain, C.R., Chao, S.-Y., Atkinson, L.P., Blanton, J.O., and de Castillejo, F., 1986, Wind-driven upwelling in the vicinity of Cape Finisterre, Spain: Journal of Geophysical Research, v. 91, p. 8470-8486.

Menard, H.W., 1969, Elevation and subsidence of oceanic crust: Earth and Planetary Science Letters, v. 6, p. 275-284.

Moore, T.C., and Heath, G.R., 1977, Survival of deep sea sedimentary sections: Earth and Planetary Science Letters, v. 37, p. 71-80.

Parrish, J.T., and Barron, E.J., 1986, Paleoclimates and economic geology: Society of Economic Paleontologists and Mineralogists Short Course 18, 162 p.

Parsons, B, and Sclater, J.G., 1977, An analysis of the variation of ocean floor bathymetry and heat flow with age: Journal of Geophysical Research, v. 82, 803-827.

Peffley, M.B., and O'Brien, J.J., 1976, A three-dimensional simulation of coastal upwelling off Oregon: Journal of Physical Oceanography, v. 6, p. 164-180.

Rainwater, E.H., 1967, Resume of Jurassic to Recent sedimentation history of the Gulf of Mexico basin: Gulf Coast Association of Geological Societies Transactions, v. 17, p. 179-210.

Roden, G.I., 1961, On the wind driven circulation in the Gulf of Tehuantepec and its effect upon surface temperatures: Geofisica Internacional, v. 1, p. 55-76.

Schmoker, J.W., 1986, Oil generation in the Anadarko Basin, Oklahoma and Texas: Modeling using Lopatin's Method: Oklahoma Geological Survey Special Publication 86-3, 40 p.

Sclater, J.G., Abbott, D., and Thiede, J., 1977, Paleobathymetry and sediments of the Indian Ocean, *in* Heirtzler, J.R., Bolli, H.M., Davies, T.A., Saunders, J.B., and Sclater, J.G., eds., Indian Ocean geology and biostratigraphy: Washington, D.C., American Geophysical Union, p. 25-59.

Sclater, J.G., Anderson, R.N., and Bell, M.L., 1971, The elevation of ridges and the evolution of the central eastern Pacific: Journal of Geophysical Research, v. 76, p. 7888-7915.

Sclater, J.G., Hellinger, S., and Tapscott, C., 1977, Paleobathymetry of the Atlantic Ocean: Journal of Geology, v. 85, p. 509-552.

Sclater, J.G., Jaupart, C. and Galson, D., 1980, The heat flow through oceanic and continental crust and the heat loss of the earth: Reviews of Geophysics and Space Physics, v. 18, p. 269-311.

Sclater, J.G., Meinke, L., Bennett, A., and Murphy, C., 1985, The depth of the ocean through the Neogene, *in* Kennett, J.P., ed., The Miocene Ocean: Geological Society of America Memoir 163, p. 1-19.

Sheridan R.E., and Grow, J.A., editors, 1988, The Atlantic continental margin: U.S.: Boulder, Colorado, Geological Society of America, The Geology of North America, v. I-2, x + 610 p.

Shor, A.N., and Uchupi, E., editors, 1984a, Eastern North American Continental Margin and Adjacent Ocean Floor, 39° to 46° N and 54° to 64° W: Woods Hole, Massachusetts, Ocean Margin Drilling Program, Regional Atlas Series, Atlas 2, Marine Science International, 24 + a-h p.

Shor, A.N., and Uchupi, E., editors, 1984b, Eastern North American Continental Margin and Adjacent Ocean Floor, 39° to 46° N and 64° to 74° W: Woods Hole, Massachusetts, Ocean Margin Drilling Program, Regional Atlas Series, Atlas 3, Marine Science International, 30 + a-h p.

Sleep, N., 1971, Thermal effects of the formation of Atlantic continental margins by continental break up: Geophysical Journal of the Royal Astronomical Society, v. 24, p. 325-350.

Southam , J.R., and Hay, W.W., 1981, Global sedimentary mass balance and sea level changes, *in* Emiliani, C., ed., The sea, Volume 7, The oceanic lithosphere: New York, Wiley-Interscience, 1617-1684.

Steckler, M.S., and Watts, A.B., 1978, Subsidence of the Atlantic-type continental margin off New York: Earth and Planetary Science Letters, v. 41, p. 1-13.

Termier, H., and Termier, G., 1952, Histoire geologique de la Biosphere: Paris, Masson & Cie., 721 p.

Tissot, B., and Welte, D.H., 1978, Petroleum formation and occurrence: Berlin, Springer-Verlag, 538 p.

Ungerer, P., Bessis, F., Chenet, P.Y., Durand, B., Nogart, E., Chiarelli, A., Oudin, J.L., and Perrin, J.F., 1984, Geological and geochemical models in oil exploration: Principles and practical examples, *in* Demaison, G., and Murris, R.J., eds., Petroleum geochemistry and basin evaluation: American Association of Petroleum Geologists Memoir 35, p. 53-77.

van Hinte, J.E., 1978, Geohistory analysis—Applications of micropaleontology in exploration geology: American Association of Petroleum Geologists Bulletin, v. 62, p. 201-222.

Veizer, J., and Jansen, S.L., 1979, Basement and sedimentary recycling and continental evolution: Journal of Geology, v. 87, p. 341-370.

Veizer, J., and Jansen, S.L., 1985, Basement and sedimentary recycling—2: Time dimension to global tectonics: Journal of Geology, v. 93, p. 625-643.

Vinogradov, A.P, 1967, Atlas of the lithological paleogeographical maps of the USSR, v. 4, Paleogene, Neogene and Quaternary: Moscow, Ministry of Geology of the USSR, 55 sheets.

Vinogradov, A.P, 1968a, Atlas of the lithological paleogeographical maps of the USSR, v. 1, Precambrian, Cambrian, Ordovician and Silurian Periods: Moscow, Ministry of Geology of the USSR, 52 sheets.

Vinogradov, A.P, 1968b, Atlas of the lithological paleogeographical maps of the USSR, v. 3, Triassic, Jurassic, and Cretaceous: Ministry of Geology of the USSR, 71 sheets.

Vinogradov, A.P, 1969, Atlas of the lithological paleogeographical maps of the USSR, v. 2, Devonian, Carboniferous and Permian: Moscow, Ministry of Geology of the USSR, 65 sheets.

Vogt, P.R., and Tucholke, B.E., editors, 1986, The western North Atlantic region: Boulder, Colorado, Geological Society of America, The Geology of North America, v. M, xiv + 696 p.

Wang, H., editor, 1985, Atlas of the Paleogeography of China: Beijing, Cartographic Publishing House, 85 p., 25 plates.

Waples, D.W., 1980, Time and temperature in petroleum formation: Application of Lopatin's method to petroleum exploration: American Association of Petroleum Geologists Bulletin, v. 64, p. 916-926.

Watts, A.B., and Ryan, W.B.F., 1976, Flexure of the lithosphere and continental margin basins: Tectonophysics, v. 36, p. 25-44.

Welte, D.H., and Yukler, M.A., 1981, Petroleum origin and accumulation in basin evolution: A quantitative model: American Association of Petroleum Geologists Bulletin, v. 65, p. 1387-1396.

Ziegler, P. A., 1982, Geological atlas of Western and Central Europe: The Hague, Shell Internationale Petroleum Maatschappij B.V. , 130 p. , 40 maps.

15

MASS-BALANCED PALEOGEOGRAPHIC MAPS: MODELING PROGRAM AND RESULTS

Christopher A. Shaw [1] and William W. Hay [2]

[1]U.S. Geological Survey, Branch of Sedimentary Processes, Denver Federal Center, Box 25046, MS 939, Lakewood, Colorado 80225 USA

[2]Museum, Cooperative Institute for Research in Environmental Sciences, and Department of Geology, Campus Box 449, University of Colorado, Boulder Colorado 80309 USA

ABSTRACT

The mass-balanced paleogeographic map program produces reconstructions of the paleotopography and paleobathymetry of a source-sink system based on mass balance principles. As discussed in the related paper by Hay, Wold and Shaw (this volume), the modeling program requires knowledge of 1) defined unit areas which make up the grid, 2) average elevation of each unit area, 3) an average stratigraphic column for each unit area with lithostratigraphic and chronostratigraphic information, 4) compilation of the mass-age distribution of the sediment in the region, 5) reconstruction of the masses of sediment deposited during each time step calculated from the modern mass-age distribution, 6) age of the ocean floor and of the contracting material beneath thinned continental margin crust, 6) specification of the position of sea level over the time steps, and 7) specification of the appropriate time step interval.

The program considers 1) changes in elevations due to bulk sediment transfer, 2) isostatic adjustments related to transfer of mass 3) compaction of sediment, 4) sea-level changes, 5) subsidence of ocean crust with age, 6) subsidence of passive continental margins with age, 7) uplift and subsidence within the continental interior, and 8) changes in paleodrainage patterns.

Using the inverse of modern erosion processes, detrital sediment is reloaded onto source areas to produce a reloaded paleotopography. By requiring that mass be conserved, that is, that the sediment yielded by erosion of the reconstructed topography be equal to the sediment which was deposited in the basin during a given time interval, the amount of uplift or subsidence within the source area is determined as an output of the program. Yields from erosion of the reloaded topography that are greater than the sediment deposited suggest that the elevations of the reloaded paleotopography are too high and that uplift of the area must have occurred subsequent to erosion and deposition of the strata in question. The reloaded topography is then reduced in elevation by the ratio of the erosional yield of the reloaded topography to the amount of sediment deposited to produce a reconstructed topography which will yield the amount of sediment deposited. The change in elevation required to achieve mass balance is "uplift" or "subsidence," the free variable of the program.

Quantitative Dynamic Stratigraphy (1989), T.A. Cross, ed., Prentice Hall, p. 277–291.

INTRODUCTION

The method of producing mass-balanced paleogeographic reconstructions described here expands on first-generation models developed by Shaw (1987). The reconstructions utilize a modification of the backstripping method of Steckler and Watts (1978) to make isostatic adjustments in response to mass transfer of the solid-phase sediment, sea-level changes, sediment compaction, subsidence of ocean crust with age, subsidence of passive continental margins with age, changes in porosity and pore fluids, and changes in paleodrainage systems. The program indicates vertical movements of the continental crust required to achieve mass balance.

The result is a stepwise hindcast of the development of a basin and its source area as a closed system with respect to detrital sediment. The model produces three-dimensional paleotopographic and paleobathymetric maps consistent with geologic data. Acceptable reconstructions can start with the configuration of the earth's surface in the past, subject the surface to processes running forward in time, and result in a series of landscapes compatible with geological knowledge ending with the modern configuration of the earth's surface. Although there are a large number of topographic configurations that will yield, upon erosion, the correct sediment mass, the reconstructed topography must be analyzed in the light of geologic data which imposes external and absolute constraints on the location of topographic highs and lows and paleoshorelines.

The assumptions of the algorithms and the data required to run the mass-balanced paleogeographic modeling programs are discussed by Hay et al. (this volume). The model has been tested on continental margins of the Gulf of Mexico and the U.S. Atlantic, and several localized source-sink systems in the conterminous United States. The computer program is written in Fortran 77 with versions compatible with both MS/DOS and VAX/VMS systems.

SPECIFICATION OF PARAMETERS

Before each series of reconstructions is run, the parameters relating to isostasy and erosion are entered by the user through a set of queries in the program's front end. For the isostatic calculations these include the densities of pore fluid, seawater and asthenosphere. For the erosion calculations, they include values of the erosional base level and erosion constant. Malcott (1928) presented a variety of definitions for base levels as applied to erosion cycles, but the term "erosional base level" is defined here as a horizontal surface above which active erosion of detrital material occurs. The detrital erosion constant represents the thickness of solid-phase sediment eroded from a source-area grid square per meter of elevation above the erosional base level. At this stage of model development, the erosion constant and base level are initially defined and remain constant throughout the entire model run, although in reality these parameters are likely to fluctuate with changes in climate. A lithology-dependent density for the solid-phase sediment is internally assigned based on measured rock and mineral densities by numerous authors.

Classification of Unit Areas

The input file data include the average elevation and an average stratigraphic column with age assignments for each unit area, as shown in Figure 1. Based on this information, the program defines five categories of unit areas—source, bypass, basin, intrasource basin and

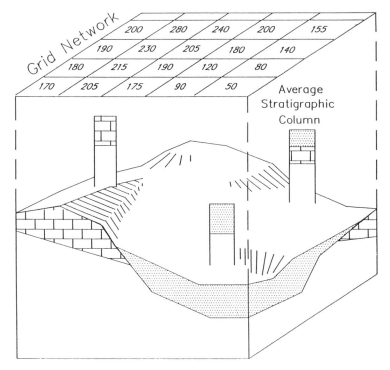

Figure 1. A grid is applied to the study area. If this is a latitude-longitude grid, the size of each unit area (degree square) decreases away from the equator and this must be taken into account in all calculations of mass, volume or elevation. Average elevation and stratigraphic data are required for each unit area of the grid. Numbers shown in the grid areas are average elevations; three average stratigraphic columns also are shown.

hiatus—based on average elevation and presence or absence of sediment deposited during the time interval being considered (Fig. 2). *Source areas* have an average elevation greater than the erosional base level and do not contain sediment deposited during the given time interval. *Bypass areas* have average elevations less than the erosional base level but greater than sea level and do not contain sediment of the time interval being considered. Bypass areas are assumed to be areas through which sediment is transported from the source area to the depositional basin and do not actively supply or consume sediment. *Basin areas* have average elevations less than the erosional base level and contain sediment deposited during the time interval being considered. *Intrasource basin areas* contain sediment deposited during the time interval being considered and have average elevations greater than the erosional base

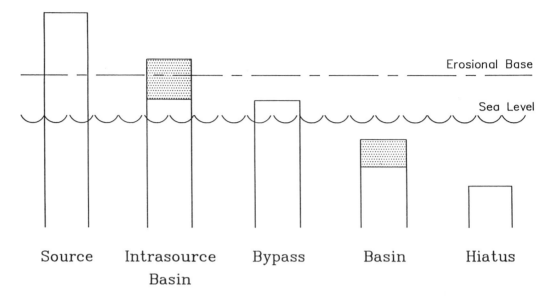

Figure 2. Schematic showing the classification of grid squares recognized by the mass-balanced paleogeographic modeling program.

level. *Hiatus areas* have average elevations below sea level but do not contain sediment of the time interval being considered. The model re-evaluates the classification of each unit area after each reconstruction.

BACKSTRIPPING SEDIMENT FROM THE DEPOSITIONAL SITES

Sediment representing the time interval under consideration is backstripped from the basin and intrasource basin areas. As a result of the removal of the sediment, the surface must be isostatically adjusted for the reduced load, and the underlying sediment decompacted.

Isostatic Adjustment

Isostatic adjustment describes the response of the lithosphere to changes in its surface load. Glacial-isostatic observations suggest that the earth's response to unloading is rapid at first but decreases nonlinearly with time to become asymptotic to an equilibrium condition. The time required for an isostatic adjustment to approximate the equilibrium condition is proportional to the mass unloaded and the duration of time over which the load was removed. In the case of sudden removal of the load, it is thought to require less than 20,000 years (Chappell, 1974; Cathles, 1980). If, as in the examples here, the model evaluates reconstructions at 1 m.y. time increments, it is assumed that isostatic adjustments are instantaneous and that the region being examined is in isostatic equilibrium.

We also assume that the outer part of the earth can be represented by an asthenosphere overlain by a solid but flexible lithosphere which includes the upper mantle and continental and oceanic crust, each of which has constant but different densities. Following Cathles (1980), the asthenosphere is assumed to be a Newtonian fluid, homogeneous with respect to density and viscosity. The assumption of a solid but flexible lithosphere overlying a Newtonian fluid does not require a critical load to initiate flow. Therefore, any change in the surface load will cause a response by the asthenosphere.

The removal of bulk sediment with water-filled pore spaces from the sea floor requires only isostatic adjustment for the solid phase because the pore water becomes part of the overlying water column. In contrast, backstripping subaerial sediments having water-filled pore spaces requires isostatic adjustment for the removal of both solid-phase and pore-fluid masses since the space occupied by pore water is replaced by air. Because the ratio of the density of air to the density of the mantle is about $1:3.3 \times 10^3$, isostatic adjustments related to changes in the thickness of the atmosphere are not considered. Equations expressing the isostatic responses to the loading and unloading of solid-phase sediment, pore fluid, and seawater are given in Watts and Ryan (1976), Steckler and Watts (1978), Ungerer et al. (1984), and Shaw (1987).

Because isostatic adjustments are instantaneous at the scale of the 1 m.y. time steps used in our reconstructions, the effects of short-term sediment loads removed by subsequent erosion are neglected. Their only lasting effect would be to compact the underlying sediment and introduce a small error into the bathymetric calculation.

Decompaction

Decompaction returns sediment to its condition prior to compaction by the load of overlying sediment. The model maintains an exponentially decreasing porosity with depth following the general porosity/depth relationship suggested by Baldwin and Butler (1985):

$$S = \{[\text{depth of burial (m)}]/6020\}^{0.1575}$$

where S is the solidity (1 − porosity) of the sediment.

More complex versions of the model could take into account different compaction curves for different sediment types, as suggested by Sclater and Christie (1980) for sand and by Ricken (1987) for carbonate. However, as Steckler and Watts (1982) noted, the increase in complexity does not significantly alter the results. Other factors, such as the magnitude of the sea-level change will overwhelm the effects of differential compaction.

Thermal Subsidence of Ocean Crust with Age

After its formation at spreading centers, ocean crust subsides and its elevation decreases with increasing age, primarily a function of heat loss with time (Sclater et al., 1971; Parsons and Sclater, 1977; Sclater et al., 1980). Reconstructions of areas underlain by ocean crust must account for subsidence with time. The model estimates the amount of thermal subsidence that ocean crust has undergone for 1 m.y. time intervals from depth-age relationships suggested by Sclater et al. (1980). The age of ocean crust beneath each square is given in the stratigraphic data-base file.

Subsidence also may be related to the thickening of ocean crust (McClain and Atallah, 1986) but this effect is extremely difficult to quantify and appears to be of only local importance. Therefore, it is not considered in the current model.

Thermal Subsidence of Continental Margins

During continental rifting, passive continental margins thin by stretching or advection of the lower, ductile part of the continental crust (McKenzie, 1978), by failure and lystric faulting of the brittle upper continental crust (Montadert and others, 1979), and by erosion following uplift (Sleep, 1971; Kinsman, 1975; Turcotte et al., 1977; Hay, 1981; Southam and Hay, 1981). Regardless of which processes are active or dominate during the rifting phase, shortly after ocean crust has formed the margin subsides as a result of thermal decay of the hot lithosphere beneath it. Sleep (1971) and Steckler and Watts (1978) have shown that a simple modification of the subsidence curve for oceanic crust (Parsons and Sclater, 1977) can be used to model the subsidence of thinned continental crust; the modified subsidence curve differs according to different thicknesses of continental crust.

To construct a subsidence curve (Steckler and Watts, 1978) for a particular site, thicknesses and paleowater depths for sediment layers must be determined and used as an input. Then the effects of compaction, sediment loading and eustatic changes must be removed so that the residual is the tectonic component of subsidence. An exponential decay curve of the form depth x $age^{1/2}$ can then be fit to the residual as an approximation of thermal subsidence. Watts and Steckler (1979) discussed a number of parameters, including specific estimates of lithospheric thickness and crustal thinning, that can be determined in order to more closely fit the thermal subsidence curve.

To reconstruct testable paleobathymetry for passive continental margins, we developed a simple, internally consistent algorithm that can be applied to a regional grid. We assume that there exists within the asthenosphere an isobaric surface that coincides with a geopotential (horizontal) surface, and that the asthenosphere is a Newtonian fluid. Hence we can write two equations that relate pressure at and depth of this geopotential surface. First, the pressure at the horizontal isobaric surface is everywhere equal to the combined weight of the overlying mantle, continental (or oceanic) crust, sediment, and water (if present). Second, the horizontal isobaric surface is everywhere the same depth below sea level, and is equal to the combined thicknesses of mantle, continental (or oceanic) crust, sediment, and water. For the present day, the values of all terms except the thickness of the mantle and continental (or oceanic) crust and the density of the mantle are known. We solve for the age-dependent density of the mantle using ocean crust subsidence curves. Then the present-day thicknesses of the mantle and continental (or oceanic) crust are determined by solving the equations for pressure at and depth of the horizontal isobar. We then assume that the thickness of the continental crust (unless exposed) has remained constant through time, and estimate the paleobathymetry/paleotopography by inserting the appropriate mantle density into the combined pressure and depth equations rearranged to solve for the depth of the water or elevation of the land surface.

Replacing Sediment on the Source Area

Although many factors such as elevation, slope, lithology, age, and climate affect weathering processes and erosion, elevation is considered the dominant influence on denudation rates (Garrels and Mackenzie, 1971; Hay and Southam, 1981; Hay, 1984). At this time, two factors are considered to control the volume of sediment returned to a source area: the average elevation of the grid square; and the absolute area of the grid square (which, if the grid squares are defined by equal latitude and longitude increments, decreases away from the equator). The program first determines the total "volume" (V) of the source area above the erosional base level. This is the sum of the products of the average elevation and absolute area for each source-area grid square (Equation 1):

where n is the number of source squares in the drainage basin, h is the average elevation of the grid square, and A is the absolute area of the grid square.

$$V = \sum_{i=1}^{n} h\,A \qquad (1)$$

The mass of reconstructed sediment is then calculated as described by Hay et al. (this volume). The volume of reconstructed solid-phase sediment is then estimated by dividing the reconstructed sediment mass by a user-defined density.

During the procedure of replacing sediment on a source area, when a source-area grid square is recognized, the fraction of its "volume" to the total source area "volume" is calculated. This fraction is then multiplied by the reconstructed solid-phase volume to determine the volume of solid-phase sediment returned to that square. The replaced bulk-sediment thickness (T_b) is thus the replaced solid-phase sediment volume (V_{sp}) divided by the area of the grid square (A). These terms are divided by one minus the porosity of rock in the source area (P_s; Equation 2).

Chemically precipitated, biogenic, and external source sediments are removed from the site

$$T_b = \frac{(V_{sp}/A)}{(1 - P_s)} \qquad (2)$$

of deposition but not returned to sources in the western-central United States.

This generation of mass-balanced paleogeographic models considers only a single, closed-system drainage basin and sediment sink. Therefore, all the sediment deposited in the northwestern Gulf of Mexico during the Cenozoic is assumed to be derived from the erosion of sedimentary rocks in a drainage basin of variable size located in the western-central United States. This simplification is required at this time since routines that model multiple drainage basins and sediment outflux from the basins are not yet completed.

Except on the global scale, erosion-deposition systems are open with respect to the dissolved load of rivers and groundwater, and may be open with respect to aeolian and glacial transport. Also, all systems contain multiple major and minor drainage divides. Recent versions of the model associate each fluvial, lacustrine and deltaic depositional site with a unique drainage basin that is either the whole drainage system or a fraction of it. The assumption for recognizing a unique drainage basin for a particular detrital deposit is that sediment is always transported downhill and generally follows a path perpendicular to topographic contours. This is true for fluvial systems, but does not apply to glacial, aeolian or subaqueous systems. As shown in Figure 3, the model recognizes and defines local source areas by examining the average elevation and sediment characteristics of the eight surrounding grid squares. Potential source areas are defined as grid squares not containing sediment deposited during the time of reconstruction and with an average elevation which is greater

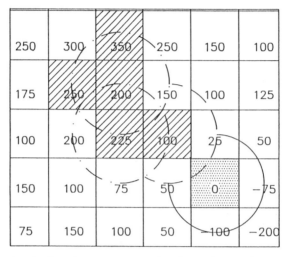

Figure 3. Schematic showing how the computer searches for a unique drainage basin for each deposit (stippled area). Hachured area represents the area of the drainage basin found by the computer. Numbers in grid squares are average elevations in meters.

than and upslope from the local deposit. If one or more of these eight surrounding areas is found to be a potential source area, each of these becomes the center for a subsequent search for additional potential sources. This process is repeated until no additional potential source areas are found.

Figure 4 shows contoured average elevation data from Figure 3. Note that the drainage basin predicted by the computer (hachured region) roughly approximates the drainage basin that would be identified by analyzing the shape of the contours. Note also that the valley directly to the north of the deposit, while at first glance appears to be a source area, has a drainage pattern that actually flows into the square directly east of the deposit.

UPLIFT AND SUBSIDENCE IN THE SOURCE AREA

Estimating Vertical Movements in the Continental Interior
In contrast to the depth-age relationships exhibited by ocean crust, tectonic movements in the continental interior are less predictable. This is at least partially because they result from more complex processes, such as collision of continental blocks, overriding of hot spots, and advection of lower continental crust.

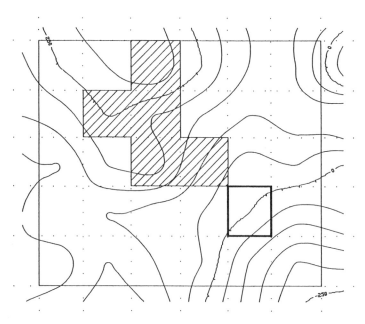

Figure 4. Contour map of elevation data shown in Figure 3. Note that the drainage system flow is normal to the topographic contours.

Because detrital erosion is thought to be elevation dependent (Garrels and Mackenzie, 1971; Hay and Southam, 1977; Hay, 1984; Hay et al., 1987), the mass of detrital sediment contained in a basin may be considered proportional to the elevation of its source area and vice versa. Hence, the reconstructed average elevation of a source area is constrained by the mass of sediment deposited in its associated basins.

In order to reconstruct paleotopography of a source area in a dynamic environment, any externally forced uplift or subsidence occurring over a given time interval must be removed. We use the term "topographic decay" to refer to the gradual reduction of topographic relief and elevation, i.e., the general loss of elevation due to erosion without tectonically induced uplift or subsidence except in response to isostasy. Topographic decay is a nonlinear function proportional to the erosion constant and initial average elevation, and inversely proportional to the elevation of the erosional base level. The rate of topographic decay decreases with time as elevation is reduced. Any departure of the reconstructed surfaces from those defined by pure topographic decay implies uplift or subsidence, which can be determined by comparing the sediment mass yielded by erosion of a hypothetical reloaded surface with the estimate of the sediment mass delivered to the basin during the same time interval. The term "reloaded surface" refers to the preliminary reconstructed elevations after sediment has been removed from the basin and returned to the source area, and after the proper isostatic adjustments have been made. It is a strictly hypothetical topography used only to compare its hypothetical sediment yield with the estimate of the true sediment mass. The hypothetical yield ratio (R) is the ratio of the eroded sediment yield from the reloaded surface divided by the estimate of the true sediment mass which existed at the time. R values greater than one imply uplift because erosion of the reloaded surface yields a sediment mass greater than actually accumulated in the basin. Because the erosion equation we use is a linear function of elevation, the final reconstructed elevation, S, of a source square is:

$$S(m) = [(\text{Reloaded Elevation (m)} - b) / R] + b \qquad (3)$$

where b is the elevation of the erosional base level in meters. Areas below the erosional base level do not contribute to the sediment yield and are not affected by the estimate of uplift or subsidence.

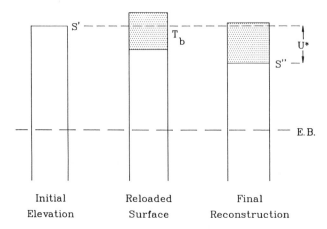

Figure 5. Continental uplift and subsidence (U^*) is estimated by evaluating the elevation of a surface at an initial (S') and final (S'') time. The real elevation of the final surface is adjusted for isostasy related to sediment loading (T_b).

The amount of uplift or subsidence (U^*) is estimated by comparing the elevations of a reference horizon within the same grid square at the beginning (S') and end (S'') of the time interval (Fig. 5) and is represented by the following equation:

$$U^* = S' - S'' + T_b \{1 - [P_s D_{pf} + (1 - P_s) D_s]\} \qquad (4)$$

where T_b is the bulk thickness of sediment returned to the square; D_s, D_{pf}, and D_m are the density of solid-phase sediment, pore fluid, and mantle, respectively; and Ps is the estimated porosity of the source rock.

ADJUSTMENT TO SEA LEVEL

After the reconstruction is completed and the final surface generated, sea level at the moment of time represented by the reconstruction must be imposed on the surface. This requires that all of the area affected by any difference between this sea level and the average sea level used in calculating erosion must again be isostatically adjusted. Changes in water depth are thus related to sea-level changes, the amount of sediment removed, decompaction of the underlying sediment, and isostatic adjustments (Fig. 6). Although asthenospheric material must move from beneath the ocean and its marginal seas to accommodate a rise in sea level, we follow the common convention and do not make a mass-conserving correction beneath the land area.

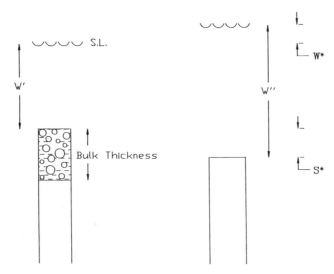

Figure 6. Backstripping involves changes in the initial water depth overlying a unit area (W') which results from sea level changes (W*) and the loss of bulk sediment after adjusting for isostasy (S*).

AN EXAMPLE OF PALEOTOPOGRAPHIC/PALEOBATHYMETRIC RECONSTRUCTION

Figures 7 and 8 show two paleotopographic/paleobathymetric reconstructions for 65 Ma. Both were run with the same erosion constants (erosional base level = +200 m; detrital erosion constant = 0.113 m of solid phase material removed per million years per meter above erosion base; dissolved erosion constant = 0, i.e., no correction for lowering of the land surface through dissolution). Figure 7 shows the result of a model run with the original Haq et al. (1987) sea-level curve. In this reconstruction, most of the continental interior is flooded. Figure 8 shows the result of a model run using a modified Haq et al. sea-level curve in which the amplitude is divided by eight. The empirical shoreline position for 65 Ma, determined from stratigraphic data by Rainwater (1964, 1967), coincides remarkably well with the modeled shoreline position calculated using the modified Haq et al. eustatic curve and imposing the resulting sea level on the topographic surface generated by the model.

The coincidence between the empirical and the modeled shoreline positions at 65 Ma, using the modified Haq et al. sea-level curve, suggests that although the trend of sea-level

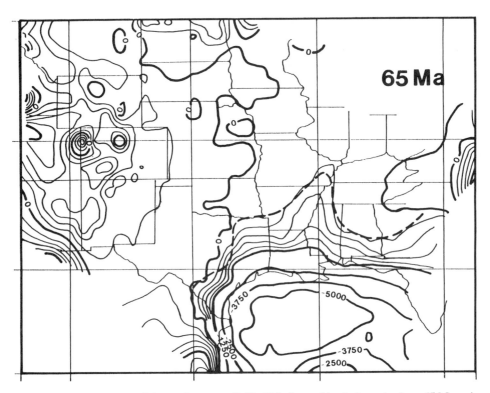

Figure 7. Reconstruction of the northwestern Gulf of Mexico and its drainage basin at 65 Ma, using third-order eustatic curve of Haq et al. (1987). Dashed line indicate position of Early Paleocene shoreling according to Rainwater (1964, 1967).

rises and falls as reported by Haq et al. (1987) may be correct, the actual amplitudes of the eustatic fluctuations may be overestimated. The actual amplitude will depend on the correction of the elevation of the land surface by replacing the material lost to dissolution. We estimate that this amounts to about 1 m/m.y. relative to sea level. It does not appear that this is solely a regional effect caused by rapid progradation of the shoreline related to high sedimentation rates. The mass-age compilations for detrital sediment in the northern Gulf margin show that rapid sedimentation has occurred primarily in the Quaternary, probably in response to rapid uplift of the High Plains and Rocky Mountains in the late Pliocene (Shaw, 1987). As has been shown by Ruddiman et al. (1986), elevated portions of the land not only affect the available supply of clastic sediment to a basin but also may have climatic implications in general by disturbing the atmospheric boundary layer and global circulation patterns. Uplift of the western part of North America may well be the trigger for continental glaciation.

It should be noted that the Cenozoic shorelines estimated by Rainwater (1967) also can be reproduced using the average sea-level values in the Haq et al. (1987) eustatic curve but with either an exceptionally high erosional base level or greatly decreased erosion constant

(Shaw, 1987). Sensitivity analysis on erosion parameters show that the reconstructions are dominated primarily by changes in sea level and that dramatic changes in the erosional parameters are required to invoke even slight changes in the predicted paleotopography.

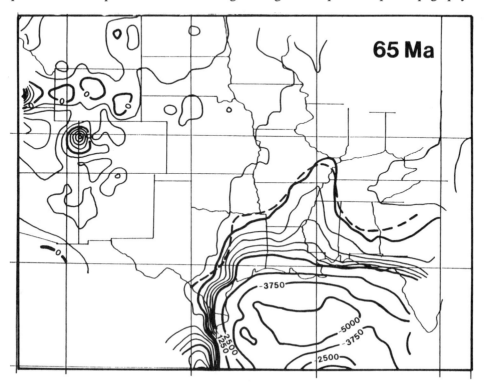

Figure 8. Reconstruction of the northwestern Gulf of Mexico and its drainage basin at 65 Ma, using modified sea-level curve of Haq et al. (1987) in which amplitudes are divided by eight. Shoreline position for 65 Ma according to Rainwater (1964, 1967) also is shown (dashed line).

CONCLUSIONS

Mass-balanced paleogeographic modeling, made possible by rapid computer imagery and decades of extensive stratigraphic, geophysical, and geochemical data base development, allows investigators to analyze the sedimentological and structural development of a basin and its source area in three dimensions. Because much of the development of a sedimentary basin is directly related to erosion cycles in the source area, understanding the evolution and decay of drainage basins is an important component when evaluating the whole erosion-deposition system. Although the current process of producing mass-balanced paleogeographic reconstructions either omits or oversimplifies many important factors, we feel that increasingly sophisticated models that restore sediment in a basin to source areas will provide relief to previously two-dimensional paleogeographic reconstructions and provide added insight to numerical oceanographic and climatic models.

ACKNOWLEDGMENTS

This work has been supported by grants OCE 8409369 and OCE 8716408 from the National Science Foundation and 19274-AC2 from the American Chemical Society's Petroleum Research Fund, by the U.S. Geological Survey, and by gifts from Texaco, Inc.. We also wish to thank Martin Perlmutter, Christopher Paola, and Timothy Cross for their reviews and comments.

REFERENCES CITED

Baldwin, B., and Butler, C.O., 1985, Compaction curves: American Association of Petroleum Geologists Bulletin, v. 69, p. 622-626.

Cathles, L.M., 1980, Interpretation of postglacial isostatic adjustment phenomena in terms of mantle rheology, in Moerner, N.-A., ed., Earth rheology, isostasy and eustasy: Proceedings of earth rheology and Late Cenozoic isostatic movements symposium, Stockholm, Sweden, July 31-August 8, 1977, Geodynamics Project: Scientific Report 49: John Wiley & Sons, p. 11-43.

Chappell, J., 1974, Late Quaternary glacio- and hydro-isostasy on a layered earth: Quaternary Research, v. 4, p. 405-428.

Garrels, R.M., and Mackenzie, F.T., 1971, Evolution of Sedimentary Rocks: New York, W.H. Norton & Co., 397 p.

Haq, B.U., Hardenbol, J., and Vail, P.R., 1987, Chronology of fluctuating sea levels since the Triassic: Science, v. 235, p. 1156-1167.

Hay, W.W., 1984, The breakup of Pangaea: Climatic, erosional, and sedimentological effects: Proceedings, 27th International Geological Congress, Moscow, v. 6 (Geology of Ocean Basins), p. 15-38.

Hay, W.W., Rosol, M.J., Sloan, J.L., II, and Jory, D.E., 1987, Plate tectonic control of global patterns of detrital and carbonate sedimentation: in Doyle, L.J., and Roberts, H.H., eds., Carbonate-clastic transitions: Amsterdam, Elsevier, Developments in Sedimentology, v. 42, p. 1-34.

Hay, W.W., and Southam, J.R., 1977, Modulation of marine sedimentation by the continental shelves, in Anderson, N.R., and Malahoff, A., eds., The fate of fossil fuel CO_2 in the oceans: New York, Plenum Press, Marine Science Series, v. 6, p. 569-604.

Kinsman, D.J.J., 1975, Rift valley basins and sedimentary history of trailing continental margins: in Fischer, A.G., and Judson, S., eds., Petroleum and global tectonics: Princeton, New Jersey, Princeton University Press, p. 83-126.

Malott, C.A., 1928, Base-level and its varieties: Indiana University Studies in Geology, v. 82, p. 37-59.

McClain, J.S., and Atallah, C.A., 1986, Thickening of the oceanic crust with age: Geology, v. 14, p. 574-576.

McKenzie, D.P., 1978, Some remarks on the development of sedimentary basins: Earth and Planetary Science Letters, v. 40, p. 25-32.

Montadert, L., de Charpal, O., Roberts, D., Guennoc, P., and Sibuet, J.-C., 1979, Northeast Atlantic passive continental margins: Rifting and subsidence processes: in Talwani, M., Hay, W., and Ryan, W.B.F., eds., Deep drilling results in the Atlantic Ocean: Continental margins and paleoenvironment: American Geophysical Union, Maurice Ewing Series, v. 3, p. 154-186.

Parsons, B., and Sclater, J.G., 1977, An analysis of the variation of ocean floor bathymetry and heat flow with age: Journal of Geophysical Research, v. 82, p. 803-827.

Rainwater, E.H., 1964, Transgressions and regressions in the Gulf Coast Tertiary: Gulf Coast Association of Geological Societies Transactions, v. 14, p. 217-230.

Rainwater, E.H., 1967, Resumé of Jurassic to Recent sedimentation history of the Gulf of Mexico basin: Gulf Coast Association of Geological Societies Transactions, v. 17, p. 179-210.

Ricken, W., 1987, The carbonate compaction law: A new tool: Sedimentology, v. 34, p. 571-584.

Ruddiman, B., Raymo, M., and McIntyre, A., 1986, Matuyama 41,000-year cycles: North Atlantic Ocean and northern hemisphere ice sheets: Earth and Planetary Science Letters, v. 80, p. 117-129.

Sclater, J.G., Anderson, R.N., and Bell, M.L., 1971, The elevation of ridges and the evolution of the central eastern Pacific: Journal of Geophysical Research, v. 76, p. 7888-7915.

Sclater, J.G., and Christie, P. A., 1980, Continental stretching: An explanation of the post Mid-Cretaceous subsidence of the central North Sea basin: Journal of Geophysical Research, v. 85, p. 3711-3739.

Sclater, J.G., Jaupart, C., and Galson, D., 1980, The heat flow through oceanic and continental crust and the heat loss of the Earth: Reviews of Geophysics and Space Physics, v. 18, p. 269-311.

Shaw, C.A., 1987, Balanced paleogeographic reconstructions of the northwestern Gulf of Mexico margin and its western-central North American source area since 65 Ma [M.S. Thesis]: Boulder, University of Colorado, 285 p.

Sleep, N.H., 1971, Thermal effects of the formation of Atlantic continental margins by continental breakup: Geophysical Journal of the Royal Astronomical Society, v. 24, p. 325-350.

Steckler, M.S., and Watts, A.B., 1978, Subsidence of the Atlantic type continental margin off New York: Earth and Planetary Science Letters, v. 41, p. 1-13.

Steckler, M.S., and Watts, A.B., 1982, Subsidence history and tectonic evolution of Atlantic-type continental margins, in Scrutton, R.A., ed., Dynamics of passive margins: American Geophysical Union, Geodynamics Series, v. 6, p. 184-196.

Turcotte, D.L., Ahern, J.L., and Bird, J.M., 1977, The state of stress at continental margins: Tectonophysics, v. 42, p. 1-28.

Ungerer, P., Bessis, F., Chenet, P.Y., Durand, B., Nogart, E., Chiarelli, A., Oudin, J.L., and Perrin, J.F., 1984, Geological and geochemical models in oil exploration: Principles and practical examples, in Demaison, G., and Murris, R.J., eds., Petroleum geochemistry and basin evaluation: American Association of Petroleum Geologists Memoir 35, p. 53-77.

Watts, A.B., and Ryan, W.B.F., 1976, Flexure of the lithosphere and continental margin basins: Tectonophysics, v. 36, p. 25-44.

Watts, A.B., and Steckler, M.S., 1979, Subsidence and eustacy at the continental margin of eastern North America, in Talwani, M., Hay, W., and Ryan, W.B.F., eds., Deep drilling results in the Atlantic Ocean: Continental margins and paleoenvironment: American Geophysical Union, Maurice Ewing Series, v. 3, p. 218-234.

16

NUMERICAL SIMULATION OF CIRCULATION AND SEDIMENT TRANSPORT IN THE LATE DEVONIAN CATSKILL SEA

Marc C. Ericksen, Didier S. Masson, Rudy Slingerland, and David W. Swetland

Department of Geosciences, The Pennsylvania State University, University Park, PA 16802 USA

ABSTRACT

A three-dimensional (x,y,z,t) hydrodynamic computer model of the Late Devonian Catskill Sea has been developed to help understand the fluid and sediment circulation processes that operated in the seaway of the Appalachian foreland basin. Input parameters to the model include estimates of the paleogeographical and paleoclimatological conditions prevailing during the Late Devonian. Given southeastern trade winds for an epeiric sea at 20° south latitude, the model predicts a dynamic circulation system with strong counterclockwise flow around an island in the vicinity of the Ozark Mountains. Shelf waters other than over the Midcontinent moved westward (Devonian coordinates) as opposed to interior flows that moved down the water surface slope. Net wind-driven circulation along the Catskill shelf across Pennsylvania and New York resulted from horizontal pressure gradients, Ekman drift, and geostrophic balance, and probably was to the paleowest. Calculations of tidal circulation also yield westward residuals on the Catskill shelf. These simulations rationalize the sedimentary facies of the Catskill complex observed in outcrop.

INTRODUCTION

Quantitative dynamic stratigraphy (QDS) is the geologic study of the character and origin of sedimentary processes and facies using numerical models that in their most complete form incorporate the fundamental physics of a sedimentary depositional system. The models are meant to be replicas of sedimentary systems that enhance understanding and encourage quantitative inferences and insight into the evolution of the systems. This general methodology, termed "analysis by synthesis" (Hut and Sussman, 1987) is an alternative to traditional reductionist methods, in that it attempts to capture all of the relevant physical processes and state variables. In complex, nonlinear systems this is particularly necessary because unforseen combinations of small, potentially overlooked, parameters often result in important phenomena.

In this paper we describe our application of QDS to the Late Devonian Catskill Sea of Eastern North America. Using a multi-layer three-dimensional hydrodynamic model and a two-dimensional bedload sediment transport model, we calculate potential circulation and

Quantitative Dynamic Stratigraphy (1989), T.A. Cross, ed., Prentice Hall, p. 293–305.

293

sediment dispersal patterns from present best estimates of the paleogeographical and paleoclimatological conditions prevailing in the Late Devonian.

Our effort has focused on simulating sedimentary events on a scale of days to seasons. We therefore do not account for basin subsidence, changes in tectonic environment, or basin evolution through geologic time. Because a great portion of the preserved rock record is thought to be the result of day- to season-long events, we view our effort as a necessary first step toward simulating the stratigraphic evolution of a basin through geologic time. Our results demonstrate that contrary to traditional thinking, tidal ranges could have been significantly enhanced in certain locations around the basin. Both wind-driven and residual circulation on the Catskill Shelf was to the paleowest, thus helping to explain aspects of Upper Devonian marginal marine facies observed in outcrop.

PREVIOUS WORK

Studies calculating the circulation of modern continental seas are numerous. Leendertse (1977, 1979), Nihoul (1982), and Heaps (1987) provide examples and illustrate that present-day models can reasonably reproduce the three-dimensional flow characteristics of complex bodies of water. By contrast, applications of circulation models to ancient systems are limited. Slater (1985) and Slingerland (1986) used numerical models to hindcast tidal circulation in the Upper Cretaceous Seaway and the North American Devonian Catskill Sea, respectively. These studies show that, contrary to traditional views, ancient continental seas probably contained appreciable tides. To our knowledge, there are no reported simulations of both wind- and tide-driven circulation in ancient seas.

Investigation of modern coastal oceans using coupled circulation and sediment transport models is only just beginning (Sheng, 1983; Sündermann, 1983; Kachel and Smith, 1986; Koutitas, 1986; and Black, 1987). The results have been sufficiently successful however, to suggest the technique can be applied profitably to ancient seas; this paper represents an initial attempt to do so.

THE MODEL

The circulation model used in this study is modified from Leendertse's (1977) three-dimensional model for turbulent flow in estuaries and coastal seas. The basic hydrodynamic and sediment transport equations, written for an incompressible fluid on a rotating earth in Cartesian coordinates with the z-axis directed positive upwards (Fig. 1), are presented in Table 1. Variable definitions are presented in Table 2. The centrifugal force is incorporated in the force of gravity, and the components of Coriolis acceleration simplify to the Coriolis parameter. The vertical equation of motion (Z-DIR) has been reduced to the hydrostatic equation by assuming that vertical accelerations may be neglected. This allows the three-dimensional flow structure to be simulated by dividing the water column into layers (Fig. 1) and integrating the system of equations over the height of each layer, a simplification that greatly reduces computational requirements. The layers are coupled using a quadratic shear stress formula. Energy dissipation equations account for sub-grid scale turbulence. The equation set is solved by a finite difference scheme written for a space-staggered grid.

Table 1: Dynamic equations and boundary conditions for circulation and sediment transport in epeiric seas.

DYNAMIC EQUATIONS FOR CIRCULATION AND SEDIMENT TRANSPORT

X-DIR Equation of Motion
$$\frac{\partial u}{\partial t} + \frac{\partial(uu)}{\partial x} + \frac{\partial(uv)}{\partial y} + \frac{\partial(uw)}{\partial z} - fv + \frac{1}{p}\frac{\partial p}{\partial x} - \frac{1}{p}\left(\frac{\partial \tau_{xx}}{\partial x} + \frac{\partial \tau_{xv}}{\partial y} + \frac{\partial \tau_{xz}}{\partial z}\right) = 0$$

Y-DIR Equation of Motion
$$\frac{\partial v}{\partial t} + \frac{\partial(vu)}{\partial x} + \frac{\partial(vv)}{\partial y} + \frac{\partial(vw)}{\partial z} + fv + \frac{1}{p}\frac{\partial p}{\partial y} - \frac{1}{p}\left(\frac{\partial \tau_{yx}}{\partial x} + \frac{\partial \tau_{yv}}{\partial y} + \frac{\partial \tau_{yz}}{\partial z}\right) = 0$$

Z-DIR Equation of Motion
$$\frac{\partial p}{\partial z} + pg = 0$$

Continuity of Fluid
$$\frac{\partial u}{\partial x} + \frac{\partial v}{\partial y} + \frac{\partial w}{\partial z} = 0$$

Continuity of Salt
$$\frac{\partial s}{\partial t} + \frac{\partial(us)}{\partial x} + \frac{\partial(vs)}{\partial y} + \frac{\partial(ws)}{\partial z} - \frac{\partial(D_x\frac{\partial s}{\partial x})}{\partial x} - \frac{\partial(D_y\frac{\partial s}{\partial y})}{\partial y} - \frac{\partial(\kappa\frac{\partial s}{\partial z})}{\partial z} = 0$$

Continuity of Heat
$$\frac{\partial T}{\partial t} + \frac{\partial(uT)}{\partial x} + \frac{\partial(vT)}{\partial y} + \frac{\partial(wT)}{\partial z} - \frac{\partial(D_x\frac{\partial T}{\partial x})}{\partial x} - \frac{\partial(D_y\frac{\partial T}{\partial y})}{\partial y} - \frac{\partial(\kappa'\frac{\partial T}{\partial z})}{\partial z} = 0$$

Equation of State
$$p = \bar{p} + p'(s,T)$$

Bed Sediment Continuity
$$g(p_p - p)(1 - \lambda)\frac{\partial h}{\partial t} + \left(\frac{\partial g_{bx}}{\partial x} + \frac{\partial g_{by}}{\partial y}\right) = 0$$

Bagnold-type Bedload Transport
$$g_{b(x,y)} = \frac{a}{\tan\alpha}(u_\star - u_{\star c})(\tau^b - \tau_{oc})$$

BOUNDARY CONDITIONS

Wind Stress
$$\tau_x^s = C^\star p_a w_a^2 \sin\psi \qquad \tau_y^s = C^\star p_a w_a^2 \cos\psi$$

Bed Stress
$$\tau_y^b = p_w g\frac{v\sqrt{u^2 + v^2}}{C^2} \qquad \tau_x^b = p_w g\frac{u\sqrt{u^2 + v^2}}{C^2}$$

Open Boundary Tide or Current

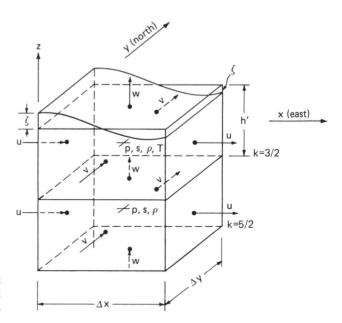

Figure 1. The layered structure of the model and relative position of variables.

The sediment transport model consists of a two-dimensional bed sediment continuity equation and a modified Bagnold bedload transport equation (Table 1). Bottom layer velocities computed by the circulation model are converted to bed stresses using the quadratic stress laws, for use in the sediment transport equations. At present, the model does not account for suspended sediment transport, wave-driven sediment transport, or heterogeneous-sized beds, and does not allow for dynamic changes in bed elevation. Copies of the source codes may be obtained from R. Slingerland at the Pennsylvania State University.

Boundary conditions to the circulation model include wind stress and open boundary water surface elevations or discharges through time (Table 1). Initial conditions specified by the user include bathymetry, basin planform, boundary tides, wind speed and direction, current inflows or outflows, water temperature and salinity, bed friction factor, and a single bed grain size. Output from the model consists of fluid velocities in three dimensions (u,v,and w components), temperature and salinity, water surface elevations, tidal ranges, residual circulation velocities, and net sediment deposition or erosion.

STRATIGRAPHIC APPLICATION OF THE MODEL

To better understand the depositional processes responsible for the Chemung and Catskill magnafacies of eastern North America we have simulated circulation and sediment transport in the Late Devonian Catskill Sea. As reconstructed by Heckel and Witzke (1979), the Sea was bounded on the south (paleocoordinates unless otherwise noted) by the Acadian Highlands, on the east by the Old Red Sandstone Continent (now Great Britain and western Europe), on the north by the Transcontinental Arch, and on the west (present day Gulf Coast)

Table 2 Variable definitions.

x, y, z	= Cartesian coordinates, positive eastward, northward, &upward, respectively (cm)
u, v, w	= respective components of velocity (cm/s)
t	= time (s)
f	= Coriolis parameter ($1.458 \times 10^{-4} s^{-1)}$
p	= pressure (dynes/cm^2)
s	= salinity (g/kg)
T	= temperature (ºC)
ρ, ρ_a, ρ_p	= density of water (g/cm^3), air (1.226×10^{-3} g/cm^3), sediment (2.65 g/cm^3)
ρ	= reference density, a constant (g/cm^3)
ρ'	= departure from ρ depending on salinity and temperature
κ	= vertical mass diffusion coefficient (set usually to 0)
κ'	= vertical thermal diffusion coefficient (set usually to 0)
τ_{xx}, τ_{xy}, τ_{yx}, τ_{yy}, τ_{xz}, τ_{yz}	= components of the stress tensor (dynes/cm^2)
D_x, D_y	= horizontal diffusion coefficients (set usually to 0)
τ^s_x, τ^s_y	= wind stress (dynes/cm^2)
τ^b_x, τ^b_y	= bed stress (dynes/cm^2)
w_a	= wind speed (cm/s)
ψ	= angle between wind direction and y axis (degrees)
C*	= wind drag coefficient : 2.6×10^{-3}
C	= Chezy coefficient (3.13×10^2 cm$^{1/2}$s^{-1})
λ	= bed porosity
h	= bed elevation (cm)
q_b	= bedload transport rate in immersed weight per unit width (Nm^{-1}s^{-1})
a/tanα	= experimental constants (usually set around 10)
u_*, τ^b	= bed friction velocity (cm/s) and shear stress (dynes/cm^2)
u_{*c}, τ_{oc}	= sediment critical friction velocity (cm/s) and shear stress (dynes/cm^2)

by the open ocean (Fig. 2). The Ozark Dome formed an island within it. For the purpose of our simulations the entire length of the Transcontinental Arch is treated as a closed boundary because intermittent seaways across the arch were shallow. The shoreline location is a time-average during the Frasnian-Fammenian based on outcrop distribution of marginal marine facies.

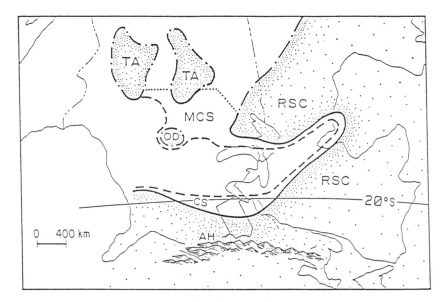

Figure 2. Paleogeography of eastern North America in Frasnian-Famennian time adapted from Heckel & Witzke (1979). The dark solid line represents the probable limit of the Catskill Sea on the craton; the dark dashed line represents the boundary between black shale and predominantly terrigenous detrital facies, taken here as the base of a clinoform. The dotted line represents a boundary to the Catskill Sea defined by shoals. TA = Transcontinental Arch; RSC = Red Sandstone Continent; MCS = Midcontinent Shelf; OD = Ozark Dome; CS = Catskill Shelf; AH = Acadian Highlands.

The bathymetry of the Catskill Sea is well constrained by the distribution of sedimentary facies (Slingerland, 1985). An extensive black shale sequence defines the deepest parts of the basin. Using the modern Black Sea as an analogue, this indicates a depth to the basin floor of 150 m or more. On the other hand, uncompacting a prograding delta sequence in eastern Kentucky yields a minimum basin depth of 230 m (Ettensohn and Barron, 1981). We used a conservative maximum depth of 150 m for the region inside the dashed line in Figure 2 to allow for maximum tidal attenuation. Our results, therefore, provide a minimum estimate of potential tidal ranges in the Catskill Sea. The shape of the seafloor was adapted from Woodrow and Isley (1983) and the facies map of Heckel and Witzke (1979). Shelf slopes everywhere rise landward from 150 m to an average shelf depth of 30 m. This bathymetry is simulated in the model by using four water layers: a surface layer 30 m deep, and three underlying layers each 40 m deep. Flows along the shelf therefore correspond to flows calculated in the surface layer of our model.

The paleolatitude of the Catskill Sea is controversial, with estimates ranging from 0° to 30° south (Woodrow, et al., 1973; Heckel and Witzke, 1979; Boucot and Gray, 1983; Kent, 1985). This study favors the intermediate latitude of 20° south consistent with Van Der Voo (1988). Over this range of latitudes, general wind patterns are not likely to change, although the wind speed decreases from 0° to 30° south latitude (Wallace and Hobbs, 1977). The Coriolis force increases away from the equator, thereby also increasing the influence of Ekman drift and geostrophic flows in the circulation in the basin.

Initial and Boundary Conditions

Three runs are presented to investigate the influence of the most probable driving mechanisms: wind-driven circulation, tide-driven circulation, and a combination of the two. Table 3 lists the initial and boundary conditions used in each model. The calculated circulation

Table 3 Input and Boundary Conditions for each of three Devonian runs.

Run Numbers	Run #1	Run #2	Run #3
Figure Numbers	3a, 3b	4	5a, 5b
Wind Direction	No Wind	from Southeast	from Southeast
Wind Speed	-	20 knots	20 knots
Tidal Amplitude at Open Boundary	60 cm	No Tide	60 cm
West Boundary Conditions	Open with M_2 Tide	Closed	Open with M_2 Tide
Maximum Thalweg Depth	150 m	150 m	150 m

patterns for these runs are presented for the surface layer only, and thus represent the average circulation over the shelves and surface layer circulation over the rest of the basin. A 20 knot southeast wind was chosen as the prevailing wind condition for the Catskill Sea based on contemporary wind circulation patterns at 20° to 30° south latitude. Variations in thickness and crystal size of the Middle Devonian Tioga Tuff supports the southeast tradewind interpretation (J.M. Dennison, personal communication, 1988). A simple semi-diurnal, lunar tide (M_2), with an amplitude of 0.6 m (similar to tidal ranges along the edges of contemporary continental shelves) is applied to the open ocean boundary. We feel this is conservative because the closer proximity of the moon during the Devonian (Lambeck, 1980) may have produced larger tidal ranges. The amount of frictional damping in the model is controlled by C, the Chezy coefficient. A constant Chezy value of 313 cm$^{1/2}$s^{-1} was used here, simulating a sandy bed.

Results and Discussion

The M$_2$ co-oscillating tides in the Catskill Sea are calculated in Run #1. Considerable tidal amplification occurs in the model (Fig. 3a) with a tidal range of over two meters predicted for parts of the Catskill Shelf. This represents more than a four-fold amplification of the tide at the open ocean boundary, confirming the earlier work of Slingerland (1986). Numerous authors (e.g., Walker and Harms, 1975; Woodrow and Isley, 1983; Dennison, 1985) have argued that the shoreline deposits of the central Appalachians indicate a low tidal range in the Catskill Sea. Others (Allen and Friend, 1968; Krajewski and Williams, 1971; Rahmanian, 1979; Slingerland and Loule, in press) concluded the opposite. This modeling suggests not only should tides have been present, but that they may have been at least of mesotidal range.

The residual velocities calculated for the surface layer (Fig. 3b) are defined as the velocities at each grid point averaged over one tidal cycle. As such they give the net drift pattern one might predict for sediment transport by tides. The residual circulation along the Catskill Shelf and in the basin thalweg is to the west; elsewhere, the residual circulation is less well-defined. These results are discussed under Run #3.

The predicted equilibrium circulation of near-surface waters resulting from a southeast wind of 20 knots is calculated in Run #2 and presented in Figure 4. Flow is generally eastward in the center of the basin, turning northeastward in the Hudson Bay Arm (easternmost extension) of the Catskill Sea. This is also true for all deeper levels in the Catskill Sea. On the Catskill Shelf circulation turns northward at the artificially closed western boundary and joins the easterly flow down the basin axis. A counterclockwise circulation gyre controls the flow around the Ozark Dome. The eastward currents are interpreted as a geostrophically balanced flow due to the wind-generated, higher water-surface elevations on the north side of the basin. The westward shore-parallel shelf flows are attributed to Ekman drift.

Run #3 combines the tide and wind forcing and represents our present best estimate of the prevailing tide and wind conditions in the Catskill Sea. Residual velocities (Fig. 5a) again show net westward (shore-parallel) transport along the Catskill Shelf. Wind-driven shelf circulation (Fig. 4) clearly dominates suggesting that, although tidal range may have been high along the Catskill coast, in the offshore wind-driven flows were more important. Halving the wind speed (not shown) results in no major changes to the general circulation pattern. In both cases, westward sediment transport is suggested, and is corroborated by a recent field study across Pennsylvania (Slingerland and Loule, in press).

Bedload transport and the resulting bed elevation changes were calculated at each node over a two-day period assuming a uniform bed grain size of 0.3 mm (Fig. 5b). Bottom velocities in the thalweg of the basin, corresponding to the outcrop belt of Devonian black shales, are not strong enough to transport sand. Zones of deposition along the coast result from horizontal expansion of the flow; zones of deposition offshore, such as the 400 km long site off the Catskill Shelf, result from vertical expansion of the flow as it passes into deeper

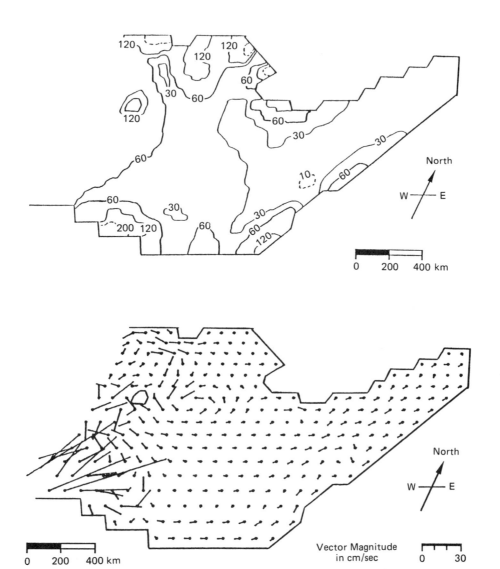

Figure 3a. (Top) Tidal range lines computed over one tidal cycle in the Devonian Catskill Sea showing the areas of tidal amplification when an M_2 tide of amplitude 60 cm is applied to the open boundary. Contour lines in cm.

Figure 3b. (Bottom) Computed residual tidal circulation in the upper layer of the Devonian Catskill Sea showing the westward drift of surface waters throughout most of the basin. Each vector represents a time-averaged value of the velocities at that node for one tidal cycle.

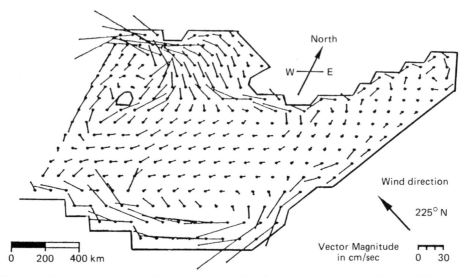

Figure 4. Computed steady state circulation pattern of surface waters (upper layer) in a closed Devonian Catskill Sea subjected to 20 knot southeast winds. Note the eastward geostrophic flow in the thalweg, and the shore-parallel flow along the Catskill shelf generated by Ekman drift. The pattern is dominated by two circulation cells: a counterclockwise gyre around the Ozark Dome, and a clockwise flow along the Catskill Shelf.

water. The large depositional regime at the west end of the basin is probably due to convergence of the two main circulation gyres and the resulting horizontal and vertical expansion as they move eastward into the basin.

Modeling results are sensitive to both bathymetry and paleogeography. Slingerland (1986) found that although tidal amplification is enhanced as bathymetry is increased to roughly 300 m, it would likely have been reduced if the Catskill shelf were less concave. A detailed sensitivity analysis of the model to bathymetry and paleogeography for the wind-driven case is still in progress. Furthermore, we did not consider the influence of density stratification in these experiments. However, analytical models (Csanady, 1983) suggest, that to a first order, the affects of stratification are most significant close to the shore, roughly within 5 km. Because the grid spacing of our models is 104 km, the general conclusions of this study will also apply to a stratified basin.

Figure 5a. (Top) Computed residual circulation in the upper layer of the Devonian Catskill Sea subjected to 20 knot southeast winds with an M_2 tide along the open boundary. Note the similarity to Figure 4 indicating the dominance of wind-driven circulation over tidal (Fig. 3a). Again the predicted flux of water along the Catskill shelf is to the west.

Figure 5b. (Bottom) Zones of deposition and erosion predicted for two days of sediment transport in a Catskill Sea with an M_2 tide of 60 cm on the open ocean boundary and a constant 20 knot southeast wind. Erosion predominates along the shelves, with pockets of deposition caused by horizontal expansion of the flows. The 400 km stretch of offshore deposition along the Catskill shelf is an example of vertical expansion of the flow as it passes into deeper water. The extensive zone of deposition at the west end of the basin results from convergence of the two major flow systems in Figure 4.

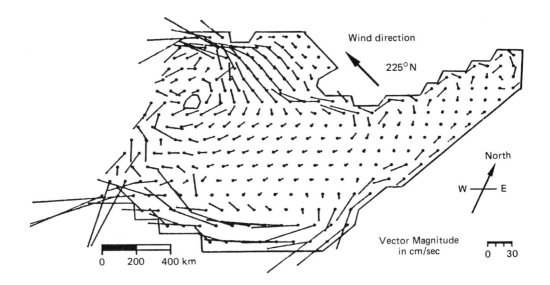

Wind direction
225° N

North
W — E

Vector Magnitude
in cm/sec

0 30

0 200 400 km

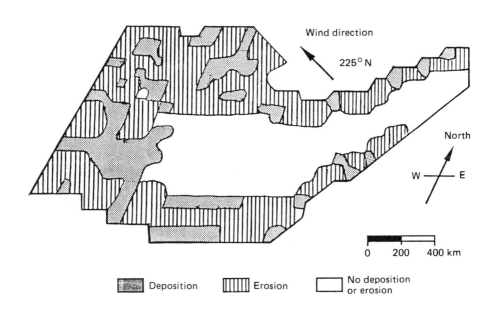

Wind direction
225° N

North
W — E

0 200 400 km

Deposition   Erosion   No deposition
or erosion

Conclusions

A three-dimensional model of circulation and sediment transport in epeiric seas has been presented as an initial step toward simulating the stratigraphic evolution of a basin through time. When applied to the Late Devonian Catskill Sea of eastern North America it predicts a wind-dominated circulation with augmented tides playing an important role in some coastal settings. The circulation was counterclockwise around the Ozark Dome and westward along the Catskill Shelf, resulting in sand transport with zones of sediment accumulation in areas of vertical or horizontal flow expansion.

Acknowledgments

We thank Jim Syvitski and Tim Cross for thoughtful reviews. This work was supported by the Societe Nationale Elf Aquitaine, and The Earth System Science Center and Department of Geoscience at The Pennsylvania State University.

References Cited

Allen, J.R.L., and Friend, P.F., 1968, Deposition of the Catskill facies, Appalachian region: with notes on some other Old Red Sandstone Basins, *in* Klein, G.deV., ed., Late Paleozoic and Mesozoic sedimentation, northeastern North America: Geological Society of America Special Paper 106, p. 21-74.

Black, K., 1987, A numerical sediment transport model for application to natural estuaries, harbours and rivers, *in* Noye, J., ed., Numerical modelling: Applications to marine systems: Holland, Elsevier, p. 77-108.

Boucot, A.J., and Gray, J., 1983, A Paleozoic Pangea: Science, v. 222, p. 571-581.

Csanady, G.T., 1982, Circulation in the coastal ocean : Dordrecht, Holland, D. Reidel, 279 p.

Dennison, J.M., 1985, Catskill delta shallow marine strata, *in* Woodrow, D.L, and Sevon, W.D., eds., The Catskill Delta: Geological Society of America Special Paper 201, p. 91-106.

Ettensohn, F.R., and Barron, L.S., 1981, Depositional model for the Devonian-Mississippian black shale sequence of North America: A tectonic-climatic approach: Morgantown, West Virginia, Morgantown Energy Technology Center Document 12040-2, 80 p.

Heaps, S.N., 1987, Three-dimensional coastal ocean models: Washington, D.C., American Geophysical Union, 208 p.

Heckel, P.H., and Witzke, B.J., 1979, Devonian world paleogeography determined from distribution of carbonates and related lithic paleoclimatic indicators, *in* House, M.R., et al., eds., Paleontological Association, London, Special Papers in Paleontology 23, p. 99-123.

Hut, Piet, and Sussman, G.L., 1987, Advanced computing for science: Scientific American, v. 257, no. 4, p. 144-153.

Kachel, N.B., and Smith, J.D., 1986, Geological impact of sediment transporting events on the Washington continental shelf, *in* Knight, R.J., and McLean, J.R., eds., Shelf sands and sandstones: Canadian Society of Petroleum Geologists Memoir 11, p. 145-162.

Kent, D.V., 1985, Paleocontinental setting for the Catskill Delta, *in* Woodrow, D.L., and Sevon, D.W, eds., The Catskill Delta: Geological Society of America Special Paper 201, p. 9-13.

Koutitas, C., 1986, A comparative study of three mathematical models for wind-generated circulations in coastal areas: Coastal Engineering, v. 10, p. 127-138.

Krajewski, S.A,, and Williams, E.G., 1971, Upper Devonian flagstones from northeastern Pennsylvania: University Park, Pennsylvania, The Pennsylvania State University, College of Earth and Mineral Sciences, Special Publication 3-71, 185 p.

Lambeck, K., 1980, The Earth's variable rotation: Geophysical causes and consequences: Oxford, Cambridge University Press, 449 p.

Leendertse, J.J., Alexander, R.C., Liu, S., 1973, A three dimensional model for estuaries and coastal seas: Vol. I, Principles of computation: Santa Monica, CA., Rand Corporation, Rand Report R-1417-OWRR, 57 p.

Leendertse, J.J., 1977, A three-dimensional model for estuaries and coastal seas: Vol. IV, Turbulent energy computation: Santa Monica, CA., Rand Corporation, Rand Report R-2187-OWRT, 59 p.

Liu, S., Leendertse, J.J., 1979, A three dimensional model for estuaries and coastal seas: Vol. VI, Bristol Bay simulations: Santa Monica, CA., Rand Corporation, Rand Report R-2405-NOAA, 121 p.

Rahmanian, V.D., 1979, Stratigraphy and sedimentology of the Upper devonian Catskill and uppermost Trimmers Rock Formations in central Pennsylvania [Ph.D. thesis]: University Park, Pennsylvania, The Pennsylvania State University, 340 p.

Sheng, Y.P., 1983, Mathematical modeling of three-dimensional coastal currents and sediment dispersion: Model development and application: Washington, D.C., U.S. Army, Office of Chief of Engineers, Technical Report CERC-83-2, 288 p.

Slater, R.D., 1985, A numerical model of tides in the Cretaceous Seaway of North America: Journal of Geology, v. 93, p. 333-345.

Slingerland, R.L., 1986, Numerical computation of co-oscillating paleotides in the Catskill epeiric sea of eastern North America: Sedimentology, v. 33, p. 487-497.

Slingerland, R., and Loule, J.P., in press, Wind/wave and tidal processes along the Upper Devonian Catskill shoreline in Pennsylvania, USA, in McMillan, N.J., ed., Proceedings of the second international Devonian symposium: Canadian Society of Petroleum Geologists Memoir.

Sündermann, J., 1983, Sediment transport modelling with applications to the North Sea, in Sündermann, J., and Lenz, W., eds., North Sea dynamics: New York, Springer-Verlag, p. 453-471.

Van Der Voo, R., 1988, Paleozoic paleogeography of North America, Gondwana, and intervening displaced terranes: Comparisons of paleomagnetism with paleoclimatology and biogeographical patterns: Geological Society of America Bulletin, v. 100, p. 311-324.

Walker, R.G., and Harms, J.C., 1975, Shorelines of weak tidal activity: Upper Devonian Catskill Formation, central Pennsylvania: in Ginsburg, R.N., ed., Tidal Deposits: New York, Springer-Verlag, p. 103-108.

Wallace, J.M., and Hobbs, P.V., 1977, Atmospheric science: An introductory survey: Orlando, Academic Press, 467 p.

Woodrow, D.L., Fletcher, F.W., and Ahrnsbrak, W.F., 1973, Paleogeography and paleoclimate at the depositional sites of the Devonian Catskill and Old Red Sandstone Facies: Geological Society of America Bulletin, v. 84, p. 3051-3064.

Woodrow, D.L., and Isley, A.M., 1983, Facies, topography, and sedimentary processes in the Catskill Sea (Devonian), New York and Pennsylvania: Geological Society of America Bulletin, v. 94, p. 459-470.

Part III

SUBBASIN-SCALE QDS MODELS

17

The Process-Response Model in Quantitative Dynamic Stratigraphy

James P.M. Syvitski
Geological Survey of Canada, Bedford Institute of Oceanography, Dartmouth, Nova Scotia, Canada, B2Y 4A2

Abstract

Process-response models simulate the input, transport and accumulation of sediment in geological basins based on equations governing the hydrodynamics of sediment transport. The response model may be constructed in a geometric, conceptual, statistical or empirical manner. The process model begins with a mathematical approach leading to a probabilistic or deterministic model, and eventually to a computer model. The best simulation models contain elements of many of these approaches, although the numerical model is most tractable and allows greatest flexibility in output resolution and constraints describing an actual basin.

Limitations of process based basin models include: (1) imperfect understanding of the laws of physics governing sediment transport, erosion and deposition; (2) the input functions must be known *a priori*; (3) the model should conserve mass; (4) the boundary conditions must contain realistic approximations to the initial basin geometry with the point sources adequately located; and (5) the transport pathways must be accurate.

Process based models can predict environmental characteristics at a given site, lithotope geometry and character. Site predictions may include both spatial and temporal rates of sedimentation, erosion and sediment accumulation. These data may be used as inputs to other models, such as slope stability models that track pore pressure, sea-floor slope and overburden stress. In addition, process based models can provide site information on catastrophic events, for example, impact pressures from episodic slides and turbidity currents. Users of lithotope predictions include the mining and petroleum industry, the military, and the geotechnical and environmental communities.

An example of a three-dimensional box model describing the filling of a deep inlet is given. The model simulates four mechanisms for the transfer of sediment from the land to the sea: (1) bedload dumping along a delta front; (2) hemipelagic sedimentation under a seaward flowing river plume; (3) proximal slope bypassing by turbidity currents; and (4) the combined effects of both short-term (wave and tidal action) and long-term (creep and slides) downslope diffusion of the accreting sediment mass.

Quantitative Dynamic Stratigraphy (1989), T.A. Cross, ed., Prentice Hall, p. 309–334.

INTRODUCTION

A new generation of predictive models is being developed to better understand the mechanics of sedimentary basin filling and the character of the resulting deposits. The models are of the "process-response" type (Whitten, 1964), wherein a specific set of process factors is linked to a specific sedimentary deposit. This paper is designed as an overview of these models, written for the nonexpert and those not desiring a comprehensive mathematical treatment. It focuses primarily on sedimentary processes of short time scales and those relating to the fill of bathymetric basins, as opposed to geologic basins where tectonic and compaction subsidence become important controls.

FROM CONCEPT TO MODEL

There are two approaches to modeling basin filling (Fig.1). The **response model** (closest to the heart of a geologist) usually begins with careful observations of the geologic record and attempts to work out the temporal and spatial distribution of sedimentary deposits (or the properties of the sediment) by trial-and-error, or with increasingly larger data sets, by best-fit approximations. The **process model** (closest to the heart of the engineer or oceanographer) relies on the rudiments of fluid/sediment interactions and the processes that govern those interactions. The characteristics of each of these approaches are described below and examples are included from the literature.

The Response Model

The first stage of the response model is to develop a **geometric model**, wherein geological information is represented in map form, fence and other diagrams. An excellent example is the 3-D model for nearshore sedimentation constructed by Davis and Fox (1972). In that example, area-time topographic data were used to visually relate responses (sedimentary facies) to changes in wave and current conditions.

The second phase is the development of a **conceptual model**, wherein relevant factors are identified and critical facets are inferred. Harbaugh and Bonham-Carter (1977) provided a conceptual model describing continental margin sedimentation as a system (Fig. 2). They defined a system as a set of dynamically interrelated components whereby changes to any part of the system would have repercussions throughout the system. Systems must have defined boundaries. In a closed system, the model neither receives input from nor provides output to the "outer world." The reverse is true for an open system and is therefore the concept of interest to geologists. For example, the open system described by Harbaugh and Bonham-Carter (1977) has endogenous components (i.e., components within the system) linked by feedback loops (Fig. 2). Exogenous components feed into the system but do not receive feedback (Fig. 2). Exogeneous inputs include sediment supply, oceanic circulation, direction and intensity of wind, and tectonic influences. The output is a simulated stratigraphic record.

Once a conceptual model is established, the original observations may be re-examined in a **statistical model** whereby relationships amongst the simultaneously varying attributes can be analyzed. Through such an analysis, relationships not previously recognized may be

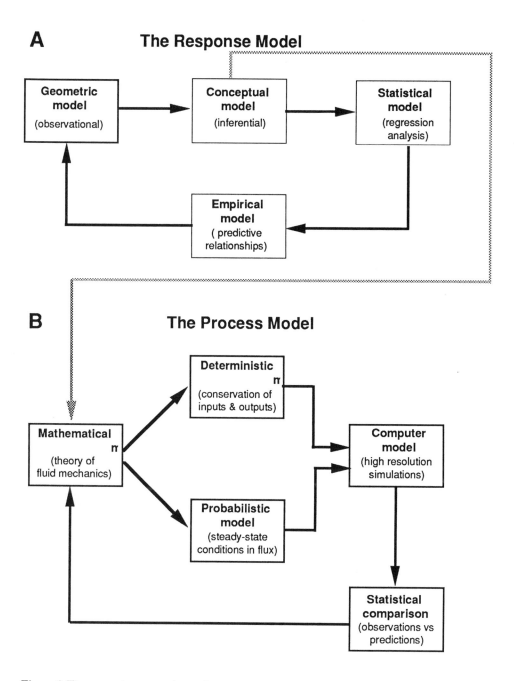

Figure 1. The two major approaches to the process-response model as related to sediment transport and the filling of sedimentary basins.

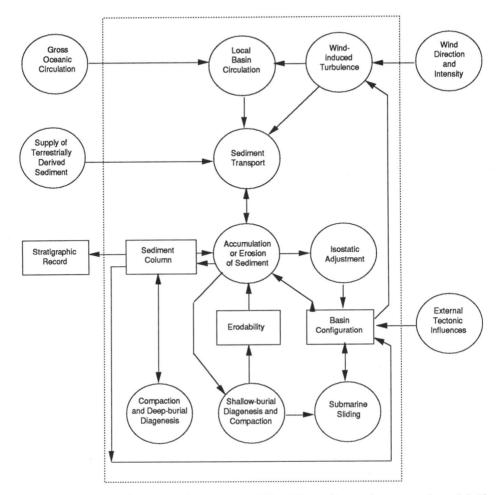

Figure 2. An example of a conceptual response model describing sedimentation on a continental shelf wherein the physics of fluid mechanics are not considered (after Harbaugh and Bonham-Carter, 1977). The within-system (endogenous) components are closely linked with each other and separated from the outer-system (exogenous) inputs and outputs.

indicated. Classic examples include the works of Klovan (1966) and Chambers and Upchurch (1979) who examined grain-size distributions in an effort to provide statistical relationships useful for unbiased predictions on the mechanism of sediment deposition.

Relationships discovered using a statistical model inevitably lead to an **empirical model** where variables can be interrelated in the form of predictive algorithms. As an example, Winters (1983) used polynomial models to estimate the annual transport of suspended sediment in an estuary with a variable salt wedge. The important variables included water flux, suspended sediment concentrations, tidal stage and river discharge. Model predictions were in agreement with observations on sedimentation rates determined from core samples (Winters, 1983).

Closing the response model loop (Fig. 1) entails: making predictions; collecting new observations to test the predictive model; modifying the conceptual model as appropriate; and statistical testing and generation of a new and refined empirical model. The applicability of empirical models, however, is usually limited to environments with similar conditions (Fox, 1978). Markov analysis of sedimentary facies transitions is an example where statistically defined relationships can be used to develop a conceptual model (e.g., Sonu and James, 1973).

The Process Model

The **process model** is not used routinely by geologists. This approach is based on some fundamental theory, such as that governing fluid mechanics and sediment transport. Like the response model described above, the theory is initially based on a conceptual model (Fig. 1), but is expressed in more detail and rigor. An example is Owen's (1977) conceptual model describing the transport, erosion and deposition of cohesive sediments (Fig. 3); the model is a hybrid of an open system in that the output (i.e., erosion or deposition of sediment) remains interactive to the system. The Owen model is more rigorous than the Harbaugh and Bonham-Carter model described above, in that primary and secondary processes and parameters are identified and the speed and size of an interaction are indicated (compare Fig. 2 with Fig. 3).

A **mathematical model** is the natural outgrowth of the process based conceptual model. It involves theoretical expressions of some basic physical laws, such as fluid mechanics. These equations can be considered mathematical approximations of reality, that when linked together can describe a physical system. They include the conservation equations of mass, momentum, and energy. The conservation of fluid mass typically is substituted for one involving volume, and is referred to as the equation of continuity. This equation usually has three terms expressed as partial differentials of the measured parameter and representing the spatial coordinates that taken together sum to nil. The continuity equation for nonconservative properties, such as sediment particles suspended in a liquid, includes an extra term defining their input (erosion) or output (deposition). Continuity equations keep track of the total volume (water, sediment, or both) and its distribution within the system being modeled. The conservation of energy equation tracks the conversion of turbulence to heat as a function of boundary friction. The conservation of momentum is based on Newton's Second Law although secondary terms relating to angular momentum typically are included. Momentum equations evaluate the operating forces at a given location and, important in sediment transport models, provide information on the boundary shear stresses.

An example of a mathematical model describing sediment transport on continental shelves is given by Smith (1977). He used both theoretical and empirical expressions and noted that some parameters are determined fairly accurately, others are merely constrained estimates, and yet others rely on untested theory (the weak link in a mathematical model). Smith divided the model into three portions: the regional physical oceanography, the boundary layer mechanics, and specific sediment transport processes. He justified the theoretical approach by commenting: "one is rarely lucky enough to measure the extreme event destined to leave an imprint in the geologic record, so the marine geologist is forced to rely on the most pertinent, best available, sediment transport model in order to couple the physical characteristics of marine strata with the processes that produce them" (Smith, 1977).

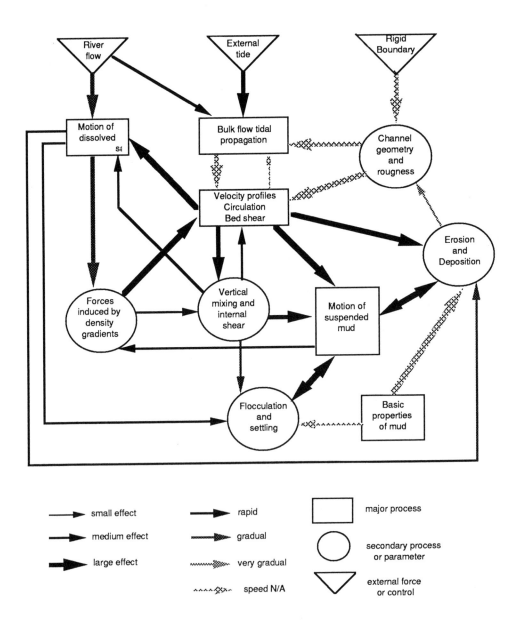

Figure 3. An example of a conceptual process model describing sediment transport in a muddy estuary and where the physics of fluid mechanics are considered (from Owen, 1977). Indicated are the known or inferred importance and speed of each linkage between components.

Two other examples of mathematical models include the work of Kranck (1986) and Ghosh et al. (1986), who provided mathematical functions useful in the prediction of grain-size distributions from the physics of erosion, transport, and deposition of sediment. Both offer the possibility of back-calculating the paleohydraulics operating within a sedimentary basin.

The **deterministic model** has the relevant response characteristics incorporated in a set of unifying equations. Such models contain no element of chance and thus their solution at any point in time is completely predetermined. Deterministic models can have a wide range of solutions depending on the theoretical expression used. For instance, there are three classes of theoretical/empirical bedload formulae describing fluvial transport (for a summary, see Syvitski et al., 1987): the DuBoys-type equations based on the mean tractive force or shear stress; the Schoklitsch-type equations based on a slope/discharge relationship; and the Einstein-type equations based upon statistical considerations of the lift forces resulting from stream line convergence. Lewis and Hwang (1986) compared five different sediment transport equations commonly employed and noted a wide variation in the predictions. Similarly, there are two different approaches to the modeling of littoral sand transport (Kamphuis et al., 1986): the wave energy flux approach, and the shear-stress modification approach.

Deterministic solutions also are dependent on the complexity of the solution evoked. For example, Van Andel and Komar (1969) did not model the energy loss during the period when the head of a turbidity current reversed and collided with its tail (Komar, 1977). A recent solution included the reverse flow in turbidity currents by modeling the role of internal solitons (Pantin and Leeder, 1987). However, neither approach includes the effect of meandering channels and thus the effect of overspill of a portion of the turbidity current. Bowen et al. (1984) modeled the effect of a flow-stripping mechanism, i.e., the result of the turbidity current head encountering a sharp bend and shedding an overspill surge over the levee crest. Hay (1987) modeled the effect of inertial overspill, i.e., a result of the curvature of the overbank in a channel bend being greater than that of the channel axis and thus the fluid columns in the current are carried up the bank and over the levee by their forward momentum.

The **probabilistic model** includes some random or stochastic components and thus no two model runs will be precisely the same. Typical sediment processes that include stochastic elements are river discharge, earthquake generated slides, and storm driven waves. For example, Harper and Penland (1987) used a standard probabilistic model for calculating the sediment transport related to wave action in the Beaufort Sea. The model uses power spectra and probability of occurrence (for both direction and wave height) to calculate the net sediment transport direction. The model is most sensitive to the rare large storm. Fox (1978) noted that probabilistic models are especially useful where portions of the underlying mathematical model are weak.

The **computer model** is usually the final product in the process model. Model simulations of key parameters are generally performed using numerical analysis. Numerical methods can be classified into finite difference methods (based on localized approximations) and finite element methods (based on global constraints of the full domain). Generally a unified differential equation is solved by one of three methods (explicit schemes, implicit schemes and method of characteristics) depending on the form of the equation (elliptic, parabolic, or hyperbolic). Numerical models typically divide the sediment transport medium into small fixed volumes or parcels that are then traced through time. As an example, Komar (1973) used numerical analysis to simulate the growth of a wave-dominated delta. The model requires: defining an initial shoreline; establishing the sediment input (rivers, beach); providing the offshore wave parameters (height, period and approach angle); and a governing equation of littoral transport of sand along the beaches. The model output is the shoreline configuration at specified time steps. Komar noted that the numerical solution allowed flexibility and variability in both sediment supply and wave parameters, including refraction/ diffraction effects not considered in the purely mathematical approach.

Computer models are normally checked against known data or observations. When a match between model output and observations is not achieved, the model may indicate changes to the supporting theory or, more usually, the boundary conditions that constrain the theory. It is important at this stage to review the assumptions used in setting up the model, so that they are clear and not buried behind algorithms. Computer models simulating sediment transport can also be verified through the use of **physical models**. For example, Komar verified his numerical model on delta progradation with wave tank experiments.

In summary, there are many approaches to the modeling of sedimentary processes within the multifaceted process-response model. Although the above discussion presented a compartmentalized description of the various components (Fig. 1), most of today's models are hybrids of one or more of these compartments. The bright future for Quantitative Dynamic Stratigraphy is through the merging of response models with process models producing unified process-response models. The level of complexity or sophistication employed will depend on requirements of specific applications and our present state of knowledge. Below we examine these aspects in more detail.

CHOICE OF MODEL

The earth scientist can take four approaches to the process-response model described above, depending on the specific objective. These include: (1) models whose predictions are at the resolution of modern sampling rates (real-time models); (2) those that predict the net product of a particular set of processes and at a time resolution beyond normal sampling schemes (event models); (3) those that principally describe only one class of processes and their responses (singular models); and (4) those that describe a complete sedimentary environment both in terms of processes and responses (unified models).

Real-time Model

Engineers and environmental scientists often are asked to understand the consequences of sediment erosion or deposition on very short time scales. Whereas a geologist is interested in the long history of deposition of turbidites within a sedimentary basin, for example, the engineer is more concerned with predicting the advent of the next turbidity current and whether that future event is capable of breaking communication cables or breaching a pipeline. In the later case, models predicting the stresses generated by a turbidity current in a real-time mode are warranted.

An example is the steady-state, spatially varied, turbidity current model proposed by Akiyama and Stefan (1986). Their model simulates the growth and the decay of turbidity currents through the use of layer-integrated equations of conservation of volume, mass and momentum. In their conclusion, they noted that the diffusion equation employed is significant because it incorporates both the erosive and depositional characters of a turbidity current. Their model can predict the behavior of the mean flow velocity, mean flow concentration, flow thickness, and rate of sediment erosion and deposition. As a consequence, their model can predict what events will break a cable from the impact pressures generated from the head of the slide, or from erosion of sediment from underneath the cable and thus breakage through loss of seabed support.

Another real-time model is given by Wilson (1979) who attempted to predict the spatial structure of suspended sediment plumes based on a solution to the advection-diffusion equation for a continuous vertical line source (i.e., a dredging barge). The discharge plume is described nicely by only two parameters: (1) one proportional to the settling velocity of the suspended material, and (2) one equal to the ratio of the diffusion velocity to the advection velocity of the ambient flow.

Let us expand on the Owen (1977) model of the movement of sediment in muddy estuaries described above (Fig. 3). In this real-time model, the change in floc size, bed shear stress, and the role of consolidation are considered. Owen used his model to determine which of the physical parameters are most sensitive to environmental changes. He found that for erosion, the critical shear stress and the rate of erosion constant are most important; for deposition, the sensitive parameters included the limiting shear stress and settling velocity of particles. Owen found surprisingly good agreement between model predictions and natural conditions, even in estuaries where flow depths, velocities, suspended sediment concentrations, and salinities all change significantly during the tidal cycle. Improvements to the Owen (1977) model, developed more from a engineering rather than an oceanographic point of view, can be found in Hayter (1986).

Whereas a real-time model can be considered the most precise means of approximating actualistic sediment transport conditions, such models are great consumers of computer time. In fact many of these models consume so much time that time allocation on a multiuser mainframe computer is difficult to acquire, and real-time models are typically run on supercomputers.

Event Model

Geologists who attempt to relate modern processes to the stratigraphic record are seldom interested in real-time modeling. Instead, event or net-product models are employed that mix and match the steady-state condition with the rare but important geologic events. These model simulations could extend over hundreds to millions of years. The interest is in the "final" deposit, not in the physics of sediment transport (even though the laws of physics may be employed to predict the final deposit).

For example, Komar (1977) modeled the morphological development of shoreline beach ridges and included the shifting in the position of a river mouth. The physics of channel abandonment are not considered; rather, the user must decide when and how far the river mouth will shift. Such an approach is especially relevant if one were to model the Yellow (Huang He) River whose river mouth has shifted eight times over a 1000 km coastline during recorded history (Wang, 1983).

Another event model describing the sediment filling of a marine basin was discussed by Harbaugh and Bonham-Carter (1977). In their model, the input parameters include: the quantity and type of sediment supplied; the initial geometry of the sedimentary basin expressed as water depth; tectonic warping (subsidence) through time and position; and the position of base level with respect to sea level. Harbaugh and Bonham-Carter demonstrated, using lag in the response time between sedimentation and subsidence, how cyclic sedimentation can occur without evoking a global change in sea level.

Singular Models

Many sediment transport models are concerned with the simulation of only the major transport/deposition processes for a particular sedimentary environment. For example, there may be an aeolian or autochthonous biogenic component to the sediment deposited in deltaic fans or submarine fans, respectively, yet few models incorporate these parameters. In these models, assumptions should be validated so that problems of mass balance are not encountered when comparing model predictions with nature. The content of organic matter can exceed 5% in some sedimentary deposits and its distribution can also be modeled (e.g., Farrow et al., 1983).

Many delta progradation models use only one land-to-sea transfer process—that of sedimentation beneath the river plume (Bonham-Carter and Sutherland, 1967; Wang and Wei, 1986). These models do not include the long-term and downslope bulk transfer of sediment by creep or small submarine slides, nor the bypassing effects of turbidity currents. To the contrary, Kenyon and Turcotte (1985) proposed a theoretical geomorphic model of a prograding river delta in which the patterns of deposition and movement of sediment on the delta-front slope are dominated by bulk transport processes. Both approaches can duplicate how the bathymetric profiles of a prograding delta, such as the Mississippi River, can change over time, yet utilize highly dissimilar processes. The coincidence of results may occur through the judicious use of constants of proportionality and empirical coefficients. So which one is right, and does it matter? In the former case, that of river plume sedimentation, the fractionation of sediment particles can be accurately predicted at rates comparable to observations based on unstable isotopes. However, the Mississippi delta continually is affected by mass flow processes supporting the second approach. In a unified approach, these two models would be linked to reflect observations of dual sediment transfer mechanisms.

Unified Model

There are few available unified process-response models that can simulate basin filling, although this is certain to change in the near future. It is the unified model, developed from the linkage of many models, that holds the greatest promise to the exploration geologist. Two recent examples include: (1) a dynamic deterministic, two-dimensional, simulation model for the transport, deposition, erosion, and compaction of clastic sediments (Bitzer and Harbaugh, 1987); and (2) a more theoretically based numerical model that simulates the growth of a river delta prograding into a fluvially dominated marine basin or graben (Calabrese and Syvitski, 1987).

The first example (Bitzer and Harbaugh, 1987) is a unified response model and therefore is not concerned with the physical laws governing sediment transport. For instance their

DEPOSIM model describes the fluid flow as being dependent on water depth within the basin: as water depth increases, velocity decreases. Their concern is *not* to simulate the real flow of water in a basin, but rather to describe a convenient means for the distribution and deposition of sediment within a basin while conserving the volume of water throughout the basin. Other models more accurately describe the physics of river water discharged into a basin as an inertial dominated effluent, a friction dominated plane jet, a buoyancy dominated buoyant jet, and a buoyancy dominated plane jet (algorithms and references can be found in Kostaschuk, 1985). However, DEPOSIM ignores these distinctions and the inherent differences on how sediment fractions might be partitioned depending on the dynamics of a river plume. The authors effectively utilize the simplifications describing sediment transport, deposition and erosion so that their model can be run with minimal effort and time on a 512 kbyte computer.

The second example is of a unified process-response model and includes the major mechanisms for the transfer of sediment from the land to the sea (Syvitski et al., 1988). They include (Fig. 4): (1) bedload dumping at the river mouth, the main hydraulic transition point; (2) sedimentation under the seaward flowing river plume that carries the suspended load; (3) proximal slope bypassing processes, such as turbidity currents, that are the result of delta-front failure; and (4) the combined effects of both short term (wave and tidal action) and longer term (creep and small slides) downslope diffusion of the accreting sediment mass, that together work to redistribute sediment downslope. Sediment accumulation is predicted from a parabolic partial differential equation that combines all four transfer mechanisms. The numerical solution employs a finite difference approximation of the parabolic equation and is solved using an explicit method. A numerical solution is necessary to maintain the ever changing source location (due to progradation), the changing sea-floor bathymetry (over rough basement topography, for example) and thus the nonlinear changes to the rate of delta-front progradation. The model, SEDFLUX, is used below to demonstrate salient points on employing unified models to describe basin filling.

The specific assumptions made about the bathymetry and dynamics of the basin that SEDFLUX is asked to simulate are given in Table 1. Where these assumptions cannot be justified, changes to an appropriate subroutine are warranted. As the requirements of a model's accuracy increase in its simulation of natural conditions, then the number of environmental inputs to the model must also increase (Table 2).

Components of SEDFLUX have been verified independently (Syvitski et al., 1988). For example, one component of SEDFLUX is a scavenging model that predicts rates of sedimentation on time scales of less than a year, and is designed to reflect the dynamics of a free, two-dimensional, buoyant jet that transports a composite size population of suspended particles. This model was verified by predicting the sedimentation rates in four basins (in Alaska, British Columbia and Norway) that differ widely in their sediment concentrations and discharge conditions. The predictions compared favorably to sedimentation rates determined from sediment traps. Model predictions on the deposition of sediment by turbidity currents compared favorably with cored lithologies in two other basins on the coast of British Columbia. The unified model was further verified through comparison of measured accumulation rates (based on ^{210}Pb, ^{137}Cs and ^{226}Ra measurements) to model predictions. SEDFLUX was able to simulate the 10,000 yr basin filling of these British Columbia inlets, favorably predicting the modern sea-floor bathymetry (Syvitski et al., 1988).

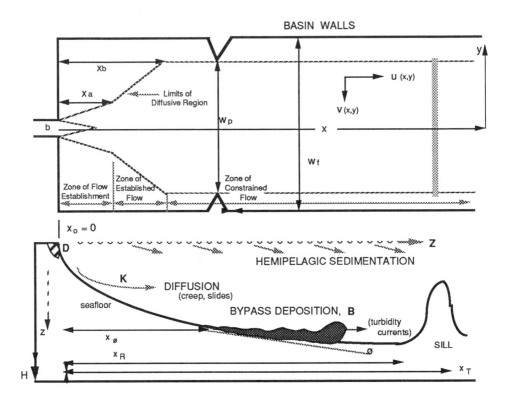

Figure 4. An example of a conceptual process-response model describing the filling of a sedimentary basin (after Calabrese and Syvitski, 1987). The upper diagram is a schematic of an areal view of an idealized two-dimensional jet issuing into a marine basin. The lower portion of the diagram is of a cross section of the basin where the major sediment transfer processes are indicated: bedload dumping (**D**); hemipelagic sedimentation (**Z**); bypass deposition (**B**); and downslope diffusion (**K**). Nomenclature on the diagram refers to parameters used in the original numerical model (see Syvitski et al., 1988).

Constraints to the Unified Process-Response Model

The process-response model must be constrained in four ways. First, unless an empirical approach is employed to simulate the movement of sediment, the principles of hydrodynamics must be observed. For example, the critical stress governing deposition of a particle from suspension must be less than the critical stress governing erosion of that same particle from the sea floor (e.g., Greenberg and Amos, 1983). As another example, the forward velocity of a sediment particle should not exceed the forward velocity of the supporting medium.

Second, the laws of conservation, i.e., those of continuity (fluid and sediment), momentum and energy, must be honored. Many times, these governing equations cannot be solved analytically, but only approximated numerically. In such cases numerical stability must be ensured. Greenberg and Amos (1983) also noted that numerical transport algorithms can lead to artificial diffusion in low-order schemes, since the suspended sediment concentration in any numerical cell is considered constant throughout the cell. [In their calculations this problem was addressed using the Flux Corrected Transport method of Zalisak (1979).]

Table 1 Assumptions made about the bathymetry and dynamics of the basin that model SEDFLUX is asked to simulate (Calabrese and Syvitski, 1987).

(1)	Lateral variations in the position of the river channel and the shape of the delta are insignificant over the time span of the simulation.
(2)	Lateral cross-sections of the basin and the river mouth are rectangular.
(3)	The delta head is perpendicular to the basin margins.
(4)	The basin has a sill.
(5)	The river plume is two-dimensional, maintaining a constant depth as it flows into a two-layer, stratified basin.
(6)	Flow velocity in the river channel is uniform.
(7)	The model distributes the total sediment deposited in any cross-section of the plume evenly across the width of the basin.
(8)	The input parameters are temporally and tidally averaged values.

Table 2 Inputs to model SEDFLUX (after Calabrese and Syvitski, 1987).

(1)	Initial height that sea level is placed above the bottom of the basin
(2)	Initial distance from the river mouth to the basin sill
(3)	Width of the basin sea floor
(4)	Distance from the river mouth over which bedload will be deposited
(5)	Sea-floor slope that deposition from turbidity currents begins
(6)	Run-out distance of the turbidity current (if less than the distance from the river mouth to the sill)
(7)	Diffusion coefficient describing the rate sediment is redistributed downslope by tides, waves, creep and landslides
(8)	Bedload and suspended load of the issuing river plume
(9)	Size distribution of the fluvial sediment, and the removal rates and densities for the various size fractions
(10)	Velocity or discharge rate at the river mouth
(11)	Dimensions of the river mouth (width and depth)
(12)	Maximum width of the river plume (usually dependent on the narrowest portion of the sedimentary basin)
(13	Critical slope for the initiation of delta-front failure
(14)	A function describing the relative fluctuation in sea level (i.e., a linear or exponential expression of sea-level height above datum with time)

Third, the boundary conditions governing the numerics must be realistic and must not contribute to numerical instabilities. For instance, a term relating to the bulk transport of sediment is employed in the SEDFLUX model described above. Modeled by a diffusion equation, bulk transport processes depend on the localized slope: downslope diffusion decreases with decreasing slope concavity. Where this term is large compared to the rate of sediment delivery to the basin, the river mouth may retreat (and without employing sea-level fluctuations). A critical boundary state (program abortion) will occur when all the deltaic sediment is removed by downslope diffusion and the rock "basement" is encountered.

The last modeling constraint is the use of realistic inputs from the external world (i.e., outside the immediate area of concern). In the Greenberg and Amos model, tides in the upper reaches of the Bay of Fundy, the area where sediment transport was to be predicted, were based on a numerical grid that included the entire Gulf of Maine, an area nearly two orders of magnitude larger. Such far-zone influences are especially important where tidal and wave processes are important. Goldsmith (1976) presented a mega-model involving 52,000 km^2 of the eastern U.S. continental shelf out to depths of 300 m, in which 19 different wave parameters were computed as output. Goldsmith demonstrated how far-zone influences control sediment transport and bottom morphology along the coast. The use of judicious inputs to the model is also very important, especially the rate sediment is delivered to a basin. For the fluvial suspended load, there are techniques that provide estimates even from ungauged basins (e.g., Mimikou, 1986). When reconstructing an ancient basin, information about the paleogeography could be used to choose a modern analog basin with known sediment discharge relationships (e.g., Milliman and Meade, 1983). First-order approximations of fluvial bedload and aeolian transport also can be estimated using information on the hinterland characteristics and appropriate transport equations (e.g., Lewis and Hwang, 1986).

SENSITIVITY ANALYSIS

An important reason for the development of process-response models in geology is their value as an educational tool. Komar (1973), for instance, was surprised to learn the speed at which a wave-dominated delta may reach an equilibrium configuration with its coastline. Sensitivity analysis is the method whereby the parameters used in a model's algorithms can be examined to determine their importance in affecting the final result. Sensitivity analysis also allows the modeler to test the numeric stability of the computer model, and the stability or validity in boundary conditions by choosing extreme values as input parameters.

Where the equations are relatively simple, the sensitivity of a particular parameter can be ascertained. For example if a model employed the algorithm $Z = [\alpha \beta^2 (\rho/t)](\mu \upsilon \omega^3)^{-1}$, the most sensitive parameter is ω, whereby small changes to this term will have the greatest affect on the dependent Z. This information can be used to direct new research towards collecting the best data, with the smallest possible error, for the term ω. Although this is simple enough to comprehend, complex models may have so many, and sometimes convoluted, interactions, that computer simulations are the only means to ascertain the sensitivity of a particular parameter.

Examples demonstrating the sensitivity of certain input parameters to the SEDFLUX model are given in Figures 5, 6, and 7. For a complete sensitivity analysis of SEDFLUX, the reader is referred to Calabrese and Syvitski (1987) who ranked the sensitivity of the many parameters affecting basin filling. They found that the fraction of bedload involved in delta-front failure, and thus turbidity currents, had a linear and inverse affect on the progradation rate of the delta front (Fig. 5A). The critical slope affecting the deposition of sediment from a turbidity current, ø, was found to have a linear affect on the rate of delta-front progradation, but only for ø < 1.2° (Fig. 5B); for ø > 1.2°, little change to the delta-front progradation rate was noted (Fig. 5B). Figure 6 illustrates how changes to the dimensions of the river mouth can affect the size fractionation of the sediment suspended in a river plume. If the river is shallow and wide, then the coarse fraction suspended in the river plume will be deposited quickly and close to the river mouth (Fig. 6A, B, and C); the distal prodelta environment would become very clay rich (Fig. 6D).

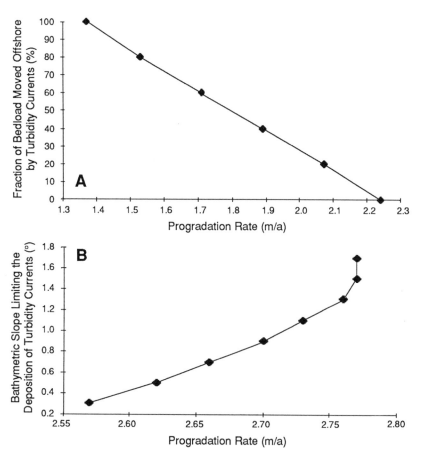

Figure 5. Sensitivity of two input parameters concerning the SEDFLUX model described in the text and as they affect the progradation rate (shoreline advance) of the delta front. A. The fraction of bedload (%) bypassed across the prodelta slope via turbidity currents. Increasing the percent of bedload moved offshore by turbidity currents decreases the rate of delta progradation. B. The critical bathymetric slope limiting the deposition of turbidity currents. Note the sensitivity of the slope decreases sharply above 1.2°.

During the simulation of the sediment filling of Knight Inlet, British Columbia, using the SEDFLUX model, actual and *a priori* input conditions were obtained from the literature. These values were used to generate the "base" case to which changes to the input parameters could be compared and their sensitivity evaluated. As the sediment input was shifted from

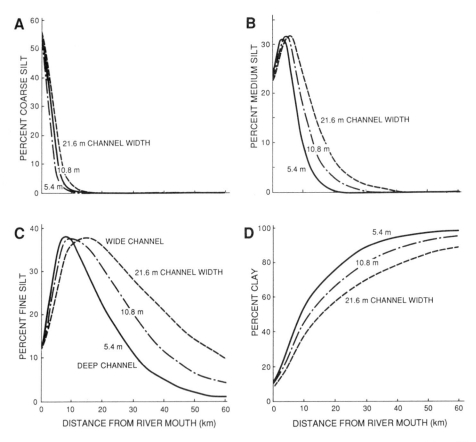

Figure 6. Sensitivity of changes in the river mouth dimensions with respect to the deposition of four size fractions. A. Coarse silt. B. Medium silt. C. Fine silt. D. Clay.

suspended load to bedload, more material accumulated in the deeper basin and the slope near the river mouth increased sharply (Fig. 7A). The impact of diffusion on accumulation rates increased in both extreme cases. When bedload accounts for the total sediment load, the proximal prodelta slopes are very steep and material dumped at the river mouth is diffused rapidly seaward. When the suspended load accounts for the total sediment load, the proximal prodelta slopes are more gentle and less sediment accumulates in the distal parts of the basin, contrary to intuition.

As expected, increasing the fraction of bedload, α, that is distributed by turbidity currents results in a greater accumulation of material in the distal parts of the basin (Fig. 7B). When $\alpha = 1.0$, the delta-front slopes become so gentle that gravity flows are not generated during 35% of the iterations, resulting in bedload dumping at the river mouth and a moderation of model sensitivity to values of $\alpha > 0.7$.

Grain size in the suspended load was varied by changing the fluvial concentration for the individual size fractions (although the total sediment delivery was held constant). Increasing the coarseness of the suspended load increases the gradient in prodelta sedimentation rates and shifts the maximum deposition landward. This results in steeper delta-front slopes (Fig. 7C).

A large coefficient of diffusion, representing a high rate of sediment transport along prodelta slopes, reduces accumulation nearshore and increases deposition in the intermediate and distal regions of the basin (Fig. 7D). The shallow slopes that result from large diffusion rates decreases the frequency of channelized flows. For $K = 25,000$ m²/a (a relatively large value; Fig. 7D), bedload was dumped entirely at the river mouth during 96% of the iterations.

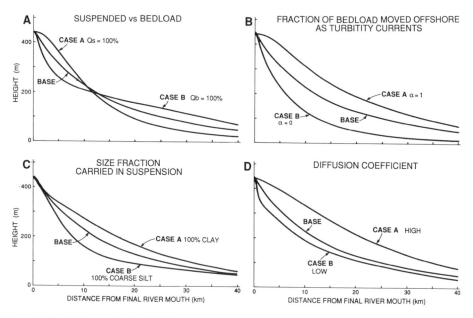

Figure 7. Sensitivity analysis of 4 of the 14 input parameters used in SEDFLUX model (after Calabrese and Syvitski, 1987). In each case extremes in the parameter are compared to a "base" case of observed input values that describe the filling of a graben-like basin on the west coast of Canada. A. Test of model sensitivity to the ratio of suspended load, Q_s, to bedload, Q_b. [Case A: Q_s = total load; Case B: Q_b = total load] B. Test of model sensitivity to α, the fraction of bedload transported by bypassing processes. [Case A: $\alpha = 1.0$; Case B: $\alpha = 0.0$] C. Test of model sensitivity to the distribution of size fractions within the suspended load. [Case A: sediment is all clay; Case B: sediment is all coarse silt] D. Test of model sensitivity to the coefficient of diffusion, K. [Case A: K = 25,000 m²/a; Case B: K = 1,000 m²/a].

SITE PREDICTION

An advantage of the process-response model is its ability to predict the flux of sediment, including the rates of sedimentation, erosion and accumulation, at a specified site in the basin being modeled. Site prediction is important to fields of applied sedimentology, engineering and environmental sciences. For instance, the distance a barge filled with dredged material must travel before it can release its (sometimes toxic) cargo into a stable environment and without subsequent erosion of the dredge spoils involves accurate site predictions with a predetermined safety factor. In such a problem, the economics of the dredging company and concern for environmental safety require highly accurate predictions. In one case study, the use of gravel to armor the dredge spoils from future erosion was considered: the dredge material contained toxic levels of cadmium and the dredging company faced a 100 km run unless the gravel-armor idea was feasible. One solution was development of an accurate simulation model with subsequent small-scale field test of the model predictions.

Another case study involved building a floating tunnel across a Norwegian fjord. The distance across the 500 m-deep fjord was 1.3 km, however the travel distance around the mountainous fjord was >100 km. One question of concern was the effect of submarine slides and turbidity currents that occasionally travelled the length of these fjords on the anchoring system for the proposed tunnel (Flaate and Janbu, 1975).

In each of these case histories detailed site predictions were required. Models capable of spatial and temporal predictions also have input/output advantages that allow them to be attached to other predictive models. For instance, a model developed to predict the discharge

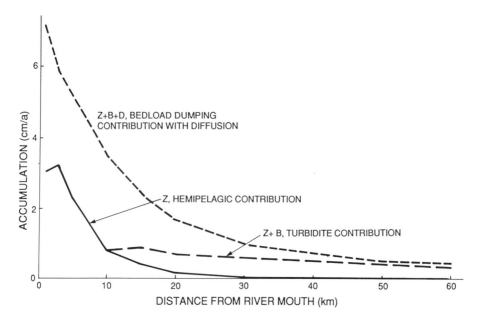

Figure 8. The contribution of sediment to the sea floor using the unified sedimentation-accumulation SEDFLUX model and as applied to the present-day situation in Knight Inlet, British Columbia (after Syvitski et al., 1988). See text and Figure 4 for detailed description and symbol identification.

of sediment based on climatic and topographic characteristics could be used as input to the SEDFLUX model discussed above. By combining the models in this way, the temporal resolution of SEDFLUX predictions conceivably could increase by an order of magnitude. Similarly the output of the SEDFLUX model, including lithology, bulk density and accumulation rates, could be used as input to a slope stability model that needs that information to track pore pressure, overburden pressure and changes in sea-floor slope over time.

Figure 8 is an example of site predictions from the SEDFLUX model. The accumulation rates depict the present-day, steady-state situation along the longitudinal axis of Knight Inlet, British Columbia (Syvitski et al., 1988). This figure shows the accumulation rate from hemipelagic sedimentation (Z), hemipelagic sedimentation plus turbidity current deposition (Z+B), and the composite model when the fraction of bedload is dumped at the river mouth and redistributed through slides into the offshore (Z+B+D with diffusion: cf. Fig. 4). The dominant processes in Knight Inlet are bedload dumping at the delta front and subsequent diffusion along the proximal slopes (Fig. 8).

BASIN-FILL PREDICTIONS

Sea Floor Geometry

A typical longitudinal seismic section through a basin filled principally by deltaic sediments shows a number of stratigraphic units consisting of genetically related, conformable reflections and bounded at their top and/or bottom by unconformities. Such units (depositional sequences) are typically characterized by sigmoid or oblique seismic reflections, that many times are considered as time lines (Vail et al., 1977). The seismic profiles of two basins in Western Turkey show classic examples of these geometries (Aksu et al., 1987a,b). Other examples based on repetitive bathymetric profiles collected off the face of a delta are described in Kenyon and Turcotte (1985). An important output of a basin-filling model is to simulate these time lines (obliquely prograding clinoforms) and thus the growth of the delta.

For example, three basin-filling simulations (SEDFLUX) of the sea-floor bathymetry in Knight Inlet, over a 10,000 year period, are represented in Figure 9. For comparative purposes, each model run has the same total sediment input to the basin and thus the mass of sediment deposited between successive time lines is equivalent. Figure 9 shows the differences in the rate of delta-front progradation and geometry of the sediment fill. A river plume is not capable of transporting sediment very far offshore, thus the rate of delta-front progradation is very high (Fig. 9A). When more of the sediment input is in the form of bedload, sediment is dumped near the river mouth (i.e., mouth or tidal bars) and also transported offshore by turbidity currents. In that case the distal basin fill accumulates more rapidly, the rate of delta-front progradation is not as fast, and very steep delta-front slopes result (Fig. 9B). The affect of moderately high rates of offshore diffusion (i.e., creep and small slides, wave and tidal action) is to the slow the rate of delta-front progradation and increase the rate of sediment accumulation at the distal portions of the basin (Fig. 9C). An important user group of this type of basin-fill simulation is engineers who are asked to calculate the life expectancy of dammed reservoirs (e.g., Frenette and Julien, 1986).

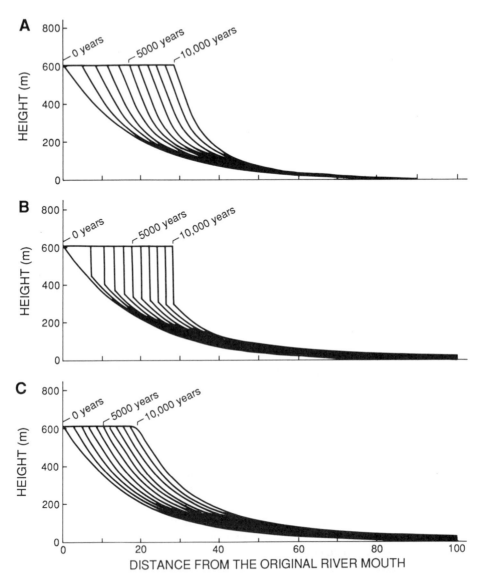

Figure 9. Simulated sea-floor positions in Knight Inlet, during progradation from a starting position approximately 30 km up-valley and to the present shoreline position (after Syvitski et al., 1988). Each model run has the same total sediment input. A. Sediment is deposited solely by hemipelagic sedimentation. b. Sediment is deposited by hemipelagic sedimentation, bedload dumping and turbidity currents, and using known levels of suspended load and bedload inputs, including 30% of the bedload being moved offshore by turbidity currents. C. Sediment deposition is the same as in the model run shown in (B) above, but as affected by downslope diffusion at K = 15,000 m^2/a.

Lithotope and Biotope Character

An important objective of a basin-fill model, especially if the model is to be useful for the exploration of hydrocarbons, is to predict the distribution of source and reservoir units, both in time and space. Unfortunately, this level of model simulation is in its infancy and rather simplistic. One recent attempt illustrates how the distribution of sand (the reservoir unit) and clay (the potential source rock), among other parameters, can be predicted for end-member cases of basin filling where the suspended load (Q_s) and bedload (Q_b) vary in extreme (Fig. 10; Syvitski and Farrow, 1988). The basin modeled in Figure 10 has only one sediment source, although the number of fluvial inputs may be varied, each with a unique Q_s:Q_b ratio (basin sediment properties are dependent on the supply level and proximity to each particular source).

If a basin receives more input from suspended load than from bedload, such that $Q_s >> Q_b$ (model 1, Fig. 10; cf. Fig. 9A), then the basin is dominated by the flux of hemipelagic particles. The accumulation rate will depend on the residence time of each type and size of suspended particle, the circulation dynamics, and the suspended load discharged from river mouths. Sand accumulates relatively close to the river mouth at high rates and thus the macrofauna population will consist mostly of mobile epifauna. Below some threshold in sedimentation rate (e.g., 6 cm/yr; Farrow et al., 1983), infaunal activity will increase. Bioturbation intensity will decrease basinward in response to a decrease in the flux of terrestrial carbon, and then increase in the distal portions of the basin through contributions from the flux of marine carbon.

Where the bedload of a river is significant (Q_b Q_s), mouth bars form and prograde onto steep foreset beds (model 2, Fig. 10; cf. Fig. 9B). This contributes to the semi-continuous failure of the sediment along the delta lip, where moderately sorted sand eventually is transported downslope along submarine channels by turbidity currents. Below some critical angle, sand is deposited as a fan over a proportion of the basin floor. As a consequence, sand bypasses the proximal prodelta muds. In the distal portion of the fan, the clay and carbon content is again high. Bioturbation normally is absent in the proximal areas that receive high rates of sedimentation, and in areas directly affected by sediment gravity flows (Farrow et al., 1983).

If the marine basin receives most of its sediment input as bedload, i.e., $Q_b >> Q_s$ (model 3, Fig. 10; cf. Fig. 9C), then the basin fill will be controlled by failure generated diffusion of the proximal prodelta slopes with some form of sediment bypassing as an invariable consequence. Liquefaction induced failures may occur anywhere along the delta foresets and these may generate cohesionless debris flows and turbidity currents. As a result, sediments within the proximal prodelta are rich in sand and the downslope bathymetry is more linear in profile. Retrogressive slide failures are common on these slopes. Bioturbation is very low for a model 3 basin, as the benthos is destroyed continually by episodic gravity flows. Carbon content within the basin sediments is highest in the distal region since there is little dilution from inorganic sedimentation (i.e., low Q_s input).

In each of the above basin models, we have tacitly assumed a benthic population that is dependent on the flux of organic detritus, marine or terrestrial, and limited by particulate loading associated with high sedimentation rates or by high-energy gravity flows. However, if the basin water is isolated occasionally, eutrophication also may effectively exclude the

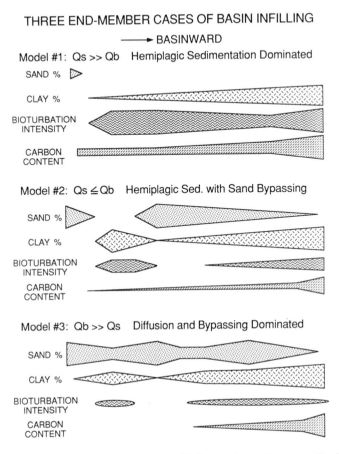

Figure 10. Model predictions of sediment parameters for three end-member cases of basin filling where the suspended load (Q_s) and the bedload (Q_b) vary in extreme (from Syvitski and Farrow, 1988). See text for details.

benthos. A basin's redox discontinuity may occur alternately within the water column, at the sediment-water interface, or within the sediment. True anoxic conditions will depend on the return interval of deep-water exchanges within these basins (for details see Syvitski et al., 1987).

In terms of petroleum potential, we are most interested in models 2 and 3—scenarios that accommodate for the action of prodelta failures and subsequent downslope transport that allows the interfingering of sand (reservoir) and mud (source) units (Fig. 11). Syvitski and Farrow (1988) pointed out that these two stratigraphic play types are mutually exclusive.

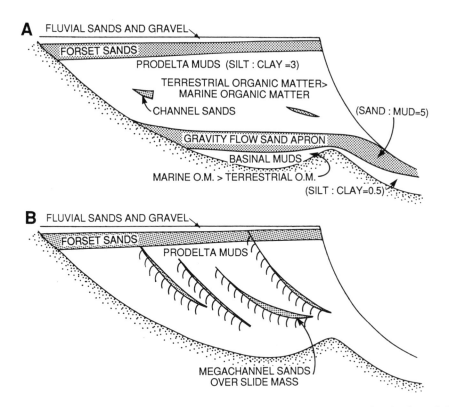

Figure 11. Cartoons based on model simulations (SEDFLUX) that illustrate two deep basin sand plays (after Syvitski and Farrow, 1988). A. A prograding delta sequence that experiences semi-continuous failure of its delta-front sands which are channelled so as to bypass the proximal prodelta muds and form a distal sand apron (cf. model 2, Fig. 10, and Fig. 9B). B. A prograding basin-fill sequence that experiences periodic and catastrophic slides, where channel sands cover the slide complex (cf. model

USERS

The users of process-response basin predictions include: (1) the mining industry in search of placer deposits, whether aggregates or heavy minerals; (2) the petroleum industry, especially where the prediction of reservoir, seal, and source rocks in small and complex hydrocarbon basins is required; (3) the military, interested in the prediction of sea-floor characteristics, and therefore acoustic signatures, in inaccessible territory; (4) the geotechnical community, whose work is centered on the analysis of sea-floor stability and the security of communication cables and pipelines; and (5) the environmental community, who often determine the long-term effects of waste disposal or evaluate point-source pollution disasters. Examples and references are scattered throughout the preceding discussion.

FUTURE

In a recent discussion on sediment transport problems, Bennett (1987) wrote, "the problem of estimating the removal time and of understanding the physical process of settling, resuspension, and burial occurs in many physical systems. The time and length scales for harbors, lakes, estuaries, rivers, and oceans may differ, but the physical questions do not." Bennett argued that we know (or can determine) the basic sediment mass balance, and how currents affect the distribution of sediments, within a basin. However, he considered that we are a long way off in our ability to accurately predict within or between systems. I agree with his first contention, but disagree with the latter. I feel that our ability to construct rigorous process-response models is improving rather quickly. By the turn of the century, the simulation of basin filling in true four-dimensional schemes should be commonplace. The SEDFLUX model outlined in this paper is just the first step. The role of bioturbation, short-term and long-term compaction, subsidence, and fluvial cannibalism of raised marine deposits are additional components that have yet to be included in the model. With the many global (earth system) projects on the horizon, such as the International Geosphere-Biosphere Project, the future will be to link the dynamic models that simulate fluvial systems with those of the continental shelf and deep ocean.

ACKNOWLEDGMENTS

I thank Felix Gradstein, Carl L. Amos, Rudy Slingerland, Dave Lawrence and Mark Ericksen for their critical comments. This is Geological Survey of Canada Contribution 51687.

REFERENCES CITED

Akiyama, J., and Stefan, H.G., 1986, Prediction of turbidity currents in reservoirs and coastal regions, *in* Wang, S.Y., Shen, H.W., and Ding, L.Z., eds., River sedimentation, Volume 3: University, Mississippi, University of Mississippi, School of Engineering, p. 26-46.

Aksu, A.E., Piper, D.J.W., and Konuk, T., 1987a, Late Quaternary tectonic and sedimentary history of outer Izmir and Candarli bays, Western Turkey: Marine Geology, v. 76, p. 89-104.

Aksu, A.E., Piper, D.J.W., and Konuk, T., 1987b, Quaternary growth patterns of Büyük Menderes and Küçük Menderes deltas, Western Turkey: Sedimentary Geology, v. 52, p. 227-250.

Bennett, J.R., 1987, The physics of sediment transport, resuspension, and deposition: Hydrobiologia, v. 149, p. 5-12.

Bitzer, K., and Harbaugh, J.W., 1987, DEPOSIM: A Macintosh computer model for two-dimensional simulation of transport, deposition, erosion, and compaction of clastic sediments: Computers and Geoscience, v. 13, p. 611-637.

Bonham-Carter, G.F., and Sutherland, A.J., 1967, Diffusion and settling of sediments at river mouths: A computer simulation model: Transactions of the Gulf Coast Association of Geological Societies, v. 17, p. 326-338.

Bowen, A.J., Normark, W.R., and Piper, D.J.W., 1984, Modelling of turbidity currents on Navy submarine fan, California borderland: Sedimentology, 31, p. 169-185.

Calabrese, E.A., and Syvitski, J.P.M., 1987, Modelling the growth of a prograding delta: Numerics, sensitivity, program code and users guide: Geological Survey of Canada Open File Report 1624, 61 p.

Chambers, R.L., and Upchurch, S.B., 1979, Multivariate analysis of sedimentary environments using grain-size frequency distributions: Mathematical Geology, v. 11, p. 27-43.

Davis, R.A., Jr., and Fox, W.T., 1972, Four dimensional model for beach and inner nearshore sedimentation: Journal of Geology, v. 80, p. 484-493.

Farrow, G.E., Syvitski, J.P.M., and Tunnicliffe, V., 1983, Suspended particulate loading on the macrobenthos in a highly turbid fjord: Knight Inlet, British Columbia: Canadian Journal of Fisheries and Aquatic Sciences, v. 40, p. 273-288.

Flaate, K., and Janbu, N., 1975, Soil exploration in a 500 m deep fjord, Western Norway: Marine Geotechnology, v. 1, p. 117-139.

Fox, W.T., 1978, Modeling coastal environments, in Davis, R.A. Jr., ed., Coastal sedimentary environments: New York, Springer-Verlag, p. 385-413.

Frenette, M., and Julien, P.Y., 1986, Advances in predicting reservoir sedimentation, in Wang, S.Y., Shen, H.W., and Ding, L.Z., eds., River sedimentation, Volume 3: University, Mississippi, University of Mississippi, School of Engineering, p. 26-46.

Ghosh, J.K., Mazumder, B.S., Saha, M.R., and Sengupta, S., 1986, Deposition of sand by suspension currents: Experimental and theoretical studies: Journal of Sedimentary Petrology, v. 56, p. 57-66.

Goldsmith, V., 1976, Wave climate models for the continental shelf: Critical links between shelf hydraulics and shoreline processes, in Davis, R.A., and Ethington, R.L., eds., Beach and nearshore sedimentation: Society of Economic Paleontologists and Mineralogists Special Publication 24, p. 24-47.

Greenberg, D.A., and Amos, C.L., 1983, Suspended sediment transport and deposition modeling in the Bay of Fundy, Nova Scotia—A region of potential tidal power development: Canadian Journal of Fisheries and Aquatic Sciences, v. 40, p. 20-34.

Harbaugh, J.W., and Bonham-Carter, G., 1977, Computer simulation of continental margin sedimentation, in Goldberg, E.D., McCave, I.N., O'Brien, J.J., and Steele, J.H., eds., The sea, Volume 6, Marine modeling: New York, Wiley Interscience, p. 623-649.

Harper, J.R., and Penland, S., 1987, Beaufort sea sediment dynamics (second edition): Woodward-Clyde Consultants 60830A, prepared for the Geological Survey of Canada, 160 p.

Hay, A.E., 1987, Turbidity currents and submarine channel formation in Rupert Inlet, British Columbia. 2. The roles of continuous and surge-type flows: Journal of Geophysical Research, v. 92, p. 2883-2900.

Hayter, E.J., 1986, Mathematical modeling of cohesive sediment transport, in Wang, S.Y., Shen, H.W., and Ding, L.Z., eds., River sedimentation, Volume 3: University, Mississippi, University of Mississippi, School of Engineering, p. 430-442.

Kamphuis, J.W., Davies, M.H., Nairn, R.B., and Sayao, O.J., 1986, Calculation of littoral sand transport rate: Coastal Engineering, v. 10, p. 1-21.

Kenyon, P.M., and Turcotte, D.L., 1985, Morphology of a delta prograding by bulk sediment transport: Geological Society of America Bulletin, v. 96, p. 1457-1465.

Klovan, J.E., 1966, The use of factor analysis in determining depositional environments from grain-size distributions: Journal of Sedimentary Petrology, v. 36, p. 115-125.

Komar, P.D., 1977, Computer simulation of turbidity current flow and the study of deep-sea channels and fan sedimentation, in Goldberg, E.D., McCave, I.N., O'Brien, J.J., and Steele, J.H., eds., The sea, Volume 6, Marine modeling: New York, Wiley Interscience, p. 603-621.

Komar, P.D., 1973, Computer models of delta growth due to sediment input from rivers and longshore transport: Geological Society of America Bulletin, v. 84, p. 2217-2226.

Kostaschuk, R.A., 1985, River mouth processes in a fjord-delta, British Columbia: Marine Geology, v. 69, p. 1-23.

Kranck, K., 1986, Generation of grain size distributions of fine grained sediments, *in* Wang, S.Y., Shen, H.W., and Ding, L.Z., eds., River sedimentation, Volume 3: University, Mississippi, University of Mississippi, School of Engineering, p. 1776-1784.

Lewis, J.M., and Hwang, G-J., 1986, An assessment study of five sediment transport models, *in* Wang, S.Y., Shen, H.W., and Ding, L.Z., eds., River sedimentation, Volume 3: University, Mississippi, University of Mississippi, School of Engineering, p. 766-775.

Milliman, J.D., and Meade, R.H., 1983, World-wide delivery of river sediment to the oceans: Journal of Geology, v. 91, p. 1-21.

Mimikou, M., 1986, Regionalization of suspended sediment rating curves, *in* Wang, S.Y., Shen, H.W., and Ding, L.Z., eds., River sedimentation, Volume 3: University, Mississippi, University of Mississippi, School of Engineering, p. 195-202.

Pantin, H.M., and Leeder, M.R., 1987, Reverse flow in turbidity currents: The role of internal solitons: Sedimentology, v. 34, p. 1143-1155.

Smith, J.D., 1977, Modeling of sediment transport on continental shelves, *in* Goldberg, E.D., McCave, I.N., O'Brien, J.J., and Steele, J.H., eds., The sea, Volume 6, Marine modeling: New York, Wiley Interscience, p. 539-577.

Sonu, C.J., and James, W.R., 1973, A Markov model for beach profile changes: Journal of Geophysical Research, v. 78, p. 1462-1471.

Syvitski, J.P.M., and Farrow, G.E., 1988, Fjord sedimentation as an anologue for small hydrocarbon bearing submarine fans, *in* Whateley, M.K.G., and Pickering, K.T., eds., Deltas: Sites and traps for fossil fuels: Geological Society of London Special Publication (in press).

Syvitski, J.P.M., Smith, J.N., Calabrese, E.A., and Boudreau, B.P., 1988, Basin sedimentation and the growth of prograding deltas: Journal of Geophysical Research (in press).

Syvitski, J.P.M., Burrell, D.C., and Skei, J.M., 1987, Fjords: Process and products: New York, Springer-Verlag, 379 p.

Vail, P.R., Mitchum, Jr., R.M., Todd, R.G., Widmier, J.M., Thompson, III, S., Sangree, J.B., Bubb, J.N., and Hatlelid, W.G., 1977, Seismic stratigraphy and global changes of sea level, *in* Payton, C.E., ed., Seismic stratigraphy—Applications to hydrocarbon exploration: American Association of Petroleum Geologists Memoir 26, p. 49-212.

Van Andel, T.H., and Komar, P.D., 1969, Ponded sediments of the Mid-Atlantic Ridge between 22° and 23° north latitude: Geological Society of America Bulletin, v. 80, p. 1163-1190.

Wang, F.C., and Wei, J.S., 1986, River mouth mechanisms and coastal sediment deposition, *in* Wang, S.Y., Shen, H.W., and Ding, L.Z., eds., River sedimentation, Volume 3: University, Mississippi, University of Mississippi, School of Engineering, p. 290-299.

Wang, Y., 1983, The mudflat system of China: Canadian Journal of Fisheries and Aquatic Sciences, v. 40, p. 160-171.

Whitten, E.H.T., 1964, Process-response models in geology: Geological Society of America Bulletin, v. 75, p. 455-463.

Wilson, R.E., 1979, A model for the estimation of concentrations and spatial extent of suspended sediment plumes: Estuarine and Coastal Marine Science, v. 9, p. 65-78.

Winters, G.V., 1983, Modeling suspended sediment dynamics of the Miramichi Estuary, New Brunswick, Canada: Canadian Journal of Fisheries and Aquatic Sciences, v. 40, p. 105-116.

Zalisak, S.T., 1979, Fully multidimensional flux-corrected transport algorithms for fluids: Journal of Computational Physics, v. 31, p. 335-362.

18

DEPO3D: A THREE-DIMENSIONAL MODEL FOR SIMULATING CLASTIC SEDIMENTATION AND ISOSTATIC COMPENSATION IN SEDIMENTARY BASINS

Klaus Bitzer and Reinhard Pflug
Geologisches Institut, Albert Ludwigs Universität Freiburg, Albertstrasse 23-B, 7800 Freiburg, Federal Republic of Germany

ABSTRACT

DEPO3D is a dynamic, deterministic, three-dimensional computer simulation model that represents interactions between moving water, and the erosion, transport, deposition and compaction of clastic sediment. DEPO3D also represents the isostatic response of the earth's crust due to changes in sediment load. DEPO3D responds geometrically by creating sedimentary deposits that may exhibit complex facies relationships. The user supplies inputs that include flow rates for water, the physical properties of sediment, and the initial configuration of the sedimentary basin that receives the sediment. DEPO3D is totally digital, so that the form of the simulated basin is represented in discretized form.

DEPO3D simulates movement of water through a basin with equations that represent steady-state potential flow. The initial water depths, and the rates at which water flows in and out of the basin are specified by the user. The user also specifies the geographic locations of "sources" of sediment and water where they enter the basin (e.g., inflowing rivers).

Transportation of sediment is a function of flow rate and direction. Sediment is transported in suspension and moves at the same rate as the water. Deposition and erosion are controlled by flow velocities. The user specifies the critical flow velocities that regulate the ability of moving water to erode, transport, and deposit sediment for each particle size.

If the velocity is sufficiently low for a given particle size, deposition occurs, with the rate of deposition dependent upon velocity, the amount of sediment in suspension, and the settling velocity of particles. Specific settling velocities are supplied for each grain size. If water velocity exceeds the critical velocity for a specific grain size, particles of that size are eroded and water depth increases. The eroded sediment reenters the water, increasing the suspended load. An increase in water depth reduces flow velocity, which in turn may lead to renewed deposition.

The load resulting from progressively accumulating sediment compacts underlying beds. DEPO3D calculates the reduction of porosity and decrease in thickness of each layer due to compaction. DEPO3D employs empirical depth-porosity functions supplied by the user for each particle-size category.

Quantitative Dynamic Stratigraphy (1989), T.A. Cross, ed., Prentice Hall, p. 335–348.

When sediment is deposited, the increased load on the crust is compensated isostatically by subsidence. The reverse is true when erosion occurs. Both compaction and isostatic compensation affect a basin's geometrical configuration, and therefore affect flow within a basin. DEPO3D's virtue is that it treats the interactions between flow, erosion, transport, deposition, compaction, and isostatic compensation in concert, producing experimental deposits that are realistic in geometrical form and facies distribution. The results of two simulation experiments are described.

OBJECTIVE

Geologists frequently make spatial or "three-dimensional" predictions about the geometry of ancient sedimentary deposits and their facies. Such predictions may involve extrapolations from observations at outcrops or in boreholes. With knowledge of sedimentation processes, predictions may be made regarding parts of the deposit not yet observed. In such predictions, comparisons are made between the "footprints" left by processes that produced the deposits, with similar footprints in modern deposits where depositional processes are observable. Interpreting processes in the past is thus an adaption of uniformitarianism, requiring observational and intuitive skills.

Our purpose is to add a quantitative component to the geologist's procedures, by simulating geologic processes and creating simulated deposits that can be compared with actual deposits. the three-dimensional model presented here, named DEPO3D, is an extension of an earlier, two-dimensional sedimentation process model, named DEPOSIM, developed by Bitzer and Harbaugh (1987).

DEPO3D requires assumptions about processes that create deposits. The assumptions are tested by using them to control DEPO3D and observing its response. Assumptions include the initial geometry of the basin to which sediment will be supplied. The user must also specify the location and magnitudes of sources of water and sediment (such as inflowing rivers) that enter the basin, as well as the physical properties of sediment, including grain sizes and sediment transport properties, such as settling rates. Given this information, DEPO3D responds according to the rules and equations with which it is constructed. The user can simulate the evolution of sedimentary deposits on a experimental basis. Of course, the experiments are constrained by DEPO3D's capabilities. The experiments allow the user to link specific geologic processes with specific responses that are realistic, yielding sedimentary sequences that can be compared with actual deposits. Most importantly, the user can interpret deposits in terms of processes that are linked together consistently and quantitatively.

In this context, computer simulation is an experimental procedure in which mathematical models are used instead of physical apparatus. Mathematical process models are not restricted to specific scales in time or space, but they are necessarily simplifications of actual processes. Furthermore, many geological processes are poorly known and can be described only by empirical or heuristic relationships. While conscious of these limitations, we suggest that experiments with DEPO3D produce results that are realistic in form and which serve to increase insight into the depositional history of sedimentary basins.

DEPO3D's Mathematical Formulation

DEPO3D incorporates fundamental equations that underlie the processes represented. The principles incorporate the conservation laws, although DEPO3D has been simplified so that only conservation of mass is satisfied. In more detailed models, energy and momentum are also conserved. Conservation of mass alone has a major influence in governing DEPO3D's performance. DEPO3D's organization of subprograms, data files, and display programs is shown in Figure 1.

Flow

Flow of water can be described by equations of fluid motion that satisfy conservation of mass, momentum and energy. In DEPO3D, flow is represented in simplified form by using stationary potential flow that accounts only for conservation of mass by means of the Laplace equation (Eq. 1). The Laplace equation states that during each increment of time, the amount of fluid entering an element volume equals the amount leaving. In DEPO3D, calculations are simplified to two horizontal dimensions because it is assumed that flow rates do not vary with depth. The Laplace equation in two dimensions is

$$\frac{\partial^2 h}{\partial x^2} + \frac{\partial^2 h}{\partial y^2} = 0 \tag{1}$$

where h represents the hydraulic head, and x and y represent the two horizontal directions, as for example, east-west and north-south.

The Laplace equation for flow is solved numerically. To obtain a numerical solution, the equation is transformed to its finite difference approximation, and then solved with a Gauss-Seidel procedure that yields a series of values of flow potential. The flow velocity components in the x and y directions are then calculated from the potential values with use of subroutine FLOW, which is based on a program of Harbaugh and Bonham-Carter (1970).

Boundary conditions for the numerical solution of Equation 1 transformed to finite-difference form are supplied as "no-flow" boundaries where the topography changes from submergent to emergent. Depths can be specified with respect to sea level. Where locations are below sea level, Neumann-type boundaries are employed and flow rates are defined.

Transport and Deposition

In DEPO3D transport and deposition of sediment include the following assumptions:
  (1)   Sediment particles are transported at the same velocity as water.
  (2)   Sediment particles are uniformly distributed within the column of water at any specific location.
  (3)   Particles of a given size have a constant sink velocity, as specified by the user.
  (4)   The velocity of water is uniform throughout a particular column of water at any specific moment.
The assumption that sink velocities remain constant is a major simplification. In actual flows,

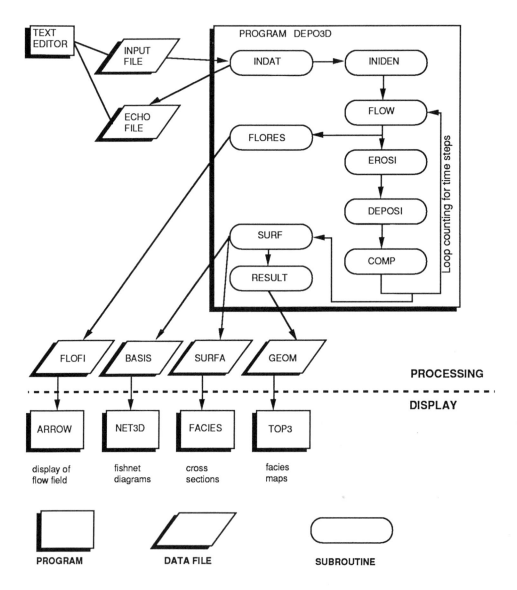

Figure 1. Principal elements of DEPO3D and graphic display programs adapted for use with DEPO3D.

shear forces induced by turbulence affect sink velocities, but turbulence is not represented in DEPO3D. Furthermore, frictional forces at flow boundaries in actual streams cause the vertical flow-velocity profile at any location to vary with depth, being slower near the bottom.

In DEPO3D the representation of sediment transport is similar to that for formulation of potential flow. Mass balance with respect to sediment suspended in an element volume of fluid is represented by the difference in the amount of sediment entering and leaving the element during an increment of time. The difference is the net amount of sediment that has been either been eroded or deposited, depending on whether the net change is positive or negative. Change in sediment concentration in a volume element can be expressed with Poisson's equation (Eq. 2):

$$\frac{\partial^2 C}{\partial x^2} + \frac{\partial^2 C}{\partial y^2} = f(S,V,G,R) \tag{2}$$

where C represents the concentration of sediment, x and y the spatial coordinates, and f(S,V,G,R) describes the net change in sediment concentration as a function of supply (S), flow velocity (V), basin geometry (G), and sink velocity of grain size (R).

DEPO3D deals with sediment whose grain sizes are specified by the user. Whether sediment will be eroded, transported or deposited depends on relationships between the critical flow velocities of specific grain sizes, and the velocity of the moving water. If the flow velocity is less than the critical flow velocity for deposition of a specific grain size, the particles will be deposited. If velocity is greater than the critical flow velocity for deposition, the particles will be transported, and they will be eroded if velocity is greater than the critical velocity for erosion. Specific data for critical flow velocities are provided by tables or by diagrams such as Hjulstrøm's (1935) diagram, in which the behavior of sediment of a given grain sizes is related to flow velocities.

In DEPO3D, the amount of sediment deposited is a function of sediment supply (S), flow velocity (V), basin geometry (G) and sink velocity of grain size (R). Such a function can be derived by considering the behavior of a single sediment particle as it passes through an elementary volume of fluid in the basin. The residence or "staying" time, t, of the particle within the fluid volume is represented by the distance it has traveled, divided by flow velocity, v, within the volume element (Eq. 3).

$$t = \frac{\sqrt{dy^2 + dx^2}}{v} \tag{3}$$

In DEPO3D, the maximum horizontal distance a particle can travel within a volume is the diagonal of an element of area (Fig. 2). Given the staying time, t, of a sediment particle, and its sink velocity, vs, the maximum vertical distance, w, which a particle travels (Eq. 4) in the volume element is

$$w = t \cdot v_s \tag{4}$$

If the vertical distance, w, is greater than or equal to water depth, then particles present in the uppermost part of the water column can settle to the basin floor, inasmuch as the staying time is sufficient. We assume that sediment particles are uniformly distributed within a water column at the beginning of each increment of time.

The function describing deposition of sediment is

$$m = \frac{w}{d} \cdot s \tag{5}$$

for which w/d must be less than 1. Equation 5 prescribes the amount of sediment to be deposited, m, as a function of sediment load, s, height of water column, d, flow velocity and sink velocity of the sediments particles (Fig. 3). Equation 5 is solved for all particle sizes and for each volume element of the basin by employing an Eulerian procedure. Accuracy of an individual solution increases with the number of iterations, but of course at the expense of increased computer time. In DEPO3D, we assume that the flow field is stationary during a time step. Therefore, the volume of flow remains constant during a time step, but the flow velocities may change because of changes in water depth caused by either erosion or deposition. A minimum water depth can be calculated that will allow a certain flow rate in a volume element so that the flow velocity will not exceed the critical velocity for deposition of a specific grain size. The minimum water depth, d_k^{min}, is calculated for each grain size, k, and its critical velocity for deposition, v_k (Eq. 6).

$$d_k^{min} = \frac{v}{v_k} \cdot d \tag{6}$$

For example, if flow velocity, v, is equal to the critical velocity, v_k, the water depth, d, at that moment will be the same as the minimum water depth, d_k^{min}, and deposition of grain size, k, will not occur.

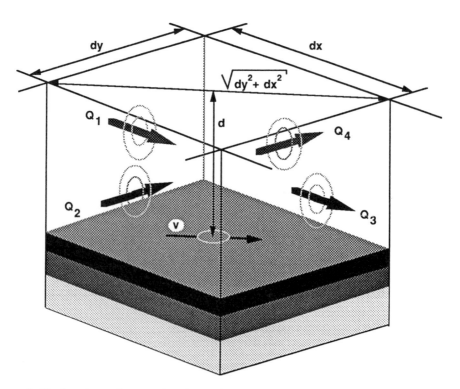

Figure 2. Single volume element of basin within which transport and deposition of sediment is calculated. Q1 through Q4 represent flow rates from adjacent cells. V represents resulting flow vector. Maximum travel distance within cell is horizontal diagonal.

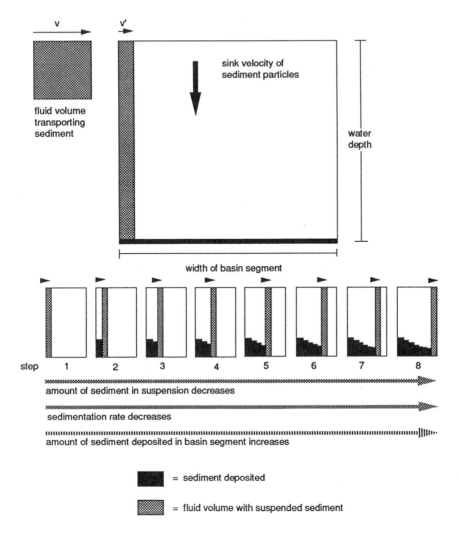

Figure 3. Sediment particles move horizontally at same velocity as water. Gravity causes particles to sink to basin floor. Horizontal and vertical velocity components regulate transport and deposition of sediment. Sediment load and deposition rate decrease with distance travelled in basin.

Erosion

DEPO3D verifies whether the calculated flow velocity exceeds the critical velocities for erosion of specifically defined particle sizes that are present in the layer of previously deposited sediment immediately below the water-sediment interface (defined as the "uppermost" layer). Three different situations can be distinguished. If flow velocity is lower than each of the critical velocities for erosion of all particle sizes present in the uppermost layer, erosion does not occur. If flow velocity exceeds the critical erosion velocities of all particle sizes in the uppermost layer, sediment will be eroded as long as the flow velocity is greater than any of the critical erosion velocities. If velocity exceeds the critical erosion velocities of one or more particle sizes, but not all of them, then sediment consisting of those sizes whose critical velocities are exceeded will be eroded in the uppermost layer, but particles of other sizes will not be eroded (Fig. 4). Erosion stops whenever the flow velocity is less than the critical erosion velocities, vc, of any particle sizes present in the uppermost layer. It is assumed that sediment consisting of those particles for which flow velocity is too low to erode, protect the sediment in layers that lie beneath the uppermost layer.

DEPO3D assumes that within a time step, flow rates remain constant but flow velocities change, depending on whether water depth increases because of erosion, or decreases because of deposition. Given this assumption, water depth E_M, at which erosion of a given grain size ceases, can be calculated (Eq. 7).

$$E_M = \frac{v}{vc} \cdot d \tag{7}$$

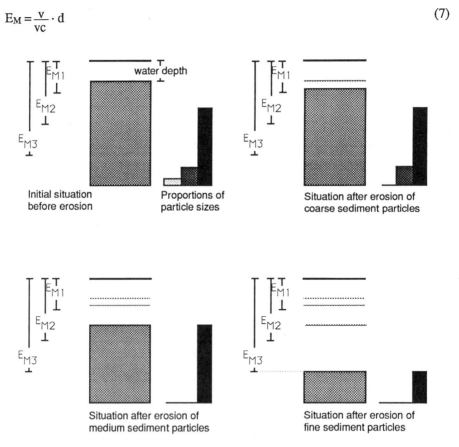

Initial situation
before erosion

Proportions of
particle sizes

Situation after erosion of
coarse sediment particles

Situation after erosion of
medium sediment particles

Situation after erosion of
fine sediment particles

Figure 4. Erosion has changed distribution of grain sizes of uppermost layer. Grain sizes 1 (coarse) and 2 (medium) were removed, the amount of grain size 3 (fine sediment) was reduced, and water depth increased with flow velocity remaining below critical erosion velocity of grain size 3.

The amount of erosion for each particle size is calculated separately. Eroded sediment reenters the moving water where it is suspended, or it may be redeposited if flow conditions are appropriate.

Compaction

DEPO3D assumes that buried sediment is compacted by the overlying sediment load, with concomitant decrease in porosity. The amount of compaction is based on an exponential function that relates changes in porosity to depth of burial. Use of an exponential function is a simplification because changes in porosity by processes occurring after loading are not considered, but an exponential porosity-depth relationship is a useful approximation.

$$P = \frac{a}{x - b} + c \tag{8}$$

where x is depth, and a, b and c are constants calculated from this function by providing three porosity-depth points. Whenever sediment is deposited, all underlying layers compact. The rate of compaction depends on the type of sediment, depth of burial, and the load of newly added sediment. DEPO3D assumes that compaction is irreversible, so that erosion does not cause decompaction, which otherwise would increase the porosity and thickness of buried layers.

In general, fine-grained sediments have high initial porosity and therefore decrease in porosity relatively rapidly with respect to depth of burial, whereas coarse-grained sediment have lower initial porosity and compact less rapidly with depth of burial. DEPO3D's representation of compaction is more or less realistic, as documented, for example, by the appearance of differential compaction that results from differences in sediment composition.

Isostatic Compensation

DEPO3D represents vertical movement where the crust responds isostatically to changes in sediment load stemming from deposition or erosion. Deposition causes an increase in load and in turn causes crustal subsidence, whereas erosion reduces load and causes crustal uplift. The isostatic response depends upon water density and depth, thickness and density of sediment, and density of the mantle. The algorithm for isostatic compensation (Eq. 9) is based on an equation of Schubert and Turcotte (1982):

$$s = \frac{\rho_m - \rho_w}{\rho_m - \rho_s} \cdot (D - d) \tag{9}$$

where ρ_m is density of mantle material, ρ_w the density of water, ρ_s the density of sediment, D the water depth before deposition, d the water depth after deposition, and s is the change of thickness of the sedimentary column to which the crust responds isostatically.

EXAMPLE APPLICATION OF DEPO3D TO SEDIMENTATION IN AN HYPOTHETICAL HARBOR

Experiment 1

Decreases in depth due to high sedimentation rates in harbors cause navigation problems. Often, the shapes of harbors are changed with the expectation that sedimentation rates will be reduced. DEPO3D can help predict the effects of such changes. Figure 5 shows a simulated harbor in Experiment 1. At the outset, the initial water depth was 9 m throughout the harbor. The inflows and outflows of water each were 180 m^3/s, yielding flow velocities of about 0.3 m^3/s across the inlets and outlets. The sediment load was 0.55 cm^3 sediment per m^3 water, which is typical for the Elbe River near Hamburg, Germany (Emeis, 1984). Each time step was 4.6 days, and the experiment ran for 40 time steps or 184 days. To minimize computer time, only three grain sizes were represented, consisting of gravel (6 mm) constituting 3% of total sediment load, sand (0.2 mm) constituting 12%, and clay (0.002 mm) constituting 85%.

Figure 6 shows the flow field at time step 1. Water entering the inlet at the left boundary flowed through the harbor, with streamlines diverging in wider parts of the harbor where flow velocities were lower. Near the outlet at the right boundary, the streamlines converged and flow velocities increased. As deposition ensued, the overall flow directions changed slightly, but flow velocities increased in most parts of the harbor because sedimentation reduced the water depth. Figure 5 provides cross sections through the deposits (the cross sections were originally drawn in color by program FACIES and are redrawn here). Thick deposits formed close to the inlet and in the central parts of the harbor, but only thin deposits formed near the outlet. Section 1 in Figure 5 shows gravel deposited in the inflow channel where depth decreased to about 5 m, which is the minimum water depth given flow rates and sediment types in the experiment. If depth decreased further, the flow velocities would have exceeded the critical velocities, and erosion would have occurred.

Sections 2 to 5 in Figure 5 show interfingering beds of gravel, sand and clay with overall increasing proportions of sand and clay in the direction of flow. Changes in sediment composition are due to lower sink velocities of fine sediment, allowing fine particles to be transported farther. Sections 6 to 9 show effects of erosion where a thin layer of gravel lies unconformably on beds of clay. Here, fine sediment was deposited earlier when flow velocities were low. But with increasing deposition, the higher velocities and decreased water depths removed previously deposited sediment, leaving the gravel lag.

Experiment 2

Experiment 1 suggests that a harbor with the form in Experiment 1 would be ineffective in reducing sedimentation rates. In Experiment 2, the harbor was changed so that a wall separated the harbor from the channel, with only a small connection between the two (Fig. 7). The flow and sediment parameters remained unchanged. The results show that most of the harbor did not receive sediment, whereas the channel was filled with gravel so that its water depth at the end of the experiment was about 5 m. Only a small part of the harbor near the

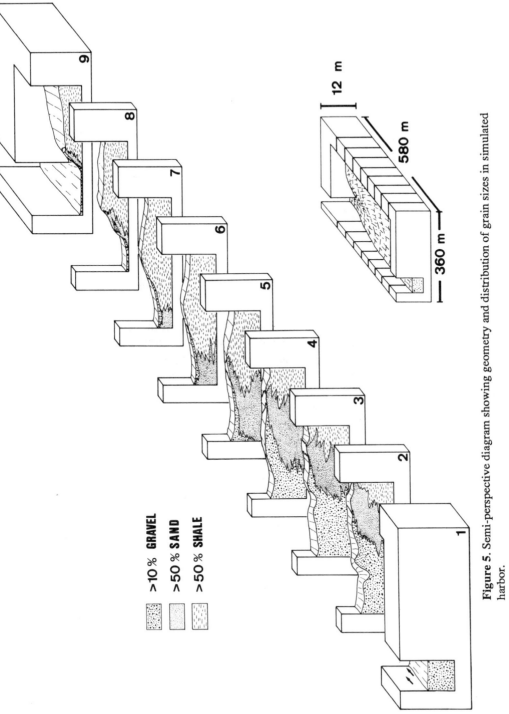

Figure 5. Semi-perspective diagram showing geometry and distribution of grain sizes in simulated harbor.

>10 % GRAVEL
>50 % SAND
>50 % SHALE

12 m

360 m
580 m

345

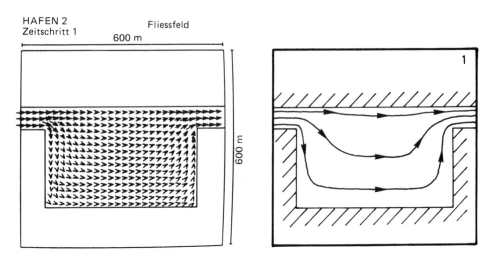

Figure 6. Harbor flow field at time step 1. Water enters at left boundary and leaves at right boundary.

connection received sediment. The flow field (Fig. 8) illustrates why the deposition rate was low in the harbor. The wall changed the flow field so that flowlines from the inlet boundary did not lead into the harbor, thereby diverting sediment from the harbor.

Although the two examples in Experiments 1 and 2 are simple, they accord with intuition, and moreover, they provide quantitative predictions about the geometry and facies distribution of deposits in the hypothetical harbors.

CONCLUSIONS

DEPO3D simulates features that are similar to actual deposits, including their overall geometry and grain-size distribution. The segregations of grain sizes that result from different sink velocities is represented realistically, as is erosion. The feedback relationships in DEPO3D link basin geometry with transport, deposition, and erosion thereby providing increased insight into the relationships between these processes and the geometry of clastic sedimentary deposits.

ACKNOWLEDGMENTS

We thank Mike Zeitlin, Texaco, and John W. Harbaugh, Stanford University, for reviews of a draft of this paper. We are especially grateful for J.W. Harbaugh's considerable assistance in revising this paper. We also benefitted from conversations with Daniel M. Tetzlaff, Western Atlas International Corporation, and Paul Martinez of ARCO Oil and Gas Corporation. Appreciation is expressed to Herbert Klein and Christoph Ramshorn of Universität Freiburg for helpful suggestions in devising the computer program.

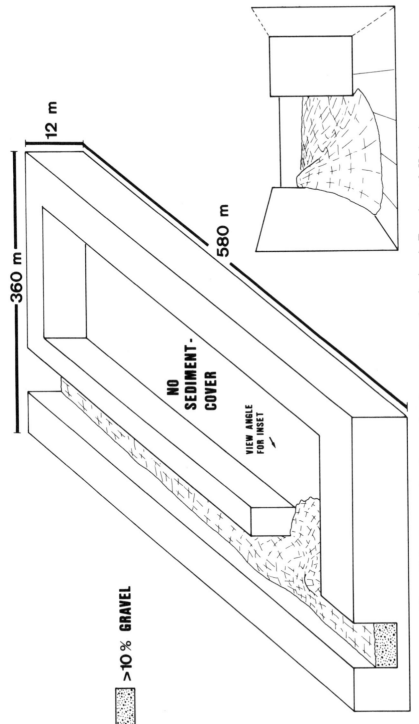

Figure 7. Semi-perspective diagram showing geometry and grain sizes in Experiment 2. Harbor is separated from channel by wall, with narrow connection near left boundary.

347

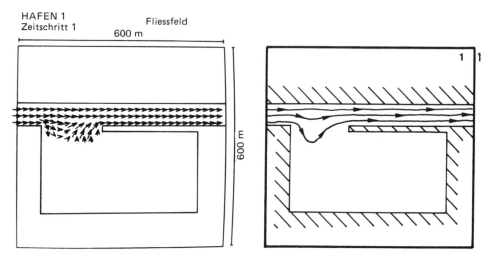

Figure 8. Harbor flow field at time step 1 at Experiment 2. Water enters at left boundary and leaves at right. Wall drastically reduces flow inside harbor.

References Cited

Bitzer, K., and Harbaugh, J.W., 1987, DEPOSIM: A Macintosh computer model for two dimensional simulation of transport, deposition, erosion, and compaction of clastic sediments: Computers and Geosciences, v. 13, p. 611-637.

Harbaugh, J.W., and Bonham-Carter, G., 1970, Computer simulation in geology: New York, Wiley-Interscience, 575 p.

Hjulstrøm, F., 1935, Studies of the morphological activity of rivers as illustrated by the river Fyris: Upsala University, Geological Institution Bulletin, v. 25, p. 221-527.

Emeis, E., 1984, Rezente Sedimentationsvorgänge in der Elbe: Schwebstoffe, *in* Paluska, A., Lammerz, U., Kleineidam, T., and Emeis, E., eds., Einige geologische Aspekte der Hamburger Gewässer, Exkursionsführer Erdgeschichte des Nordsee- und Ostseeraumes: Selbstverlag des Geologisch-Paläontologischen Instituts der Universität Hamburg, p. 1-45.

Turcotte, D.L., and Schubert, G.,1982, Geodynamics: New York, John Wiley & Sons, 450 p.

19

INTERACTION OF WEATHERING AND TRANSPORT PROCESSES IN THE EVOLUTION OF ARID LANDSCAPES

Robert S. Anderson [1] and Neil F. Humphrey [2]
[1]Earth Sciences Board, University of California Santa Cruz, Santa Cruz, California 95064 USA
[2]Division of Geological and Planetary Sciences, California Institute of Technology, Pasadena, California 91125 USA

ABSTRACT

Sediment delivered to depositional basins derives ultimately from the hillslopes on adjoining topographic highs. Progress in modeling basin sedimentation therefore requires a knowledge of the weathering and sediment transport processes operative on these hillslopes. Although in general the rate of sediment delivery depends upon a kaleidoscope of variables, we choose here to address a simple situation in which there is both hope of making progress, and a significant arena for detailed checking of model predictions in the real world.

We present a two-dimensional finite difference model of topographic evolution of an arid, internally drained landscape. A weathering algorithm that produces transportable material is combined with a diffusive transport algorithm that operates on this material, and an angle of repose algorithm to handle oversteepened slopes of loose debris. The weathering transforms bedrock into transportable material—"soil" in the engineering sense—as a specifiable function of the local soil depth. The transport model is diffusive in the sense that the downslope flux of debris is proportional (to the first power) to the local topographic gradient. Importantly, however, the downslope rate of increase of flux may not exceed the available soil; this leads to the important weathering-limited case pertinent to arid regions.

Two end-member cases are illustrated. First, when the initial soil depth is great and the rate of soil production is sufficient to counter denudation due to transport, any initial topographic relief is diffused, leading to smoothing of all initial roughness. The results of earlier studies of such landscape elements as fault scarps and shoreline terraces in the semi-arid western U.S. are reproduced. Second, when the soil production rate is very low (e.g., on the order of 0.001 mm/yr) the evolution of topographic highs is characterized by approximately parallel slope retreat, most initial roughness is retained, and topographic lows are filled and smoothed. Note that, in this latter case, the diffusivity that scales the transport rate is not the controlling parameter; rather, the weathering rate provides the control.

Other problems associated with treating larger scales confront the stratigrapher. Among these problems is the issue of deciding in which cases a two-dimensional model is appropriate, and the issue of scaling the diffusivity that controls the sediment transport.

Quantitative Dynamic Stratigraphy (1989), T.A. Cross, ed., Prentice Hall, p. 349–361.

INTRODUCTION

A great deal of attention is now being paid to the evolution of crustal scale geologic structures (e.g., King et al., 1988; Stein et al., 1988; Isaaks, 1988). The evolution of basins adjacent to growing mountain fronts, the stratigraphy from which is used to interpret the timing of tectonism (e.g., Johnson et al., 1986; Jordan et al., 1988), is driven not only by plate tectonic forcing, but by thermal (Kominz and Bond, 1986) and sediment loading histories (Stein, et al., 1988; King, et al., 1988). Although perhaps taking a back seat to coseismic and interseismic deformations due to large scale tectonic forces, sediment loading history was stressed by Stein et al. and King et al. as a significant determinant of the final basin structure. The sediment loading history of such basins may be constrained by detailed study of the basin stratigraphy (e.g., Jordan et al., 1988). Modeling of basin evolution, however, requires model components that are capable of predicting the sediment flux from the mountain front to the adjacent basin. This flux varies in time and represents an integral over time and space of the weathering and sediment transport processes in the mountains.

The ball is now in the geomorphologists' court. Current geomorphological studies often are concerned with problems limited either in temporal or spatial scale, or both. Although many geomorphologists are loathe to lose touch with the detailed physics of the processes acting to modify the landscape, they ultimately must model at appropriately large scales (i.e., those relevant to stratigraphers and geodynamicists) the delivery rates of sediment of various grain sizes to basins adjacent to tectonically active mountain fronts. [See Ritter, 1988, for a historical discussion of the process/landscape-development conflict within the geomorphological community.] To accomplish this the weathering and transport processes active on the mountain fronts must be understood and modeled at a large range of time and spatial scales. Ideally, such models should include spatial and temporal variations of both tectonic and climatic forcings.

We start by presenting a simple model of landscape development, which is two dimensional, small in scale, and has constant climatic forcing. Explicit connection is therefore maintained between the active local processes and the resulting form of both the deposits and the topography, allowing easy interpretation of results. We then proceed to the larger scales, where a different suite of processes must be addressed, and simple extension of the mathematics is warranted, if at all, only with extreme caution, and for particular climatic and tectonic settings.

As an illustration of larger scale geomorphologic models, we treat the case of an initially imposed topography that then evolves under the action of a constant, arid climate. Although this scenario ignores entirely the time dependent tectonic forcing, the validity of this approach as a first step is based upon the assumption that the time scales for endogenic (constructive or tectonic) processes are much shorter than those for exogenic (destructive) processes. For the case of arid climate considered here, this assumption is probably a good one, as subduction related uplift rates far outstrip weathering rates in arid regions (Isaaks, 1988). The arid case avoids the potential for transport of dissolved load out of the local (i.e., model) system and the attendant problems of tracking chemistry as well as mass (see Brimhall and Dietrich, 1987, for such an attempt). The lack of rainfall also restricts the development of through-going drainage networks and promotes the formation of closed basins, where again mass is conserved within the basin/range pair of concern.

   Topographies in the real world are complex and three dimensional, with variability over a large range of spatial scales. In addition, the boundary conditions and transport processes may change through time. Therefore, no attempt is made to maintain analytic simplicity. We use a numerical method that not only allows treatment of complex initial conditions, but that may be altered to accommodate complex temporal and spatial changes in the climatic and tectonic forcings. In addition, the numerical technique allows for treatment of nonlinear surficial processes, including on-off or switching interplay between two or more processes.

PROCESS MODELS

Surface Transport Processes

The sediment transport process is considered as diffusive in the sense that the volume flux per unit width of slope, Q, is directly proportional to the local slope,

$$Q = -\alpha \frac{\partial z}{\partial x} \tag{1}$$

where both Q and the proportionality constant, α, have units L^2/T. Such a slope dependent law results from rainsplash, creep, frost heave, and animal induced disturbances (e.g., Selby, 1985). It does not allow for slope-length dependent processes (where Q goes as x) such as overland flow (Carson and Kirkby, 1972). Conservation of mass requires that the rate of change of elevation at a point is proportional to the divergence of the volume flux,

$$\frac{\partial z}{\partial t} = -\frac{1}{1-p} \frac{\partial Q}{\partial x} \tag{2}$$

where p is the porosity of the soil. Combination of (1) and (2) yields the one-dimensional diffusion equation

$$\frac{\partial z}{\partial t} = \frac{1}{1-p} \frac{\partial}{\partial x}\left(\alpha \frac{\partial z}{\partial x}\right) \tag{3}$$

If the coefficient, α, relating transport to local slope is considered independent of x, this collapses to the more familiar form

$$\frac{\partial z}{\partial t} = \kappa \frac{\partial^2 z}{\partial x^2} \tag{4}$$

where the porosity and the transport coefficient have been combined to produce a diffusivity, κ. The erosion or deposition rate depends only upon the local curvature of the surface.

Angle of Repose

Transport equation (1) does not apply to very steep slopes. We have utilized a simple algorithm that mimics downslope transport of debris. The flux equation for this process is taken to be

$$Q = Q_o \left(1 + \left(\frac{\partial z/\partial x}{\partial z/\partial x_c}\right)\right)^\varepsilon \qquad (5)$$

where Q_o is the flux as calculated from equation (1), $(\partial z/\partial x)_c$ is the critical slope or the "angle of repose" (taken to be roughly 40°), and ε is an arbitrary but high power, here chosen to be 6. This additional highly nonlinear term in the transport law, although generally negligible for low slopes, causes a large increase in transport as the chosen angle of repose is approached. Slopes steeper than the angle of repose are allowed as initial conditions, but the transport rate across them is essentially infinite.

Weathering

If the entire hillslope were composed of transportable material, the original profile would diffuse, short wavelength roughnesses decaying most rapidly (e.g., Culling, 1960; Farr and Anderson, 1987), resulting in broadening, smoothing, and flattening through time. Yet in all but the simplest of cases, principally the scarp and terrace degradation cases (e.g., Andrews and Hanks, 1985; Andrews and Bucknam, 1987; Pierce and Colman, 1986), the material to be transported first must be produced *in situ* by weathering processes. We talk loosely, then, of "soil" (in the engineering sense) as the products of the weathering processes that are of small enough grain size to be transported by surficial processes.

The weathering process is modeled as a simple mass-conservative transfer of material from the underlying bedrock into the soil profile. For the arid case under consideration, this is taken to be a mechanical process driven by freeze/thaw, wetting/drying, and thermal cycling at the bedrock surface. As noted by Gilbert (1880), the rate of soil production in an arid environment should decay as the soil thickens, as this layer reduces the frequency and intensity of the weathering mechanisms operating on bedrock. We have therefore taken the weathering rate to be a simple function of soil depth,

$$\frac{\partial S}{\partial t} = -S_a \, e\left\{-S_b\left[z(x) - s(x)\right]\right\} \qquad (6)$$

where $S(x)$ is the topography at the soil/bedrock interface, $z(x)$ is the surface topography, S_a is the weathering rate on bare rock, and S_b scales the decay of weathering rate with soil depth (Fig. 1). Similar treatments have been utilized by Ahnert (1970, his mechanical weathering end member) and discussed in Carson and Kirkby (1972). Ahnert (1970) argued that there should be a peak in the weathering rate at some non-zero soil depth, resulting from the soil's ability to retain moisture necessary for the freezing and wetting processes. Although this is easily incorporated in our model, we wish first to treat the simplest cases.

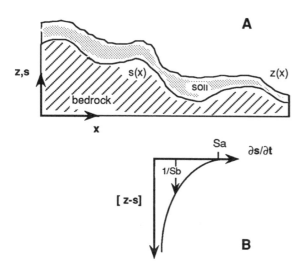

Figure 1. Definition sketch. (A) Topography, z(x), and soil/bedrock interface, s(x). Both surfaces are allowed to evolve through time, the bedrock surface moving monotonically downward at a weathering rate prescribed by a simple function of soil depth, as shown in (B); the topographic surface moving either up (deposition) or down (erosion) depending, in the diffusive case presented here, only upon the local curvature of the surface.

SMALL-SCALE EXAMPLES

We treat first two small-scale cases of glacial moraines and fault scarps in which the assumptions of two dimensionality and of uniform process (i.e., uniform climatic forcing and lack of vegetational or lithologic variation over the scale of the system) are justified.

Geomorphologists, glacial geologists, and climatologists are interested in determining the ages of glacial moraines because these ages provide dates of maximum glacial advances. Cross-sectional profiles of moraines have been used to infer relative ages, where the more rounded broader-crested moraines strewn with highly weathered boulders are older, and steep-crested moraines with fresh boulders are younger (e.g., Porter and Denton, 1967). The assumed age relations implicitly depend on diffusive decay of topography. Their usefulness in providing absolute ages is now being explored (e.g., Kaufman, 1987; M. Bursik, personal communication, 1987). In general, however, the moraine profiles we have seen do **not** obey in detail the behavior expected from a simple diffusive processes. They look younger (less rounded) along their crests than along their flanks. We believe that this is because, although moraines are sedimentary deposits, their hard-packed cohesive material must be weathered before individual grains can be moved by surficial transport processes.

Our model results show that youthful appearing crests are expected when the weathering rate is sufficiently low and/or the initial "soil" depth is sufficiently small to produce a weathering-limited situation (contrast Fig. 2, pure diffusion, with Fig. 3). In the weathering-limited case (Fig. 3) soil is transported downslope as fast as it is produced, leaving behind a bare "bedrock" surface. We see that the crest eventually is rounded, but at a much later time than the rounding of the contact between the moraine and the valley floor which occurs immediately upon initiation of erosion.

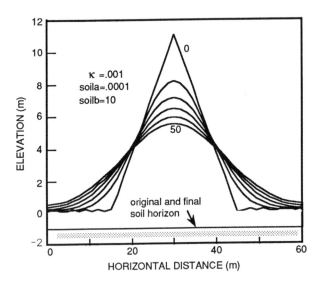

Figure 2. Small scale example: a glacial moraine evolving by "pure diffusion"—i.e., the transport-limited case, wherein the original soil profile s(x)= -1m for all x is sufficient to ensure unlimited soil supply throughout the evolution of the profile. Lines represent 0, 10, 20, 30, 40, 50 kyr profiles of surface, z(x), and soil/bedrock interface, s(x). Diffusivity, κ; and weathering rate parameters, S_a ("soila") and S_b ("soilb") as indicated, in mks units. Note that initial irregularities in the topographic profile of the moraine flanks are buried rapidly, and are not reflected in the final profiles. Bedrock is patterned.

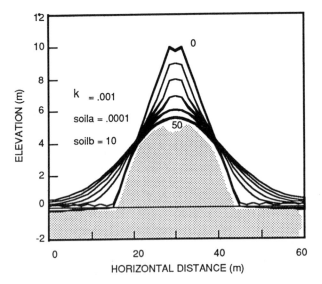

Figure 3. Glacial moraine evolving under weathering-limited conditions. Original soil depth s_o=0.2m, in contrast to Figure 2. Both soil and surface profiles are shown for each time step. Note bare "bedrock" upper portion of profile maintained until roughly 40 kyr, and marked contrast of curvature at crest vs. flanks throughout this interval, in accord with observed Sierra Nevadan glacial moraine profiles. Final bedrock profile patterned.

Figures 4 and 5, respectively, represent thickly and thinly soil-mantled fault scarps in bedrock that evolve through a combination of weathering and transport processes through a 50 kyr interval. Note the almost bare bedrock surface on the upper portion of the scarp maintained throughout the interval in Figure 5, and the asymmetry of the resulting scarp profile about the inflection point. [While the profile in the depositional zone becomes rounded very quickly, the abrupt break in slope representing the original scarp crest is slow to decay.] This contrasts with results from either linear diffusion (e.g., Andrews and Hanks, 1985; Pierce and Colman, 1986) or nonlinear diffusion (Andrews and Bucknam, 1987) scenarios, both of which maintain symmetry throughout.

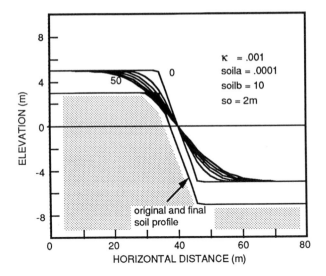

Figure 4. Evolution of small fault scarp entirely within the soil profile (s_0=2m), resulting in purely diffusive evolution. Note 4x vertical exaggeration. Parameters as shown, in mks

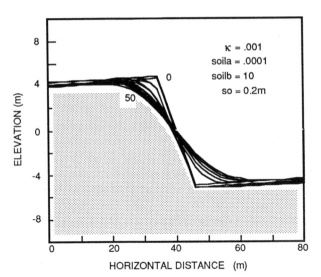

Figure 5. Evolution of small scale fault scarp with thin (0.2m) original soil mantle over bedrock. Note 4x vertical exaggeration. Parameters same as Figure 4. Times 0, 10, 20, 30, 40, 50 kyr.

SCALING UP

Our initial interest in the large-scale basin sedimentation problem resulted from parallelism between our model results for moraines and the general appearance of mountain fronts in arid environments, where rough, rocky upper slopes with little soil cover loom over broad, smoothly alluviated valley floors. It seemed the model results had produced an analog of the arid mountain front in several respects. In Figure 3 we see that the bare upper slopes are succeeded downslope by a relatively straight reach of thinly mantled bedrock with a convex-up profile, as has been argued (e.g., Lawson, 1915) should be the geometry of mountain-flanking pediments.

In addition, our model could be adapted to treat numerous large scale problems. For example, a diffusion model recently has been used to approximate sediment supply in stratigraphic models (Flemings and Jordan, 1987, and in review; Jordan and Flemings, this volume; Karner, 1986). Similarly, predictive models of fault related geologic structures (King et al., 1988; Stein et al., 1988) and lithospheric behavior (Stephenson, 1984) must rely on sediment supply models.

Simply increasing the length scales of our model moraine to represent the cross-sectional envelope of a mountain mass, and maintaining the weathering and transport processes as calibrated on surfaces of known age (e.g., through the work of Hanks and Andrews, 1985; Hanks and Bucknam, 1987; and Pierce and Colman, 1986), results in disappointingly little transport. Even when the weathering rate is increased such that the soil depths are great, the apparent time scale for decay of the mountain mass is much too long (order 10^7 yrs). Associated with the chosen diffusivity is an implicit relation between the length and time scales over which surficial processes are active. In the case of moraines and small fault scarps in colluvium and alluvium, the rainsplash, creep, and animal disturbance processes act with transport lengths and depths on the order of centimeters and time scales of less than one year, resulting in diffusivities of order $0.001 m^2/yr$. Given even geologically long periods of time to operate, these processes are not capable of reducing a mountain mass to a smooth pile of waste.

However, on the scale of a mountain front, the transport processes include scale-dependent processes such as fluvial activity. The scales are sufficiently large to allow for channelization of flows, down which are routed floods and associated debris flows that deliver sediment distances on the order of kilometers into the basin below. We encounter a basic dichotomy of geomorphological systems due to the nonlinearity of the sediment transport law (see Shreve, 1979)—that real landscapes are composed of hillslopes and of channels. As we move to larger scales, then, we must address not two processes and their relative rates, but (at the least) three: weathering, hillslope transport, and channel transport. The system is coupled in the sense that the channels form lower boundary conditions for the hillslopes. [In fact, the implicit boundary condition in our small-scale models is that all material delivered to the toe of the slope remains there—no out-of-plane sink for sediment, representing channel removal, is allowed.] The system of channels and hillslopes is inherently three dimensional. There are, however, important circumstances under which 2-D models may continue to be of use. Much hinges upon the relative efficiencies of the three processes, and in particular upon which is the rate-limiting process. If the processes act with similar strengths, 3-D modeling is necessary.

If channel transport is the rate-limiting process, the sole determinant of the flux at the basin edge is the sediment transport capacity of the exiting stream systems. When fluvial processes are very efficient in removing material delivered to them by the hillslopes, then there are two possibilities: (1) the hillslope transport rate is supply limited, implying that the entire system is supply limited and the determinant of the basin edge flux is the spatial mean weathering rate in the basin; and (2) the diffusion transport on hillslopes is rate limiting (as would be the case perhaps in humid deep-soil regions), in which case one may treat the entire mountain mass as being diffusive, with an appropriately chosen effective diffusivity, as discussed below.

We believe an heuristic argument may be made that the rate-limiting process in most active mountain belts is the weathering rate, implying that the transport rates—both hillslope and channel—effectively remove sediment from the system as fast as it is produced. This is based upon the observation that there is little evident storage on hillslopes and in valley trains in such landscapes, especially when contrasted with the material missing between the valley walls. It is therefore probable that the decay of a large number of mountain systems may be assumed to be weathering limited. This most likely reflects the fact that the weathering process is **not** scale dependent, whereas the transport processes become more efficient at progressively larger scales. The weathering-limited case therefore becomes more typical of the larger scales relevant to the problem of basin deposition. This is especially the case in arid climates.

We may place this argument in a mathematical context. If the quantity we wish to predict is the flux at the edge of a mountain front, we may write a general expression for mass conservation in the mountain mass, which is the integrated form of equation (2), as

$$Q_L(t) = \int_0^L \frac{\partial z}{\partial t} \, dx \qquad (7)$$

where the zero of x is the topographic/erosion divide such that $Q_0=0$ for all time (a zero flux boundary condition that implies in general no slope at the divide), and L is the distance between the mountain crest and the mountain front/basin edge. In general, z/t is a function of the transport and weathering processes. Since in general these processes depend on the rest of the basin, and are not just point values, the kernel of equation (7) is a very complex function that may be approached only numerically.

There are two cases in which equation (7) is particularly tractable, and that correspond to two recognizable weathering/transport scenarios. First, if the transport processes are completely efficient, such that all weathered debris is quickly removed from the mountain slopes and from the channels, then $z/t = <S_a>$, the weathering rate on bare rock, averaged over the mountain front, and equation (7) becomes

$$Q_L(t) = L <S_a> (t) \qquad (8)$$

Second, if the weathering process is completely efficient so that there is always "soil," the channel process is sufficiently effective to remove sediment delivered to it. And, if the transport process is linear with slope, that is equation (4) holds, then equation (7) becomes

$$Q_L(t) = \kappa \left[\left(\frac{\partial z}{\partial x} \right)_L \right] (t) \qquad (9)$$

In the first case, the temporal history of the sediment delivery rate should be controlled by the climatically modulated (spatial) mean of the weathering rate, $<S_a>$, and by (probably minor) progressive lengthening of the basin. In the second case, the flux is apparently controlled by the local slope at the exit from the mountain front, and by the effective diffusivity of the processes involved. Note, however, that the controlling slope is actually the basin-wide mean value of the hillslopes as they intersect the channels, which will be much greater than the slope of any channel as it exits the system.

If, as a first step toward solving large-scale problems (and following the lead of Flemings and Jordan, 1987 and in review; Jordan and Flemings, this volume), we retain the diffusive transport law in a large-scale 2-D model in which the detailed 3-D topography is represented by some cross-sectional envelope, the effective diffusivity must be chosen with care. This diffusivity will be greater than that determined from small-scale studies of fault scarps and moraines for several reasons. The main feature of the real topography not included in the 2-D representation is the complex of valleys interconnected with a fluvial system. Yet the diffusive transport processes act only on the local hillslopes of those valleys. This has two ramifications. First, the diffusive transport in the model will be driven by the slope of the cross-sectional envelope, which in general is considerably smaller than these local slopes. From equation (1), in order to calculate the correct flux from the system, Q, the effective diffusivity must be enhanced by the ratio between the local and the envelope slopes. Second, and more importantly, because the diffusive processes act on a length scale of the local hillslope, and the diffusion equation scales as the square of the linear scale in the problem, the correction for doing the calculation using the larger scale topographic envelope is the square of the ratio between the half-width of the mountain range and the half-width of a valley within the range. These effects can be quite large, and increase dramatically with the scale of the problem treated. The slope correction introduces a factor of about 5 to 50. For a typical range in the Great Basin of the western U.S. of approximately 10 km half-width, characterized by valleys 1 km wide, the length-scale correction is on the order of 400x. Combining these effects, we expect that the effective diffusivities used in 2-D large-scale models should be enhanced 3 to 6 orders of magnitude over those reported from work on fault scarps. As expected, Flemings and Jordan (in review) found that the stratigraphic record of foreland basins can be fit with diffusivities of approximately 10^3 m^2/yr, as contrasted with the 10^{-3} m^2/yr diffusivities typical of the small scale examples treated above.

In addition, the processes acting to spread the sediment derived from the mountain front out over the surface of the basin are different at these larger scales, and generally involve channelized distributary networks that act to disperse sediment long distances from its point of exit from the mountain front. If the sediment transport algorithm to be used remains diffusive, then the effective diffusivity must be chosen so as to match the expected mean fluxes on the associated surface slopes. As an example, we illustrate a case in which we have chosen κ' such that the debris-flow dominated fan environment of arid regions is crudely mimicked. If, for instance, the typical debris flow delivers debris of mean thickness 1 m and mean width 10 m out 1000 m onto the fan surface, with a recurrence interval of 10 years, and the spacing between debris flow lobes along the mountain front was 1000 m, then the volume flux, Q, from the mountain front is roughly 1 m^2/yr. [Note that in the arid, weathering-limited case, this constrains the mean weathering rate within the uplands.] If the slope driving the debris across its fan is of the order 0.1°, then the effective diffusivity would be $\kappa'=10$ m^2/yr [$= Q/(\partial z/\partial x)$]. Given the same slope, this process fluxes five orders of magnitude more sediment than the small scale creep processes dominating the fault scarps and glacial moraines described earlier. Using this diffusivity throughout the mountain-front/basin system, the two-dimensional profiles resulting from the decay of a 40°, 3 km high scarp on a tilted fault-block mountain front are shown in Figure 6. The evolution of the scarp is clearly weathering limited, and hence retains a barren, rocky, steep appearance, while the basin below fills with the sediment slowly produced by the weathering process. The mountain front retreats approximately parallel, at a rate commensurate with the weathering rate on bare rock, S_a. The roughly equal thickness slabs of soil thus derived in each time step are distributed over the basin in thinning and lengthening wedges whose shape is determined by the local surface slope of the growing "fan."

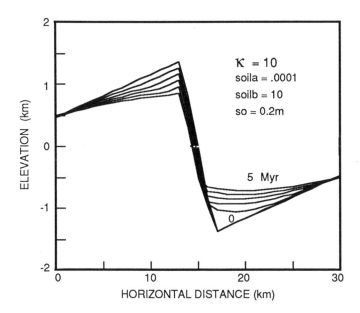

Figure 6. Large scale example of tilted (4 degrees) fault block mountain front with 3 km original relief. Note 7.5x vertical exaggeration. Scales and parameters in mks units. Times 0, 1, 2, 3, 4, 5 Myr. Original slope (40°) is maintained throughout the time interval, showing only parallel retreat, while being buried at base and truncated by back-directed transport at crest. Soil thickness too thin to show at this scale.

DISCUSSION

In this preliminary model we have assigned a single diffusivity, representing the efficiency of sediment transport on a given slope, to the entire system. In reality, the processes involved in spreading sediment over the basin may have very different efficiencies than do the processes moving sediment down hillslopes to the channels in the uplands. It is evident that as we increase the temporal and spatial scales of the questions we ask of the landscape, many aspects of the landscape taken to be constants in the small-scale process models will become important variables of the larger system. If, as we have argued, the weathering-limited system is common, perhaps most important in the mountain-front/basin problem is the altitudinal dependence of the weathering rate. At any given time, the mean annual rainfall will vary strongly with altitude. Higher elevations will be subjected to the freeze/thaw processes of the periglacial environment. Above the freezing line, the possibility exists for permanent ice and glacial erosion whereby valley bottoms are deepened and local sideslopes steepened. The spatial distribution of these processes will change through time due to global or regional changes in climate and to tectonic motions, both of which force variations in the relative position of the freezing line and in the statistics (mean, frequency distribution, and correlation) of precipitation through time (see Wolman and Miller, 1960). All of this will be reflected in the variations of the sediment flux to the adjacent basins.

The challenge remains, however. Correct geomorphological models of weathering and transport are needed to help stratigraphers who are attempting to gain insight into the tectonic history of the mountain range through the sedimentary record of the adjacent basins. From the above discussion, it is obvious that the inversion of the sedimentary record will result in nonunique answers for the tectonic forcing. It is our feeling at present that the best we can hope to do through forward modeling is identify the sensitivity of the resulting sedimentary record to realistic choices of the climatic and tectonic forcings.

ACKNOWLEDGMENTS

We gratefully acknowledge a critical review of the manuscript provided by Terry Jordan and Peter Flemings, and the careful scrutiny of Tim Cross.

REFERENCES CITED

Ahnert, F., 1970, Brief description of a comprehensive three-dimensional process-response model of landform development: Zeitschrift für Geomorphologie, Supplement, v. 24, p. 11-22.

Andrews, D.J., and Hanks, T.C., 1985, Scarps degraded by linear diffusion: Inverse solution for age: Journal of Geophysical Research, v. 90, p. 10193-10208.

Andrews, D.J., and Bucknam, R.G., 1987, Fitting degradation of shoreline scarps by a nonlinear diffusion model: Journal of Geophysical Research, v. 92, p. 12857-12867.

Brimhall, G.H., and Dietrich, W.E., 1987, Constitutive mass balance relations between chemical composition, volume, density, porosity, and strain in metasomatic hydrochemical systems: Results on weathering and pedogenesis: Geochemica et Cosmochimica Acta, v. 51, p. 567-587.

Carson, M.A., and Kirkby, M.J., 1972, Hillslope form and process: Cambridge, Cambridge University Press, 475 p.

Culling, W.E.H., 1960, Analytical theory of erosion: Journal of Geology, v. 68, p. 336-344.

Farr, T.G., and Anderson, R.S., 1987, Simulation of surface modification processes in arid regions: Applications to remote sensing and the study of climate change: Geological Society of America Abstracts with Programs, v. 19, p. 659.

Flemings, P.B., and Jordan, T.E., 1987, Sedimentary response in a foreland basin to thrusting: A forward modelling approach: Geological Society of America Abstracts with Programs, v. 19, p. 664.

Flemings, P.B., and Jordan, T.E., in review, A synthetic stratigraphic model of foreland basin development: Journal of Geophysical Research.

Gilbert, G.K., 1880, The Geology of the Henry Mountains: Washington, D.C., U.S. Geographical and Geological Survey of the Rocky Mountain Region, 169 p.

Isaaks, B.L., 1988, The Altiplano-Puna and the Bolivian orocline: Journal of Geophysical Research, in press.

Johnson, N.M., Jordan, T.E., Johnsson, P.A., and Naeser, C.W., 1986, Magnetic polarity stratigraphy, age, and tectonic setting of fluvial sediments in an eastern Andean foreland basin, San Juan Province, Argentina, in Allen, P.A., and Homewood, P., eds., Foreland basins: International Association of Sedimentologists Special Publication 8, p. 63-75.

Jordan, T.E., and Flemings, P.B., 1989, From geodynamic models to basin fill—A stratigraphic perspective, in Cross, T.A., ed., Quantitative dynamic stratigraphy: New Jersey, Prentice-Hall (this volume).

Jordan, T.E., Flemings, P.B., and Beer, J.A., 1988, Dating thrust-fault activity by use of foreland-basin stratigraphy, in Kleinspehn, K., and Paola, C., eds., New perspectives in basin analysis: New York, Springer-Verlag, p. 307-330.

Karner, G.D., 1986, On the relationship between foreland basin stratigraphy and thrust sheet migration and denudation: EOS (American Geophysical Union Transactions), v. 67, p. 1193.

Kaufman, D., 1987, Morphometric analysis of Pleistocene glacial deposits in the Kigluaik Mountains, northwest Alaska [M.S. Thesis]: Seattle, Washington, University of Washington, 50 p.

King, G.C.P., Stein, R.S., and Rundle, J.B., 1988, The growth of geological structures by repeated earthquakes: 1, Conceptual framework: Journal of Geophysical Research, in press.

Kominz, M.A., and Bond, G.C., 1986, Geophysical modelling of the thermal history of foreland basins: Nature, v. 320, p. 252-256.

Lawson, A.C., 1915, Epigene profiles in the desert: California University Department of Geology Bulletin 9, p. 23-48.

Pierce, K.L., and Colman, S.M., 1986, Effect of height and orientation (microclimate) on geomorphic degradation rates and processes, late-glacial terrace scarps in central Idaho: Geological Society of America Bulletin, v. 97, p. 869-885.

Porter, S.C., and Denton, G.H., 1967, Chronology of neoglaciation in the North American cordillera: American Journal of Science, v. 265, p. 177-210.

Ritter, D.F., 1988, Landscape analysis and the search for geomorphic unity: Geological Society of America Bulletin, v. 100, p. 160-171.

Selby, M.J., 1985, Earth's changing surface: Oxford, Clarendon Press, 607 p.

Shreve, R.L., 1979, Models for prediction in fluvial geomorphology: Journal of the International Association of Mathematical Geologists, v. 11, p. 165-174.

Stein, R.S., King, G.C.P., and McCarthy, J., 1988, The growth of geologic structures by repetitive earthquakes: 2, Field examples of continental dip-slip faults: Journal of Geophysical Research, in press.

Stephenson, R., 1984, Flexural models of continental lithosphere based on the long-term erosional decay of topography: Geophysical Journal of the Royal Astronomical Society, v. 77, p. 385-413.

Wolman, M.G., and Miller, J.P., 1960, Magnitude and frequency of forces in geomorphic processes: Journal of Geology, v. 68, p. 54-74.

20

A Simple Basin-Filling Model for Coarse-Grained Alluvial Systems

Chris Paola
Department of Geology and Geophysics, University of Minnesota, Minneapolis, MN
55455 USA

Abstract

Modeling sediment filling of basins may be approached by coupling the equations of sediment transport with those governing basin subsidence. Subsidence directly induces changes in the fluvial system by forcing deposition, which causes readjustment of the transport system away from its nondepositing (graded) configuration. Averaged over sufficiently long times this produces a configuration that is in equilibrium with the distribution of subsidence. Modeling this equilibrium configuration requires using sediment-transport equations in a way that is exactly the reverse of common practice in engineering and geomorphology: normally one knows the flow and uses it to calculate the deposition rate, but in basin modeling one knows the deposition rate and uses it to calculate the flow. An additional important effect of deposition induced by subsidence is sorting due to selective deposition of less-transportable sediments, usually gravel.

These ideas are used to construct a very simple model of the filling of alluvial basins with coarse-grained alluvium. The model accounts for downstream fining by selective deposition. It is different from most aggradation models in that it conserves excess bed shear stress rather than channel width. For uniform subsidence the model predicts only limited downstream variation in channel depth, even in the presence of grain-size changes of a factor of four or more. This prediction is confirmed by data from modern rivers and Lower Cretaceous gravels of the U.S. Western Interior. Concentrating subsidence near the source causes much more rapid downstream fining and larger depth variation.

Introduction

Currently available geodynamic models for basin evolution provide, among other things, reasonably accurate predictions of subsidence and thus sediment thickness through time as the basin evolves. In formulating these models it is generally assumed that sedimentation in the basin keeps pace with subsidence, so that the basin remains filled. Thus the output of the geodynamic models is a series of cross sections in which the basin sediments appear as white space. The ultimate goal of basin-filling models is to replace the white space with useful information, such as the locations of boundaries between major facies types, and variations in important properties of the sedimentary fill (e.g., grain size and sand-body geometry). This paper describes a simple dynamic model for the filling of basins in which coarse-grained sediment is transported and deposited by rivers.

Although the model uses relatively straightforward variables, including subsidence rate, depth, slope, and grain size, it is strikingly difficult to find data sets in the literature that can be used to test it. In part, this is because it is difficult to gather enough data to define the time-averaged behavior of a complex and (usually) partially exposed alluvial system. But the time-averaged behavior must be known, at least in general, before the effects of fluctuations in climate, sediment supply, and so on can be understood; and for the most part the sedimentary record is the only place where measurements can be made over the appropriate time spans. The theory developed in this paper is an example of how the long-term behavior of alluvial systems can be modeled. It will have served a useful purpose even if it does nothing more than inspire someone to collect the data necessary to refute it.

THE MODEL

The model presented in this paper starts with the assumption that, averaged over sufficiently long times, subsidence and sedimentation are locally in balance. It is easy to show that the governing equations discussed below tend toward such a balance under geologically reasonable conditions, and the existence of sections of alluvial sediment many kilometers thick shows that sedimentation can balance subsidence in real basins. This assumption has important implications for the way the governing equations are solved. In common engineering or geomorphic practice, the properties of the flow and channel (e.g., velocity, depth and slope) are known and are used to find the spatial distribution of sedimentation rate. In the basin-filling case the distribution of sedimentation rate is known because it is determined by the distribution of subsidence. The sedimentation-rate distribution is then used to determine changes in flow properties. The basin-filling problem is the inverse of the geomorphic problem, with respect to both independent and dependent variables, and to the order of solution of the equations.

Governing Equations

Four basic equations govern the motion of water and sediment in river systems. In two-dimensional, steady, width-averaged form these are: conservation of mass for the water,

$$Q = bhu \tag{1}$$

and the sediment,

$$\frac{\partial \eta}{\partial t} = \frac{-1}{C_0} \frac{\partial q_s}{\partial x} \tag{2}$$

and conservation of momentum for the water,

$$\frac{\partial(bhu^2)}{\partial x} = -ghb \frac{\partial(h + \eta)}{\partial x} - b\tau$$

where u is average velocity, x downstream distance, b channel width, h channel depth, Q water discharge, η bed elevation, t time, C_0 sediment concentration in the bed (here assumed = 0.6), q_s sediment discharge per unit width, g gravitational acceleration, and τ kinematic bed shear stress. The kinematic shear stress is T/ρ_w where T is the dynamic stress (force/unit area) and ρ_w is the fluid density. The stress and velocity are related by $\tau = c_f u^2$ where c_f is a nondimensional friction factor.

Finally there is conservation of momentum for the sediment, which in practice must still be expressed empirically. A simple but typical transport law is that of Meyer-Peter and Muller (1948):

$$q_s = \frac{8\,(\tau - \tau_c)^{3/2}}{g\,(s-1)} \tag{4}$$

where τ_c is the critical shear stress for moving the bed sediment, and s is the sediment specific gravity.

At a minimum, the information that must be specified to solve these equations comprises the discharges (in units of L^2/T) of water and sediment supplied to the basin, and the distribution of subsidence across the basin. If the crude approximation of constant c_f is made, a typical value for this parameter (about 0.01) is required; otherwise c_f is a function of the bed roughness, which in general is determined by flow conditions and requires an additional equation. In addition, the grain size and specific gravity of the input sediment must be specified.

There are a number of fundamental problems in applying this system of equations to sedimentary basins. First, the equations as given above are appropriate for describing the behavior of the bed over relatively short times. Over long time intervals, the equations have to be applied in some time-averaged sense; since they are nonlinear the time-averaged forms are likely to differ from those given above. No general solution for this problem has been proposed. A very simple one is used in this paper, but more refined schemes for time averaging will be required in future basin-filling models.

Second, the independent variables of the above set of equations are the bed elevation η, the average velocity u, the depth h, the sediment-transport rate q_s, and the flow width b. There are only four equations. A relation between the width and the other variables is needed to close the system. A variety of theoretical approaches have been used, including variational principles (Chang and Hill, 1977; Yang and Song, 1979) and force-balance models like that of Parker (1978). There is still a good deal of controversy about the width-closure problem even for equilibrium conditions, and there is no general formulation that can handle dynamic width variation in systems far from equilibrium.

Third, the set of equations given above is for systems with constant grain size, but of course this is almost never the case across a sedimentary basin. A good deal of downstream facies variation is directly or indirectly driven by downstream grain-size changes, so the grain size D, should be included as another independent variable in the system. This requires an additional equation that relates downstream sorting to the other variables, as a function of the distribution of grain sizes in the input sediment.

Approximations

ELIMINATION OF NONUNIFORM TERMS   In Equation (3), the terms involving spatial derivatives of depth and velocity represent local nonuniformity in the flow. In equilibrium ("graded") rivers the flow usually varies gradually enough that these terms are small and the flow may be regarded as quasiuniform, that is,

$$\tau = -gh \frac{\partial \eta}{\partial x} \tag{5}$$

In aggrading systems retention of the nonuniform terms is associated with large-scale wave-like behavior of the bed. Ribberink and van der Sande (1985) have shown that small-amplitude bed waves broaden as they migrate so that the nonuniform terms become negligible beyond a distance of the order of 30 times the backwater length h/S, where S is the channel slope. For coarse-bedded streams slopes are typically of the order of 10^{-3} and depths of the order of a meter, so the backwater length is of the order of 1 km. It is thus reasonable to assume that the space and time scales of interest are large enough that the quasiuniform approximation applies. However finer-grained rivers with high discharges can have slopes of less than 10^{-4} and depths of the order of several tens of meters, giving backwater lengths in the hundreds of kilometers. In this case the nonuniform terms should be retained, at least in linearized form.

This broadening of aggradational waves may not occur if the disturbance is nonlinear (i.e., if the transport rate changes abruptly). In this case the quasiuniform approximation might be inaccurate even over large spatial scales.

TIME SCALES   As mentioned above, Equations (1) through (4) apply over the time scales of individual flows (e.g., a single flood). To couple sedimentation and subsidence, the deposition term (the left-hand side of Equation [2]) has to be set equal to the average subsidence rate, which in general is several orders of magnitude lower than instantaneous deposition rates during floods (Sadler, 1981). The simplest possible resolution of this problem is to introduce an intermittency, I, the average fraction of geologic time during which deposition actually takes place. For deposition to be occurring in a given location the location must have a channel on it and the channel has to be active (i.e., in flood). Thus the intermittency is the product of the average flood frequency and the average channel-return frequency, both represented nondimensionally as fractions of time. The model presented below produces an internal estimate of a characteristic sediment-transport rate during floods. The intermittency, whch does not vary spatially, is the long-term rate of sediment supply to the basin divided by this short-term transport rate.

WIDTH VARIATION   In a simple long-term average model, the total channel width per unit width of basin, here denoted as β, is a more useful parameter than the width(s) of individual channels. The equations are thus normalized to unit basin width. The actual number of channels depends on internal instabilities within the channels (e.g., Fredsoe, 1978) and on the spatial distribution of water supply to the system.

It has been observed that many gravel-bed rivers with noncohesive banks attain shear stresses during floods that are only fractionally greater than the critical stress needed to move the bed. Parker (1978) showed that this is because, in the absence of vegetation or cohesive material to stabilize the banks, bank erosion and channel widening will occur if the stress rises substantially above critical. Thus rivers with noncohesive banks are self-regulating. This self-regulation clearly cannot occur if the width is held constant. In the present model the stress is assumed to be given by

$$\tau = (1+e)\tau_c$$

where e is the fractional excess stress, typically in the range 0.2 - 0.6 (Parker, 1978). The constant value e = 0.4 is used here, and τ_c is given by

$$\tau_c = 0.05g\ (s\text{-}1)\ D$$

which is strictly valid only for well sorted gravel, but is reasonably accurate for mixtures of sizes if the median diameter D_{50} is used for D (Komar, 1987).

The discussion in this and the preceding sections allows us to write Equations (1), (2) and (5) respectively as:

$$\langle q_w \rangle = \frac{I\,\beta h}{c_f^{1/2}}\ \left[(1+e)\,\tau_c\right]^{1/2} \tag{6}$$

$$\sigma = \frac{-8Ie^{3/2}}{C_0\,g\,(s-1)}\ \frac{\partial\left(\beta\tau_c^{3/2}\right)}{\partial x} \tag{7}$$

$$(1+e)\,\tau_c = -\,gh\,\frac{\partial \eta}{\partial x} \tag{8}$$

where σ is the subsidence rate and $\langle q_w \rangle$ is the long-term average water discharge per unit width of basin (Fig. 1). Note that because the assumed form of the channel stress requires that banks be at most weakly cohesive, these equations apply only to basins or sections of basins in which fine-grained and/or vegetated floodplains are not well developed.

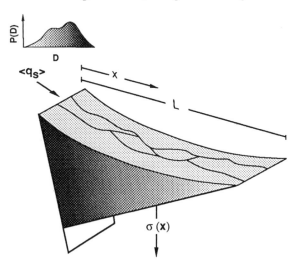

Figure 1. Definition sketch. D is grain size, P(D) the probability-density function for grain size in the supply, $\langle q_s \rangle$ the long-term average rate of sediment supply, x downstream distance, L the length of the basin, and σ the long-term average subsidence rate. Shading indicates grain-size variation.

SORTING Downstream fining and associated changes in channel geometry are a major factor in causing facies changes in fluvial deposits. Thus even a simple model for alluvial basins should account for downstream fining. Fining is thought to result from two main processes: abrasion and weathering, and selective deposition of large clasts (e.g., Shaw and Kellerhals, 1982). For present purposes I will ignore abrasion, thus restricting the model to relatively durable clasts and/or relatively rapidly subsiding basins. For further discussion and a simple abrasion model, see Paola (1988).

Despite its fundamental importance in both geomorphology and sedimentology, downstream fining by selective deposition is not well understood; some of the problems are discussed by Komar (1987). In view of this and in the interest of simplicity, I assume that the sorting is perfect; that is, at each point the rivers deposit the coarsest grain size available from the sediment supply until it runs out, then the next coarsest, and so on. This assumption is an extreme oversimplification. What makes it a reasonable place to start is that the predictions of any sorting model are strongly limited by the accuracy to which the grain-size distribution in the sediment supply can be specified. There is at present no general method for predicting the size distribution of the debris supplied to river systems, nor do available studies of alluvial basin fill generally provide enough information to reconstruct the supplied distribution. Thus the input distribution must be determined from the distribution of sizes observed in the basin, and this is often known only to the level of variation in mean size or (worse) some form of average largest clast size. Given this uncertainty, more sophisticated sorting models may not produce substantially better results than the simple one adopted here.

With perfect sorting, mass balance for each grain-size fraction can be written as

$$\frac{dD}{dx} = \frac{-\sigma}{\langle q_s \rangle P(D)} \tag{9}$$

where $\langle q_s \rangle$ is the long-term average rate of sediment supply and $P(D)$ is the grain-size probability-density function in the supply (in units of $1/D$).

A SIMPLE TEST Many field studies have shown downstream fining that follows Sternberg's exponential law:

$$D = D_0 \exp\left(\frac{-x}{L_a}\right) \tag{10}$$

where D_0 is the grain size at the upstream end, and L_a is a length scale determined from the data. One such study, that of Schlee (1957), showed uniform aggradation and included data on downstream variation in mean size as well as the distribution of the supplied sediment. In logarithmic (phi) form, this distribution is fairly uniform over the range of sizes that appear as mean sizes in the deposit (this excludes the coarse and fine tails of the supply). Using a logarithmically uniform input distribution and uniformly distributed subsidence in Equation (9) gives precisely the exponential downstream fining defined by Equation (10), which may help explain why this exponential form is so commonly observed.

The selective-deposition process discussed so far occurs mainly within river channels. Sorting by selective retention in flood plains, such as is implied in the channel-stacking models of Allen (1978, 1979), Leeder (1978) and Bridge and Leeder (1979), is a separate process that applies mainly to the suspended fraction of the sediment load. Including it would not make sense in a model directed primarily at systems with limited flood-plain development.

DISCUSSION

Major Features of Solutions

Equations (6) through (9) above can be solved easily given the following input: subsidence rate as a function of downstream distance, rate of sediment supply (long-term average), and the grain-size distribution of the input sediment. If absolute estimates of the dependent (output) variables are desired then a value for the long-term average supply of water to the basin is required as well. It is not clear how such values could be obtained for most ancient basins, but fortunately estimates of the dependent variables relative to their values at the upstream end of the basin are sufficient for many purposes. In this case the water discharge is not needed.

The method of solution is straightforward. First Equation (9) is solved to give the grain-size variation. The grain size fixes the distribution of τ_c so that Equation (7) can be solved for β. Both differential equations are solved using standard finite-difference methods. Finally Equations (6) and (8) are solved algebraically to obtain the depth and slope.

As mentioned above, until some independent method is available, the input size distribution must be determined from what is found in the basin. The first set of calculations shown below is intended for comparison with observations in the basal Cretaceous gravel of the western interior of the United States (Cloverly Formation and equivalents) made by the author and Paul L. Heller (University of Wyoming) over some 600 km between the Wyoming-Utah thrust belt and the Black Hills, South Dakota. The results of the calculations are shown in Figure 2, with grain-size and depth variation shown in Figure 2a. The agreement between observed and predicted values for grain size results from basing the grain-size input on grain sizes measured in the basin and is not to be taken as a test of the model. On the other hand, the channel-depth calculations are quite striking in that the depth is predicted to vary only weakly across most of the basin despite the four-fold variation in grain size. The data shown are paleodepth estimates based on thickness of gravel-sand fining-upward sequences. They affirm this prediction remarkably well except for the anomalous point at $x/L = 0.15$, which represents only one measurement (the other points represent 3 to 5 measurements each). The error bars on the points give the average geometric standard deviation at each site.

Figure 2b shows the calculated variation in normalized width and slope. The normalized width β varies by about a factor of 3 down the basin, indicating that in the presence of aggradation and downstream fining, stream systems with noncohesive banks cannot be expected to maintain constant width. The strong linear decrease in slope results in the classical concave profile associated with depositional alluvial systems.

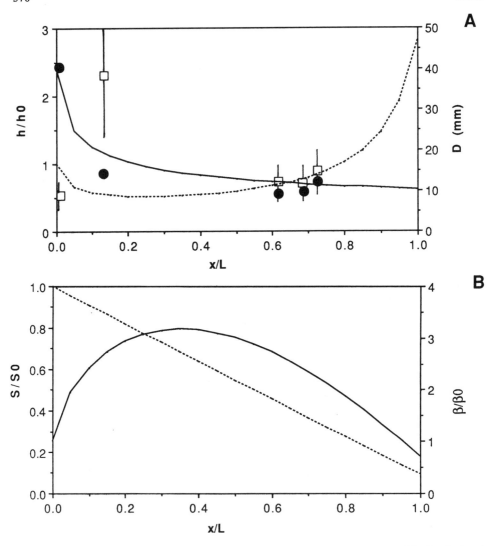

Figure 2. (A) Comparison of grain size (D, solid line) and channel depth (h, dashed line) calculated using Equations (6) - (9) with measurements in the Cloverly Formation (Albian-Aptian) and equivalents in Wyoming and Utah. Paleodepth data are shown as open squares, grain size as filled circles. As discussed in the text, only the depth variation is a true test of the model. Depth is normalized by its value at the upstream end (h_0). L is the length of the basin (about 600 km) and x is downstream distance. (B) Variation in fractional width (β, solid line) and in slope (S, dashed line) for the same input parameters. Both parameters are normalized by their values at the upstream end.

The predicted tendency for rivers with noncohesive banks to respond to downstream fining by changing their width rather than their depth can also be seen in modern rivers. For example, in the Peace River in western Canada, fining from 41 to 0.31 mm is accompanied by a width increase of 50% and a depth increase of only 16%. In the Chilliwack River, British Columbia, fining from 140 to 32 mm is accompanied by a width increase of nearly 500% and a depth decrease of 45%. Both examples also show substantial downstream increases in discharge (data from Church and Rood, 1983).

As discussed in the previous section the model generates an internal estimate of the intermittency, I, needed to balance short-term and long-term sediment-transport rates. For a given input grain-size distribution, I is proportional to the subsidence rate which is not well constrained for the Cloverly Formation and equivalents. For a subsidence rate of 0.01 km/10^6 yr, which is probably the correct order of magnitude, I is about 3×10^{-4}. Assuming an annual flood that lasts for a few days, the flood contribution to I is of the order of 10^{-2}. Thus the intermittency due to channel switching is also of the order of 10^{-2}, and is directly proportional to the sedimentation rate, as would be expected. [Note that a higher value for I means more frequent switching.]

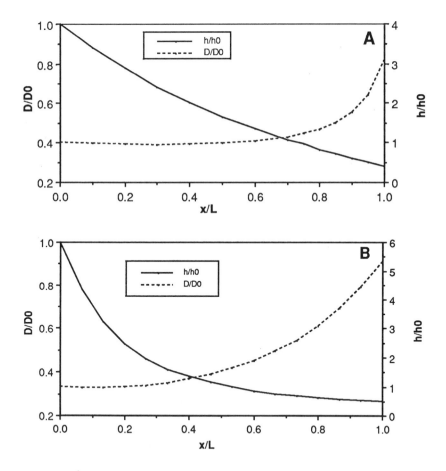

Figure 3. Variation in depth (h, dashed lines) and grain size (D, solid lines) for (A) uniform subsidence and log-uniform grain size input, and (B) exponentially decreasing subsidence rate and log-uniform grain size input. In both graphs h and D are normalized to their upstream values, h_0 and D_0 respectively. L is the length of the basin and x is downstream distance.

Figure 3 shows the results of changing the distribution of subsidence on grain-size and relative-depth variation. The input grain-size distribution in this case was uniform in logarithmic size units. For uniform subsidence (Fig. 3a) the size variation is exponential, as discussed in the previous section, and the depth remains nearly constant across most of the basin. The effect of an exponentially decreasing rate of subsidence is shown in Figure 3b: the grain size varies rapidly near the upstream end of the basin in response to the rapid subsidence. This has the effect of sharply reducing the distance over which the depth remains constant. In general, concentrating subsidence near the source region has the effect of compressing the uniform-subsidence results near the source and stretching them farther away.

Comparison with Other Models

Most proposed models for aggradation in alluvial systems conserve channel width and allow the bed shear stress to vary freely (e.g., Soni, 1981; Begin, 1987; Wyrwoll, 1988; and Snow and Slingerland, 1987, who allowed width to vary only as a function of discharge). The model described here allows width to vary freely but conserves excess bed shear stress. Clearly both types of model are extreme cases, and each applies best to a different type of alluvial system. The one proposed here is for coarse-grained, bedload dominated systems with weak flood-plain development; on the other hand, conserving width requires strongly cohesive banks that are stable under a wide range of bed shear stresses. Stable banks are generally associated with well developed, vegetated flood plains and abundant suspended mud. Bank stability is also enhanced by the presence of bed forms. If the banks are stable, shear stresses developed on the bed during floods can greatly exceed the critical values for moving sediment in the channel. With constant width and $\tau \gg \tau_c$, Equations (1), (2), (4) and (5) reduce to the diffusion equation, which is probably the most widely applied aggradation model proposed to date (e.g., Soni, 1981; Begin, 1987; Wyrwoll, 1988; Jordan and Flemings, this volume). The diffusion equation also governs the evolution of the bed topography, η, for the system, Equations (6) through (8), although the predicted variations in h and β are much different. Also, the diffusion coefficient is larger in the constant-width model by a factor of e/(1+e).

CONCLUSIONS

It is possible to formulate models for the filling of basins by rivers, based on fundamental laws of fluid flow and extensively tested sediment-transport laws. The model proposed here is computationally simple, has no adjustable parameters, and requires a minimum of input about the system; specifically, the rate and size distribution of sediment supply, and the distribution of subsidence. The model predicts mean grain size and paleochannel depth across the basin. Both of these can be tested in the field with conventional sedimentologic techniques.

In contrast to most other models of fluvial aggradation, the one proposed here allows for free variation of channel width but conserves excess bed shear stress. It is most applicable to coarse-grained systems with limited flood-plain development. The other major class of model conserves width but allows for free variation in bed stress. This class is most applicable to river systems with well developed flood plains and strongly cohesive banks. Clearly a more general model would behave like each in the appropriate limit.

Acknowledgments

Many of the ideas in this paper were developed and sharpened during discussions with Paul Heller, Charles Angevine and Gary Parker. In addition, I thank Karen Kleinspehn, Rudy Slingerland, and Bob Anderson for insightful reviews of the original version of this paper. Financial support was provided by the Atlantic Richfield Corporation junior-faculty support program and NSF grant EAR-87-07041.

References Cited

Allen, J.R.L., 1978, Studies in fluviatile sedimentation: An exploratory quantitative model for the architecture of avulsion-controlled alluvial suites: Sedimentary Geology, v. 21, p. 129-147.

Allen, J.R.L., 1979, Studies in fluviatile sedimentation: An elementary geometrical model for the connectedness of avulsion-related channel sand bodies: Sedimentary Geology, v. 24, p. 253-267.

Begin, Z.B., 1987, ERFUS 6—A FORTRAN program for calculating the response of alluvial channels to baselevel lowering: Computers and Geosciences, v. 13, p. 389-398.

Bridge, J.S., and Leeder, M.R., 1979, A simulation model of alluvial stratigraphy: Sedimentology, v. 26, p. 617-644.

Chang, H.H., and Hill, J.C., 1977, Minimum stream power for rivers and deltas: Journal Hydraulics Division, American Society of Civil Engineers, v. 103, p. 1375-1389.

Church, M., and Rood, K., 1983, Catalogue of alluvial river channel regime data: Vancouver, BC, University of British Columbia, Department of Geography, 99 p.

Fredsoe, J., 1978, Meandering and braiding of rivers: Journal of Fluid Mechanics, v. 84, p. 609-624.

Jordan, T.E., and Flemings, P.B., 1988, From geodynamic models to basin fill—A stratigraphic perspective, in Cross, T.A., ed., Quantitative dynamic stratigraphy: Prentice-Hall (this volume).

Komar, P.D., 1987, Selective grain entrainment by a current from a bed of mixed sizes: A reanalysis: Journal of Sedimentary Petrology, v. 57, p. 203-211.

Leeder, M.R., 1978, A quantitative stratigraphic model for alluvium, with special reference to channel deposit density and interconnectedness, in Miall, A.D., ed., Fluvial sedimentology: Canadian Society of Petroleum Geologists Memoir 5, p. 587-596.

Paola, C., 1988, Subsidence and gravel transport in alluvial basins, in Kleinspehn, K.L. and Paola, C., eds., New Perspectives in Basin Analysis: New York, Springer-Verlag, p. 231-243.

Parker, G., 1978, Self-formed straight rivers with equilibrium banks and mobile bed. Part 2. The gravel river: Journal of Fluid Mechanics, v. 89, p. 127-146.

Ribberink, J.S., and van der Sande, J.T.M., 1985, Aggradation in rivers due to overloading—Analytical approaches: Journal of Hydraulic Research, v. 23, p. 273-284.

Sadler, P.M., 1981, Sediment accumulation rates and the completeness of stratigraphic sections: Journal of Geology, v. 89, p. 569-584.

Schlee, J., 1957, Upland gravels of southern Maryland: Geological Society of America Bulletin, v. 68, p. 1371-1430.

Shaw, J., and Kellerhals, R., 1982, The composition of Recent alluvial gravels in Alberta river beds: Alberta Research Council Bulletin 41, 151 p.

Snow, R.S., and Slingerland, R.L., 1987, Mathematical modeling of graded river profiles: Journal of Geology, v. 95, p. 15-33.

Soni, J.P., 1981, Unsteady sediment transport law and prediction of aggradation parameters: Water Resources Research, v. 17, p. 33-40.

Stanley, K.O., and Wayne, W.J., 1972, Epeirogenic and climatic controls of Early Pleistocene fluvial
    sediment dispersal in Nebraska: Geological Society of America Bulletin, v. 83, p. 3675-3690.
Wyrwoll, K.-H., 1988, Determining the causes of Pleistocene stream-aggradation in the central coastal
    areas of western Australia: Catena, v. 15, p. 39-51.
Yang, C.T., and Song, C.S.S., 1979, Theory of minimum rate of energy dissipation: Journal Hydraulics
    Division, American Society of Civil Engineers, v. 105, p. 769-784.

21

The Nickpoint Concept and Its Implications Regarding Onlap to the Stratigraphic Record

Samuel W. Butcher

Department of Geological Sciences, Brown University, Providence, Rhode Island 02912 USA

Abstract

An understanding of processes involved in river profile alteration is fundamental to understanding alluvial sedimentation and onlapping sequences found in the stratigraphic record. Tectonism, isostasy and climate may cause either alluviation or incisement of the river profile. Relative sea-level rise will cause flooding and alluviation, whereas sea-level fall will most frequently cause incisement.

Coastal onlap and headward migration of the nickpoint are two responses of the river profile to relative sea-level fall. Coastal onlap during relative sea-level fall is theoretically possible where the coastal plain is not well developed and when the earth has no continental ice budget. However, in areas where the coastal plain is well developed and during times of glacio-eustatic sea-level fall, nickpoint migration and river profile lowering is the dominant response. Two-dimensional forward models of coastal clastic sedimentation require that the discrepancies between the two scenarios of river profile response to relative sea-level lowering be resolved.

Introduction

Accurate interpretation of alluvial stratigraphy and onlap sequences requires a better understanding of the processes and causal mechanisms that alter river profiles and thereby control sediment accumulation or removal in alluvial environments. Numerous studies have concluded that climatic variations may change river profiles, either locally or regionally. For example, an increased flux of sediment through a river system may cause alluviation, whereas an increased flux of water (runoff) without change in sediment flux may increase concavity and cause local incisement (Snow and Slingerland, 1987). Other studies have concluded that base-level (sea-level) fluctuations cause river profile alteration in specific locations (e.g., Fisk, 1944), but there is no consensus as to how the profile will vary with change in base level. Some investigators have concluded that sea-level fall will cause river incisement, at least locally (e.g., Leopold and Bull, 1979; Begin et al., 1980). Others have concluded, based on what may be a geometrical argument, that sea-level fall may cause alluviation (e.g., Pitman, 1986). Accurate stratigraphic process-response modeling is predicated upon resolving these discrepancies.

Quantitative Dynamic Stratigraphy (1989), T.A. Cross, ed., Prentice Hall, p. 375–385.

375

This paper discusses changes in the river profile in response to sea-level fluctuation with emphasis on response to relative sea-level fall. The validity of onlap during sea-level fall (Vail et al., 1984) is examined in light of the nickpoint concept. Recent models of clastic sedimentation on a subsiding margin have been made which incorporate sediment erosion and deposition with fluctuating sea level (Turcotte and Kenyon, 1984; Helland-Hansen et al., 1988). An understanding of the dynamic adjustments of river profiles to fluctuating sea level (base level) may help define parameters for these models.

Basic Concepts Concerning River Profile Alteration

Morphology of the river profile may be considered a function of four processes: tectonic, isostatic, climatic (including changes in vegetation, sediment and water discharge rates, and sediment type), and base level (Zeuner, 1959). Tectonic processes, such as thermal subsidence, often occur at a rate which is an order of magnitude slower than changes in sea level due to ice-volume fluctuation (Pitman, 1978). Therefore, tectonic processes of this type are considered negligible for the purposes of this discussion. Isostatic processes have been documented as occurring at rates comparable to glacio-eustatic sea-level fluctuation (Pinter et al., 1987); however, for the purpose of focusing on river profile responses to base-level changes, they too are not considered in this discussion.

Climatic variation is also important in determining the shape of the river profile. However, it is often difficult to determine how climatic change has altered the river profile because with climatic variation comes potential changes in vegetation patterns, sediment types, sediment yields, and water yields, to name a few complex interactions. In studies of alpine regions, river aggradation occurs in part due to climatically induced changes in sediment yield (Penck et al., 1909; Eberl, 1930; Bryan and Ray, 1940; Zeuner, 1959). The same climatic variation which may change sediment yield in one portion of the river profile may cause changes solely in water discharge rates in another part of the profile. Increased water discharge rates increase the concavity of a river profile (Snow and Slingerland, 1987) and cause local incisement.

Rivers have aggraded in response to climate variations of a lesser degree than glaciation (Leuninghoener, 1947; Schumm, 1968; Leopold, 1976; Haible, 1980). In these cases, climatic variations change the sediment yield as well as the water discharge rates, resulting in aggradation of the river profile. Although climatic variation is often closely linked to glacio-eustatic sea-level fluctuation, this paper assumes that river profile response to relative sea-level fluctuation may be examined independently.

Equilibrium Profile

An equilibrium profile, or graded stream, is one in which slope, velocity, depth, width, roughness, pattern, and channel morphology delicately and mutually adjust to provide the power and efficiency necessary to transport the load supplied from the drainage basin without aggradation or degradation on the channels (Leopold and Bull, 1979). Rivers that are out of equilibrium will aggrade or incise in an attempt to achieve a graded profile which, as previously noted, is a function of tectonic, isostatic, climatic variation (with associated elements), and base-level controls (Hack, 1957; Zeuner, 1959; Fairbridge, 1968).

Changes in Base Level

In response to a rise in base level (sea level), the total height of the river profile is reduced, and the river loses energy. This loss of energy, in addition to flooding of the coastal plain, causes rivers to aggrade. Onlap of the coastal plain has been shown to occur in response to relative sea-level rise (Leverett, 1921; Fisk, 1944; Trowbridge, 1954; Fisk and McFarlan, 1955). Snow and Slingerland (1987) have shown that if all other forces are held constant, then theoretically, this aggradation would take place throughout the river profile. However, Leopold and Bull (1979) concluded that alluviation in response to base-level rise is more local in nature.

In the event of base-level (sea-level) fall, a nickpoint develops on the river profile (Fig. 1). The nickpoint represents the point below which the profile is adjusting to the newest base level and above which the profile is not adjusting to the most recent base level (Begin et al., 1980; Gardner, 1983). Below the nickpoint, incisement occurs down to a new graded profile and downcutting processes dominate the profile. Above the nickpoint, incisement may or may not occur as the river responds independently to the most recent base-level fall.

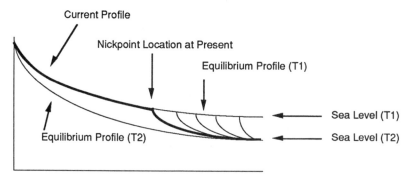

Figure 1. Diagram indicating the nickpoint concept. The drop in sea level from Time 1 (T1) to Time 2 (T2) initially caused cliffing at the shoreline. This cliffing eventually erodes in a headward direction along the river profile until a new equilibrium profile is achieved which was related to the T2 base level. Below the nickpoint, the river profile is determined by the T2 sea level. Above the nickpoint, the river profile acts independently of the change in sea-level. The bold line represents the river profile at some time when the nickpoint had not yet migrated all the way up the T1 river profile.

Early discussions of river profile response to base-level fall considered nickpoints to be more or less a single point or location in the river profile which migrated up the profile with incisement occurring downstream of the nickpoint (e.g., Zeuner, 1959). More recently, it has been shown that the response to base-level lowering is not a simple nickpoint, but may incorporate a complex system of oversteepened reaches, headcuts, and plunge pools (e.g., Begin et al., 1980); that multiple nickpoints may exist on a single river profile due to successive base-level falls (Gardner, 1983); and that river profile behavior may incorporate periodic alluviation within the overall trend of incisement below the nickpoint (Slingerland and Snow, 1988). Theoretically the nickpoint (i.e., the response to base-level lowering) should migrate to the head of the profile (Zeuner, 1959; Begin et al., 1981; Snow and Slingerland, 1987). However, recent work has shown that nickpoints may in fact migrate only part of the distance up the profile (Leopold and Bull, 1979).

Nickpoint migration is affected by both global and local changes in base level. For example, temporary base levels such as lakes or lagoons may not immediately respond to a drop in sea level (Fig. 2). The nickpoint will migrate from its shore position up to the temporary base level and expose the temporary base level to the new height of sea level. River profiles controlled by the base level of the lake or lagoon conceivably could have aggraded during times when sea level was falling. It will be only after some finite length of time when the nickpoint has migrated headward up to the lake, or lagoon, that the level would be affected by the lower base level.

Competing Processes

Because the river profile morphology is a function of a number of different, primarily independent, processes, there is the potential for competition among these processes in determining whether the river profile will alluviate or incise. In this case, the relative magnitude of these processes, and therefore their relative influence, may vary (Fairbridge, 1968).

In situations where climatic changes may be causing alluviation at the same time that sea-level lowering may be causing incisement, the nickpoint might be considered the boundary between aggradation in the upper portions of the profile and incisement in the lower portions. Stephens and Synge (1966) claimed that in northern Europe it is possible to trace river terraces formed as a result of alluviation during glaciation to areas where there is evidence for incisement due to lowering of sea level. In a different example, Fisk (1944) showed that the Mississippi River profile was lowered during the last glacial maximum, and that the extent of profile lowering was nearly to the point of maximum glaciation. In this example, profile lowering may have been caused by nickpoint migration, however, climatic variation resulting in increased water discharge rates may have caused increased concavity and profile lowering (Snow and Slingerland, 1987). Therefore, base-level lowering and climatic change may have acted in concert to lower the profile of the Mississippi River.

DISCUSSION

Alternative Responses to Changes in Base Level

Pitman (1986) proposed that sea-level fall results in onlap of sediments (i.e., aggradation of the river profile). He also proposed that in areas where the steeply sloping uplands grade directly into the continental shelf (i.e., the nearly flat coastal plain is not well developed), sea-level fall would cause fluvial systems to deposit a portion of their sediment load, thus raising the river profile (Fig. 3). This argument is based on the premises that: 1) erosion and deposition maintain the equilibrium profile; 2) the uplands have a steeper slope than the continental shelf; and 3) rates of sea-level fluctuation are proportional to thermal subsidence rates, that is, no high frequency glacio-eustatic sea-level component is considered (Pitman, 1978; Pitman, 1986). Although the scenario proposed by Pitman (1986) is theoretically reasonable, it requires a change in concavity of the river profile (Fig. 3) which in turn requires the river profile to respond to an external force other than base-level fall (e.g., decreased

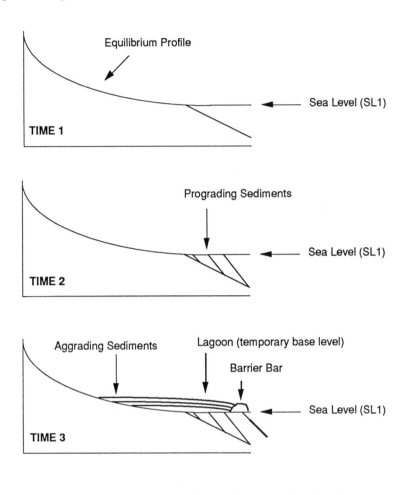

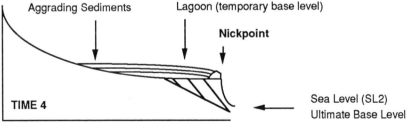

Figure 2. A hypothetical sequence of events, the last of which would appear to be aggradation in the face of sea-level fall. The gradual development, at constant sea level, of a progradational sequence and back-barrier lagoon is shown in Panels 1, 2, and 3. At T4, there is a drop in sea level to SL2 causing cliffing and migration of a nickpoint. However, the upper portion of the river is controlled by the temporary base level (SL1), not sea level. For this reason, the river would appear to be aggrading in the face of sea-level fall. Eventually the nickpoint would migrate up to the temporary base level and aggradation would cease.

runoff or change in sediment type). This paper presumes that a significant ice volume has existed throughout major portions of the Earth's history and that this ice budget and sea level are expected to fluctuate with Milankovitch perturbations and frequencies (Matthews and Poore, 1980; Matthews, 1984). Additionally, a nearly flat coastal plain is frequently developed between the steeply sloping uplands and the shallowly sloping continental shelf (e.g., the Mississippi River Valley). In this case, relative sea-level fall would cause incisement of the river valley and nickpoint migration rather than alluviation (Fig. 3).

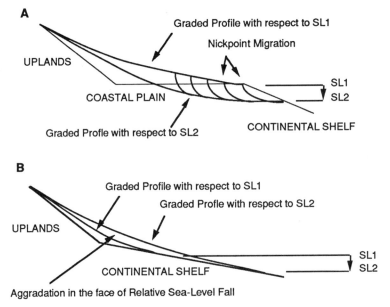

Figure 3. Location of the nickpoint migration in relation to the river profile. In relation to the overall river profile, nickpoint migration is most commonly initiated on the flat coastal plain when relative sea level falls onto the more steeply sloping continental shelf (Fig. 3A). In this example, the uplands grade into the nearly flat coastal plain. The coastal plain in turn grades into the relatively steeply sloping continental shelf. Incisement, due to nickpoint migration, occurs primarily in the coastal plain.
In the less likely event that the coastal plain is not developed and that the steeply sloping uplands grade directly into the less steeply sloping continental shelf, onlap or aggradation may occur during periods of relative sea-level fall (Fig. 3B). Although the potential for this geometry exists, the more frequent geometry is one in which a coastal plain is developed. Therefore, response of the river profile to relative sea-level fall is most frequently incisement.

Evaluation of Onlap Relations from Seismic Stratigraphy

Based on seismic and well log data, Vail et al. (1984) proposed alluvial onlap (aggradation) during times of relative sea-level fall. Relative sea-level fall occurs when absolute sea level is falling faster than regional subsidence (Vail et al., 1977).

In light of the nickpoint concept and the previous discussion, alluviation during periods of relative sea-level fall is difficult to justify. Although Vail et al. (1984) diagrammed onlapping sediments during periods of relative sea-level fall (their Figures 3 and 7), text describing scenarios in which this situation might occur has not been presented (Vail et al., 1977; Vail et al., 1979; Vail et al., 1980; Hardenbol et al., 1981; Vail et al., 1981). This author recognizes that the shoreline will migrate basinward, as diagrammed by Vail et al. (1984); however, there is not a physically based argument as to why alluvial onlap should encroach on the pre-existing surface, solely as a response to base-level fall. Figure 4 outlines a possible explanation of a Vail et al. (1984) methodology. The presumption of Figure 4 is that sea level is fluctuating at a rate beyond the resolution of seismic stratigraphy, in which case features from this high-frequency sea-level curve may be hidden from the seismic signal.

In cases other than the ice-free world scenario of Pitman (1986), relative sea-level fall would result in incisement of the river profile to a lower profile rather than alluviation.

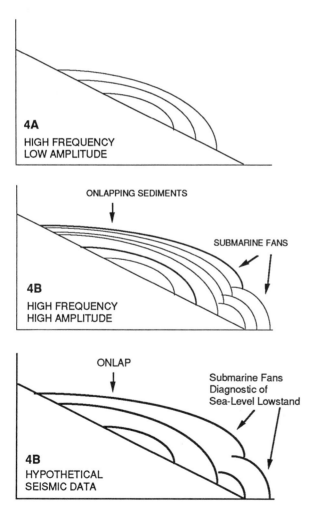

Figure 4. A schematic diagram of a possible misinterpretation which might lead to the conclusion of onlap in the face of sea-level fall (Matthews, pers. comm.). In Figure 4A, sea-level fluctuations have a high frequency and low amplitude creating a common onlapping sequence for a subsiding margin. At some later time, the amplitude of the sea-level fluctuations increases (Fig. 4B). The lowstand portions of the fluctuation create submarine fans (diagnostic of lowstand), the highstand portions of the fluctuations create alluvial and coastal onlap. Because these fluctuations may be of higher frequency than can be resolved by seismic techniques Figure 4B might be misinterpreted as indicated by Figure 4C. The lower portion contains seismic structure diagnostic of sea-level lowstand, therefore, the alluviation might be misinterpreted by others to have occurred during sea-level lowstand.

Application of Nickpoint Concept to Forward Modeling

Recent models of clastic sedimentation on a subsiding margin have been made with the assumption that an erosional environment exists landward of the shoreline and that deposition occurs on the seaward side of this boundary (e.g., Turcotte and Kenyon, 1984). In some cases, the erosion/deposition boundary is located landward of the shoreline to account for alluvial deposition (e.g., Helland-Hansen et al., 1988). Nonetheless, sea-level fall results in the migration of the erosion/deposition boundary to a new location, basinward of its previous position, causing some previously deposited material to be eroded.

Current models calculate erosion solely as a function of time over the entire subaerial exposure surface (Helland-Hansen et al., 1988). Erosion is considered a process that works in a vertical sense, eroding the entire exposure surface down to some new elevation.

In the nickpoint concept, however, erosion is both a downcutting process as well as a headward-cutting process. Vertical incisement as well as headward-migration rates could be parameterized in these model runs as a function of time using previously derived calculations (e.g., Leopold and Bull, 1979; Snow and Slingerland, 1987). In some areas, headward migration of the nickpoint is probably so rapid that it would be considered instantaneous on geologic time scales, and vertical incisement might just as well be assumed (e.g., Fisk, 1944).

During periods of relative sea-level fall, the alluvial valley acts as a sediment source, just as it acts as a sediment sink during periods of relative sea-level rise. Since nickpoint migration rates determine to a large extent the additional volume of sediment being introduced from this source to a sedimentary basin, these rates will affect the resulting stratal geometry. For this reason nickpoint migration rates in lithologically distinct areas are an interesting tunable parameter that should be incorporated into basin modeling efforts.

CONCLUSIONS

Prior to interpretation of the riverine or alluvial component of the stratigraphic record, it is essential that those processes affecting clastic sediment deposition be understood. Only after we understand the processes involved in alluvial sedimentation (aggradation and onlap) can we use the stratigraphic record to draw conclusions about the past.

Depending on the nature of tectonic, isostatic and climatic variations, alluviation or incisement may occur along portions of the river profile (Zeuner, 1959). Sea-level rise will cause flooding in the lower course of the river profile which will cause alluviation (onlap), whereas sea-level fall will cause incisement (e.g., Fisk, 1944; Leopold and Bull, 1979) and the development of a nickpoint. The nickpoint represents the point below which the profile is adjusting to the lower sea level. Whether or not the nickpoint will eventually migrate to the head of the river profile is unclear (e.g., Leopold and Bull, 1979; Snow and Slingerland, 1987).

Tectonic, isostatic, climatic and sea-level controls may act independently or as competing processes on the river profile (Stephens and Synge, 1966). During glaciation, for example, climatic changes may cause the river profile to alluviate whereas sea-level fall will cause incisement. In this case, the nickpoint may be considered the separation point between the dominance of these two independent processes.

Pitman (1986) proposed that onlap occurs during sea-level fall when three conditions exist: 1) erosion and deposition maintain equilibrium profiles on the river profile; 2) the coastal plain is not well developed; and 3) the rates of sea-level fluctuation are of the same magnitude as thermal subsidence rates (i.e., no global ice volume). This author assumes that the coastal plain is more frequently developed than not, and that a significant ice volume has existed throughout major portions of earth history (Matthews and Poore, 1980; Matthews, 1984). Therefore, sea-level fluctuation rates have been much higher than thermal subsidence rates. Given these assumptions, and in light of the nickpoint concept, onlap during periods of sea-level fall is difficult to justify.

Development of quantitative dynamic stratigraphic models of clastic sedimentation on a subsiding margin is in its initial stage (e.g., Helland-Hansen et al., 1988). Currently, erosion is parameterized as a downcutting process which acts over a limited subaerial exposure surface. However, nickpoint migration (incisement) is both a downcutting process as well as a headward migrating process over the entire alluvial valley. Therefore nickpoint migration is a tunable parameter that should be incorporated into these forward models in order to accurately represent the source/sink nature of the alluvial valley. Incorporation of the nickpoint concept to basin modeling efforts will increase the accuracy of predicted sedimentary geometries.

ACKNOWLEDGMENTS

I would like to thank R.K. Matthews for the introduction to the concepts presented in this paper, as well as M. Jervey, R. Slingerland and T.M. Quinn for their helpful reviews and suggestions. I would also like to thank S.M. Cannon for editorial assistance.

REFERENCES CITED

Begin, Z.B., Meyer, D.F., and Schumm, S.A., 1980, Knickpoint migration due to base level lowering: Journal of the Waterway Port Coastal and Ocean Division, v. WW3, p. 93-102.

Begin, Z.B., Meyer, D.F., and Schumm, S.A., 1981, Development of longitudinal profiles of alluvial channels in response to base-level lowering: Earth Surface Processes and Landforms, v. 6, p. 49-68.

Bryan, K., and Ray, L.L., 1940, Geologic antiquity of the Lindenmeier site in Colorado: Smithsonian Miscellaneous Collections, v. 99, 59 p.

Eberl, B., 1930, Die Eiszeitenfolge im nordlichen Alpenvorlande: Augsberg, 427 p., 2 pls.

Fairbridge, R.W., 1968, The encyclopedia of geomorphology: Stroudsburg, Pennsylvania, Dowden, Hutchinson and Ross, pp. 49-51, 891-893, 1124-1138.

Fisk, H.N., 1944, Geological investigation of the alluvial valley of the lower Mississippi River: Mississippi River Commission U.S. Army Corps of Engineers, 75 p.

Fisk, H.N., and McFarlan, E., Jr., 1955, Late Quaternary deltaic deposits of the Mississippi River: Geological Society of America Special Paper 62, p. 279-302.

Gardner, T.W., 1983, Experimental study of knickpoint and longitudinal profile evolution in cohesive, homogeneous material: Geological Society of America Bulletin, v. 94, p. 664-672.

Hack, J.T., 1957, Studies of longitudinal stream profiles in Virginia and Maryland: U.S. Geological Survey Professional Paper 294B, 97 p.

Haible, W.W., 1980, Holocene profile changes along a California coastal stream: Earth Surface Processes, v. 5, p. 249-264.

Hardenbol, J., Vail, P.R., and Ferrer, J., 1981, Interpreting paleo-environments, subsidence history and sea-level changes of passive margins from seismic and biostratigraphy: Oceanologica Acta, No. SP, p. 33-44.

Helland-Hansen, W., Kendall, C.G.St.C., Lerche, I., and Nakayama, K., 1988, A simulation of continental basin margin sedimentation in response to crustal movements, eustatic sea level change and sediment accumulation rates: Mathematical Geology, v. 20, p. 777-802.

Leverett, F., 1921, Outline of Pleistocene history of Mississippi Valley: Journal of Geology, v. 29, p. 615-626.

Leopold, L.B., 1976, Reversal of erosion cycle and climatic change: Quaternary Research, v. 6, p. 557-562.

Leopold, L.B., and Bull, W.B., 1979, Base level, aggradation and grade: Proceedings of the American Philosophical Society, v. 123, p. 168-202

Leuninghoener, G.C., 1947, The Post-Kansan geologic history of the lower Platte Valley area: University of Nebraska Studies (2), Lincoln, 82 p.

Matthews, R.K., and Poore, R.Z., 1980, Tertiary $\delta^{18}O$ record and glacio-eustatic sea-level fluctuations: Geology, v. 8, p. 501-504.

Matthews, R.K., 1984, Oxygen isotope record of ice-volume history: 100 million years of glacio-eustatic sea-level fluctuation, in Schlee, J.S., ed., Interregional unconformities and hydrocarbon accumulation: American Association of Petroleum Geologists Memoir 36, p. 97-107

Penck, A., and Bruckner, E., 1909, Die Alpen im Eiszeitalter: Leipzig, 3 vol. 1189 + xxxvi p., 31 pls.

Pinter, N., Gardner, T.W., Slingerland, R.W., Wells, S.G., and Bullard, T.F., 1987, Late Quaternary uplift rates, Osa Peninsula, Costa Rica: Geological Society of America Abstracts with Programs, v. 19, p. 806.

Pitman, W.C., III, 1978, Relationship between eustasy and stratigraphic sequences of passive margins: Geological Society of America Bulletin, v. 89, p. 1389-1403.

Pitman, W.C., III, 1986, Effects of sea level change on basin stratigraphy: American Association of Petroleum Geologists Bulletin, v. 70, p. 1762.

Schumm, S.A., 1968, River adjustment to altered hydrologic regimen: Murrumbidgee River and paleochannels Australia: U.S. Geological Survey Professional Paper 598, 65 p.

Slingerland, R.L., and Snow, R.S., 1988, Stability analysis of a rejuvenated fluvial system: Zeitschrift für Geomorphologie N.F. 67 p. 93-102.

Snow, R.S., and Slingerland, R.L., 1987, Mathematical modeling of graded river profiles: Journal of Geology, v. 95, p. 15-33.

Stephens, N., and Synge, F.M., 1966, Pleistocene shorelines, in Dury, G.H., ed., Essays in geomorphology: New York, American Elsevier, p. 1-52.

Trowbridge, A.C., 1954, Mississippi River and Gulf Coast terraces and sediments as related to Pleistocene history—A problem: Geological Society of America Bulletin, v. 65, p. 793-812.

Turcotte, D.L., and Kenyon, P.M., 1984, Synthetic passive margin stratigraphy: American Association of Petroleum Geologists Bulletin, v. 68, p. 768-775.

Vail, P.R., Mitchum, R.M., Todd, R.G., Widmier, J.M., Thompson, S., Sangree, J.B., Budd, J.N., and Hatelid, W.G., 1977, Seismic stratigraphy and global changes of sea-level, in Seismic stratigraphy: Applications to hydrocarbon exploration: American Association of Petroleum Geologists Memoir 26, p. 49-212.

Vail, P.R., and Hardenbol, J., 1979, Sea-level changes during the Tertiary: Oceanus, v. 22, p. 71-79.

Vail, P.R., Mitchum, R.M., Shipley, T.H., and Buffler, R.T., 1980, Unconformities of the North Atlantic: Philosophical Transactions of the Royal Society of London, v. 294, p. 137-155.

Vail, P.R., and Todd, R.G., 1981, Northern North Sea Jurassic unconformities, chronostratigraphy and sea-level changes from seismic stratigraphy: Petroleum geology of the continental shelf of northwest Europe, Institute of Petroleum Geology, p. 216-235.

Vail, P R., Hardenbol, J., and Todd, R.G., 1984, Jurassic unconformities, chronostratigraphy, and sea-level changes from seismic stratigraphy and biostratigraphy, *in* Schlee, J.S., ed., Interregional unconformities and hydrocarbon accumulation: American Association of Petroleum Geologists Memoir 36, p. 129-144.

Zeuner, F.E., 1959, The Pleistocene Period: Its climate, chronology and faunal successions: London, Hutchinson and Co., 447 p.

22

Modeling Base-Level Dynamics as a Control on Basin-Fill Geometries and Facies Distribution: A Conceptual Framework

William C. Ross
Marathon Oil Co., P.O. BOX 269, Littleton, CO 80160 USA

Abstract

Basin-fill geometries, represented by stratal surfaces, can be numerically modeled by systematically varying the relative rise and fall of laterally continuous base-level surfaces. These base-level surfaces represent the upper limit to which sediments can be deposited in marine and nonmarine environments for a given profile of equilibrium. Together, the nonmarine and marine profiles of equilibrium represent a continuous base-level surface across the topset portion of a basin (i.e., the environments near base level).

Sediments transported along a given facies tract are deposited and preserved as this base-level surface rises relative to an initial surface of deposition (i.e., during a relative rise in base level). When this base-level surface is falling (or stationary) relative to an initial surface of deposition, the surface becomes nondepositional or erosional as sediments are bypassed basinward.

Facies modeling is conducted within the framework of a changing relative base-level surface. Shoreline positions, shelf widths, and the occurrence of basinal sands are determined by budgeting sand–shale distributions in fluvial and coastal-plain environments under varying conditions of relative base-level rise.

Introduction

Progress in the disciplines of sedimentology, stratigraphy, and basin analysis is achieved when we improve our ability to reconstruct and understand the geologic processes which have produced the stratigraphic record. Success in this area is limited by our ability to integrate observational data sets into existing qualitative process-response or analog models (typically qualitative) for the various physical and biological processes controlling depositional system development. Quantitative stratigraphic modeling procedures represent a powerful tool for the integration of observational data by attempting to numerically simulate the major controls on the development of depositional sequences. This paper describes the relationship of quantitative stratigraphic modeling techniques to traditional approaches, reviews the conceptual framework upon which a series of quantitative models were

Quantitative Dynamic Stratigraphy (1989), T.A. Cross, ed., Prentice Hall, p. 387–399.

387

developed, and highlights the power of these models in providing new insights into stratigraphic problems.

Quantitative approaches to stratigraphic analysis provide the geologic model builder with the means to systematically evaluate the relative importance and interdependency of the various geologic processes which are *believed* (by the modeler) to influence the development of depositional sequences. The evaluation process, performed through a series of model experiments, affords the interpreter-geologist the opportunity to modify the model so that it is consistent with the physical laws (e.g., conservation of mass) and modeling assumptions considered most important.

Geologic interpretations and numerical models that are constrained to be internally consistent are no more "unique" than qualitatively described models. Both kinds of models are subject to the same pitfalls, such as poor input data and an incorrect conceptual framework guiding the model assumptions. All quantitative stratigraphic models must apply some simplification or approximation strategy in the model design which reflects the biases of the model builder. This paper describes the conceptual framework that was used to guide the construction of one such model.

The forward models under consideration were designed to simulate the development of basin-fill geometries, stratal surfaces and gross facies distributions over time intervals of several thousand years or greater per time step. A given model does not attempt to deal with the depositional processes operating within individual depositional environments over periods of days, hours, or minutes. Rather, the model assumes that the first-order controls on gross basin-fill geometries and facies distributions can be approximated over long time intervals by considering a few key variables. The models presented in this paper were designed and implemented on a Macintosh computer. Although the algorithms are not shown, the figures were derived from the model runs.

ASSUMPTIONS AND LIMITATIONS OF ONE-DIMENSIONAL MODELS

Approaches to stratigraphic modeling can range from simple, one-dimensional to complex, three-dimensional models with increasing numbers of variables required for consideration in the model design. In one-dimensional models, vertical "space" is created or destroyed for sediment preservation or erosion through the relative rise or fall of a sea-level or "base-level" surface, respectively. Base level, as defined by Sloss (1962), is ". . .an equilibrium surface above which deposition is temporary and below which preservation is possible." The relative rise and fall of this surface is modeled by summing the effects of a constant tectonic subsidence function (amplified by isostatic sediment loading) and a sinusoidal sea-level curve (Fig. 1). The resulting curve, referred to as a relative sea-level or base-level curve, describes the creation and destruction of vertical space available for sediment deposition.

If we assume that sufficient sediment is available to fill all of the space as it is created, then a simple, one-dimensional "aggradation" model can be used to simulate a vertical stratigraphic column (Fig. 1, horizontal axis). No lithologies are implied in this stratigraphic column, but the spacing of time lines and unconformities indicate variations in the rate of aggradation and/or nondeposition, respectively. These concepts are not new, having been discussed originally by Barrell (1917).

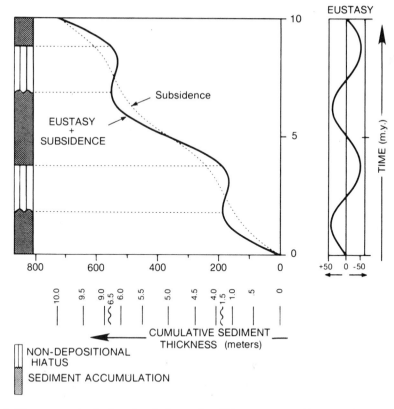

Figure 1. Time versus depth plot of a relative sea level history generated by summing the effects of a symmetrical, sine-wave eustasy curve and a constant tectonic subsidence function. The effects of sediment loading amplify the tectonic subsidence component while relative sea level or base level is rising. Periods of potential sediment aggradation (gray shaded) and nondepostion (vertical lines) are indicated on the left side of the vertical (time) axis. A synthetic stratigraphic column, with million year age-labels and unconformities, is shown along the horizontal (depth) axis.

By varying the rate of subsidence for a given sea-level curve, one can examine the effects of subsidence in contributing to the completeness of the stratigraphic record (Fig. 2). Subsidence, when summed with sinusoidal eustasy, produces asymmetric relative sea-level curves with periods of relative sea-level rise becoming progressively longer than falls as rates of subsidence are increased (Morrow, 1986). When the rate of subsidence exceeds the maximum rate of sea-level fall, a completely preserved stratigraphic section is possible (curve A, Fig. 2).

Figure 3 illustrates the differences in synthetic stratigraphic columns generated under slow versus rapid subsidence conditions. Correlations of stratal surfaces across this hypothetical basin illustrate the time-transgressive nature of unconformities that develop under differential subsidence conditions. The offlapping and onlapping stratal patterns occur during periods of accelerating and decelerating rates of sea-level *fall*, respectively.

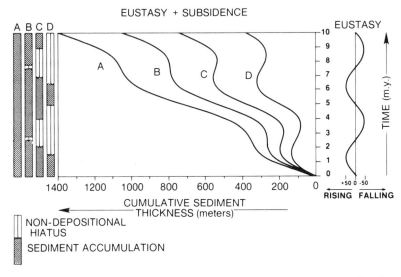

Figure 2. Plots of four relative sea level histories generated from the same eustasy function and four different subsidence functions. The shaded columns at the left of the diagram show periods of potential sediment aggradation when relative sea level is rising. With increased rates of subsidence the periods of relative sea level rise become longer and the synthetic stratigraphic columns become more complete.

This one-dimensional model can be applied to stratigraphic column modeling only by assuming that sufficient sediment is available to fill the space generated at each time step. When the rate of "space" creation exceeds the rate of sediment supply a two-dimensional model is more appropriate.

CONCEPTS IN TWO-DIMENSIONAL MODELING

The extension of one-dimensional "aggradation" modeling to two-dimensional basin-fill or sequence modeling requires conceptual models that address two important issues: 1) the determination of base level in two-dimensional space; and 2) the manner in which lithologies are distributed and facies boundaries determined below this base-level surface.

Base-Level Modeling

In the one-dimensional aggradation model, the upper limit to which sediments could be deposited (i.e., base level) was defined as sea level. In nature, base level is coincident with the sea-level surface only at the shoreline position. Landward of this position, base level is represented by the graded profile of equilibrium (Mackin, 1948) for the system of rivers crossing the alluvial plain (Fig. 4). The slope of this surface is controlled locally by characteristics of the river, including discharge, grain size, and valley slope (Lane, 1937). However, the average regional slope will be determined by the relative positions of two important measured distances: 1) the length (L) of the river drainage from the first nickpoint (i.e., abrupt change in river gradient) to the shoreline; and 2) the difference in elevation between this nickpoint and sea level (Fig. 4; see also Butcher, this volume).

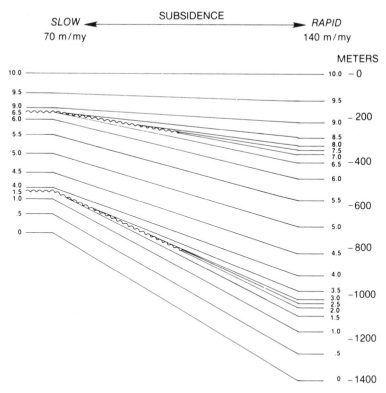

Figure 3. A correlation between two synthetic stratigraphic columns generated with the same eustasy function and two different subsidence functions.

Base level in the marine environment is defined as a graded profile of equilibrium across the shelf that is controlled by a variety of marine processes. The concept of the graded shelf, originally described in qualitative terms (e.g., Johnson, 1919; Swift, 1970), has been described more recently by a number of authors in quantitative terms (e.g., Jago and Barusseau, 1981; Niedoroda et al., 1984). The quantitative models relate the slope and depth of a position upon a "graded" shelf to the effective wave energy and grain-size distribution across that shelf. Using this relationship, a shelf profile of equilibrium can be constructed given a knowledge of 1) the grain-size distribution reaching and redistributed within the marine environment, and 2) the effective wave power influencing the system under consideration. Effective wave power can be estimated if a knowledge of the size of the ocean basin or water body controlling the wind fetch is known or assumed (Pickrill, 1983). The distribution of grain sizes entering the marine environment can be derived within the model as output from the facies modeling routine described subsequently.

Thus, as discussed by Wheeler (1964), base level can be considered as a continuous surface extending across the nonmarine and marine portions of the system (i.e., topset deposits, Fig. 5). As applied to two-dimensional models, the creation and destruction of vertical space across the entire topset facies tract is modeled as the rise and fall of a continuous base-level surface.

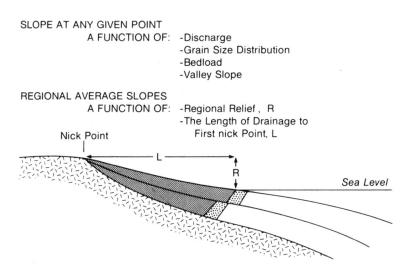

SLOPE AT ANY GIVEN POINT
        A FUNCTION OF: -Discharge
                           -Grain Size Distribution
                           -Bedload
                           -Valley Slope

REGIONAL AVERAGE SLOPES
        A FUNCTION OF: -Regional Relief , R
                           -The Length of Drainage to
Nick Point   First nick Point, L

L

R

Sea Level

Figure 4. Parameters controlling the non-marine profile of equilibrium.

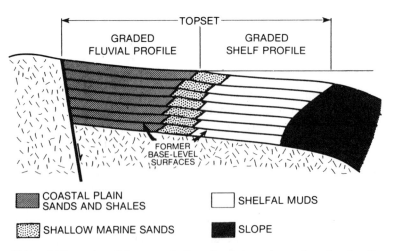

TOPSET

GRADED
FLUVIAL PROFILE

GRADED
SHELF PROFILE

FORMER
BASE-LEVEL
SURFACES

COASTAL PLAIN
SANDS AND SHALES

SHELFAL MUDS

SHALLOW MARINE SANDS

SLOPE

Figure 5. Illustration of the concept of base level. A continuous base-level surface extends across the topset portion of the basin and represents the upper limit to which sediments can be preserved. Over geologic time the relative rise of this surface, due to subsidence and/or eustatic change, controls the geometry of preserved topset sediments.

Facies Modeling

Modeling the relative rise and fall of a continuous base-level surface generates the vertical space in two dimensions necessary for sediment accumulation and preservation in the topset portion of a system. In simplistic terms, if the sediment supply (in two-dimensional space) exceeds the area created during that time interval, then some fraction of the sediments will bypass the topset setting, accumulate in the foreset or bottomset settings, and the system will prograde. Conversely, if the new space created exceeds the sediment supply, then the system will retreat. A more detailed method for distributing sediment, required for facies modeling, is possible by considering the differential rate of accumulation and preservation of sand versus silt and clay across an entire facies tract.

SHORELINE MODELING   Perhaps the most important environmental or facies boundary to understand for modeling purposes is the boundary between the marine and nonmarine systems. It seems clear that the shoreline position should represent a balance between rates of base-level change (i.e., subsidence plus sea level), rates of sediment input, and rates of erosion (Curray, 1964). However, it is difficult to determine just how to balance these important variables.

Figure 6 illustrates the geometric relationships that develop with a relative rise of base level (or sea level) prior to the addition of sediments. Depending on the rate at which sediment is supplied and the manner in which it is distributed, the new shoreline can theoretically occupy any position along this new sea-level surface (e.g., positions 1, 2, or 3 in Fig. 6).

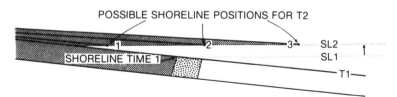

Figure 6. Geometric configuration developed after a relative sea level rise between time 1 (T1) and time 2 (T2). The determination of the shoreline position along this new sea-level surface (eg., positions 1, 2, or 3) will depend on the total volume of sediments entering the basin and the manner in which the constituent sands and muds are distributed within the available space.

The model strategy for determining the position of the shoreline is based on the assumption that the progradation of the coastal-plain and shoreface couplet is *limited* by the volume (area in 2-D) of "sand" available to the system for each time step (i.e., km² sand/unit time). During a relative rise in base level, a certain quantity of sand is preserved or "consumed" in the nonmarine environment to accommodate the new "space" that is created. The shoreface environment also requires a certain volume of sediment to buttress the shoreline system against erosion and removal of sediment by marine currents and waves. For a given time step, the basinward extent of the shoreline is determined by the distance the coastal-plain and shoreface couplet progrades before the available sand volume is exhausted.

In two-dimensional models, the effects of an imbalanced addition or removal of sediment from out of the plane of the model, for example by long-shore drift, either should be simulated or considered unimportant and ignored. This model assumes a strike-oriented, wave-dominated system with many river systems crossing the coastal plain. Utilizing this assumption, any sand removed from the shoreface by longshore transport is balanced by sand entering the system from upcurrent.

Figure 7 illustrates two important considerations used to determine the position of the shoreline based on the sand–mud budgeting approach. The first consideration is the ratio of sand to mud in the system. For systems with low sand–mud ratios, the basinward extent of shoreline progradation is restricted by the limited availability of sand to "fill" the nonmarine portion of the system (Fig. 7a). Conversely, drainage systems carrying equal total volumes

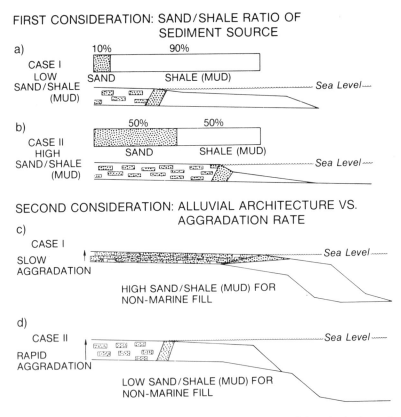

Figure 7. Two major considerations which are important in determining the basinward extent of shoreline facies during each model time step. The first consideration is the original sand-shale (mud) ratio of sediments entering the basin (Figures 7a and b). For a given quantity of sediment, low sand-mud ratios limit the progradation of the shoreline (a), while shorelines in systems with high sand-mud ratios extend further into the basin (b). The second consideration is the style of fluvial architecture (Figures 7c and d). During periods of slow aggradation, fluvial channels rework floodplain shales leaving high sand-shale ratios in the preserved, non-marine sediment (c). Rapid aggradational episodes (d) are characterized by low sand-shale ratios of preserved non-marine fill (see text).

of sediment, but with higher sand–mud ratios, are able to prograde much farther (Fig. 7b).

The second consideration in determining how rapidly sands are "consumed" or deposited in the nonmarine setting is the variation in fluvial architecture which affects the preservation rate of sands versus shales. Figures 7c and d show two end members of fluvial architecture which reflect different rates, low and high respectively, of aggradation. The relationship between variations in fluvial architectural styles and aggradation rates has been discussed and modeled by Allen (1978) and Bridge and Leeder (1979). The general model relates the rate of base-level rise (i.e., the aggradation rate) to rates of channel-belt migration and avulsion to determine the stacking geometries of fluvial sandstones preserved within flood-plain shales.

Although not tested, their models suggest a general decrease in channel sand interconnectedness and sand–shale ratios with increasing rates of aggradation, and increasing connectedness and sand–shale ratios with slowing rates of aggradation (Figs. 7c and d). For a given volume of sediment, the extent of basinward progradation of the coastal-plain and shoreface couplet is partially controlled by the rate of sand preservation, as a function of aggradation rate.

THE TWO-DIMENSIONAL MODEL   To utilize the alluvial architecture model in two-dimensional facies modeling, a general relationship was assumed between the rate of sand preservation in the nonmarine setting and the rate of aggradation (Table I, column 3). Given this relationship, the volume (or area in 2-D) of coastal-plain sediment required to "exhaust" the available sand supply for a given sand–mud ratio can be calculated (Table I, column 4). Given the coastal-plain volume and the geometry of new space created in the topset setting, a shoreline distance (measured from the left edge of the model) can be calculated for each time step (Table I, column 5). The shoreline positions calculated according to this procedure are termed "sand-limited" shorelines in Figure 8. The distance of progradation of these shorelines was limited by the total volume of sand (km^2/Myr), the rate of preservation of sands in the nonmarine setting (sand–mud ratio), and the aggradation rate (km/Myr).

The procedure for calculating the position of a sand-limited shoreline assumes progradation across a shallow platform. When the system is prograding into a "deep-water" basin, the capacity of the shoreline system to prograde is limited by the availability of shale (silt and clay) to first build a shallow platform. "Shale-limited" shorelines shown in Figure 8 develop when the water depth, into which the shelf–slope system is prograding, exceeds the volume (area in 2-D) of shale required to fill it. When this occurs (time steps 5, 6, and 7 in Table I and Fig. 8), some fraction of the available sand supply is "required" to fill the basin before the shoreline system can continue to prograde.

Based on these simple sand–shale budgeting considerations, the model predicts that during intervals with shale-limited shorelines significant quantities of sand will bypass the coastal-plain and shoreface environments and be deposited in deep-water (below wave base) environments. Although the model calculates the amount of sand that must be deposited in deep-water environments, no conceptual or quantitative strategy has been included for distributing these sands within slope and basinal settings. Rather, for graphical reasons, the bypassed sands are shown as deposited at the base of slope (Fig. 8). Examples of analogous delta-fed turbidite systems in the ancient have been reviewed by Heller and Dickinson (1985).

396

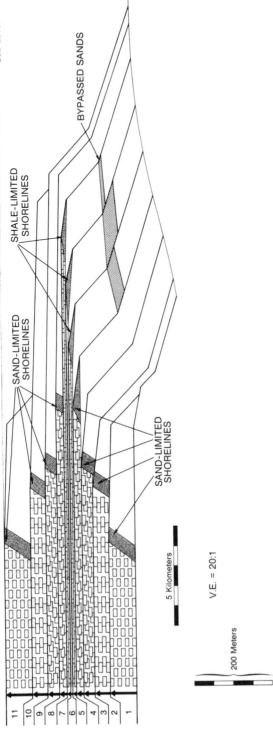

Sea Level

BYPASSED SANDS

SHALE-LIMITED
SHORELINES

SAND-LIMITED
SHORELINES

SAND-LIMITED
SHORELINES

11
10
9
8
7
6
5
4
3
2
1

5 Kilometers

V.E. = 20:1

200 Meters

Figure 8. Two-dimensional basin-fill geometry and facies model generated by varying the rate of relative base level rise from rapid (case 1, .07 km/Myr) to slow (case 6, .005 km/Myr) to rapid again (Table I). A constant volume of sand and shale (mud) is maintained for each model calculation. The basinward extent of the sand-limited shoreline systems is determined by the distance the coastal-plain and shoreface couplet can prograde before the available sand volume (area in 2-D) is exhausted (see text and Table I). Shale-limited shorelines develop when the water depth required for shelf-slope progradation exceeds the volume of shale (mud) required to fill it (see text). The model predicts that the progradation of shale-limited shoreline systems will bypass significant quantities of sands into deep-water (below wave base) environments.

TABLE I

SHORELINE MODELLING DATA

GIVEN: TOTAL SEDIMENT AVAILABLE = 1 Km^2/my

  TOTAL   SAND   AVAILABLE = .2 Km^2/my

| Case | MODEL INPUT | | CALCULATED | | |
	Aggrad. Rate (Km/my)	Sand/Mud in Non-Marine	Amount of Non-Marine Sediment Required to Exhaust Sand Supply (Km^2)	Kms to Shoreline*	Km^2 of Sand Dep. Below Wave Base
1	.070	.4	.5	7.14	——
2	.040	.5	.4	10	——
3	.030	.6	.33	11	——
4	.020	.7	.28	14	——
5	.010	.8	.25	25	0.64
6	.005	1.0	.2	40	.1
7	.010	.8	.25	25	0.16
8	.020	.7	.28	14	——
9	.030	.6	.33	11	——
10	.040	.5	.4	10	——
11	.070	.4	.5	7.14	——

* Assuming a shallow platform

$$\text{Amount of Non-Marine Sediment Required to Exhaust Sand Supply (Km}^2) = \frac{\text{Total Sand Available (Km}^2)}{\text{Sand/Mud in Non-Marine}}$$

$$\text{Kms to Shoreline} = \frac{\text{Non-Marine Sediment (Km}^2)}{\text{Aggrad. Rate (Km/my)}}$$

As the rate of base-level rise increases in the model (times 8, 9, 10 and 11 in Table I and Fig. 8), there is a return to sand-limited shorelines which retreat landward across the shallow platform. In contrast to the regressive shorelines and shelves, which require proportionally larger volumes of shale (mud) to prograde into deep water, the transgressive systems retreat across a shallow platform and are oversupplied with respect to shale. As a result, the model predicts the development of broad, wave-graded shelves, some of which actually prograde while the shoreline moves landward (e.g., times 8 and 9, Fig. 8). A possible modern analog for the progradation of a wave-graded muddy shelf is the subaqueous Amazon delta system (Nittrouer et al., 1986).

Summary

The simple stratigraphic models described in this paper emphasize the concept of base level and its control on sediment deposition and preservation in the stratigraphic record. Modeling the variables that control the relative rise and fall of this surface over time provides the geometric framework within which facies modeling can be performed.

Quantitative stratigraphic models are not inherently different from qualitative models in the respect that both are sensitive to inaccuracies in input data and model assumptions. However, the formal description of a natural system in quantitative terms allows for a rigorous, systematic evaluation of the assumptions built into a given model. In addition, models often produce nonobvious (but intuitively plausible) results that provide insight into natural systems which is not produced by traditional approaches to stratigraphic analysis (e.g., the sand-limited versus shale-limited shoreline systems in Fig. 8). As a continuing process, the results of modeling runs can be field tested to verify the model or supply information necessary for its refinement. In this way the model should not be viewed as an end to itself but as a tool to aid in the development of better conceptual models in sedimentology, stratigraphy and basin analysis.

Acknowledgments

This paper benefitted greatly from the constructive comments and criticisms of Tim Cross, Fred Schroeder, Daniel Tetzlaff, Jeff May, and Beverly Hazlett. Jeff Melton provided the drafting services. I thank Marathon Oil Company for permission to publish this paper.

References Cited

Allen, J.R.L., 1978, Studies in fluviatile sedimentation: An exploratory quantitative model for the architecture of avulsion-controlled alluvial sites: Sedimentary Geology, v. 21, p. 129-147.

Barrell, J., 1917, Rhythms and the measurement of geologic time: Geological Society of America Bulletin, v. 28, p. 745-904.

Bridge, J.S., and Leeder, M.R., 1979, A simulation model of alluvial stratigraphy: Sedimentology, v. 26, p. 617-644.

Butcher, S.W., 1988, The nickpoint concept and its implications regarding onlap to the stratigraphic record, *in* Cross, T.A., ed., Quantitative dynamic stratigraphy: New Jersey, Prentice-Hall (this volume).

Curray, J. R., 1964, Transgressions and regressions, *in* Miller, R. L., ed., Papers in marine geology: New York, Macmillan, p. 175-203.

Heller, P.L., and Dickinson, W.R., 1985, Submarine ramp facies model for delta-fed, sand-rich turbidite systems: American Association of Petroleum Geologists Bulletin, v. 69, p. 960-976.

Jago, C F., and Barusseau, J.P., 1981, Sediment entrainment on a wave-graded shelf, Rousillon, France: Marine Geology, v. 42, p. 279-299.

Johnson, D., 1919, Shore processes and shoreline development (second edition, 1938): New York, John Wiley & Sons, 585 p.

Lane, E.W., 1937, The importance of fluvial morphology in hydraulic engineering: American Society of Civil Engineering Transactions, v. 63, p. 123-142.

Mackin, J.H., 1948, Concept of the graded river: Geological Society of America Bulletin, v. 59, p. 463-512.

Morrow, D.W., 1986, The sea-level rise staircase on continental margins and origin of upward-shoaling carbonate sequences: Bulletin of Canadian Petroleum Geology, v. 34, p. 284-285.

Niedoroda, A.W., Swift, D.J.P., Hopkins, T.S., and Chen-Mean Ma, 1984, Shoreface morphodynamics on wave-dominated coasts: Marine Geology, v. 60, p. 331-354.

Nittrouer, C.A., Kuehl, S.A., DeMaster, D.J., and Kowsmann, R.O., 1986, The deltaic nature of Amazon shelf sedimentation: Geological Society of America Bulletin, v. 97, p. 444-458.

Pickrill, R.A., 1983, Wave-built shelves on some low energy coasts: Marine Geology, v. 51, p. 193-216.

Sloss, L.L., 1962, Stratigraphic models in exploration: American Association of Petroleum Geologists Bulletin, v. 46, p. 1050-1057.

Swift, D.J.P., 1970, Quaternary shelves and the return to grade: Marine Geology, v. 8, p. 5-30.

Wheeler, H.E., 1964, Base level, lithosphere surface, and time-stratigraphy: Geological Society of America Bulletin, v. 75, p. 599-610.

23

SED0: A Simple Clastic Sedimentation Program for Use in Training and Education

Daniel M. Tetzlaff
Western Atlas International, P.O. Box 1407, Houston TX 77251 USA

Introduction

SED0 is a simple computer program that simulates the processes of erosion, transport and deposition of clastic sediments by water in vertical cross section. It is a simplified version of a broad family of two-dimensional sedimentation models introduced by Harbaugh and Bonham-Carter (1970), and further exemplified by Bitzer and Harbaugh (1987). SED0 is written in FORTRAN 77, and does not require any graphics hardware other than a character screen or conventional printer. It is transportable and easy to modify. Yet it illustrates many of the principles that are present in more advanced models.

Principles of Operation

SED0 is based on the concept of a fluid that moves along the surface of a two-dimensional section according to the pull of gravity and the effect of friction (or bed drag). Changes in the velocity of the fluid along the section produce changes in its transport capacity, which in turn cause erosion or deposition. The rates of erosion and deposition also depend on the sediment type that underlies the flow or is being transported.

It is possible to modify a number of parameters associated with the physical properties of the fluid and the sediments, as well as the shape and sediment distribution in the original basin and the initial velocity and sediment content of the flow. The parameters have arbitrary units and cannot necessarily be compared to physical quantities. The program operates in cycles represented by a full trajectory of the fluid along the section. An output on a disk file or line printer can be obtained after any number of cycles. The output file on disk can be used again as input after optionally modifying any of the parameters that describe the current state of the system. This provides a large number of possibilities for running experiments with the model.

Topography and sediment type

The sedimentary basin is simulated by a two-dimensional integer array of cells (LCELLS(I,J)), each cell containing one of five possible sediment types. The sediment types are read in as symbols (+ = basement, o = gravel, . = sand, : = silt, | = clay) and then converted to integers

Quantitative Dynamic Stratigraphy (1989), T.A. Cross, ed., Prentice Hall, p. 401–415.

(1 to 5) to facilitate their handling inside the program. They are reconverted to the corresponding symbols when they are printed. In the printed output the section should be viewed sideways. The depositional surface of the basin may slope continuously in one direction, or it may contain irregularities or slopes in either direction.

Fluid motion

The flow along the surface of the basin has been represented by a single "fluid element" which is assumed to enter the section at the left. The fluid is assumed to possess an initial velocity (VI), and as it progresses towards the right its velocity on each column (V) is calculated from its velocity on the previous column, the difference in surface elevation between the columns, gravitational acceleration (A), and bed drag or friction.

Friction was assumed to be proportional to the square of the velocity times a friction coefficient (CF). An alternative approach, which would make the solution much simpler, would be to consider the friction to be proportional to the velocity (instead of its square). However, the first approach agrees better with the laws of fluid mechanics (Vennard and Street, 1975). Therefore the equation used to represent the movement of the fluid element is

$$\frac{\partial^2 x}{\partial t^2} + k\left(\frac{\partial x}{\partial t}\right)^2 = a$$

where x = horizontal position of fluid; t = time; k = friction coefficient; a = acceleration (due to slope and gravity; for low slopes a = g s where s is the tangent of the slope).

This equation does not have a simple analytical solution. An approximate method is used to calculate the velocity at each column from the preceding velocity. The new velocity is calculated as if there were no friction, then the change in velocity due to friction is calculated as the old velocity squared times the coefficient of friction. This method yields a solution that can be shown to approach the true solution as the width of each column approaches zero.

The velocity increment ignoring friction depends only on the difference in surface elevation, DV, between two adjacent columns and the acceleration of gravity, A, in the following manner. If DV is positive, the velocity increment is SQRT(2*DV*A); if DV is negative, the velocity increment is -SQRT(-2*DV*A).

The formula for calculating the new velocity without friction (VN) therefore becomes VN = V+SQRT(2*DV*A) if DV is positive, and VN = V-SQRT(-2*DV*A) if DV is negative. The term representing friction must then be subtracted. This term is assumed to be the square of the geometrically averaged velocity (without friction) over the current cycle, multiplied by a factor (CF). This expression reduces to VN*V*CF. The program listing (Appendix A) clarifies the use of these formulas in SED0. Note that VN may eventually become negative. Since it is assumed that the fluid always moves from left to right, V is arbitrarily made 0 whenever it becomes negative. In that case the cycle stops at that point but this does not affect the subsequent evolution of the system.

Transport

The ability of a certain volume of fluid to transport sediments depends, among other factors, on the power dissipated by turbulence. This power is proportional to the velocity squared times the friction coefficient. The total transport capacity, TT, is therefore assumed to be TT=CF*V*V*CT where CF is the coefficient of friction, V is the velocity and CT is the coefficient of transport. Hence for a given CF and CT the transport capacity is proportional to the square of the velocity.

At any point the fluid element is assumed to be carrying a certain load of each sediment type (S(K)). The sum of these loads is the total sediment load SS. At any column the transport capacity of the fluid may be greater than, equal to or less than the total sediment load. If it is greater, erosion occurs; if it is equal, nothing happens and the fluid moves on to the next column; if it is less, sediment is deposited.

Erosion

When the fluid carries less load than its total transport capacity, then part of the underlying sediment is eroded and incorporated into the flow. How much is eroded depends on the lithology or sediment type. Each type is assumed to have an associated coefficient of erosion (CE(K)). The total amount of rock or sediment eroded (ST) is assumed to be equal to the difference between the total sediment load and the transport capacity, multiplied by this coefficient: ST=(TT-SS)*CE(K).

Since the surface elevation in each column is an integer, the amount to be transferred is rounded off to the nearest integer. This amount is subtracted from the rock column elevation and added to the load of the corresponding sediment type. If the "sediment type" underlying the flow is "basement," then it is assumed that the sediment eroded from it is sand. The program can also be modified to cause the eroded sediment to be a mixture of certain proportions of gravel, sand, silt and clay.

Deposition

When the transport capacity of the fluid is too low for the total load of sediment being carried, then part of the load is deposited. The amount of sediment deposited is equal to the difference between load and transport capacity times a coefficient of deposition (CD). This amount is also rounded off to the nearest integer and is subtracted from the coarsest sediment in the fluid load (coarse sediments are deposited deposited first) and added to the elevation of the corresponding column. The kind of sediment being deposited is assigned to the new cell or cells located between the old new elevations of the column).

PROGRAM ORGANIZATION

Two programs, listed in Appendix A, were used in the simulation. Program INIT is used to generate the initial state of the system. By calling subroutine GEN it produces a wedge of a given sediment type and initializes all other variables to zero. It then calls subroutine WRITE to write out this system. By editing the written file, the numerical parameters or cells representing sediment can be changed.

Program SED0 is the main program. It calls subroutine READ to read in the state of the system from a disk file. Then it calls subroutine CYCLE, which causes the system to go through one cycle or complete trajectory of one fluid element through the section. This subroutine is called inside a DO loop once for each of the specified number of cycles. The following subroutines are called by INIT and/or SED0.

Subroutine GEN(ICOLS, ITOP, LITH) generates an input file consisting of a wedge of rock that has a constant slope towards the right side of the section. The number of columns is ICOLS. The highest (left most) column has a height of ITOP cells. LITH is the sediment type of the whole wedge (a number from 1 to 5).

Subroutine READ reads the input file. It first reads the physical parameters of the fluid and sediments. Then it reads the sediment type of the cells and converts the symbols to numbers. It also generates the array containing the elevation of each column (LEVEL (I)).

Subroutine WRITE writes the output file. Numbers representing sediment types are converted again into characters, and headings are written to explain the meaning of each variable. The file can also be read in again by subroutine READ without any modifications.

Subroutine CYCLE performs one cycle of the system. Its main steps are: (1) Initialize variables. (2) Start the cycle by entering a DO loop that scans the columns from left to right. For each column: calculate level difference, velocity, and transport capacity; then compare transport capacity to actual load and according to their relative values go to the set of steps simulating erosion or to the set of steps simulating deposition. (3) Simulate erosion. Calculate the amount of material to be eroded, subtract it from the sediment column and add it to the load of the flow. (4) Simulate deposition. Calculate the amount and type of material to be deposited, subtract it from the load of the flow and add it to the sediment column.

Sample Runs

Several runs were tried simulating different conditions. The output from these runs are listed in Appendix B. The following cases were simulated. (1) Basement wedge with no sediment input and initial flow velocity 1. (2) Basement wedge with a large input of gravel, sand, silt and clay, run for 100 cycles. (3) Layered sequence with no sediment input to see the effect of differential erosion. (4) Sedimentation in a partially closed basin. (5) Sedimentation in an environment where hypothetical wave motion limits the height of the columns to 25 units. One additional step was added to subroutine CYCLE to make this possible. This step is indicated in the program listing.

EXTENSIONS AND FURTHER EXPERIMENTS

The model seems to present a plausible way to simulate the processes of erosion and deposition of clastic sediments in a two-dimensional sedimentary basin. It is possible to simulate many more conditions than the ones shown here. Some of these would possibly require extensive file editing between runs to adjust the conditions after every cycle. For example, the progressive displacement along a vertical fault can be simulated by editing the output file after each cycle to "lift" the raising edge of the fault, and then reentering the file as input. However, the program could be modified to produce these changes automatically.

The resulting topography in most of the cases shown is rough and irregular, and the simulated deposits often appear to have been generated "at random" even though the model is completely deterministic. This is due to the "chaotic" nature of the model. If one of these examples is repeated with slightly different initial conditions, the irregularities in the output may vary, but the overall appearance of the output should not change significantly. This behavior is also characteristic of physical experiments in sedimentation, and is ubiquitous in nature.

It is interesting to note that in some of the sample runs the topography tends to look similar after a large number of cycles (e.g., compare cycle 20 of cases 3 and 5 which start with a totally different topography but similar initial fluid velocities and sediment loads). In general the topography tends to look like an exponential curve. This suggests that this may be an "equilibrium profile" for the given conditions of the flow.

Although SED0 only represents four sediment types plus "basement," more sediment types can be added, and continuous mixtures of sediment types also could be represented. In the output file, one could combine a given number of adjacent cells into one larger cell whose sediment type could be classified into a compositional diagram. The resulting larger cells would be, for example, "silty clay" and "gravely sand." The effect would be the same as increasing the unit of observation.

Another interesting development would be to carry out a Markov chain analysis (or other statistical analysis) on different columns of the resulting section. Example 5 (Appendix B) may already be suitable for such a study.

Possibly the most interesting extension would be to develop a three-dimensional model based on some of these principles. This would require a much larger and complex program, but some of the results obtained here make it look feasible, especially considering that the longest runs presented in this report, which ran through 100 cycles, took only a few seconds on a personal computer.

Other possible areas of experimentation include the following. (1) Study of the effects faults of different attitudes and speeds of displacement. (2) Simulate deltaic deposition by assuming that below a certain level (i.e., sea level) the transport capacity drops, and possibly also the equations that govern flow and transport change. (3) Simulate a cyclic change in the input velocity and sediment load. (4) Adjust the input parameters to simulate the migration of sand dunes on a river bed. (5) Simulate isostatic adjustment and sediment compaction. (6) Simulate salt domes protruding into the sequence. (7) Study the effects of a sudden release of a large amount of sediment in order to simulate a turbidity current, and compare this deposit to the one produced by a slow gradual accumulation. (8) Incorporate some extra steps in the program to provide for the effects of "landslides" due to slope instability when the slope surpasses a certain value. (9) Study the equilibrium profile under a number of different conditions. (10) Modify the program to produce a section showing sediment age (for example in the form of single digits representing the cycle during which the sediment was deposited).

References Cited

Bitzer, K., and Harbaugh, J.W., 1987, DEPOSIM: A Macintosh computer model for two-dimensional simulation of transport, deposition, erosion, and compaction of clastic sediments: Computers and Geosciences, v. 13, p. 611-637.

Harbaugh, J.W., and Bonham-Carter, G., 1970, Computer simulation in geology: New York, Wiley Interscience, 575 p.

Vennard, J.K., and Street, R.L., 1975, Elementary fluid mechanics (fifth edition): New York, J. Wiley and Sons, 740 p.

Appendix A: Program Listing

```
C       PROGRAM INIT
C...THIS PROGRAM GENERATES AN INPUT FILE REPRESENTING INITIAL
C... STATE OF THE SYSTEM
C       SUBROUTINES CALLED: GEN AND WRITE
C       FILE GENERATED MUST BE EDITED TO ASSIGN NON-ZERO PARAMETERS.
C       FILE MUST ALSO BE RENAMED TO USE AS INPUT
            COMMON  A,CF,CT,CD,CE(5),SI(5),VI,NCYE,NCYD,NCY,NCO,
        &       LEV(180),LCELL(180,132)
C...OPTIONAL OPEN STATEMENT FOR OUTPUT (UNIT 21)
C       CAN BE INSERTED HERE
            CALL   GEN(20,20,1)
            CALL   WRITE
            STOP
            END


C       PROGRAM SEDO
C...THIS IS THE MAIN PROGRAM TO SIMULATE EROSION-TRANSPORT-
C       DEPOSITION
C...VARIABLES USED ARE:
C       A=ACCELERATION
C       CF=COEFFICIENT OF FRICTION
C       CT=COEFFICIENT OF TRANSPORT
C       CD=COEFFICIENT OF DEPOSITION
C       CE(K)=COEFFICIENT OF EROSION OF SEDIMENT TYPE K
C       SI(K)=AMOUNT OF SEDIMENT INPUT OF TYPE K
C       VI=INITIAL VELOCITY
C       NCYE=NUMBER OF ENDING CYCLE
C       NCYD=NUMBER OF CYCLES BETWEEN DISPLAYS
C       NCY=NUMBER OF CURRENT CYCLE
C       NCO=NUMBER OF COLUMNS OF ROCK OR SEDIMENT
C       LEV(I)=LEVEL OF THE TOP OF COLUMN I
C       LCELL(I,J)=SEDIMENT TYPE IN CELL J OF COLUMN I
C...SUBROUTINES CALLED BY MAIN PROGRAM:
C       READ: READS IN CURRENT STATE OF THE SYSTEM FROM DISK
C       WRITE: WRITES OUT CURRENT STATE OF THE SYSTEM ON DISK
C       CYCLE: SIMULATES ONE CYCLE OF FLOW AND SEDIMENTATION
            COMMON  A,CF,CT,CD,CE(5),SI(5),VI,NCYE,NCYD,NCY,NCO,
        &       LEV(180),LCELL(180,132)
C...OPTIONAL OPEN STATEMENTS FOR INPUT (UNIT 20) AND OUTPUT
C       (UNIT 21) CAN BE INSERTED HERE
            CALL   READ
            CALL   WRITE
            DO  1   I=1,NCYE-NCY
                CALL   CYCLE
    1           IF (NCY/NCYD*NCYD.EQ.NCY) CALL WRITE
```

```
            STOP
            END
C...SUBROUTINES:
            SUBROUTINE   GEN(ICOLS,ITOP,LITHN)
C...THIS SUBROUTINE GENERATES THE INITIAL STATE OF THE SYSTEM BY
C        PRODUCING A SLOPING WEDGE OF SEDIMENT.
C...VARIABLES PASSED TO SUBROUTINE ARE:
C      ICOLS = NUMBER OF COLUMNS
C      ITOP = LEVEL OF TOP OF LEFTMOST COLUMN
C      LITHN = SEDIMENT TYPE NUMBER (1 TO 5)
            COMMON   A,CF,CT,CD,CE(5),SI(5),VI,NCYE,NCYD,NCY,NCO,
          &      LEV(180),LCELL(180,132)
            NCO=ICOLS
            DO 2 I=1,NCO
               LEV(I)=(I.-1)*(ITOP-1.)/(NCO-1.)+ITOP+0.5
               DO 1 J=1,LEV(I)
      1           LCELL(I,J)=LITHN
      2        CONTINUE
            RETURN
            END
C
            SUBROUTINE   READ
C...THIS SUBROUTINE READS IN INITIAL STATE OF THE SYSTEM
C...VARIABLES USED ARE THE SAME AS IN MAIN PROGRAM
            COMMON   A,CF,CT,CD,CE(5),SI(5),VI,NCYE,NCYD,NCY,NCO,
          &      LEV(180),LCELL(180,132)
            CHARACTER*1   LITH(6),LSYMB(132)
            DATA   LITH/'+','o','*',':','|',' '/
            READ   (20,1)   A,CF,CT,CD,(CE(K),K=1,5),(SI(K),K=2,5),VI,
          &      NCYE,NCYD,NCY,NCO
      1     FORMAT   (4(22X,F10.3/)//,22X,5F7.3/27X,4F7.3/
          &       22X,F10.3/4(22X,I6/))
            IF (NCYD.LE.0)  NCYD=1
            DO 5 I=1,NCO
               LEV(I)=0
               READ  (20,2,END=6)   (LSYMB(J),J=1,132)
      2        FORMAT (1X,132A1)
               DO 4 J=1,132
                  IF(LSYMB(J).EQ.LITH(6))   GOTO  5
                  DO 3 K=1,5
      3              IF(LSYMB(J).EQ.LITH(K)) LCELL(I,J)=K
      4           LEV(I)=LEV(I)+1
      5        CONTINUE
      6     RETURN
            END
C
            SUBROUTINE   WRITE
C...THIS SUBROUTINE WRITES OUT CURRENT STATE OF THE SYSTEM
C...VARIABLES USED ARE THE SAME AS IN MAIN PROGRAM
            COMMON   A,CF,CT,CD,CE(5),SI(5),VI,NCYE,NCYD,NCY,NCO,
          &      LEV(180),LCELL(180,132)
            CHARACTER*1   LITH(6)
            DATA   LITH/'+','o','*',':','|',' '/
            WRITE   (21,1)   A,CF,CT,CD,(CE(K),K=1,5),(SI(K),K=2,5),VI,
          &      NCYE,NCYD,NCY,NCO
      1 FORMAT (' ACCELERATION             =',F10.3/
          &' COEFF OF FRICTION      =',F10.3/
          &'  COEFF OF  TRANSPORT    =',F10.3/
          &'  COEFF OF  DEPOSITION =',F10.3/
          &' LITHOLOGIES',9X,'= BASEM. GRAVEL    SAND    SILT    CLAY'/
```

```
        &'  SYMBOLS          ',9X,'= ++++++ oooooo ****** :::::: ||||||'/
        &'  COEFF OF EROSION       =',5F7.3/
        &'  INIT SEDIMENT INPUT    =',7X,4F7.2/
        &'  INIT FLUID VELOCITY    =',F10.3/
        &'  ENDING CYCLE #         =',16/
        &'  DISPLAY EVERY          =',16/
        &'  CURRENT CYCLE #        =',16/
        &'  TOTAL # OF SED COLS    =',16/)
          DO 3 I=1,NCO
              IF(LEV(I).EQ.0)   WRITE (21,2)
     2      FORMAT()
     3          IF(LEV(I).NE.0) WRITE (21,4)
     &          (LITH(LCELL(I,J)),J=1,LEV(I))
     4      FORMAT (1X,132A1)
          RETURN
          END
C
          SUBROUTINE  CYCLE
C...THIS SUBROUTINE EXECUTES ONE CYCLE OF THE SEDIMENTARY SYSTEM
C...VARIABLES USED ARE THE SAME AS IN MAIN PROGRAM PLUS:
C     V=VELOCITY
C       PF=POWER DISSIPATED BY FRICTION=CF*V**2
C       S(K)=SEDIMENT LOAD OF TYPE K
C       SS=TOTAL SEDIMENT LOAD=SUM OF S(K) FOR ALL K'S
C       ST=TOTAL AMOUNT OF SEDIMENT TO BE MOVED FROM BED TO FLOW
C       DV=DIFFERENCE IN LEVEL TO NEXT COLUMN=LEV(I)-LEV(I+1)
C       TT=TOTAL TRANSPORT CAPACITY=PF*CT
C       TD=TRANSPORT CAPACITY DIFFERENCE=TT-SS
          COMMON  A,CF,CT,CD,CE(5),SI(5),VI,NCYE,NCYD,NCY,NCO,
     &      LEV(180),LCELL(180,132)
          DIMENSION  S(5)
C...INITIALIZE
          NCY=NCY+1
          V=VI
          SS=0.
          DO 1 K=2,5
              S(K)=SI(K)
     1        SS=SS+S(K)
C...CYCLE
          DO 5 I=1,NCO-1
              DV=LEV(I)-LEV(I+1)
              IF(DV.GE.0.)   VN=V+SQRT(2*DV*A)
              IF(DV.LT.0.)   VN=V-SQRT(-2*DV*A)
              V=VN-V*VN*CF
              IF(V.LT.0.)   V=0.
              PF=V*V*CF
              TT=PF*CT
              TD=TT-SS
              IF(TD)   3,5,2
C...ERODE
     2      TD=TT-SS
              IF(LEV(I).EQ.0) GOTO 5
              K=LCELL(I,LEV(I))
              ST=TD*CE(K)
              DV=LEV(I)-LEV(I+1)
              IF(ST.LT.0.5) GOTO 5
              LEV(I)=LEV(I)-1
              IF(K.EQ.1)K=3
              S(K)=S(K)+1.
```

```
                SS=SS+1.
                GOTO  2
C...DEPOSIT
      3      TD=TT-SS
             ST=TD*CD
             IF(ST.GT.-0.5)  GOTO  5
C...USE  NEXT  LINE  TO  LIMIT  DEPOSITION  LEVEL
C            IF  (LEV(I).GE.25) GOTO 5
             DO  4  K=2,5
                IF  (S(K).LE.0.5)  GOTO  4
                S(K)=S(K)-1.
                SS=SS-1.
                LEV(I)=LEV(I)+1
                LCELL(I,LEV(I))=K
                GOTO  3
      4         CONTINUE
      5      CONTINUE
          RETURN
          END
```

APPENDIX B: SAMPLE INPUT AND OUTPUT

Several sample input files and their corresponding output files are shown below. Only the last cycle is shown for the output.

Example 1 (input): Basement wedge with no sediment input and initial flow velocity 1.0.

```
ACCELERATION              =      0.500
COEFF  OF  FRICTION       =      0.080
COEFF  OF  TRANSPORT      =      5.000
COEFF  OF  DEPOSITION  =      0.500
LITHOLOGIES               = BASEM. GRAVEL   SAND    SILT    CLAY
SYMBOLS                   = ++++++ oooooo  ******  ::::::  ||||||
COEFF  OF  EROSION        =   0.200   0.600   0.700   0.800   0.900
INIT SEDIMENT INPUT =              0.00    0.00    0.00    0.00
INIT FLUID VELOCITY =      1.000
ENDING CYCLE #            =      20
DISPLAY EVERY            =      4
CURRENT CYCLE #          =      0
TOTAL # OF SED COLS =      48
```

Example 1 (output): After 20 cycles, basement has been eroded. Product of erosion is sand that has been transported downstream.

```
ACCELERATION              =      0.500
COEFF  OF  FRICTION       =      0.080
COEFF  OF  TRANSPORT      =      5.000
COEFF  OF  DEPOSITION  =      0.500
LITHOLOGIES               = BASEM. GRAVEL   SAND    SILT    CLAY
SYMBOLS                   = ++++++ oooooo  ******  ::::::  ||||||
COEFF  OF  EROSION        =   0.200   0.600   0.700   0.800   0.900
INIT SEDIMENT INPUT =              0.00    0.00    0.00    0.00
INIT FLUID VELOCITY =      1.000
ENDING CYCLE #            =      20
DISPLAY EVERY            =      4
CURRENT CYCLE #          =      20
TOTAL # OF SED COLS =      48
```

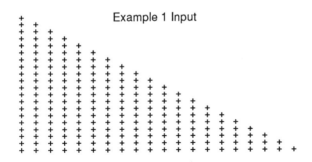

Example 1 Input

Example 1 Output

Example 2 (input): Basement wedge with input of gravel, sand, silt, and clay.

```
ACCELERATION              =        0.500
COEFF OF FRICTION         =        0.080
COEFF OF TRANSPORT        =        5.000
COEFF OF DEPOSITION       =        0.300
LITHOLOGIES               = BASEM.  GRAVEL    SAND     SILT     CLAY
SYMBOLS                   = ++++++  oooooo   ******   ::::::   ||||||
COEFF OF EROSION          =  0.200   0.500    0.600    0.700    0.800
INIT SEDIMENT INPUT  =                2.00     2.00     2.00     2.00
INIT FLUID VELOCITY  =        4.000
ENDING CYCLE #            =      100
DISPLAY EVERY            =       20
CURRENT CYCLE #          =        0
TOTAL # OF SED COLS =            48
```

Example 2 (output): After 100 cycles, deposition has produced an approximately exponential profile. Coarse sediment has been deposited near the source, while finer sediment
has been carried out farther.

```
ACCELERATION              =        0.500
COEFF OF FRICTION         =        0.080
COEFF OF TRANSPORT        =        5.000
COEFF OF DEPOSITION       =        0.300
LITHOLOGIES               = BASEM.  GRAVEL    SAND     SILT     CLAY
SYMBOLS                   = ++++++  oooooo   ******   ::::::   ||||||
COEFF OF EROSION          =  0.200   0.500    0.600    0.700    0.800
INIT SEDIMENT INPUT  =                2.00     2.00     2.00     2.00
INIT FLUID VELOCITY  =        4.000
ENDING CYCLE #            =      100
DISPLAY EVERY            =       20
CURRENT CYCLE #          =      100
TOTAL # OF SED COLS =            48
```

Example 2 Input

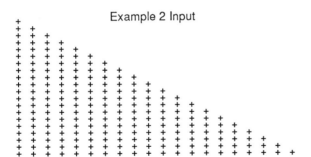

Example 2 Output

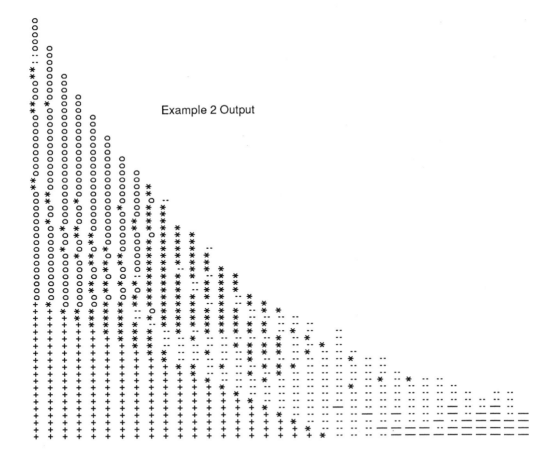

Example 3 (input): Initial wedge consisting of layers with high erosivity contrast.

```
ACCELERATION          =       0.500
COEFF  OF  FRICTION   =       0.080
COEFF  OF  TRANSPORT  =       5.000
COEFF  OF  DEPOSITION =       0.300
LITHOLOGIES           = BASEM. GRAVEL   SAND    SILT    CLAY
SYMBOLS               = ++++++ oooooo  ******  ::::::  ||||||
COEFF  OF  EROSION    =   0.050   0.700   0.800   0.900   1.000
INIT SEDIMENT INPUT   =           0.00    0.00    0.00    0.00
INIT FLUID VELOCITY   =       2.000
ENDING CYCLE #        =      20
DISPLAY EVERY         =       4
CURRENT CYCLE #       =       0
TOTAL # OF SED COLS   =      48
```

Example 3 (output): Differential erosion has produced a step profile, and erodible sediment has been transported downstream.

```
ACCELERATION          =       0.500
COEFF  OF  FRICTION   =       0.080
COEFF  OF  TRANSPORT  =       5.000
COEFF  OF  DEPOSITION =       0.300
LITHOLOGIES           = BASEM. GRAVEL   SAND    SILT    CLAY
SYMBOLS               = ++++++ oooooo  ******  ::::::  ||||||
COEFF  OF  EROSION    =   0.050   0.700   0.800   0.900   1.000
INIT SEDIMENT INPUT   =           0.00    0.00    0.00    0.00
INIT FLUID VELOCITY   =       2.000
ENDING CYCLE #        =      20
DISPLAY EVERY         =       4
CURRENT CYCLE #       =      20
TOTAL # OF SED COLS   =      48
```

```
+                                   Example 3 Input
+  +
+  +  +
+  +  +  +
+  +  +  +  +
-- -- -- -- -- --
-- -- -- -- -- -- --
-- -- -- -- -- -- -- --
-- -- -- -- -- -- -- -- --
+  +  +  +  +  +  +  +  +  +  +
+  +  +  +  +  +  +  +  +  +  +  +
+  +  +  +  +  +  +  +  +  +  +  +  +
+  +  +  +  +  +  +  +  +  +  +  +  +  +
-- -- -- -- -- -- -- -- -- -- -- -- -- -- --
-- -- -- -- -- -- -- -- -- -- -- -- -- -- -- --
-- -- -- -- -- -- -- -- -- -- -- -- -- -- -- -- --
-- -- -- -- -- -- -- -- -- -- -- -- -- -- -- -- -- --
```

```
+      +      +      +      +      Example 3 Output
--     --     --     --     --
--     --     --     --     --
--     --     --     --     --
--     --     --     --     --
+      +      +      +      +  +  +  +  +  +  +
+      +      +      +      +  +  +  +  +  +  +  +
+      +      +      +      +  +  +  +  +  +  +  +  +
+      +      +      +      +  +  +  +  +  +  +  +  +  +
-- +   --  +  --  +  --  +  -- -- -- -- -- -- -- -- -- --
-- +   --  +  --  +  --  +  -- -- -- -- -- -- -- -- -- --      -- -- -- --        -- -- -- -- -- --
```

Example 4 (input): Initial topography consists of a slope partly closed by a sill, forming a basin.

```
ACCELERATION              =        0.500
COEFF  OF  FRICTION       =        0.080
COEFF  OF  TRANSPORT      =        5.000
COEFF  OF  DEPOSITION  =           0.300
LITHOLOGIES            = BASEM.  GRAVEL   SAND    SILT    CLAY
SYMBOLS                = ++++++  oooooo  ******  ::::::  ||||||
COEFF  OF  EROSION     =   0.200   0.500   0.600   0.700   0.800
INIT SEDIMENT INPUT =               2.00    2.00    2.00    2.00
INIT FLUID VELOCITY =             4.000
ENDING CYCLE #            =       20
DISPLAY EVERY            =        4
CURRENT CYCLE #         =         0
TOTAL # OF SED COLS  =           48
```

Example 4 (output): Basin has filled with coarse sediment near the source, and finer sediment distally. Sill has been partly eroded and coarse sediment has been deposited downstream.

```
ACCELERATION              =        0.500
COEFF  OF  FRICTION       =        0.080
COEFF· OF  TRANSPORT      =        5.000
COEFF  OF  DEPOSITION  =           0.300
LITHOLOGIES            = BASEM.  GRAVEL   SAND    SILT    CLAY
SYMBOLS                = ++++++  oooooo  ******  ::::::  ||||||
COEFF  OF  EROSION     =   0.200   0.500   0.600   0.700   0.800
INIT SEDIMENT INPUT =               2.00    2.00    2.00    2.00
INIT FLUID VELOCITY =             4.000
ENDING CYCLE #            =       20
DISPLAY EVERY            =        4
CURRENT CYCLE #         =        20
TOTAL # OF SED COLS  =           48
```

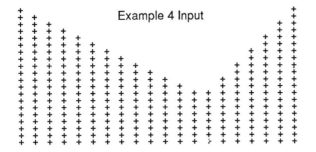

Example 4 Input

Example 4 Output

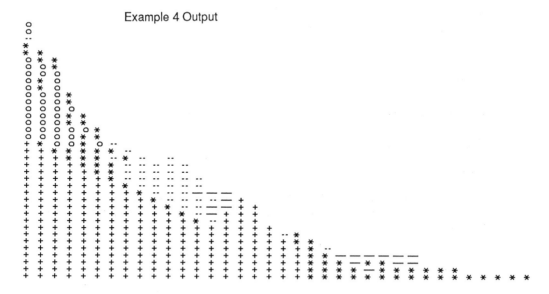

Example 5 (input): Initial state similar to Example 2, but subroutine CYCLE has been modified to forbid erosion above a certain level representing the "wave-base" level in the submarine part of a delta (see comments in subroutine CYCLE, Appendix A).

```
ACCELERATION          =      0.500
COEFF OF FRICTION     =      0.080
COEFF OF TRANSPORT    =      5.000
COEFF OF DEPOSITION   =      0.300
LITHOLOGIES           = BASEM. GRAVEL   SAND    SILT    CLAY
SYMBOLS               = ++++++ oooooo  ******  ::::::  ||||||
COEFF OF EROSION      =      0.200   0.500   0.600   0.700   0.800
INIT SEDIMENT INPUT   =              2.00    2.00    2.00    2.00
INIT FLUID VELOCITY   =      4.000
ENDING CYCLE #        =    100
DISPLAY EVERY         =     20
CURRENT CYCLE #       =      0
TOTAL # OF SED COLS   =     48
```

Example 5 (output): Sediment has been deposited forming typical coarsening upwards sequence.

```
ACCELERATION          =      0.500
COEFF OF FRICTION     =      0.080
COEFF OF TRANSPORT    =      5.000
COEFF OF DEPOSITION   =      0.300
LITHOLOGIES           = BASEM. GRAVEL   SAND    SILT    CLAY
SYMBOLS               = ++++++ oooooo  ******  ::::::  ||||||
COEFF OF EROSION      =      0.200   0.500   0.600   0.700   0.800
INIT SEDIMENT INPUT   =              2.00    2.00    2.00    2.00
INIT FLUID VELOCITY   =      4.000
ENDING CYCLE #        =    100
DISPLAY EVERY         =     20
CURRENT CYCLE #       =    100
TOTAL # OF SED COLS   =     48
```

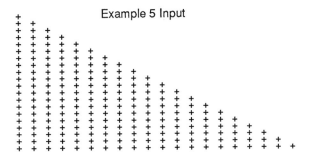

Example 5 Input

Example 5 Output

24

QUANTITATIVE MODELING OF CARBONATE STRATIGRAPHY AND WATER-DEPTH HISTORY USING DEPTH-DEPENDENT SEDIMENT ACCUMULATION FUNCTION

Raymond F. Gildner [1] and John L. Cisne [2]

[1]Department of Geological Sciences and the Institute for the Study of the Continents, Cornell University, Ithaca, NY USA
[2]Department of Geological Sciences, the Institute for the Study of the Continents and the Division of Biological Sciences, Cornell University, Ithaca, NY USA

ABSTRACT

Stratigraphic architecture and water-depth history that are expected in response to a sea-level rise can be determined from a simple mathematical model. In the past, most such models have used a description of depth-dependent carbonate accumulation appropriate only for reefs. We here propose a new general function (the epeiric sea function) appropriate for other carbonate environments, and compare it to a formulation appropriate for reef environments (the reef function). Numerical methods produce solutions of the model for discrete time intervals; a graphical method of solution allows a better intuitive understanding of the model.

Both the reef function and the epeiric sea function predict reasonable stratigraphic architectures. However, the reef function predicts that water depth would rapidly decrease and the reef surface be maintained at sea level—conditions necessary for the formation of many syndepositional features observed in ancient reefs, and that the shoreline will prograde monotonically into the basin in what is generally referred to as a regression. The epeiric sea function does not predict shoaling, but that water depth will be maintained at a shallow level. It predicts spiked water depth versus stratigraphic position curves in response to a smooth, sinusoidal sea-level rise rather than requiring a spiked variation in sea level. Both functions predict the occurrence and conditions necessary for keep-up, catch-up and give-up response patterns.

INTRODUCTION

Numerous authors have commented on the response of carbonate strata to increases in sea level. Schlager (1981) noted the paradox between the existence of drowned carbonate platforms and observed rates of sea-level rise and carbonate production. The paradox states that observed maximum rates of carbonate sediment production are so much greater than any plausible rate of sea-level rise that all reefs should be able to "keep-up" with sea-level rise and no reefs should drown. Kendall and Schlager (1981) extended the paradox to include carbonate platforms. These discussions were qualitative due to the lack of precise data for carbonate production as a function of water depth.

Stratigraphic responses of carbonates to sea-level rise have been grouped into three classes: keep-up, catch-up and give-up (Neumann and Macintyre, 1985). Keep-up carbonates maintain a relatively constant and shallow water depth throughout the duration of a sea-level rise. Give-up carbonates are unable to maintain their water depth throughout a sea-level rise and their water depth increases through time. In the case of reefs or shallow water carbonate platforms, they "drown" and become relict features. Catch-up carbonates behave in an intermediate manner; although they initially may accumulate at a rate less than that of sea-level rise and thus deepen, they recover, begin to out-pace the sea-level rise, and become more shallow.

A corollary to the paradox is that not only should carbonate banks not drown, they should always reach sea level. The tendency of carbonates in lagoonal environments to remain below sea level has been ascribed to environmental factors in the lagoon such as hypersalinity, elevated temperatures, nutrient poisoning and turbidity. Such adverse conditions would have been especially prevalent in epeiric seas (Shaw, 1964), and the geologic record has many examples of carbonate strata that accumulated in shallow water for millions of years, in effect keeping-up but not shoaling. In fact, most carbonates that accumulated following a transgression (that is, on depositional surfaces that were initially above sea level) do not record intertidal or supratidal environments, but conditions of greater water depth. In strata of the Galena Group of the upper Mississippi River Valley, shallow water depths were maintained without subaerial exposure over the 8 to 13 Myr interval represented by the upper Middle Ordovician (Sloan, 1987; Delgado, 1983). Superimposed on the shallow water depth are several stratigraphically thin, "short" intervals of shallowing, giving the water-depth curve a spiked shape (Fig. 1). This spiked nature of the curve is especially noticeable at the base of the section from Guttenberg, Iowa (Cisne and Gildner, in press).

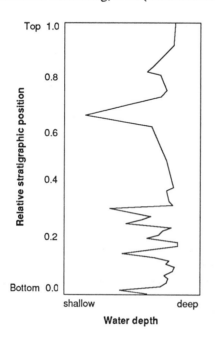

Figure 1. Plot of the water-depth history for the Galena Group (upper Middle Ordovician) at Guttenberg, Iowa (Cisne and Gildner, accepted). Depths were determined using the ordination technique of reciprocal averaging (Cisne and Gildner, accepted; Cisne et al., 1984; Cisne and Rabe 1978). Note the spiked nature of the stratigraphically short shallowing events.

During extended periods in the Silurian, strata maintained relatively constant, shallow water depths without shoaling. Water depth increases and decreases in patterns which can be recognized between distant areas both within the North American continent and among continents, indicating that eustasy was an important control (Johnson, 1981; Jia-yu et al., 1984).

We are, then, presented with two different keep-up responses of carbonates to a sea-level rise: with shoaling, appropriate for reefs; and without shoaling, appropriate for other environments such as epeiric seas. Although previous mathematical models have adequately described the former response (e.g., Turcotte and Bernthal, 1984; Strobel et al., this volume), few have addressed the latter.

We apply a numerical model to describe the relationship between water-depth history as preserved in the stratigraphic record and sea-level history throughout the time of deposition. The approach used in this study is a two-part mathematical model. We mathematically represent the relationship between water depth, sea-level rise and sediment accumulation, utilizing functions that relate water depth to sediment accumulation rate. We substitute different functions into the model and compare the predicted stratigraphic architecture and stratigraphically recorded water-depth history to the geologic record. Although precise information on absolute water depth, sea level and time is difficult to collect, and presently available observations from the geologic record cannot support a rigorous test of the model for exact values of the function parameters, modeling can be used to distinguish the nature and behavior of sediment accumulation with sufficient resolution to allow comparisons between functions.

A possible point of confusion is the distinction between a period of sea-level rise and a transgression. In this paper, and following the strict definitions, the terms "transgression" and "regression" refer to the change in shoreline position landward and seaward, respectively. As such, they are related to sea-level change, but are not equivalent (Vail et al., 1977).

MODEL AND FUNCTIONS

Modeling carbonate sediment accumulation must be approached with a different conceptual base than that of clastics due to inherent differences in depositional processes. Clastic sedimentation is essentially a problem of mass balance: the mass of sediment entering an area must equal the sum of the mass of the sediment deposited or eroded and of the sediment leaving the area. Carbonate sediments, on the other hand, are biogenic and autochthonous. For this reason, the mass balance and fluid mechanics methods which have proven so powerful for modeling clastic sedimentation (e.g., Turcotte and Kenyon, 1984) are largely irrelevant to the modeling of carbonate sediment accumulation. Instead, water depth is usually considered the primary factor in carbonate sediment accumulation (e.g., Turcotte and Bernthal, 1984; Bice, 1986; Read et al., 1986; Bova and Read, 1987).

A sketch of the progress of a transgression at two times (Fig. 2) shows that the change in water depth, w, is the difference between the amount of relative sea-level rise, L (including both eustatic changes and tectonic subsidence), and the amount of sediment accumulated, S. The terms α and β are constants of isostatic adjustment to the load of water and sediment, with densities ρ_m, ρ_s, and ρ_w for the asthenosphere, sediment and sea water, respectively. It should be noted that the values of α and β do not affect the behavior of the solution, only the magnitude of the results, as long as $\rho_m > \rho_s, > \rho_w$. From Figure 2 and adding terms for isostatic adjustment,

$$\Delta w = \alpha \Delta L - \beta \Delta S \tag{1a}$$

$$\alpha = \frac{\rho_m}{\rho_m - \rho_w} \tag{1b}$$

$$\beta = \frac{\rho_m - \rho_s}{\rho_m - \rho_w} \tag{1c}$$

or, in instantaneous terms and using dot notation, writing $\frac{dx}{dt}$ as $\dot{x}$,

$$\dot{w} = \alpha \dot{L} - \beta \dot{S} \tag{2}$$

Equations 1a and 2 are different representations of the same model (see Cisne et al., 1984, for a complete derivation).

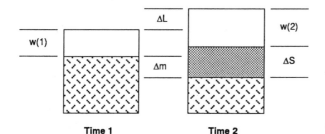

Figure 2. Cartoon depicting the origins of Equation 1. (The term Δm describes the lithospheric deflection and cancels out in the derivation of Equation 1. See Cisne et al., 1984, for a complete derivation.)

The model points out that it is the *rate* of sea-level change, and not its magnitude, which determines the water-depth history. Deepening-upwards does not indicate highstands of sea level, but periods of sea-level rise at rates higher than sediment accumulation. Similarly, shallowing-upwards does not necessarily indicate a sea-level fall, but only that sediment is accumulating faster than sea level is rising.

In practice, the derivative with respect to time of an arbitrarily defined driving function representing sea level, L(t), and an assumed depth-dependent carbonate sediment accumulation function, $\dot{S}$ (w), are inserted into Equation 2. This study uses two different sea-level functions to demonstrate the responses to each as recorded in the synthetic stratigraphic column. The first represents a linear sea-level rise and is used to demonstrate the graphic solution of the model:

$$L(t) = A\, t \tag{3a}$$

$$\dot{L}(t) = A \tag{3b}$$

where A is an amplitude variable.

The second driving function represents a sinusoidally increasing sea-level rise and is used to demonstrate the numerical solution of the model:

$$L(t) = A\, t + B \sin (\omega t) \tag{4a}$$

$$\dot{L}(t) = A + B\omega\cos(\omega t) \qquad\qquad\qquad (4b)$$

where A and B are amplitude variables, ω is the period of the oscillations, and t is time. Others (e.g., Turcotte and Bernthal, 1984; Read et al., 1986) have used the linear term, A, to represent tectonic subsidence, and the oscillating term, $B\sin(\omega t)$, to represent eustatic changes. Alternatively, the linear term may represent a small segment of a low-order eustatic cycle, and the oscillating term may represent several higher-order eustatic cycles.

One depth-dependent carbonate sediment accumulation function chosen for this study is from the literature and the other is introduced in this paper. Some authors have directly translated Schlager's (1981) carbonate production curves as a depth-dependent carbonate sediment accumulation function (herein referred to as the reef function) of a form similar to:

$$\dot{S}(w) = \dot{S}_M \qquad\qquad 0 < w < w_M \qquad\qquad (5a)$$

$$\dot{S}(w) = \dot{S}_M e^{(-w/w_M)} \qquad\qquad w > w_M \qquad\qquad (5b)$$

That is, the maximum sediment accumulation rate, $\dot{S}_M$, is reached at all water depths more shallow than the critical water depth, w_M. Below the critical water depth, the sediment accumulation rate decreases monotonically and possibly exponentially (Neumann and Macintyre, 1985). Some models have assumed a constant sediment accumulation rate independent of time (e.g., Bice, 1986; Turcotte and Bernthal, 1984; Read et al., 1986; Bova and Read, 1987). Furthermore, a constant carbonate sediment accumulation rate is an essential part of the graphic correlation method (Shaw, 1964; Miller, 1977). Whether or not claiming to follow a sediment accumulation model of the form of Equation 5, the use of a constant sediment accumulation rate is *de facto* subscription to the concept of the reef function.

Cisne et al. (1984) proposed a function specifically to describe sediment accumulation of shallow water carbonates, that is, those deposited in water depths more shallow than w_M as defined above:

$$\dot{S}(w) = c\,w \qquad\qquad 0 < w < w_M \qquad\qquad (6a)$$

$$c = \left(\frac{\dot{S}_M}{w_M}\right) \qquad\qquad 0 < w < w_M \qquad\qquad (6b)$$

In this function, sediment accumulation rate increases linearly from zero at sea level to its maximum at w_M. This formulation was created to describe depth-dependent carbonate sediment accumulation only in shallow water, and cannot be extended into deeper water beyond the depth of maximum sediment accumulation rate, w_M, where the function would predict impossibly high sediment accumulation rates. Cisne et al. (1984) determined and discussed the expected stratigraphic responses to a sea-level rise for this function.

To extend the function to sediment accumulation in deeper water, we propose a function which incorporates the exponential tail of the function from the reef function. A continuous formulation of the function (herein referred to as the epeiric sea function) is:

$$\dot{S}(w) = \dot{S}_M\left(\frac{w}{w_M}\right)e^{(1 - w/w_M)} \qquad\qquad (7)$$

The sediment accumulation rate increases with water depth to the maximum at w_M, followed by an exponential decrease of the sediment accumulation rate with increasing water depth. Bova and Read (1987) used a depth-dependent sediment accumulation function with three intervals of constant sediment accumulation rate which shows a similar trend: a maximum sediment accumulation rate in the shallow subtidal, with decreasing sediment accumulation rates in both more shallow and deeper waters.

Whereas the reef function describes reef environments very well, with ideal conditions for carbonate production and accumulation, the epeiric sea function is intended to describe carbonate sediment accumulation in less favorable conditions. The two functions are complementary and may represent two points along a continuum of functions. If true, the existence of the continuum argues strongly that factors other than water depth are necessary for a general carbonate sediment accumulation function.

The functions used here could easily be replaced with any number of formulations with similar behavior. To avoid the possibility that differences in the model results reflect our choice of formulation rather than the functions' behaviors, we reformulate the reef function in Equation 5 to more closely parallel the epeiric sea function.

$$\dot{S}(w) = \dot{S}_M \qquad\qquad\qquad 0 < w < w_M \qquad\qquad (8a)$$

$$\dot{S}(w) = \dot{S}(w) = \dot{S}_M \left(\frac{w}{w_M}\right) e^{(1 - w/w_M)} \qquad w > w_M \qquad\qquad (8b)$$

Both depth-dependent sediment accumulation functions are plotted in Figure 3. Note that both can be written in a general form reflecting the depth-dependent nature of our modeling approach.

$$\dot{S}(w) = \dot{S}_M \, f\left(\frac{w}{w_M}\right) \qquad\qquad\qquad\qquad (9)$$

Equation 1 is a first-order differential equation into which is substituted an arbitrary driving function for sea-level change and the depth-dependent carbonate sediment accumulation function whose behavior we wish to study. The reef and the epeiric sea formulations of the depth-dependent carbonate sediment accumulation function result in differential equations which can only be solved with numerical methods. The use of numerical solutions makes it difficult to gain an intuitive understanding for the equations; in an attempt to provide the reader with this intuition we present a graphical solution.

Relative sedimentation rate

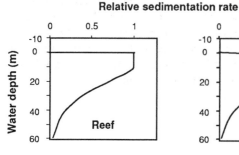 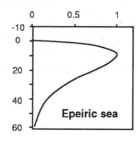

Figure 3. Plots of depth-dependent carbonate sediment accumulation functions used in the model. Ordinate is the sediment accumulation rate relative to $\dot{S}_M$. 3a) reef function based on observations of carbonate production presented by Schlager (1981). 3b) epeiric sea function presented in this study.

GRAPHICAL SOLUTION

For keep-up carbonates, the rate of change of water depth with respect to time is zero, such that the left-hand side of Equation 1 equals zero. The isostatically adjusted rates of carbonate sediment accumulation and relative sea-level rise must be equal.

$$\alpha \dot{L} = \beta \dot{S} \tag{10}$$

For one point in time, $\dot{L}$, being a variable only of time, is a constant and can be presented as a fraction of $\dot{S}_M$ (e.g., $\dot{L} = k \dot{S}_M$), such that

$$\frac{\dot{L}}{\dot{S}_M} = k = \frac{\beta}{\alpha} f\left(\frac{w}{w_M}\right) \tag{11}$$

Thus, the keep-up water depth is a function of the ratio between the rate of sea-level rise and the maximum sediment accumulation rate and this ratio can be shown graphically as a plot of water depth versus the two rates (Fig. 4). Since the rate of sea-level change, $\dot{L}$, does not vary with water depth, it plots as a horizontal line (Fig. 4a). Water depth decreases wherever the sediment accumulation rate is greater than the rate of sea-level rise; that is, wherever the curve representing the sediment accumulation rate lies above the line representing the rate of sea-level rise. Conversely, water depth increases wherever the sediment accumulation rate is less than the rate of sea-level rise; that is, wherever the curve representing the sediment accumulation rate lies below the line representing the rate of sea-level rise. The depth-dependent sediment accumulation rate, $\dot{S}$, is depicted with the curved line (Fig. 4b); it is intended to be a general curve and not related to any particular model. Where the two lines intersect, water depth remains constant with time.

The keep-up condition is reached when the line representing the sediment accumulation rate crosses the line representing the rate of sea-level rise with a positive slope. Consider the case where the water depth initially is greater than the water depth corresponding to the point of intersection between the two curves. In this circumstance, water depth decreases until the depth corresponding to the point of intersection is reached. Similarly, in the case where the water depth initially is less than the depth corresponding to the point of intersection of the two curves, water depth increases to this depth. This depth is an asymptotically stable solution of Equation 9; it is the keep-up depth. The same stable water depth was found by Cisne et al. (1984) in the exact solution to their linear function, and is shown in the following numerical section for the epeiric sea function and in an unusual form for the reef function.

If the sediment accumulation rate at sea level is less than the maximum sediment accumulation rate and the rate of sea-level change is greater than zero but less than the maximum, isostatically adjusted, carbonate sediment accumulation rate, a keep-up water depth will exist. The major distinction between the reef and epeiric sea function in this regard is that the reef function predicts the keep-up water depth will be at sea level, whereas the epeiric sea function predicts the keep-up water depth will vary with the rate of sea-level rise, but the depositional surface will remain below sea level.

It is also possible that the curve representing the depth-dependent sediment accumulation function will intersect the line representing the rate of sea-level change where it has a negative slope. Under these circumstances the reverse argument from the stable water depth may be applied. Should the water depth initially be slightly deeper than the depth corresponding to the point of intersection, the water depth will become still deeper, moving away from that depth. In the case where the initial water depth is more shallow than that of the point of intersection, water depth will decrease. This depth is an unstable solution of Equation 9. This intersection point marks the boundary between those depths which become more shallow and those which become deeper. It is the boundary between the catch-up and the give-up conditions.

There are two conditions where the two curves will not intersect. If the rate of sea-level rise is greater than the maximum sediment accumulation rate, the give-up condition is prescribed and water depth will increase. If the rate of sea-level rise is less than zero, the location is exposed and elevation will increase.

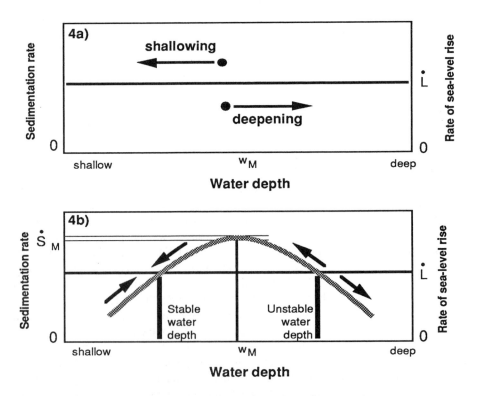

Figure 4. The graphic method of solution of Equation 1. 4a) Should sediment accumulation rate exceed the rate of sea-level rise (horizontal line), the water depth at the site decreases. conversely, should the accumulation rate be less than the rate of sea-level rise, the water depth increases. 4b) Using a depth-dependent sediment accumulation function (grey curve), the sediment accumulation rate follows the curve to the left if the accumulation rate is greater than the rate of sea-level rise, and to the right if it is less. The accumulation rate will converge to the rate of sea-level rise at the keep-up water depth (convergent double arrow) and diverge from the unstable water depth (divergent double arrow). See text for a more detailed explanation.

To demonstrate the graphical solution, numerical solutions were found for the model using the two sediment accumulation functions and a constant rate of sea-level change. Water-depth histories of points at various initial water depths, as determined using numerical methods, are shown (Fig. 5, right side) with the stable and unstable solutions found using the graphical method (Fig. 5, left side). The vertical line indicates the isostatically adjusted rate of sea-level rise; horizontal gray lines between the two plots indicate positions of stable and unstable water depths. The plot of the numerical solutions shows the asymptotic nature of the stable, keep-up water depth.

One consequence of the fact that the keep-up water depth is a function of the ratio between the rate of sea-level rise and the maximum sediment accumulation rate is the interpretation that periods of different keep-up water depths indicate periods of differing rates of sea-level rise. Even when superimposed upon different local subsidence rates, such shifts would indicate broadly contemporaneous periods and be correlative, as studies of Silurian strata have shown (Johnson, 1981; Jia-yu et al., 1984).

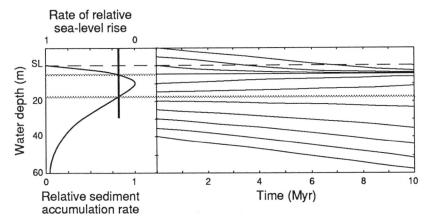

Figure 5. Demonstration of the graphic solution for a constant rate of sea-level rise using the epeiric sea function for depth-dependent sediment accumulation presented in the text. The left plot shows the depth-dependent sediment accumulation function, sediment accumulation rate (x-axis) versus water depth (y-axis), with the stable and unstable solutions shown by horizontal gray lines and rate of sea-level rise shown by the vertical line. The right plot shows the water-depth history, time (x-axis) versus water depth (y-axis, at the same scale as the left hand plot). Each curve presents the history of a location with an initial water depth as indicated by its point of intersection on the y-axis. The dashed horizontal line indicates sea level.

NUMERICAL SOLUTION

Substituting $\dot{L}(t)$ and $\dot{S}(w)$ into Equation 1 produces a first-order linear differential equation without exact solutions. There are many different methods available to determine solutions to such equations. We used a four-term Runga-Kutta method, using 1,000 iterations for time steps of 0.1 Myr (for an introduction to the Runga-Kutta method, see Braun, 1975; for a more thorough discussion, see Press et al., 1986). The Runga-Kutta method was chosen for its simplicity, power and familiarity. The discontinuous nature of the reef function presented a complication for the method (the discontinuity of the depth-dependent sediment accumulation at w = 0 causes the solution to converge poorly), but this did not seriously affect the numerical results.

Each sediment accumulation function was tested using a sinusoidally increasing sea level with the parameters listed in Table 1. Those working in modern carbonate environments may be surprised at the low values we chose for both the maximum sediment accumulation rate, $\dot{S}_M$, and the rate of sea-level change. Part of the apparent disagreement lies in the difference between carbonate *sedimentation* rates and carbonate *sediment accumulation* rates (sedimentation less erosion). We obtained the values from observation of a major outcrop of the Galena Group at Guttenberg, Iowa, whose water-depth history is shown in Figure 1. There, approximately 30 m of strata accumulated in about 10 Myr, giving us a rough measure of the rates.

Table 1: Model and function parameters used for demonstration and discussion.

Isostatic terms α, β	Sea level $\dot{L}(t)$	Sediment accumulation rate $\dot{S}(w)$
$\rho_m = 3300$ kg/m^3	A = 2.16 m/Myr	$\dot{S}_M = 7.21$ m/Myr
$\rho_s = 2100$ kg/m^3	B = 2.16 m/Myr	$w_M = 10$ m
$\rho_w = 1030$ kg/m^3	$\omega = 0.2$ Myr^{-1}	

MODEL RESULTS AND DISCUSSION

Results of the numerical solution of the model for both functions are shown in Figure 6. The solution for the reef function demonstrates the paradox stated earlier in this paper and its corollary: all sites at and above sea level at the beginning of the transgression are never submerged more than is necessary for the initiation of carbonate sediment accumulation, after which they keep-up with sea level. Furthermore, those sites with an initial water depth above the unstable water depth, as defined previously, rapidly become more shallow and soon reach sea level. The solution for the epeiric sea function shows the convergence of water depths to the stable water depth.

In both functions, fluctuations of sea level are reflected in those sediments deposited below sea level, as indicated by the undulations in the curves in Figure 6. However in the reef function, after sediments reach sea level, water depth no longer fluctuates. After shoaling, the stratigraphic record no longer contains any information about sea level change.

Inverting Equation 1 and substituting water depth and sea-level values from the solution of the models allows us to predict stratigraphic architecture. To convert the one-dimensional model results to two-dimensional synthetic stratigraphic architecture, we use a conceptual devise that we refer to as a "pseudo-cross-section" (Fig. 7). Instead of using horizontal location as the ordinate, as one would in the two-dimensional case, we use the initial water depth. The resulting plot of elevation versus initial water depth would be a true synthetic cross section only if a constant slope is assumed. A rough idea of facies is implied by shading the strata that were deposited in the shallow subtidal, defined as a water depth of less than 3 m.

The "pseudo-cross-section" constructed using the reef function predicts the development of a platform at sea level which progrades into the basin, recording the period of sea-level rise as a regression. Again, this is a demonstration of the paradox stated earlier in this paper. Coupled with water depth information from the numerical solutions, we see that the entire prograding platform is deposited at sea level. The "pseudo-cross-section" shows toplap at the edge of the platform at every instance of constant relative sea level (note toplap between the third and fourth isochrons from the base at an initial water depth of 5 m, and between the seventh and eighth isochrons at an initial water depth of 10 m). The toplap is accompanied by exposure of the platform's surface, and presumably the formation of observable lithologic effects such as erosion, dolomitization, karstification, evaporite deposition and mineralization. These features are often found in reef environments, but are less common in other carbonate environments.

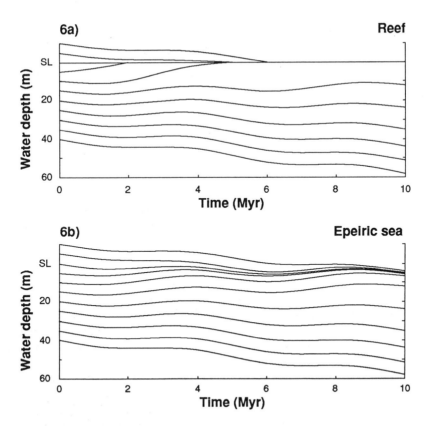

Figure 6. Numerically derived water-depth histories for a sinusoidally increasing sea-level rise using 6a) reef function and 6b) epeiric sea function.

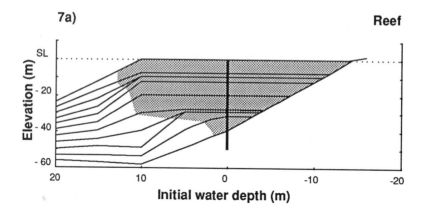

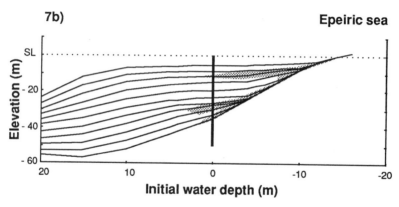

Figure 7. "Pseudo-cross-sections" derived from the water-depth histories for a sinusoidally increasing sea-level rise (Fig. 6). 7a) reef function and 7b) epeiric sea function. Solid curves represent equally spaced isochrons, dotted horizontal line indicates sea level and the vertical solid lines indicate the stratigraphic columns shown in Figure 8. Gray areas are those strata deposited in less than 3 m of water.

The "pseudo-cross-section" constructed using the epeiric sea function resembles that of the reef function, with the exception that the platform is subtidal and records the period of sea-level rise as a transgression, rather than a regression. The bathymetry across the platform edge is less sharp, but this might not be detectable in the geologic record. The major differences between the two lie in the predicted internal geometry and facies patterns. Unlike the "pseudo-cross-section" constructed using the reef function, that of the epeiric sea function shows no toplap seaward of the initial shoreline and predicts that submarine conditions are maintained throughout the section. The facies pattern simulated by the epeiric sea function shows that shallow intertidal facies are restricted to a few thin but laterally widespread layers within the stratigraphic section. However, a constant or decreasing relative sea level might be expected to cause the cessation of sediment accumulation long enough for submarine hardgrounds and corrosion surfaces to form.

The widespread and long term uniformity of conditions predicted by the epeiric sea function is reminiscent of "layer cake" stratigraphy. Something very like "layer cake" stratigraphy occurs in the Galena Group of the upper Mississippi River Valley where facies persist across the outcrop area, individual beds abundant in one taxon can be traced for hundreds of kilometers, and isochronous volcanic ash beds separate identical beds from outcrop to outcrop. Corrosion surfaces in the Galena Group have been attributed to submarine origin (Delgado, 1983), and may be a reflection of constant relative sea level for a period of time.

Water-depth histories simulated by sediment accumulation under the reef and epeiric models, are shown in Figure 8 as water depth versus stratigraphic position, the same format in which we presented the water-depth history of the Guttenberg, Iowa, data (Fig. 1). The information presented is for the point initially at sea level in Figure 6, except that water depths are now plotted against stratigraphic accumulation rather than time. A site that was at the shoreline at the beginning of the period of sea-level rise has distinctly different histories using the two functions. Sediments accumulated in the reef function were at sea level, except for two brief intervals above sea level which would be recorded in the stratigraphic column by the post-depositional features mentioned above. Sediment accumulation in the epeiric sea function would result in a water-depth curve with spikes corresponding to periods of constant relative sea level or sea-level fall.

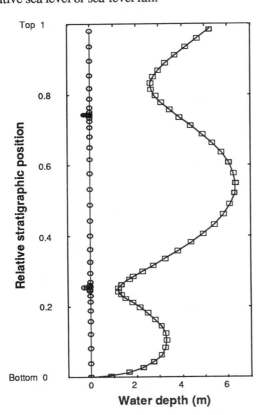

Figure 8. Predicted water depth as recorded in a stratigraphic section at a location that was along the shoreline at the beginning of the period of sea-level rise, using the reef function (circles, above) and the epeiric sea function (squares, below). Model and function parameters are the same as those in Figures 6 and 7, and shown in Table 1.

Although the spikes in Figure 8 are broader, they resemble the spikes in the water-depth curve for the Galena Group at Guttenberg, Iowa (Fig. 1). The spikes in the Ordovician could be interpreted to be responses to brief eustatic falls (e.g., Gildner and Cisne, 1987; Cisne and Gildner, accepted). However, given the model results of the epeiric sea function, the Ordovician spikes may record a longer time period than would first be assumed by the thickness of strata involved, and may record a smooth, less catastrophic eustatic change.

The model presented here does not incorporate lag-times, the delay between submergence and the initiation of carbonate sediment deposition. In the geologic case of a rapid eustatic rise associated with the melting of glaciers, the rate of sea-level change is very high, and water depth increases rapidly. Often in such cases sediment accumulation is initiated in the deeper subtidal, indicating a delay between submergence and the onset of sediment accumulation. This delay has been successfully modeled using a lag-time between submergence and the initiation of sediment accumulation (Bova and Read, 1987).

SUMMARY

Using a mathematical model for the response of water depth to sediment accumulation and sea-level change, we have compared two simple functions which describe the depth-dependent variation in the carbonate sediment accumulation rate. The model points out that it is the rate, not the magnitude, of sea-level change that determines water-depth history. We have developed a simple graphical technique to predict the keep-up water depth and the boundary between the give-up and catch-up conditions using any depth-dependent sediment accumulation function. Numerical methods, however, are necessary to predict the water-depth history.

The usual formulation of depth-dependent carbonate sediment accumulation (reef function) predicts features that are supported by the geologic record for reef environments, but are uncommon in other carbonate environments: syndepositional features reflecting subaerial exposure and the ubiquity of regressive stratigraphic sequences and of shallow subtidal and intertidal sediments.

The epeiric sea formulation of depth-dependent sediment accumulation predicts stratigraphic architectures and facies which are in reasonable agreement with the geologic record for carbonate environments other than reefs. The predicted water-depth history for a sinusoidally increasing sea level is similar in form to the observed water-depth history of an Ordovician stratigraphic section in the upper Mississippi River Valley. By increasing or decreasing the rate of sea-level rise, the water depth which can be maintained will vary. The pattern recorded from Silurian strata in various locations around the world (Johnson, 1981; Jia-yu et al., 1984) is of this nature. Deeper and more shallow water depths may reflect differing rates of sea-level change rather than higher and lower eustatic levels.

It remains to be seen whether the two depth-dependent sediment accumulation functions are descriptions of two distinctly different processes or two endpoints of a continuum of environmental conditions.

ACKNOWLEDGMENTS

This paper grew from poster sessions presented at the Geological Society of America Annual Meeting in Phoenix, Arizona and the QDS Workshop. The authors would like to thank our visitors for their discussion and suggestions. K. McGuirk, Dr. Teresa Jordan, S. Ackerly, J. Nicholson and C. Caruso of Cornell University reviewed early versions of the manuscript. R.K. Matthews, Tim Cross, and Chris Kendall reviewed the submitted manuscript. The extent to which this paper is comprehensible stems from their helpful criticisms and suggestions, for which the authors are very grateful. Chris Kendall also unknowingly made a comment which greatly illuminated some points to the first author. This work was completed with assistance of NSF grant EAR–8407723.

REFERENCES CITED

Bice, D.M., 1986, Computer simulation of prograding and retrograding carbonate platform margins: Geological Society of America Abstracts with Programs, v. 18, p. 540.

Bova, J.A., and Read, J.F., 1987, Incipiently drowned facies within a cyclic peritidal ramp sequence, Early Ordovician Chepultepec interval, Virginia Appalachians: Geological Society of America Bulletin, v. 98, p. 714-727.

Braun, M., 1976. Differential equations and their applications: New York, Springer-Verlag, 518 p.

Cisne, J.L., 1985, Depth-dependent sedimentation and the flexural edge effect in epeiric seas: Measuring water depth relative to the lithosphere's flexural wavelength: Journal of Geology, v. 93, p. 567-576.

Cisne, J.L., and Gildner, R.F., in press, Measurement of sea level change in epeiric seas: The Middle Ordovician transgression in the North American Midcontinent, *in* Wilgus, C., ed., Sea level change—An integrated approach: Society of Economic Paleontologists and Mineralogists Special Publication 42.

Cisne, J.L., Gildner, R.F., and Rabe, B.D., 1984, Epeiric sedimentation and sea level: Synthetic ecostratigraphy: Lethaia, v. 17, p. 267-288.

Cisne, J.L., and Rabe, B.D., 1978, Coenocorrelation: Gradient analysis of fossil communities and its applications in stratigraphy: Lethaia, v. 11, p. 341-364.

Delgado, D.J., 1983, Deposition and diagenesis of the Galena Group in the Upper Mississippi Valley, *in* Delgado, D.J., ed., Ordovician Galena Group of the Upper Mississippi Valley—Deposition, diagenesis, and paleoecology: Guidebook, 13th Annual Field Conference Great Lakes Section, Society of Economic Paleontologists and Mineralogists, p. A1-A17.

Gildner, R.F., and Cisne, J.L., 1987, Correlation of the Galena Group by crosscorrelation of sea level curves: Geological Society of America Abstracts with Programs, v. 19, p. 200.

Jia-yu, R., Johnson, M., and Xue-chang, Y., 1984, Early Silurian (Llandovery) sea-level changes in the upper Yangzi region of Central and Southwestern China (in Chinese and English): Acta Palaeontologica Sinica, v. 23, p. 672-698.

Johnson, M.E., 1981, Correlation of Lower Silurian strata from the Michigan Upper Peninsula to Manitoulin Island: Canadian Journal of Earth Sciences, v. 18, p. 869-883.

Kendall, C.G.St.C., and Schlager, W., 1981, Carbonates and relative changes in sea level: Marine Geology, v. 44, p. 181-212.

Miller, F.X., 1977, The graphic correlation method in biostratigraphy, *in* Kauffman, E.G., and Hazel, J.E., eds., Concepts and methods of biostratigraphy: Stroudsberg, Pennsylvania, Dowden, Hutchinson and Ross, p. 165-186.

Neumann, A.C., and Macintyre, I., 1985, Reef response to sea-level rise: Keep-up, catch-up and give-up, *in* Proceedings of the Fifth International Coral Reef Congress: Tahiti, v. 3, p. 105-110.

Press, W.H., Flannery, B.P., Teukolsky, S.A., and Vetterling, W.T., 1986, Numerical recipes: The art of scientific computing: New York, Cambridge University Press, 818 p.

Read, J.F., Grotzinger, J.P., Bova J.A., and Koerschner, W.F., 1986, Models for generation of carbonate cycles: Geology, v. 14, p. 107-110.

Schlager, W., 1981, The paradox of drowned reefs and carbonate platforms: Geological Society of America Bulletin, v. 92, p. 197-211.

Shaw, A.B., 1964, Time in stratigraphy: New York, McGraw-Hill, 365 p.

Sloan, R.E., 1987, Tectonics, biostratigraphy and lithostratigraphy of the Middle and Late Ordovician of the Upper Mississippi Valley, *in* Sloan, R.E., ed., Middle and Late Ordovician lithostratigraphy and biostratigraphy of the Upper Mississippi Valley: Minnesota Geological Survey Report of Investigations 35, p. 7-20.

Turcotte, D.L., and Kenyon, P.M., 1984, Synthetic passive margin stratigraphy: American Association of Petroleum Geologists Bulletin, v. 68, p. 768-775.

Turcotte, D.L., and Bernthal, M.J., 1984, Synthetic coral-reef terraces and variations of Quaternary sea level: Earth and Planetary Science Letters, v. 70:121-128.

Vail, P.R., Mitchum, R.M., Jr., and Thompson, S., III, 1977, Seismic stratigraphy and global changes of sea level. Part 3: Relative changes of sea level from coastal onlap, *in*, Payton, C.E., ed., Seismic Stratigraphy—Applications to hydrocarbon exploration: American Association of Petroleum Geologists Memoir 26, p. 63-82.

25

INTERACTIVE SIMULATION (SED-PAK) OF CLASTIC AND CARBONATE SEDIMENTATION IN SHELF TO BASIN SETTINGS

John Strobel [1], Freddy Soewito [2], Christopher G. St. C. Kendall [1], Gautam Biswas [3], James Bezdek [4], and Robert Cannon [2]

[1]Department of Geological Sciences, University of South Carolina, Columbia SC 29208 USA
[2]Department of Computer Science, University of South Carolina, Columbia SC 29208 USA
[3]Department of Computer Science, Vanderbilt University, Nashville, TN 37235 USA
[4]Boeing High Technology Center, PO Box 24969, MS 7J-24, Seattle WA 98124 USA

ABSTRACT

SED-pak is an interactive computer simulation program that tracks the sedimentary geometries produced by the filling of a two-dimensional basin from both sides, with a combination of clastic sediment and *in situ* carbonate growth. The simulation program is implemented in "C" on an Apollo DN3000 workstation using graphical and plotting functions. The modeled geometries of clastic and carbonate sediments evolve through time and respond to depositional processes that include tectonic movement, eustasy, and sedimentation. Clastic modeling includes sedimentary bypass and erosion, and sedimentation in alluvial and coastal plains, marine shelf, basin slope and basin floor. Carbonate modeling includes progradation, the development of hardgrounds, downslope aprons, keep-up, catch-up, back-step and drowned reef facies, as well as lagoonal and epeiric facies. Also included in the model are extensional vertical faulting of the basin, sediment compaction and isostatic response to sediment loading. Sediment geometries are plotted on the graphics terminal as they are computed so the user can immediately view the results. Then, based on these observations, parameters can be changed repeatedly and the program rerun until the user is satisfied with the resultant geometry.

To improve the user friendliness of the interactive program the user can conveniently enter data, including the initial basin configuration, local tectonic behavior, sea-level curves, the amount and source direction of clastic sediment and the growth rates of carbonates as a function of water depth.

INTRODUCTION

The first version of this simulation program, called SEDFIL, was developed at the Gulf Corporation Research Center in Pittsburgh, Pennsylvania, by Rande Burton and Christopher Kendall (Burton et al., 1987). This version was then improved by Elvira Camino and Richard

Quantitative Dynamic Stratigraphy (1989), T.A. Cross, ed., Prentice Hall, p. 433–444.

433

Slater, also of Gulf Research in Pittsburgh. Further work at the University of South Carolina led to many more improvements (Helland-Hansen et al., 1988). Recently, the algorithm was extended to include carbonate deposition, faulting, and sediment deposition from two sides of the basin and was called SEDBSN (Scaturo et al., 1988). In this paper we discuss in detail further additions and improvements that have been made to the clastic deposition algorithm. The original computer code was written in FORTRAN. It has now been rewritten in the C programming language. It has been restructured to speed up execution and reduce space requirements, and renamed SED-pak.

The objective of SED-pak is to model clastic and carbonate sediment accumulation in various depositional settings. It tracks evolving sedimentary geometry as a basin is filled by clastic sediments and the *in situ* production of carbonates. The program is designed to let the particular processes that are effective in specific depositional settings dictate sediment accumulation. Further, the user can conveniently change the input parameters to study the effects changes in these parameters have on the evolving geometries.

The program closely mimics geological processes involved in clastic and carbonate sediments deposition over large spans of time. Some of the significant clastic depositional phenomena incorporated include clastic filling of topographic depressions, creation of sediment wedges or fans, draped fill over topography, procedures to ensure that clastic sediments penetrate a minimum distance into the basin, and devices to monitor the volume and areal distribution of the sediments being deposited. The effects of eustatic changes on clastic sediment deposition also are modeled. Important carbonate phenomena modeled include the influence of eustasy on the accumulation rate and the rates of progradation or retreat of carbonates.

Clastic geometries are primarily influenced by the volume of sediments delivered to and redistributed within a basin. Carbonate geometries, in contrast, are influenced by the amount of *in situ* carbonate accumulation and by the volume of transported carbonate talus that accumulated over that same time interval. The initial and evolving shape of the basin surface directly affects the geometries of the sediments as they are deposited. Additional influences on sediment geometries include the amount of regional and hinged subsidence, faulting of the basin, water depth, erosion of previously deposited sediments, submarine slumping, compaction of sediments, and isostatic subsidence of the basin responding to sediment loading. All these influences are incorporated into our simulation model and the corresponding computer program.

Siliciclastic sediment deposition is treated differently from *in situ* carbonate sediment accumulation. It is assumed that the source of the sand and shale sediments is not located within the boundaries of the region simulated. In contrast, the source for carbonate sediments is assumed to be located within the boundaries of the simulation. In addition, the program simulates deposition of both nonmarine and marine sediments.

In the case of sand and shale deposition, it is assumed that the user has good estimates of both the volume (expressed as cross-sectional areas) of clastics involved, and the distance that each grain-size type penetrates the basin over the time intervals modeled. Within each time step, no attempt is made to recreate interbedded relationships between the sand and shale. Instead, the total amount of sediment deposited is modeled, with the areal distribution of each sediment type shown as a ratio of the total quantity of sediment deposited at each time step. The sedimentary ratios deposited at each location and time step are preserved.

Figure 1 illustrates the overall flow of control of the simulation program. The program is iterative and deposits sediments layer by layer for a user-specified number of time steps. Each time step corresponds to a user-defined number of real years (usually a few million years).

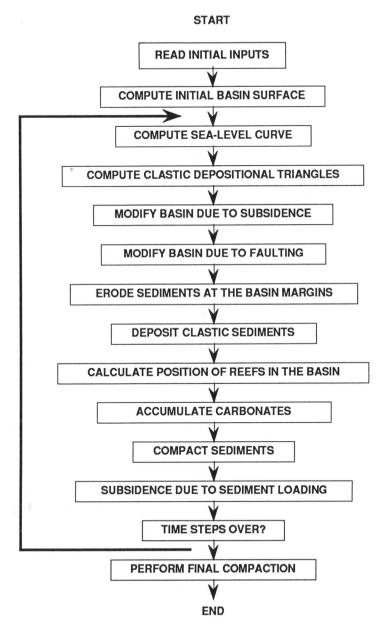

Figure 1. Flow diagram for computer program SED-pak.

FUNCTIONAL COMPONENTS OF THE SIMULATION PROGRAM

Initialization of the Basin Surface and Absolute Sea Level

The transverse axis of the basin being simulated can be subdivided into as many as 300 evenly spaced vertical columns. The program simulates sediment deposition column by column. The user defines the initial basin surface by specifying an initial height for each column. For convenience, the user may provide heights for some of the columns, and the simulation program will compute the initial heights of the remaining columns by linear interpolation. Similarly, the absolute sea level for each time step is specified as an input parameter. Again the user may specify sea level for some time steps and the simulation program uses linear and sinusoidal interpolation to determine the sea level at each of the remaining time steps. Sea level is defined by a first-order linear curve on which a second-order sinusoidal curve is superimposed. Both the first- and second-order curves are user defined.

Structural Movement

For each time step the simulation program first computes the basin's response to structural movement. Linear or non-linear subsidence or uplift are computed first, and then any changes in the basin surface due to faulting are computed. The location of faults and the slip rate along each are input parameters to the program. Users may set the slip-rate parameter as a time-dependent function. To simplify the computations involved with faulting, it is assumed that all faults extend to the bottom of the basin and only the vertical effects of faulting are modeled by the simulation program. The latter is, in general, a reasonable approximation because sediment geometries in the simulated area are more sensitive to vertical than to horizontal changes. This is because vertical distances are measured in fractions of feet or meters, whereas horizontal distances are of the order of hundreds of feet or meters. For example, if the basin modeled is in reality 150 km wide, subdividing it into 300 equally spaced columns results in column width of 500 m. Therefore, for a horizontal dislocation to be large enough to cause significant changes, it would have to exceed 250 m, or one-half the column width. This simplification is justified for high angle faults, but fails for low angle or listric faults.

Sediment Erosion

Before each time step the simulation checks for any erosion that may occur at the margins of the basin. Sediments above sea level with a surface slope greater than a user-supplied stable sediment repose angle are subject to erosion; the amount of sediment eroded is directly proportional to the repose angle. This reflects the greater instability of sediments on steeper slopes, which makes them more prone to erosion. The model treats all sediment types equally. Any sediment eroded from the basin margins will be added to the amount of clastic sediments being deposited in the basin for the current time step.

Clastic Deposition

The two-dimensional simulation represents the volume of sediment to be deposited as an area in the form of a right triangle. The area of this triangle is a measure of the volume of sediment available for deposition, and the base of the triangle represents the maximum distance the sediment penetrates into the basin from its margin. Sediments are deposited column by column and the amount of sediment deposited is subtracted from the cross-sectional areas of the triangles. Deposition continues until the sediment triangle's cross-sectional area becomes zero.

As illustrated in Figure 2, sediment deposition on the left side of the basin begins with the first column on the left that lies below sea level and progresses to the right. Similarly, deposition from the right margin starts with the first column on the right that is below sea level, and proceeds to the left. The user may set a parameter that activates deposition from either or both sides.

When depositing sediments column by column, the program first checks the surface to ascertain whether it is stable enough for deposition. This is accomplished by determining the height of the next column and calculating whether the slope between the current and next column exceeds a pre-defined angle of stability (Fig. 3). If the slope exceeds the angle of stability, no sediment is deposited in that column. If the slope of maximum stability is not exceeded, the simulation will deposit clastic sediment on the column up to the stable angle, provided sufficient volume of sediment is available (Fig. 4).

The maximum sediment height that the simulation will deposit on any given

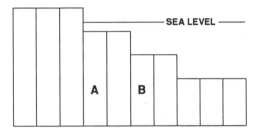

Figure 2. Clastic deposition begins at the first column whose elevation lies below sea level (column A) and proceeds basinward towards column B.

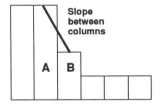

Figure 3. During clastic deposition the program checks to determine if the slope between columns A and B exceeds the user-defined stable slope. If it does, as in this case, no clastic sediments will be deposited on column A.

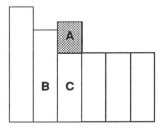

Figure 4. If the maximum stable slope between columns B and C is not exceeded, and if sediment is available, it is deposited on column B. The maximum quantity of sediment that can be supported on column B is the combined heights of columns C and A. The amount A is derived from the user-defined maximum stability slope.

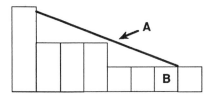

Figure 5. "A" is the slope up to which sediment will accumulate before becoming unstable. In other words it is the maximum stable slope, a surface that defines the top of the alluvial fan or submarine wedge for column B.

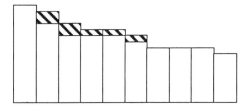

Figure 6. The shaded area is the sedimentary tail that extends over several columns into the basin.

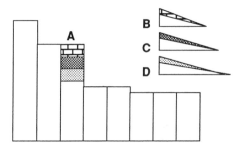

Figure 7. After sediment is deposited on a column, such as A, the heights of the sedimentary triangles are reduced, reflecting the area (and so volume) of sediment deposited on that column. A carbonate triangle "B" is included to represent the deposition of talus.

column depends not only on the slope of stability, but also on the present cross-sectional area of the sediment triangles. If the areas of the combined triangles cause sediment to exceed the stability slope, then the simulation deposits sediments only up to that slope. This slope, if it is above sea level, defines the surface of the alluvial plain; if it is below sea level, it defines the surface of a submarine slope (Fig. 5). By contrast, if the areas of the combined triangles cause sediment accumulation to be less than the stability slope, then the simulation deposits the available sediments onto the current column, as part of the sediment "tail" (Fig. 6).

The term sediment "tail" refers to the sediment wedge deposited in the area of the basin seaward of the shore (Fig. 6). Here the combined heights of the triangles are less than the angle of repose of the alluvial fan. In this case the simulation deposits sediments in columns equaling the combined heights of the sediment triangles. As each column is deposited, the heights and lengths of the sediment triangles are reduced to reflect the area (and so volume) of sediment just deposited (Fig. 7).

Sediment fans are constructed whenever the combined height of the clastic sediment triangles exceeds the height of sediments that can be deposited in a given column of the simulation. An alluvial fan or wedge is created when sediments are deposited above sea level. A submarine fan or wedge is created when the maximum sediment height to which a sediment may be deposited lies below sea level. Each type of fan or wedge has a different user-defined stability slope. The sediment wedge extends from the current column away from the center of the basin (Fig. 6). To create the sediment wedge, clastics are deposited in each column up to the user-defined slope for that particular wedge. Sediment wedges terminate upslope at the point at which the present surface of the basin exceeds the stable slope (Fig. 8).

Once the simulation has deposited sediment onto the column as a sedimentary wedge, the area of each clastic triangle is reduced to match the area of clastic material deposited and thus reduce the amount of sediment available for further deposition (Fig. 9). Since the length of a triangle represents the distance the sediment penetrates into the basin, the triangle's length is reduced to reflect the penetration of sediments into the basin for the current time step.

At every time step, the program checks to see if clastic sediments are being deposited upslope on some topographical feature. This is done by keeping track of the lowest elevation upon which clastic sediments have been deposited for that time step. The maximum topographic height that sediments may be deposited upslope is computed by summing the lowest elevation and a user-specified fraction of the height of the combined sediment triangles. The second factor signifies that some deposition (drape) upslope is possible when the program encounters a local topographic high. However if there is still sediment available for deposition, the program fills sediments backwards from the topographic high. If sufficient sediment remains, the program then causes the sediment to top and cross the offshore relief (Fig. 10), and continue deposition beyond the relief in a basinward direction.

Carbonate Deposition

Deposition of carbonates is initiated after clastic deposition for that time step is complete. The main influence on the amount of

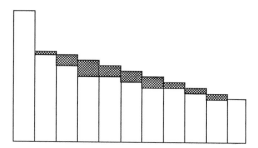

Figure 8. Method for determining the upslope termination of the sedimentary wedge (shaded). The sediment wedge extends marginward from column B and terminates at column A. No sediments are deposited on column A because of the unstable slope between A and the next column seaward.

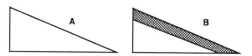

Figure 9. After deposition of a sedimentary wedge ("A"), the triangle's height and length are reduced ("B"). This reduction in height and length of the sediment triangle is shown in "B," and reflects the quantity of sediment being deposited. Thus the ability of sediment to penetrate further into the basin is reduced. The shaded area represents sediment removed during deposition.

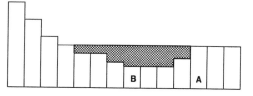

Figure 10. Sediment will fill a topographic depression before crossing the enclosing topographic high. In this case the simulation determines that column A is above the maximum depositional height for sediments (as determined from column B). If sediments are available, the simulation then fills in the topographically low area up to the height of column A. In cases of abundant sediment, column A will be overtopped.

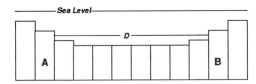

Figure 11. Line "D" shows the user-defined depth to which waves limit carbonate growth potential. In this case the simulation determines that columns A and B are where waves start affecting carbonate production. The zone between line "D" and sea level is the zone of wave damping on the seaward sides of reefs and buildups.

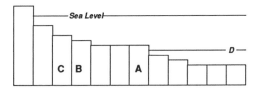

Figure 12. The carbonate growth potential is reduced from columns A through B by wave damping. Column C is not affected since it exceeds the user-defined distance from where waves start breaking (column A). Line "D" represents the depth below which waves do not damp the growth of carbonates.

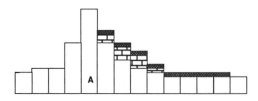

Figure 13. Simulation of talus extending away from an isolated reef (column A). Note separation of talus into "near-reef" (limestone symbol) and "turbidite" (shaded). The user provides the extent of each type of talus and the percentage of carbonates available for each type.

carbonate to be deposited in each column is the depth of water for that column. The modeled rate of accumulation mimics the response of carbonate-producing organisms to photosynthesis. Other factors that influence the rate of carbonate accumulation in each column are: 1) wave energy damping of the rates of carbonate accumulation on build-ups or reef margins; 2) the presence of reefs or build-ups on the margins of lagoons that limit aerating waves or nutrient supply into the lagoons and within epeiric seas, so reducing accumulation rates; and 3) the amount of clastic material deposited in the column, which damps carbonate growth and mimics the oversupply of nutrients to the area. Transported carbonate talus is modeled by assuming that talus deposition follows rules similar to those for clastic materials.

Carbonate accumulation is suppressed by the effects of wave energy over a specific range of depths some distance below the water surface. The locations at which waves start affecting carbonate accumulation on the depositional surface are defined in the following manner. The program determines where the sea floor lies above a user-defined water depth (Fig. 11), above which waves begin to break. Wave energy is then modeled as affecting the sea floor, suppressing carbonate accumulation at a user-defined distance shoreward from the position at which waves start breaking (Fig. 12). Locations of the build-up crests or reefs are determined by testing, from deep to shallow water, the columns on which carbonate growth, suppressed only by wave energy, are the first to reach sea level. Once these locations are determined, the areas between them are examined. If the depth of the sea does not exceed a user-supplied maximum depth for lagoons and epeiric settings, then that area is considered to be a lagoon or epeiric shelf and its rate of accumulation damped accordingly; otherwise, the area is designated as being either subaerial or open ocean.

Deposition of carbonates begins after the columns are classified as open ocean, carbonate build-up crest (or reef), lagoon (or epeiric shelf), or a subaerial surface. First, build-ups or reefs are modeled and their growth is constrained by water depth, wave energy and the presence of clastics in the water column. The program limits carbonate growth of the build-up crests to sea level. Excess carbonate accumulation, which would cause the build-ups to rise above sea level, is stored as talus, transported off the build-up and deposited. The user defines what percentage of the talus is to be transported as backreef facies into lagoons or over the epeiric shelf, and what percentage is to be transported downslope off the carbonate platform into the basin. The user also defines the distance to which the talus extends away from the reef into the ocean as either near-reef debris or turbidite fans (Fig. 13). The talus is then deposited. After the build-up growth, backreef, and talus are modeled, carbonate accumulation is modeled for the lagoonal, epeiric shelf, and open ocean settings. Carbonate sediment accumulations in lagoon and epeiric shelf settings are damped by user-defined amounts, so that these environments can be modeled to not reach sea level during the time step. As before, carbonate growth in these areas is suppressed by the presence of clastics in the water column and, in the case of the open ocean, by wave energy.

Subsidence and Compaction

Deposition of carbonates completes sediment additions to the basin for the current time step. As a next step the simulation program considers alteration phenomena produced by the sediments just deposited. This includes compaction of sediments and subsidence of the basin due to sediment loading, or conversely an isostatic rise in the basin surface due to erosion of sediments.

After compaction, subsidence or uplift, the program increments the time-step parameter and repeats the entire set of procedures. This iterative process continues for a user-specified number of time steps. At this point, a final compaction is performed. This is based on a user-designated parameter that specifies the depth the basin is buried beneath later deposits. The geometries of these deposits are not modeled, but their thickness is assumed.

DISCUSSION

The SED-pak simulation program, written in the C programming language, currently runs on a DN3000 series Apollo workstation. The graphics interface is implemented using system-specific graphics routines (the GPR package), however we hope to reimplement this interface using the X-windows system in the near future. This should make the entire package transportable to any UNIX-based engineering workstation (and quite a few PC-based systems).

The program is run interactively by the user. At present the program is provided with a set of initial numerical inputs. At a later time we plan to be able to add these inputs in a graphical mode. Once the inputs have been implemented, the user runs the simulation. After each time step the simulation can display sediment geometries. The program may be run to completion, or it may be stopped and reentered to change values of input variables and then executed again or terminated. Input data sets may be saved for later use.

Clastic deposition values, which are related to the area and so heights and lengths of the sediment triangles, may be determined from local well logs, core and seismic data. The lengths of the triangles are derived from the horizontal extent the sediment penetrates the basin from the shelf break. The triangle's areas are determined from the data set by examining the maximum cross-sectional area of the sediment in the area of interest.

The clastic simulation algorithm creates sediment fans, erodes and deposits sediment in a natural fashion. In particular, clastic sediments are prevented from being deposited directly upslope so that the filling of topographic lows within a basin model appears realistic.

The simulation's carbonate algorithm handles carbonate accumulation plausibly; however it requires that users provide a great deal of information which may not be readily available. We realize that in many locations information may be incomplete, inaccurate or simply lacking (Scaturo et al., 1988) partly because of unavailability of published data on carbonate production rates as a function of depth, carbonate production rates in lagoons or epeiric settings, damping of carbonate growth due to the presence of clastics, and the depth to which carbonate production is suppressed by wave energy (Scaturo et al., 1988). As a consequence, some of the default values used are no more than educated guesses.

The simulation algorithm treats the basin's response to sediment loading as if the basin were perfectly elastic, rather than incorporating an element of crustal rigidity in its isostatic

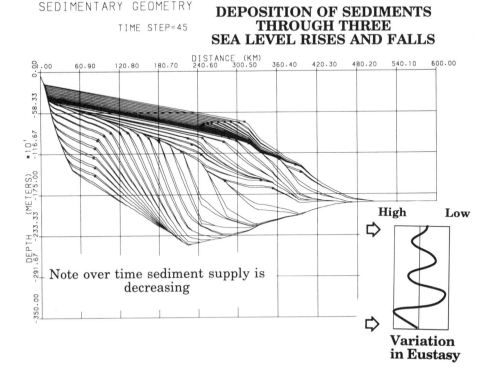

Figure 14. The product of SED-pak simulating clastic fill from a one-sided basin over three sea-level cycles.

component. However, users can provide an approximation of crustal rigidity indirectly by increasing the density provided for the mantle, either uniformly across the basin, or as a function of geographic position. Increases in mantle density cause the basin surface to be less responsive to subsidence produced by sediment loading, and this essentially matches the effects of the crust's rigidity. Tests have shown that increasing the mantle density by a factor of 10 almost eliminates crustal subsidence.

While the simulation algorithm handles the reduction of porosity due to sediment loading it does not yet address the migration of fluids within the pore space and hence burial diagenesis. In order to extend its functionality to include the fluid history of the basin, the program needs to be altered to allow the over-pressuring of fluids and the occurrence of sedimentary seals for fluids within the area being modeled. Finally, to address the question of whether the evolving sediment geometry of the basin allows the existence of hydrocarbons, the thermal history, rock type and organic content of the basin needs to be incorporated into the simulation model.

Experimental results confirm that the simulation SED-pak provides a good two-dimensional picture of the sedimentary geometries created by filling a basin with clastic (Fig. 14) and carbonate sediments (Fig. 15). Various tectonic features and the basin's response to

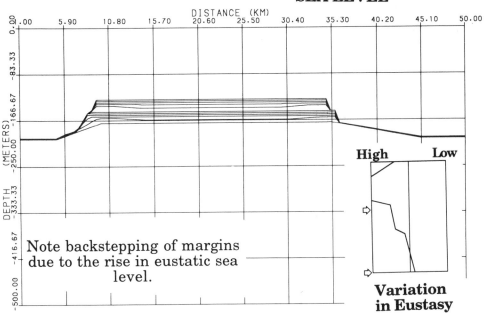

Figure 15. The product of SED-pak simulating carbonate accumulation on an isolated carbonate platform during a series of abrupt relative sea-level rises. The response of the carbonate margin is to back step. Notice that the response to lagoonal sediment accumulation has been damped during some of the time steps.

sediment filling are also incorporated into our model. *In situ* carbonate sediments are deposited in a process driven fashion and the carbonates grow directly as a result of the basin's shape and water depth. Clastic sediments require users to provide information on the total volume of the sediment type and the extent to which the sediment extends into the basin. After the amount of clastic sediment is known, the deposition of this sediment follows general geologic rules.

ACKNOWLEDGMENTS

We express our appreciation to the Japanese National Oil Corporation, Union Petroleum of California, AMOCO, and the Standard Oil Company for partial financial support for development of the SED-pak simulation. We appreciate the thoughtful and constructive criticisms and suggestions of Bob Goldhammer and Matt Matthews in their reviews of the original manuscript.

REFERENCES CITED

Burton, R., Kendall, C.G.St.C., and Lerche, I., 1987, Out of our depth: On the impossibility of fathoming eustasy from the stratigraphic record: Earth Science Reviews, v. 24, p. 237-277.

Helland-Hansen, W., Kendall, C.G.St.C., Lerche, I., and Kazuo, N., 1988, A simulation of continental basin margin sedimentation in response to crustal movements, eustatic sea level change and sediment accumulation rates: Mathematical Geology, v. 20, p. 777-802.

Scaturo, D.M., Strobel, J.S., Kendall, C.G.St.C., Wendte, J.C., Biswas, G., Bezdek, J.C., and Cannon, R.C., 1988, Judy Creek: A case study for a two dimensional sediment deposition simulation: Society of Economic Paleontologists and Mineralogists Special Publication (in press).

Vail, P.R., Mitchum, R.M., Jr., Todd, R.G., Widmier, J.M., Thompson, S., III, Sangree, J.B., Bubb, J.N. and Hatelid, W.G., 1977; Seismic stratigraphy and global changes of sea level, *in* Payton, C.E., ed., Seismic Stratigraphy—Applications to hydrocarbon exploration: American Association of Petroleum Geologists Memoir 26, p. 49-212.

26

FORWARD MODELING OF BANK-MARGIN CARBONATE DIAGENESIS: RESULTS OF SENSITIVITY TESTS AND INITIAL APPLICATIONS

Terrence M. Quinn
Department of Geological Sciences, Brown University, Providence, RI 02912-1846 USA

ABSTRACT

The incorporation of deductive principles into quantitative stratigraphic models is an important first step toward a better understanding of the geological processes affecting the dynamic stratigraphies of carbonate sequences. An updated version of a one-dimensional forward model (Matthews and Frohlich, 1987) that simulates the interaction of bank-margin subsidence, sea-level fluctuation, and near surface carbonate diagenesis and outputs a synthetic stratigraphic sequence is presented. The results of model sensitivity experiments indicate that the interaction of model approximations of geological processes produce complex diagenetic stratigraphies.

The results of specific model sensitivity experiments indicate that (1) subsurface stratigraphic traps are generated in response to differential basin subsidence and simple model rules governing porosity enhancement and destruction, and (2) relatively thick sequences of dolomite are generated by the superposition of thin mixing zones in response to rapid and numerous fluctuations in sea level. Lastly, forward model output is compared to the diagenetic stratigraphy of Enewetak Atoll. There are several discrepancies as well as similarities between the model output and the observed data. Hiatuses at 1.4 and 3.0 Ma are well represented in both model and observed data. The sea-level highstand at 2.1 Ma in the model is in relatively good agreement with the amplitude of the sea-level rise suggested by the data, but is displaced by 0.6 m.y.

INTRODUCTION

A forward model is based on a specific set of input variables and assumptions that attempt to describe, often by simplification, the physical laws controlling geologic processes (e.g., Steckler et al., 1988). In this light, an important function of forward modeling is to gain insight into geological processes and stratigraphic responses. Stratigraphic applications of forward modeling clastic and carbonate systems focus on gaining insight into the processes of subsidence, sea-level change, and sedimentation, and the products of their interaction. Over the last couple of years, several quantitative stratigraphic forward models have been presented for clastic systems (e.g., Kendall et al., 1986) and carbonate systems (e.g., Read et al., 1986; Matthews and Frohlich, 1987).

Quantitative Dynamic Stratigraphy (1989), T.A. Cross, ed., Prentice Hall, p. 445–455.

Sensitivity experiments are commonly used to evaluate the interaction and significance of model parameters. In a sensitivity experiment, a single model parameter is perturbed while all other model parameters are kept constant in an effort to isolate cause and effect in the model. I present the results of sensitivity experiments using an enhanced version of the one-dimensional forward model originally presented by Matthews and Frohlich (1987). I also present the initial results and comparisons of applying the model to the diagenetic stratigraphy of Enewetak Atoll.

The Forward Model

A one-dimensional forward model of bank-margin carbonate diagenesis is implemented on an Apple Macintosh 1 Mbyte personal computer. The computer program, written in FORTRAN77, simulates the interaction of sea-level fluctuation, basin subsidence, and meteoric diagenesis in a bank-margin carbonate environment.

In the model simulation, when a subaerially exposed surface is flooded by the sea, new bank-margin carbonate sediments accumulate up to the new sea level. All newly deposited sediment is arbitrarily defined as 100% aragonite with a user-defined initial porosity value. The explicit assumption of the model is that accumulation rates of aragonite sediments are sufficient to fill available space created by the interaction of basin subsidence and sea-level fluctuation.

Upon subaerial exposure, sediments are subject to meteoric diagenesis. Mineralogic stabilization occurs in the vadose, phreatic, mixed-water, and marine environments at environmentally specific rates. The thickness of the vadose environment is determined dynamically by the program at each time step and is defined as the distance between the top of the subaerially exposed bank margin and sea level. Thicknesses of the phreatic lens and the mixed-water environment are user-defined prior to the start of the model run. The marine environment extends downward from the base of the mixing zone to the base of the sedimentary column.

Time steps are defined as the time difference (in kyr) between successive absolute sea-level highstands and lowstands. Diagenetic alteration of aragonite to calcite is a function of cumulative residence time in vadose, phreatic, and marine environments. Residence time is accumulated in each diagenetic environment during the transition from one sea-level stand to another and during the user-defined duration of sea-level stillstand. The mixed-water environment is chosen by the user as either a site of dolomitization or calcification. Mixed-water dolomitization replaces all available aragonite first, then replaces calcite. Vadose and phreatic lens calcites are defined to be equally susceptible to dolomitization.

Model Input

Input parameters include sea-level history, linear rate of subsidence, rate of isostatic compensation in response to sediment loading, rates of mineralogical stabilization, thicknesses of diagenetic environments, linear rate of erosion, initial porosity, and linear rate of cementation. The amplitude and stillstand duration of the input sea-level history can be further modified by the user at the start of the model run. Table 1 is a compilation of input data for the model control case to which all other model experiments are compared. The input sea-level record for all sensitivity experiments is the Late Cenozoic (6 Ma) tropical planktonic $\delta^{18}O$ composite (Prentice, 1988), shown in Figure 1, which is converted to meters of sea-level equivalence (e.g., Matthews and Frohlich, 1987).

Table 1 Forward model input parameters for control case.

Sea Level Flag (if 1, sea level in sigma; 0 if in per mil)	0
Subsidence Rate (m/kyr)	0.100
Isostatic Compensation (m/m; 1.0 = none)	1.00
Marine Diagenetic Clock (kyr)	5000.00
Mixing Zone Diagenetic Clock (kyr)	5000.00
Phreatic Lens Diagenetic Clock (kyr)	40.00
Vadose Diagenetic Clock (kyr)	1000.00
One Standard Deviation of Sea Level (sigma)	0.33
Isotopic Calibration (per mil/m)	0.011
Mean Sea Level Position in Meters (xbar)	-60.00
Mean Aragonite d18O (per mil)	-2.00
Mean Calcite $\delta^{18}O$ (per mil)	-6.00
Mean Dolomite $\delta^{18}O$ (per mil)	+2.00
Rate of Subaerial Erosion (m/kyr)	0.000
Thickness of Phreatic Lens + Mixing Zone (m)	5.00
Thickness of Phreatic Lens (m)	5.00
Position of Phreatic Lens Top Relative to Sea Level	0.00
Number of Sea-Level Stillstands	269
Stillstand Duration Flag (if 1, difference between SS)	0
Mixing Zone Flag (if = 0, mixing zone = marine)	0
Sea-Level Amplitude Compression Factor (0 = none)	0
Stillstand Duration (kyr)	5.00
Initial Water Depth of Rock Column (m subsea)	60.00
Initial Porosity of Aragonite Sediment (%)	30.00
Cement Flag (if = 0, cementation occurs in stillstands)	0
Vadose Cementation Clock (+ number = dissolution)	1000.00
Phreatic Lens Cementation Clock (kyr)	-50.00
Mixing Zone Cementation Clock (kyr)	0000.00
Marine Cementation Clock (kyr)	0000.00

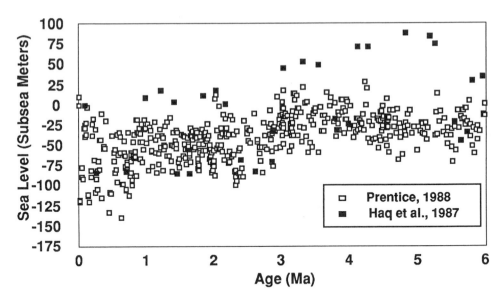

Figure 1. Two versions of the last 6 m.y. of sea-level variation. The low frequency curve with the solid black squares is taken from Haq et al. (1987), whereas the high frequency curve with the solid white squares is modified from Prentice (1988).

Model Output

Results of the forward model are displayed in a variety of user-requested formats. The standard outputs are mineralogy (in the form of a lithologic column), position within the stratigraphic column of subaerial exposure surfaces, and sediment age versus depth (Fig. 2). For reasons of spatial economy, the diagenetic bed/sea-level matrix (see Matthews and Frohlich, 1987, Fig. 3), which is another optional output format, is not presented here.

RESULTS OF SENSITIVITY EXPERIMENTS

Sensitivity analysis involves perturbation of one model parameter while holding all other model parameters constant. In this method, cause and effect relationships among parameters may be isolated and analyzed. Results of sensitivity analysis may also provide insight into the relative contribution of individual geological processes to the overall stratigraphic response of the system. In this section, I present the results of two sensitivity experiments comparing rate of mineralogical recrystallization and thickness of diagenetic environment.

Rate of Mineralogical Recrystallization

Mineralogical recrystallization of metastable precursors to stable mineralogies in the near surface diagenetic environment is driven primarily by solubility contrasts between coexisting carbonate minerals (e.g., Matthews, 1974). The rate of mineralogical recrystallization is

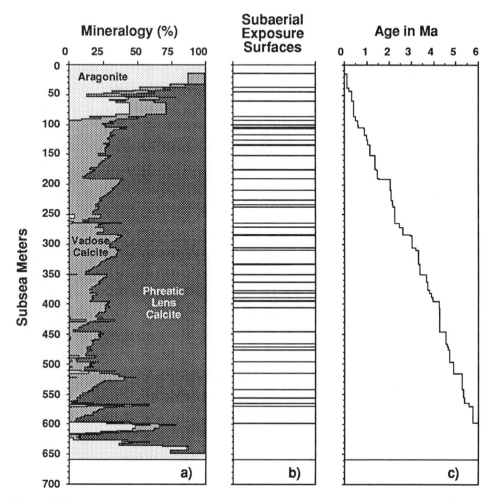

Figure 2. Forward model output of control case. Panel a) mineralogic data in the form of a lithologic column. Panel b) depth of predicted subaerial exposure surfaces. Panel c) sediment age versus depth plot. This output is defined as the control case. Input parameters responsible for this output are summarized in Table 1.

environmentally specific and is a function of several parameters including hydrology, climate, and topography. In the model simulation, mineralogical recrystallization occurs as a function of lapsed residence time in a specific diagenetic environment and a user-defined rate of recrystallization for that environment.

In the first sensitivity experiment, the rate of mineralogical recrystallization in the vadose zone is decreased from 1000 kyr (control case) to 200 kyr, and all other model parameters are unchanged. The model output for this experiment, shown in Figure 3a, may be compared with model output from the control case (Fig. 2). Note that the degree of mineralogical stabilization has markedly increased from the control case.

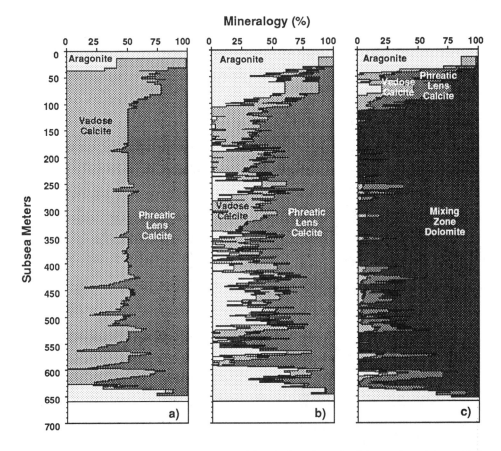

Figure 3. Mineralogic outputs (in the form of lithologic columns) of forward model sensitivity experiments. In panel a), the rate of mineralogical recrystallization in the vadose environment is decreased from 1000 kyr (Fig. 2) to 200 kyr. Note that the amount of vadose calcite has increased markedly from Figure 2. In panel b), the thickness of the phreatic lens environment is decreased from 5 m (Fig. 2) to 1 m. Note that with the phreatic lens only 1 m thick, primary aragonite is preserved throughout the sedimentary column. In panel c), the thickness of the mixed-water environment is increased from 0 m (Fig. 2) to 4 m. Note that the superposition of thin mixing zones in response to rapid and numerous sea-level variations has produced a thick dolomite-rich sedimentary sequence.

Thickness of Diagenetic Environment

In the second sensitivity experiment, the thickness of the phreatic lens is decreased from 5 m (control case) to 1 m. The interaction of thickness of the diagenetic environment and high-frequency sea-level fluctuations increases the probability that strata in the sedimentary column are affected by meteoric diagenetic processes. Note that by simply decreasing the thickness of the phreatic lens diagenetic environment, primary aragonite is preserved throughout the sedimentary column (Fig. 3b).

APPLICATION OF THE FORWARD MODEL

Results of model experiments that simulate stratal geometries of dolomite bodies and predict the occurrence of subsurface stratigraphic traps are presented in this section. Thick sections of dolomite are produced by the interaction of mixed-water dolomitization and high-frequency sea-level variation. Subsurface stratigraphic traps are produced by varying subsidence rates across a basin.

Prediction of Dolomite Stratal Geometries

In the last decade or so, dolomite formation in fluids of mixed marine and fresh water has gained favor among carbonate workers despite the recent arguments of Machel and Mountjoy (1986) and Hardie (1986). In particular, Humphrey (1987) concluded that not only has dolomite of mixed-water origin formed in an uplifted terrace of Barbados (Golden Grove), but dolomitization occurred within one Late Pleistocene sea-level fall (<10 kyr).

In this sensitivity experiment, the rate of 100% recrystallization to dolomite in the mixed-water environment is set at 40 kyr and the thickness of the mixed-water environment is set at 4 m. Figure 3c is the output of this model experiment and clearly shows that thick sections of dolomite are formed in the mixed-water environment by the superposition of thin mixing zones in response to rapid and numerous sea-level variations.

Prediction of Subsurface Stratigraphic Traps

In this series of model experiments, the rate of basin subsidence was set at 10 cm/kyr, 5 cm/kyr, and 1 cm/kyr. All sediments have an arbitrarily defined initial porosity value of 30%. Cementation or dissolution occurs as a function of lapsed residence time in each diagenetic environment at environmentally specific rates. In this model run, 100% occlusion of porosity in response to stillstand phreatic lens cementation is arbitrarily defined to occur in 50 kyr, whereas all other diagenetic environments have no affect on cementation. In the model simulation, updip porosity pinchouts are generated in response to differential basin subsidence (Fig. 4).

MODEL/DATA COMPARISON

The carbonate cap on mid-ocean atolls offers a unique opportunity to examine the record of sea-level variations because atolls having a relatively simple tectonic history (i.e., those without a significant geoid elevation anomaly) allow constraints to be placed on subsidence history. If the assumption is made that carbonate sediment aggradation keeps pace with sea-level rise, then sea-level history can be deduced (Major and Matthews, 1983; Quinn, 1987; Halley and Ludwig, 1987). Thus, an atoll setting is a logical first choice to test carbonate stratigraphic forward models.

Sea-level records deduced from the carbonate cap on mid-ocean atolls are an independent check on sea-level records deduced from seismic stratigraphy (e.g., Haq et al., 1987) and tropical planktonic $\delta^{18}O$ data (e.g., Prentice, 1988). Figure 1 is a plot of two versions of the

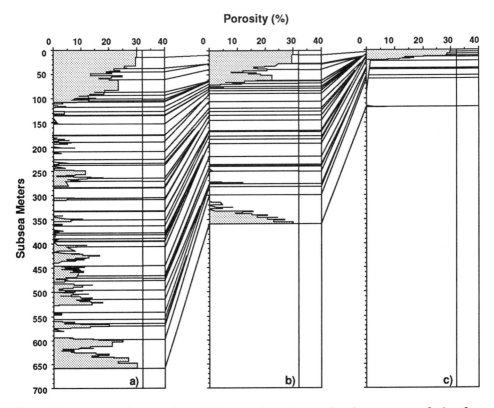

Figure 4. Forward model output of a sensitivity experiment that predicts the occurrence of subsurface stratigraphic traps in response to differential basin subsidence. Subsidence rate varies from 10 cm/kyr in panel a) to 5 cm/kyr in panel b) to 1 cm/kyr in panel c). Model predicted subaerial exposure surfaces (dashed lines) are correlated to subaerial exposure surfaces of equivalent age in each of the panels. 100% occlusion of porosity in response to stillstand phreatic lens cementation is arbitrarily defined to occur in 50 kyr. Note updip pinchout of porous zones.

last 6 m.y. of sea-level variation. Note that the two curves are discrepant in period, frequency, and amplitude of sea-level variations. Studies of the carbonate cap on mid-ocean atolls provide important data for resolving these discrepancies.

Detailed lithologic, petrographic, stable isotopic, geochemical, and Sr-isotopic data on carbonate sediments from Core KAR-1 from Enewetak Atoll were used to subdivide the upper 200 m of section into individual unconformity-bounded sediment packages (Henry and Wardlaw, 1986; Quinn, 1987; Ludwig et al., 1988; Figs. 5a,c). Figure 5d is the result of an initial attempt to forward model the record of subaerial exposure surfaces observed in Core KAR-1.

A subsidence rate of 3.8 cm/kyr provided the best qualitative fit to the data. There are several discrepancies as well as similarities between the model and the observed data. Hiatuses at 1.4 and 3.0 Ma are well represented in both model and observed data. The sea-level highstand at 2.1 Ma in the model is in relatively good agreement with the amplitude of

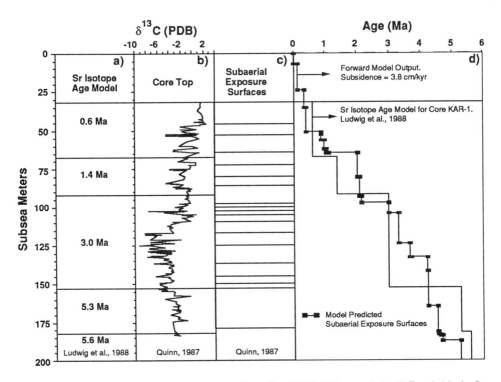

Figure 5. Initial comparison of model with data from Core KAR-1, Enewetak Atoll. Panel a) is the Sr-isotope chronostratigraphy (from Ludwig et al., 1988). Panel b) is δ^{13}C stratigraphy (from Quinn, 1987). Panel c) displays the recognized subaerial exposure surfaces selected on the basis of stable isotopic and petrographic data (Quinn, 1987). Panel d) is model/data comparison of age versus depth. Note that in several places there is relative agreement in both the timing and the amplitude of sediment aggradation and subaerial exposure between model output and data from Core KAR-1.

the sea-level rise suggested by the data, but is displaced by 600 kyr. Unconformity-bounded sediment packages selected on Sr-isotope criteria alone (Fig. 5a) appear to agree somewhat with the Haq et al. (1987) sea-level curve; detailed stable isotopic and petrographic data indicate that numerous unconformity-bounded sediment packages exist below the resolution of Sr-isotopes and the Haq et al. (1987) sea-level curve (Quinn, 1987). These higher resolution unconformity-bounded sediment packages are in relatively good agreement with the sea-level record deduced from tropical planktonic δ^{18}O data (Prentice, 1988).

DISCUSSION

Quantitative stratigraphic models must make estimates of three fundamental model parameters: subsidence, sedimentation, and sea level. Subsidence rate is frequently parameterized as a linear function of time in stratigraphic forward models, whereas sedimentation rate in models of carbonate depositional sequences is parameterized as a function of time and space—usually depth below mean sea level (e.g., Read et al., 1986; Matthews and Frohlich, 1987; Goldhammer et al., 1987). Although there is general agreement regarding input parameterization of subsidence and sedimentation rates among the different carbonate stratigraphic forward models, there is little agreement about the frequency and amplitude of sea-level variations. Presently in different models, sea-level variations are input as 1) third-order cycles (Haq et al., 1987) deduced from seismic stratigraphy; 2) sinusoids with Milankovitch-like characteristics (e.g., Read et al., 1986; Goldhammer et al., 1987); or 3) high-resolution tropical planktonic $\delta^{18}O$ data (a sea-level proxy; Matthews, 1984, 1989) that are converted to meters of sea-level equivalence (e.g., Matthews and Frohlich, 1987; Fig. 1). Fundamentally different stratal geometries are produced in the various models as a result of the temporal and spatial differences between the different records of sea level.

Accurate prediction of stratal geometries is a fundamental goal in stratigraphic applications of forward modeling. However, herein lies a paradox because forward modeling is extremely sensitive to input parameterization. "Pattern matching" between model output and observed stratal geometries (based on seismic data) is not necessarily an indication of model fidelity. Little insight into geological processes is gained when geologically "reasonable" output is produced by geologically "unreasonable" estimates of model parameters.

Forward models also are susceptible to over-parameterization, and it is important to keep in mind that the number of model parameters, their complexities, and their interdependencies are not necessarily correlated with the quality of the model. In fact, an overabundance of complex model parameters can obscure the relationship between interactive geological processes, which is a *goal* of the forward modeling exercise.

CONCLUSIONS

Initial results of forward modeling of bank-margin carbonates indicate that complex, multistage diagenetic stratigraphies are the expected result of the interaction between glacio-eustatic sea-level fluctuations and basin subsidence. Application of the computer based forward model to real carbonate bank margins resulting in convergence between synthetic diagenetic stratigraphy produced by the model and data from real rocks will greatly facilitate interpretation of the evolution of carbonate bank margins by providing information on the timing and nature of diagenesis. Model-aided recognition of paleophreatic lenses and paleosubaerial exposure surfaces may also provide independent constraints on the history of eustatic changes.

ACKNOWLEDGMENTS

I would like to thank R.K. Matthews and C. Frohlich for their insights into model development and forward modeling in general. I would also like to thank K.E.A. Al-Saqri, S.W. Butcher, and J.D. Humphrey for constructive criticism throughout this project and S. Choh for performing numerous sensitivity experiments. This work is part of my Ph.D. dissertation under the direction of R.K. Matthews and is supported in part by National Science Foundation Grant EAR-8607357. I thank C.G.St.C. Kendall and T.A. Cross for helpful reviews and suggestions.

References Cited

Goldhammer, R.K., Dunn, P.A., and Hardie, L.A., 1987, High-frequency glacio-eustatic sea-level oscillations with Milankovitch characteristics recorded in Middle Triassic platform carbonates in Northern Italy: American Journal of Science, v. 287, p. 853-892.

Halley, R.B., and Ludwig, K.R., 1987, Disconformities and Sr-isotope stratigraphy reveal a Neogene sea-level history from Enewetak Atoll, Marshall Islands, Central Pacific: Geological Society America Abstracts with Programs, v. 19, p. 691.

Haq, B.U., Hardenbol, J., and Vail, P.R., 1987, Chronology of fluctuating sea-level since the Triassic: Science, v. 235, p. 1156-1167.

Hardie, L.A., 1987, Dolomitization: A critical view of some current views: Journal of Sedimentary Petrology, v. 57, p. 166-183.

Henry, T.W., and Wardlaw, B.R., 1986, Pacific Enewetak Atoll Crater Exploration (PEACE) Program Enewetak Atoll, Republic of the Marshall Islands, Part 3: Stratigraphic analysis and other geologic and geophysical studies in the vicinity of KOA and OAK craters: U. S. Geological Survey Open-File Report 86-555, p. 4.3-4.49.

Humphrey, J.D., 1987, Processes, rates, and products of early near-surface carbonate diagenesis: Pleistocene mixing zone dolomitization and Jurassic meteoric diagenesis [Ph. D. thesis]: Providence, Rhode Island, Brown University, 263 p.

Kendall, C.G.St.C., Lerche, I., and Nakayama, K., 1986, Simulation of continental margin sedimentation: American Association of Petroleum Geologists Bulletin, v. 70, p. 606.

Ludwig, K.R., Halley, R.B., Simmons, K.R., and Peterman, Z.E., 1988, Strontium-isotope stratigraphy of Enewetak Atoll: Geology, v. 16, p. 173-177.

Machel, H.-G., and Mountjoy, E.W., 1986, Chemistry and environments of dolomitization—A reappraisal: Earth Science Reviews, v. 23, p. 175-222.

Major, R.P., and Matthews, R.K., 1983, Isotopic composition of bank-margin carbonates on Midway Atoll: Amplitude constraint on post-early Miocene eustasy: Geology, v. 11, p. 335-338.

Matthews, R.K., 1974, A process approach to diagenesis of reefs and reef associated limestones, *in* Laporte, L.F., ed., Reefs in time and space: Society of Economic Paleontologists and Mineralogists Special Publication 18, p. 234-256.

Matthews, R.K., 1984, Oxygen isotope record of ice volume history: 100 million years of glacio-eustatic sea-level fluctuation, *in* Schlee, J.S., ed., Interregional unconformities and hyrdrocarbon accumulation: American Association of Petroleum Geologists Memoir 36, p. 97-107.

Matthews, R.K., 1989, Quaternary sea-level change, *in* Revelle, R., ed., Sea-level change: National Research Council Studies in Geophysics, National Academy of Sciences (in press).

Matthews, R.K., and Frohlich, C., 1987, Forward modeling of bank-margin carbonate diagenesis: Geology, v. 15, p. 673-676.

Prentice, M.L., 1988, The deep sea oxygen isotopic record: Significance for Tertiary global ice volume history, with emphasis on the Latest Miocene/Early Pliocene [Ph. D. thesis]: Providence, Rhode Island, Brown University, 567 p.

Quinn, T.M., 1987, Late Cenozoic limestones on Enewetak Atoll: A 5 Ma record of glacio-eustasy and diagenesis: Geological Society America Abstracts with Programs, v. 19, p. 811.

Read, J.F., Grotzinger, J.P., Bova, J.A., and Koerschner, W.F., 1986, Models for generation of carbonate cycles: Geology, v. 14, p. 107-110.

Steckler, M.S., Watts, A.B., Thorne, J.A., 1988, Subsidence and basin modeling at the US Atlantic passive margin, *in* Sheridan, R.E., and Grow, J.A., eds., The Atlantic continental margin, US: Geological Society of America, The Geology of North America, v. I-2, p. 399-416.

Part IV

STRATIGRAPHIC RESOLUTION AND MODEL VERIFICATION

27

CLASTIC FACIES MODELS, DEPOSITIONAL SYSTEMS, SEQUENCES AND CORRELATION: A SEDIMENTOLOGIST'S VIEW OF THE DIMENSIONAL AND TEMPORAL RESOLUTION OF LITHOSTRATIGRAPHY

William E. Galloway
Department of Geological Sciences, The University of Texas at Austin, Austin, Texas
78713 USA

ABSTRACT

Basin fills consist of assemblages of genetic stratigraphic units that are bounded by or contain correlative stratigraphic surfaces, facies, and marker beds. Because most basins are filled episodically, stratigraphic units defined by hiatal surfaces provide fundamental elements for both regional and local correlation and analysis. Depositional sequence analysis emphasizes the importance of erosional unconformities produced by fluctuating sea level. In contrast, the concept of depositional episodes and resultant genetic stratigraphic sequences uses transgressions and subsequent maximum-marine-flooding surfaces to delineate units created by the interplay of sediment supply, subsidence, and base-level change. The two models suggest somewhat different stratigraphic architectures but rely on correlation of surfaces to define both the lithostratigraphic and chronostratigraphic framework. Correlation of sequences extends regionally within basins, and may encompass several basins, depending upon causal mechanism. Resolution is on the order of 10^6 years.

Each sequence typically contains several depositional systems, which provide the three-dimensional building blocks of the basin fill. Depositional systems, in turn, consist of process-related facies assemblages deposited in more localized environments. Deposition within systems and their constituent environments is commonly punctuated by a series of depositional events, which create discrete facies sequences. Such events provide a relatively high-resolution temporal framework, but correlation is commonly limited to a single depositional system. Causal mechanisms are classified as intrinsic, tectonic/geomorphic, climatic, and eustatic. Intrinsic mechanisms are autogenic and include alternating channel incision and backfilling, channel avulsion, and delta/fan lobe switching. Time scales are on the order of 10^3 years. Tectonic/geomorphic events include tectonic triggering of large-scale gravity resedimentation and recurrent channel piracy. Time scales probably extend from 10^2 to 10^4 years. Climatic mechanisms include induced changes in runoff and sediment supply as well as periodic storm events. Suggested periods of climatic cycles range from 10^4 to 10^5 years. Storm events are more frequent but erratic, and stratigraphic resolution is limited by physical continuity of resultant beds or surfaces. Finally, eustatic fluctuations create widely

Quantitative Dynamic Stratigraphy (1989), T.A. Cross, ed., Prentice Hall, p. 459–477.

correlative facies shifts. However, correlation is complicated by the complex response of different depositional systems to modest base-level changes and by the interplay of other variables.

Correlation is highly interpretative. Correlations should always be tested against alternative hypotheses, including the hypothesis that physically correlative surfaces or strata do not exist. Further, correlations should be made within the context of the depositional system(s) under analysis.

INTRODUCTION

> *In classifying together distant unknown formations under one name, in giving them a simultaneous origin, and determining their date, not by the organic remains we have discovered, but by those we expect to hypothetically hereafter to discover in them; we have given one more example of the passion with which the mind fastens on general conclusions, and of the readiness with which it leaves the consideration of unconnected truths.*

> — Adam Sedgwick, 1831, Presidential Address, Geological Society of London, recanting the diluvian theory.

Sedimentary basins have been described as tape recorders of geologic history. They contain a stratigraphic record of the progression of events, large and small, that have influenced erosion and transport of sediment from adjacent land areas and the chemistry and circulation of waters within the basin itself. Their depositional and erosional record reflects the interplay of tectonism, climate, and fluid dynamics. From this interplay emerges a variety of physical stratigraphic features—beds, facies, bounding surfaces, and repetitive sequences or cycles—that can be defined and correlated.

Development of a consistent lithostratigraphic and chronostratigraphic correlation framework is a most basic, yet challenging task for the basin analyst. My basic tenet is that correlation is most effective and accurate when it is based upon an understanding of the depositional and tectonic processes that created the basin fill. This discussion is best considered as a set of personal observations and working concepts that are grounded in the literature of process sedimentology and tested by two decades of application to regional and local problems in stratigraphic interpretation in terrigenous clastic basin fills.

CORRELATIVE ELEMENTS OF BASIN FILLS

Clastic basin fills contain three fundamental stratigraphic components: depositional systems, hiatal surfaces, and bedding architecture. Depositional systems (Fisher and McGowen, 1967) are the building blocks of the sedimentary pile. Each depositional system consists of an assemblage of process-related facies and is the stratigraphic record of a major paleogeomorphic system. Terrigenous clastic depositional systems are commonly recognized on the basis of the geometry of their framework sand units. They grade laterally into adjacent systems, forming logical associations of paleogeographic elements. They may, however, be separated stratigraphically from underlying and overlying systems by unconformity or other hiatal surfaces. Analysis of depositional systems has focused on the three-dimensional sedimentary framework of the basin and has become a major facet of basin analysis (e.g., Galloway and Hobday, 1983; Miall, 1984).

Hiatal surfaces are created by periods of nondeposition, very slow deposition, or erosion at many temporal and spatial scales. They include unconformities, disconformities, hardgrounds, and condensed sections. Hiatal surfaces have several origins, and may themselves be part of a migratory facies tract (Curray, 1964; Swift, 1968) or the direct result of active sedimentary or tectonic loading and subsidence of the basin margin (Quinlan and Beaumont, 1984). Although hiatal surfaces commonly approximate time lines, like facies boundaries they may sometimes cross time lines and separate time-equivalent deposits in some geologic settings (Christie-Blick et al., in press). Condensed intervals are thin sedimentary veneers that record very slow deposition. The purist may correctly argue that a bed, however thin, is not a surface. However, condensed sections are little more than bundles of diastems bound by a gossamer of sediment. Semantic arguments only obscure the important genetic relationship between unconformities and condensed intervals. In terrigenous clastic basin fills, condensed intervals may be recognized by a variety of compositional or paleontologic attributes and commonly provide time-stratigraphic markers (Busch, 1971; Loutit et al., in press). Thin, widespread, highly fossiliferous mudstones, hemipelagic drapes, marls, and limestone beds reflect terrigenous sediment starvation. Widespread, thin marker beds of unusual sediments, such as glauconite beds (green sand), phosphate horizons, and siliceous, radioactive (hot) shales are also indicators of slow rates of deposition. In subaerial environments laterally extensive, mature paleosoils or coaly zones record slow rates of clastic sediment accumulation (Fisher and McGowen, 1967; Retallack, 1983; Kraus, 1987). Hardgrounds and downlap surfaces indicate long periods of nondeposition. Unconformities are created by subaerial erosion, such as valley incision, and by submarine erosion, such as ravinement and submarine canyon incision.

Bedding architecture is a general term used here to describe the geometric relationships of stratification or bedding surfaces within depositional systems and at bounding surfaces. Terminology (onlap, downlap, offlap, toplap, baselap, etc.) is derived largely from seismic stratigraphy (Mitchum et al., 1977), but also has been used in facies analysis (e.g., Galloway and Brown, 1973; Cant, 1984) where the contrasting geometries of progradational, aggradational, and laterally accreted sedimentary units have long been recognized. The unique capability of reflection-seismic data to resolve surfaces and the geometry of stratification within the framework of such surfaces has greatly expanded our understanding and use of bedding architecture in interpreting depositional processes and history.

In summary, an analysis of a sedimentary basin fill must incorporate and reconcile the geometric, genetic, and temporal relationships among three-dimensional lithofacies distribution (depositional systems), bedding geometries, bounding surfaces, and condensed intervals. The initial requirement for regional correlation and mapping is the definition of operational stratigraphic units that group all sedimentary facies which record an episode of basin filling within a common paleogeographic framework. Delineation, mapping, and interpretation of such genetic packages will then define the principal depositional systems preserved within the basin fill, and provide a framework for detailed facies interpretation and correlation. Busch (1971) provided an early, integrative model for such correlation within delta systems.

Stratigraphic Organization of Basin Fills

Early in the history of stratigraphic analysis, geologists noticed that deposition within many sedimentary basins appeared to be cyclic or episodic. Recurrent facies organization or depositional motifs characterized thick stratigraphic intervals. With the increasing availability of subsurface data, which allowed three-dimensional stratigraphic synthesis, episodic depositional patterns were found to occur at a variety of temporal and geographic scales.

In a seminal paper, Frazier (1974) developed a conceptual model for defining genetic stratigraphic units that was based on the pervasive episodicity observed in the stratigraphic fabric of marine basins. Using three-dimensional studies of the Quaternary depositional systems of the northern Gulf Coast basin as a field laboratory, Frazier distilled several sedimentologic principles that form a foundation of genetic stratigraphic correlation and analysis. At the regional scale, marine clastic basin filling occurs as a series of *depositional episodes*—pulses of sediment input that prograde the basin margin and are separated from older and younger episodes by periods of regional or interbasinal transgression and coastal submergence (Fig. 1). Each episode deposits a *depositional complex*, an assemblage of related sedimentary facies derived from common sources and separated from deposits of underlying and overlying complexes by marine tongues, hiatal surfaces, or condensed intervals recording periods of transgression and submergence. Within the depositional complex, genetic facies and depositional architectures display predictable, ordered relation-

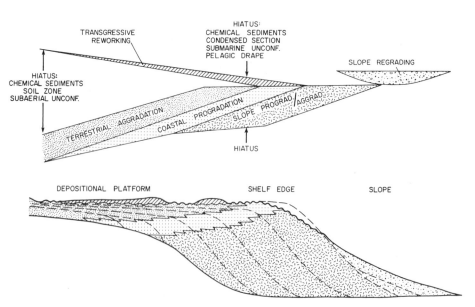

Figure 1. Schematic representation of a simple depositional episode (upper time-space diagram) and its resultant genetic stratigraphic sequence (lower cross section). The genetic stratigraphic sequence contains an orderly arrangement of facies of related depositional systems bounded by marine hiatal surfaces produced by shelf/basin sediment starvation following transgression and flooding of the basin margin. From Galloway (1987); original modified from Frazier (1974).

ships and distributions (Fig. 1). Together, depositional episodes and complexes constitute the principal genetic time- and rock-stratigraphic subdivisions of basin history and fill.

Within major depositional episodes, Frazier recognized short-term depositional events, which created subregional and/or high-frequency stratigraphic units termed facies sequences. Depositional events have several possible origins, including autogenic processes (such as channel avulsion and delta-lobe switching) and secondary base-level or climatic fluctuations. Correlation commonly is limited within a particular depositional system, even when the cause is external, because of the complex response of depositional systems to climatic, eustatic, and other external variables (Schumm, 1981; Cant, 1984; Morton and Price, 1987).

At the shortest time scale, catastrophic events, such as storms, floods, or earthquakes, may punctuate the stratigraphic record. Recognition of tempestites (Aigner, 1985) or catastrophites provides the finest scale of physical stratigraphic correlation and is one basis of fine-scale event stratigraphy. Response of a depositional system to such catastrophic events is environment-specific. Physically correlative beds or surfaces are likely to have areal distributions that are much less than those of the depositional system effected.

Depositional Complexes and Sequences

Three approaches to the stratigraphic analysis of sedimentary basins have been developed and applied in recent years. All share, to varying degrees, a common heritage in the conceptual framework of depositional episodes and complexes as articulated by Frazier (1974). The *seismic sequence* was introduced and described in the well-known AAPG Memoir 26. The concept of an unconformity-defined stratigraphic unit, termed a *depositional sequence*, evolved from the strengths of seismic records in displaying surfaces and bedding architecture and from an assumed control of depositional patterns by rapid global sea-level change (Vail et al., 1984). An alternative sequence-stratigraphic unit incorporating both unconformity and maximum-marine-flooding surfaces and three-dimensional depositional systems—the *genetic stratigraphic sequence* (Galloway, 1987 and in press)—has been utilized in analysis of Cenozoic deposits of the Gulf basin (Galloway et al., 1986; Galloway et al., in press). Several key differences distinguish the genetic stratigraphic sequence from the Vail et al. sequence model.

First, a depositional sequence is "the succession of sediments deposited during a complete sea-level cycle, that is, from a sea level fall to subsequent rise and ending with the next fall" (Haq et al., 1987). The genetic stratigraphic sequence, building upon Frazier's concept of the depositional episode and complex, is an assemblage of related depositional systems derived from common sources and deposited between periods of maximum marine flooding of the coastal plain.

Second, as a consequence of this selection of contrasting interpreted relationships between major sedimentary pulses and relative base-level stability, proposed sequence-bounding surfaces are shifted in phase by 180°. Vail et al. (1984) equated maximum progradation of the basin margin with periods of eustatic fall and lowstand. The result of such eustatic fall is development of a widespread subaerial erosion surface (type 1 or 2 unconformity) that constitutes a time-equivalent depositional sequence boundary. By contrast, the genetic stratigraphic sequence incorporates the pulse of sediment input and progradation (regardless of cause) as the principal depositional event and separates such events by periods of transgression and flooding of the depositional platform.

Thirdly, the two sequence models place different emphases on the timing, process, and role of shelf-margin erosion and retrogradation. The Vail et al. model restricts canyon cutting and submarine fan deposition to periods of low sea level, subaerial valley incision, and deposition directly on the upper slope (Vail et al., 1984; Mitchum, 1985). In the genetic sequence model, shelf margin and slope erosion and retrogradation are ongoing processes controlled by the interplay of inherent instability of the platform margin, temporal and spatial variation in rate of sediment supply, basin hydrography, coastal and platform morphology, and base-level change (Galloway et al., 1988).

The temporal framework and facies stratigraphy of a genetic stratigraphic sequence is shown schematically in Figure 1. The time-space diagram (top) and lithostratigraphic cross section (bottom) compare the time- and rock-stratigraphic organization of a simple depositional episode and resultant sequence. For simplicity, no depositional events (*sensu* Frazier) are shown, and no additional sediment is supplied during transgression. The sequence consists of offlap components, transgressive components, and bounding hiatal intervals reflected in the stratigraphic record as disconformities or condensed intervals. Offlap components consist of a variety of environments and depositional facies, including terrestrial aggradational facies of the coastal plain and progradational deposits of the shore zone. Progradation initially builds across the earlier, flooded depositional platform, and then beyond the foundered margin onto mixed progradational and aggradational deposits of the slope. Transgressive components consist of reworked shore-zone and shelf facies deposited during and soon after shoreline retreat, and an onlapping apron of gravitationally resedimented upper-slope and shelf-margin deposits at the toe of the slope. In reality, shoreline retreat often is accompanied by substantial sediment accumulation, recording an extended period of onlap (Curray, 1964). The term retrogradation has been used to describe such long-term onlap. Finally, the genetic depositional sequence is bounded on its basinward periphery by two stratigraphic surfaces that record the relative sediment starvation of the outer shelf and slope during maximum marine flooding. Thus the seaward part of a genetic stratigraphic sequence is bounded by a veneer of reworked transgressive deposits (perhaps including a ravinement surface) and a regional marine condensed interval or submarine unconformity (Fig. 1). An internal hiatal surface may also occur within the updip portion of the sequence. As the shore zone, which is the locus of deposition, shifts basinward, the inner coastal plain becomes a sediment bypass zone. Minor relative falls in base level or peripheral uplift due to load-induced subsidence of the basin margin (Quinlan and Beaumont, 1984) readily lead to nondeposition or even valley incision and low-angle truncation of older strata (see Winker, 1979; Galloway, 1981; and Galloway et al., 1986, for discussions).

A generalized genetic stratigraphic sequence model applicable to sediment-rich offlap continental margins, such as the northwest Gulf of Mexico (Fig. 2) incorporates an additional discontinuity surface produced by gravitational deformation of the thick sedimentary prism (Galloway, 1987). The resultant stratigraphic architecture provides an abundant, and complex array of facies sequences, erosional surfaces, and condensed intervals for both regional and local correlation. The principal genetic stratigraphic sequences remain, how-

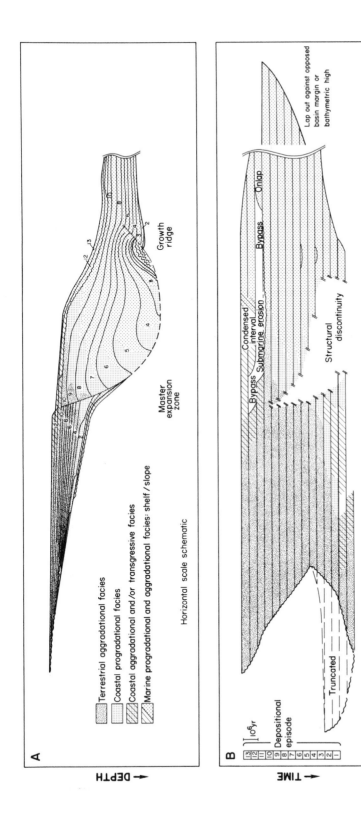

Figure 2. Chronostratigraphic diagram of a depositional episode (B) and stratigraphic dip cross section (A) of an ideal genetic depositional sequence. The genetic sequence is bounded by sediments or surfaces produced during transgressional flooding of the basin margin and contains both erosional and structural discontinuity surfaces resulting from load-induced uplift and gravity deformation. From Galloway (in press).

465

ever, the fundamental units of the basin fill and are readily recognized, correlated, and mapped with outcrop, well, or seismic data.

Regardless of the choice of sequence models, physical stratigraphic correlation relies in recognition and tracing of hiatal surfaces. Both subaerial-unconformity and marine-flooding-surface boundaries have advantages in different paleogeomorphic settings.

In plan view, sequences typically consist of the deposits of several individual depositional systems. In marine basins, a common pattern consists of one or more major extrabasinal fluvial and associated delta systems (forming coastal and shelf-margin headlands) and flanking shore-zone, shelf, and coastal-plain complexes of the interdeltaic coastal bight (Fig. 3). If water depths are adequate, slope systems also will extend seaward and underlie the paralic facies sequences. Deltas form the principal depocenters, particularly for sand. Marine reworking or gravitational remobilization disperses sediment to the shore-zone, shelf, and slope systems. The resultant mosaic of depositional environments limits the geographic extent and correlativity of facies, lithosomes, marker beds, and secondary surfaces. This limitation is mitigated by the interdependence of many different environments within the larger sediment-dispersal system and base-level framework, and leaves much potential for intersystem correlation of large and even small depositional events. For example, lobe switching of the Mississippi delta system influences the episodic progradation of the adjacent interdeltaic chenier plain (Frazier, 1974). The delta system is, by contrast, much less sensitive to the minor changes in Holocene sea level that influence interdeltaic shore zones.

Both the depositional sequence and genetic sequence paradigms apply best to marine (or lacustrine) basins that have filled by marginal progradation. Modified or alternative models applicable to nonmarine and deep, tectonically bounded marine basins remain to be developed. Widespread bounding surfaces and repetitive episodes of filling appear to have tectonic origins (Gloppen and Steele, 1981) and may provide the basis for sequence definition in such basin settings.

In summary, clastic deposition commonly displays repetitive or episodic patterns of deposition at a hierarchy of scales. Correlation is possible at each level, allowing for a high degree of stratigraphic resolution in many settings. However, the areal extent of the correlative unit or surface typically decreases in proportion to the frequency of its development. Interpretive correlation of logical process-response coupling between related (contiguous) depositional systems may extend the range of the physical stratigraphic framework. At least it is a challenging alternative to correlations based, by default, upon assumed synchroneity of local events.

DEPOSITIONAL SYSTEMS, FACIES MODELS, AND CORRELATION

The number of process-defined clastic depositional systems is limited. A complete sediment dispersal complex might include alluvial-fan, fluvial, delta, shore-zone, shelf, and slope/basin systems. In addition, independent eolian and lacustrine systems are found in some basins. Knowledge of the processes and facies that characterize each system provides the basis for developing an interpretive correlation framework. It also has demonstrated the dangers of basing correlations on apparent lithosome continuity or similarity of vertical sequence. For example, similarity of vertical facies sequence (and resultant log response) for channel mouthbar sandstones more likely reflects the inherent succession of process

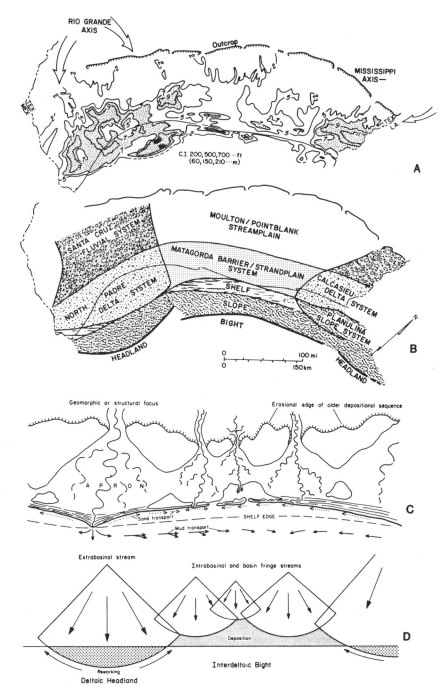

Figure 3. Depositional systems comprising the lower Miocene genetic sequence and their paleogeographic relationships. A. Net sandstone isolith map defining principal lithofacies trends and depocenters. B. Interpreted depositional systems. C. Simplified paleogeography. D. Schematic coastal plain depositional elements. Modified from Galloway et al. (1986).

environments typical of delta progradation than contemporaneity or physical continuity of the sand bodies.

The depositional styles and extent of our three-dimensional documentation of well-described examples determine the degree of stratigraphic resolution that may be achieved by physical correlation within a particular depositional system. Depositional process and facies models are well developed for fluvial, alluvial-fan, delta, and shore-zone systems. Highly detailed correlation at the subregional scale is commonly possible in delta and shore-zone deposits. Markers are abundant, and environments shift dramatically in response to autogenic processes or to minor extrinsic controls on base level or sediment supply. Deposits of storm events may be recognized.

Though fluvial and fan systems are equally well understood, they provide a more difficult correlation problem. Facies and surfaces are discontinuous and similar in appearance. Internal markers are few or nonexistent. At best, packages of stacked channel fills can be differentiated from muddier interchannel facies. Interbedded lacustrine or organic-rich swamp facies may provide the most continuous beds for establishing correlation form lines. Because large areas of the alluvial plain or fan surface may be far removed from the active sediment dispersal path at any one time, mature paleosoils (hiatal surfaces) provide particularly useful stratigraphic markers (Retallack, 1983; Kraus, 1987).

The subaqueous deposits of shelf systems commonly contain an array of subtle marker horizons and surfaces, although shelf depositional processes and syntheses of facies models are active fields of investigation (McCave, 1985). Submarine-slope and adjacent basin-plain systems offer some of the most complex problems of correlation. Both process and facies models are in a state of flux (e.g., Mutti, 1985; Stow, 1985; Bouma et al., 1985; Mutti and Normark, 1987). Gravity-resedimentation processes typically leave a heterogeneous strati-graphic record containing many discontinuous surfaces and sediment bodies. Potentially useful correlation markers include very large, catastrophic turbidites (Mutti, 1985) and widespread pelagic drapes (Bouma et al., 1985). Physical continuity of individual facies is highly variable and generally unpredictable.

Eolian systems, though characterized by their comparative textural uniformity, contain a hierarchy of surfaces. These include super surfaces that correlate regionally and reflect response of ergs to major geological changes (Kocurek, 1988).

Resolution of Physical Correlation

Given the availability of a sufficient data base, the hierarchy of features preserved within terrigenous stratigraphic successions commonly allows the development of detailed physical stratigraphic correlation frameworks. However, except in certain geologically special circumstances (e.g., rapid Quaternary climatic and resultant large-scale sea-level fluctuations), abundance, and therefore temporal resolution, of correlatable features typically decrease in proportion to the areal extent of the feature. Correlatable depositional events and/or surfaces can be classified conveniently as sequence scale, depositional system scale, or depositional environment (facies-association) scale. In reality, many of the processes and resultant stratigraphic features found at the system scale also occur at the facies-association scale. Thus the latter two scales will be discussed together.

DEPOSITIONAL SEQUENCE RESOLUTION   Regionally correlative genetic stratigraphic sequences commonly exhibit time spans of 10^6 years, as do the third-order eustatic depositional sequences of Vail et al. (1977). Tertiary depositional episodes of the northwest Gulf of Mexico extended from 1 to 8 m.y. in duration (Galloway, 1987). Comparable Cretaceous depositional sequences of the Rocky Mountain foreland basin were deposited over intervals ranging from 3 to 7 m.y. (Weimer, 1984). In contrast, presumed glacio-eustatic sequences of the late Paleozoic were shorter, ranging from 0.3 to 0.8 m.y. in duration (Driese and Dott, 1984; Heckel, 1986).

Because genetic stratigraphic sequences are, by definition, regional to interregional features of basin fills, they provide a basic and typically obvious physical-stratigraphic correlation framework with a resolution of about 1 m.y. Finer subdivision may be possible where prominent secondary events of regional extent, such as transgressive pulses, can be recognized within the sequence (e.g., Galloway et al., 1986; Armentrout, 1987). Alternatively, genetic stratigraphic sequences may be subdivided into operational units based on regional correlation of approximately equivalent horizons or of the lower progradational and upper retrogradational parts of the cycle (e.g., Galloway et al., 1982).

DEPOSITIONAL SYSTEM RESOLUTION   The degree of temporal resolution by physical stratigraphic correlation within depositional systems is highly dependent upon the type of depositional processes that characterize the system, the basin setting, and extrinsic variables such as climatic or sea-level stability.

Autogenic processes are particularly important at the system scale. They are best understood in systems (fluvial, delta, and submarine fan) in which channelized flow is periodically diverted by avulsion and the locus of deposition displaced. Although data allowing the dating of avulsion events are sparse, Leeder (1978) compiled existing information and estimated major channel avulsions in aggrading fluvial systems to recur at an average period of 10^3 years. Detailed magnetostratigraphy allowed Raynolds and Johnson (1985) to estimate that Neogene fluvial depositional events occurred at any particular position on the Himalayan foredeep on the order of every 10^4 to 10^5 yr. Thus recognition and correlation of the stratigraphic products of avulsion have the potential to provide a resolution of depositional events within a particular depositional system at time scales of thousands to tens of thousands of years. However, correlative markers—widespread paleosoil horizons, coal beds, interchannel lake shales, delta destructional facies, geochemically or paleontologically condensed intervals, or hemipelagic drapes—are necessarily restricted to those parts of the depositional system that were temporarily isolated from active sediment supply.

Less well understood is the evidence for inherent episodicity that has been well documented by Schumm (1977) in scale model as well as natural fluvial and fan systems. These studies have demonstrated patterns of repetitive, alternating incision and aggradation of channels under stable environmental conditions. The consequence of this episodicity would be pulses of sediment supply entering the depositional basin at different times through each major and minor fluvial system. The stratigraphic consequences of such episodicity in large fluvial systems remains speculative. However, fluvial systems are demonstrably characterized by episodic behavior and complex response to geologic change (Schumm, 1981), features typical of chaotic systems.

Tectonic and related geomorphic events may impose correlative structure within specific depositional systems. Two examples illustrate the possibilities. Within slope depositional systems, tectonic events commonly trigger large-scale gravity resedimentation. Mapped continental margin slump and debris flow deposits cover as much as thousands of square kilometers. Type I turbidites of Mutti (1985) may well have such an origin. Recurrence rates of major earthquake events at 10^2 to 10^3 years (Vita-Finzi, 1986) likely define the maximum degree of resolution obtainable by detailed correlation of such catastrophites.

Tectonism or longer-term geomorphic evolution within the zone of fluvial sediment transfer can shift the position at which a trunk stream enters a basin, or even the ultimate basin itself. The Yellow River has alternated within historic time between the Yellow Sea margin and the Gulf of Bohai at a rate of approximately 10^3 yr (Shen, 1979). Galloway (1981) noted that both the Catahoula (Oligocene) and Oakville (Early Miocene) paleo Rio Grande fluvial systems alternated throughout their history between two principal depositional axes located more than 100 km apart on the northwest Gulf coastal plain (Fig. 4). Maximum time intervals between such large-scale axis switching are calculated to have been on the order of 10^5 yr. The delta systems fed by these fluvial systems show high-frequency correlative depositional events with dimensions of 100 to 300 km along strike (Fig. 4).

Climatic changes influence runoff, sediment yield, and the nature and frequency of storm events, and thus may impart correlative stratigraphic features within depositional systems. Relationships between climate and sediment yield are complex and not well understood (Walling and Webb, 1983). The most dramatic responses seem to be associated with periods of climatic change. Depositional effects are most direct within fluvial systems but may initiate complex responses in related deltaic and coastal systems. Climatic fluctuations have predictable spans of 10^3 to 10^6 years. Storm events and resultant tempestites (Aigner, 1985) likely are environmentally specific. Periods of 10^0 to 10^2 years offer the potential for a high degree of temporal resolution, but usually within restricted areas of shelf and shore-zone systems. Regional correlation usually relies on tracing distinctive packages of multiple storm beds.

Finally, variations in the oceanography affect the patterns of sediment distribution and deposition within marine basins, sometimes over large areas. Shifting current systems both influence and respond to deposition. A common result is the development of many locally to widely correlative electrical or gamma-ray log markers that reflect slight differences in the texture or composition of suspended load deposited in shelf and shallow-basin settings. A high-resolution correlation network can be constructed within marine shales using such log markers. In addition, minor eustatic changes (arbitrarily defined as less than 10 m) influence shelf and shore-zone facies sequences. However, in clastic coastal-plain, shore-zone and nearshore depositional systems, their signature may be difficult if not impossible to distinguish from that of other, less predictable mechanisms. Large-scale, rapid eustatic changes, such as those characterizing the Quaternary, produce well-defined, regional cyclostratigraphic markers and surfaces that may be correlated regionally within and between depositional systems (e.g., Frazier, 1974; Armentrout, 1987; COSOD II Working Group 1, 1987)

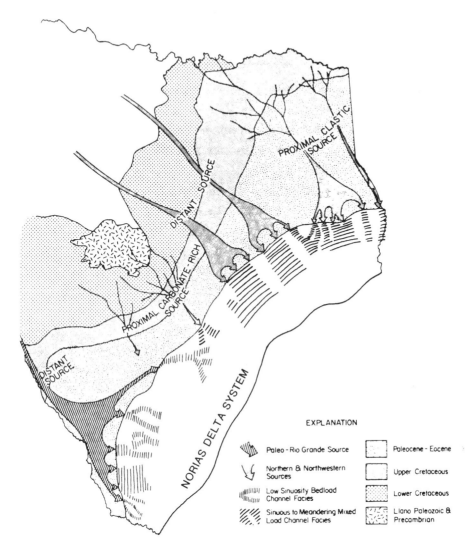

Figure 4. Interpretive paleogeographic map of the late Oligocene Catahoula fluvial systems of the Texas coastal plain. The map is based on detailed lithofacies mapping (Galloway, 1977). The major paleo Rio Grande fluvial system shows repeated shifts among valley axes dispersed more than 300 km north to south along the South Texas coastal plain. Shifts occurred landward of the Oligocene depositional coastal plain, indicating a paleogeomorphic or tectonic control of valley axes position. This fluvial system was the principal source of sediment for the late Oligocene depocenter, the Norias delta system (Galloway et al., 1982), which shows repeated lobe shifts caused by repeated relocation of the fluvial axis. Similar but more localized shifts characterized other major fluvial systems of the northeastern coastal plain and associated delta system.

Examples: Reality and a Caution

Figures 5 and 6 illustrate two examples of physical stratigraphic correlation at a hierarchy of scales. A subregional strike cross section (Fig. 5) of a portion of the upper and lower Wilcox genetic depositional sequences (representing about 10 m.y.) illustrates the variable degree of

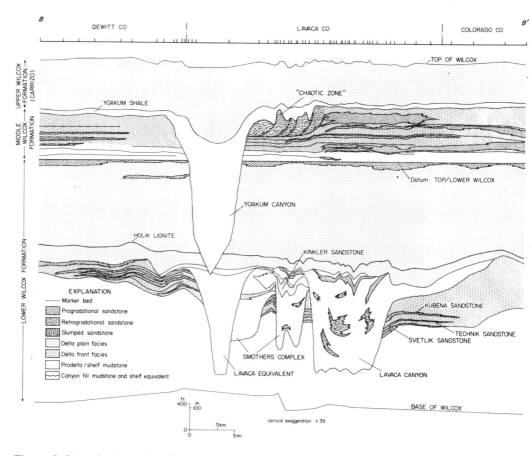

Figure 5. Subregional stratigraphic strike cross section through the downdip Wilcox (Paleocene-Eocene) genetic sequences. The lower/middle and upper Wilcox sequences are separated by the fossiliferous, glauconitic Yoakum shale -- a regional condensed section associated with large-scale submarine canyon excavation (Dingus, 1987; Galloway et al., 1988). Distal delta front and shoreline progradational and retrogradational sand facies show moderate continuity and are bounded by thin, low-resistivity marine-shale marker beds. These facies sequences record short-term depositional events associated with autogenic delta-lobe switching. Delta-plain facies sequences contain few correlative features, as do submarine canyon fills. An exception is the Holik lignite, which is a widespread (ca. 100 km along strike) marker probably deposited during abandonment and transgression of the western lobes of the Lower Wilcox Rockdale delta system.

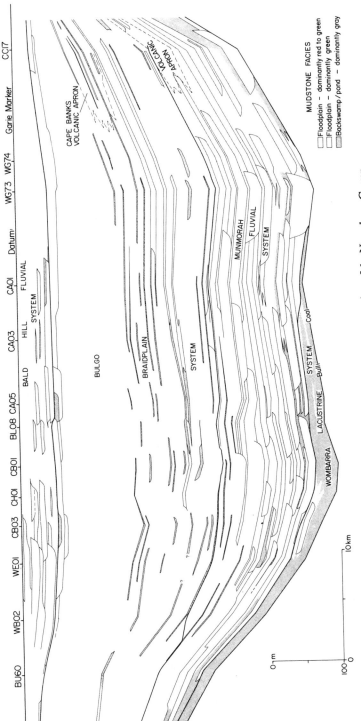

Figure 6. Correlation and interpreted facies cross section of the Narrabeen Group (Early Triassic) of the Sydney Basin, Australia. Correlative facies include lacustrine mudstones and laminites (Wombarra lacustrine system), which record widespread lacustrine flooding of the basin, and paleosoil horizons within floodplain mudstones, which record periods of nondeposition across areas of the floodplain removed from active channel overflow. Limited areal extent of the paleosoils indicates the dominance of autogenic processes within the fluvial systems.

473

detail possible in physical-stratigraphic correlation of shelf, delta-front/prodelta, delta-plain, and fluvial facies complexes. In addition, several large submarine canyon fills transect the paralic deposits. Numerous correlative units allow detailed correlation within marine shelf and prodelta mudstone, revealing complexities in depositional and structural architectures. Progradational and transgressive/retrogradational sandstone facies show a high degree of continuity at the scale of the section. Although time-transgressive in detail, they constitute facies sequences of local depositional events of relatively short duration. In contrast, aggradational delta-plain facies contain few features that can be traced confidently for more than a few kilometers. Both canyon-forming erosional and slump (tectonic) surfaces are delineated by careful physical stratigraphic correlation. Canyon fills (a variety of slope system) are dominated by discontinuous, gravity-resedimented deposits that defy all efforts at correlation between even the most closely spaced wells.

Triassic fluvial and lacustrine deposits of the Sydney Basin (Fig. 6) illustrate the more difficult correlation problem in nonmarine systems. The section traverses a single family of depositional systems, and correlations shown generally are limited within a system. Three levels of correlation can be recognized. Major stratigraphic units are delineated by regional, muddy lacustrine deposits (Bulli coal and overlying Wombarra Shale at the base; Garie Formation forming the stratigraphic datum for the section). Within the interval defined by these regional "maximum-flooding" units, correlation of five operational units delineated by regional changes in fluvial depositional style (base of the Bald Hill Claystone) or gamma-ray log markers (such as the green marker) was accomplished. Mapping of operational units delineated four depositional systems in this area. These systems, in turn, are the products of three tectonically and climatically defined depositional episodes. The closely spaced continuous cores allowed recognition and correlation of paleosoil horizons. The paleosoil correlations, in turn, provided a basis for a tentative correlation of discrete channel fill units in the lower part of the section. However, lateral discontinuity of all facies of the upper Bulgo operational unit (readily confirmed by outcrop study) precluded further correlation and subdivision of this facies sequence. The total section illustrated represents about 5 m.y. Obviously, time resolution is limited in such a dominantly nonmarine basin fill.

A final caution is in order. Zeller (1964) performed an experiment in geopsychology that demonstrated a critical and sobering fact. The most important decision in correlation is not the selection of features to correlate or the methodology to be used. Rather, it is the decision that correlation exists at all. Having made that decision (reinforced by a model that suggests an expected resolution and extent of correlative events), correlation is almost inevitable, even when the "stratigraphy" correlated was created from a random number table. In short, correlation is too easily in the eye of the beholder, regardless of its physical stratigraphic reality.

Real depositional systems are complex, reflecting the interplay of many variables. Correlation is required to unravel the interplay; however, at the same time, correlation must continually be critically tested and judged against realistic process models and independent lines of evidence. At best, physical stratigraphic correlations are approximations of reality. At worst, they lead to misinterpretation of fundamental attributes of basin fill stratigraphy and geologic history.

ACKNOWLEDGMENTS

I thank Ashton Embry and Doug Cant for critical and helpful reviews of the manuscript.

REFERENCES CITED

Aigner, T., 1985, Storm depositional systems: Berlin, Springer-Verlag, Lecture notes in earth sciences 3, 174 p.

Armentrout, J.M., 1987, Integration of biostratigraphy and seismic stratigraphy: Pliocene-Pleistocene, Gulf of Mexico: Proceedings of Gulf Coast Section, Society of Economic Paleontologists and Mineralogists Foundation Eighth Annual Research Conference, p. 6-14.

Bouma, A.H., Normark, W.R., and Barnes, N.E., 1985, eds., Submarine fans and related turbidite systems: New York, Springer-Verlag, 351 p.

Busch, D.A., 1971, Genetic units in delta prospecting: American Association of Petroleum Geologists Bulletin, v. 55, p. 1137-1154.

Cant, D.J., 1984, Development of shoreline-shelf sand bodies in a Cretaceous epeiric sea deposit: Journal of Sedimentary Petrology, v. 54, p. 541-556.

Christie-Blick, N., Mountain, G.S., and Miller, K.G., in press, Seismic stratigraphic record of sea-level change, in Revelle, R., ed., Sea-level change: National Research Council Studies in Geophysics, National Academy of Sciences.

COSOD II Working Group I, 1987, Scientific goals of an ocean drilling program designed to investigate changes in the global environment: Report of the second conference on scientific ocean drilling, JOIDES, p. 15-46.

Curray, J.R., 1964, Transgressions and regressions, in Miller, R.L., ed., Papers in marine geology, Shepard commemorative volume: New York, McMillan Co., p. 175-203.

Dingus, W.F., 1987, Morphology, paleogeographic setting, and origin of the Middle Wilcox Yoakum canyon, Texas coastal plain [M.A. Thesis]: Austin, Texas, University of Texas at Austin, 78 p.

Driese, S.G., and Dott, R.H., Jr., 1984, Model for sandstone-carbonate "cyclothems" based on upper member of Morgan Formation (Middle Pennsylvanian) of northern Utah and Colorado: American Association of Petroleum Geologists Bulletin, v. 58, p. 574-597.

Fisher, W.L., and McGowen, J.H., 1967, Depositional systems in the Wilcox Group of Texas and their relationship to occurrence of oil and gas: Gulf Coast Association of Geological Societies Transactions, v. 17, p. 105-125.

Frazier, D.E., 1974, Depositional episodes: Their relationship to the Quaternary stratigraphic framework in the northwestern portion of the Gulf basin: The University of Texas at Austin, Bureau of Economic Geology Geological Circular 71-1, 28 p.

Galloway, W.E., 1977, Catahoula Formation of the Texas coastal plain: Depositional systems, composition, structural development, ground-water flow history, and uranium distribution: The University of Texas at Austin, Bureau of Economic Geology Report of Investigations 87, 59 p.

Galloway, W.E., 1981, Depositional architecture of Cenozoic Gulf coastal plain fluvial systems: Society of Economic Paleontologists and Mineralogists Special Publication 31, p. 127-155.

Galloway, W.E., 1987, Depositional and structural architecture of prograding clastic continental margins: Tectonic influence on patterns of basin filling in the Gulf of Mexico: Norsk Geologisk Tidsskrift v. 67, p. 237-251.

Galloway, W.E., in press, Genetic stratigraphic sequences in basin analysis: Sequence architecture and genesis: American Association of Petroleum Geologists Bulletin.

Galloway, W.E., Bebout, D.G., Dunlap, J.B., Jr., and Fisher, W.L., in press, Cenozoic stratigraphy and depositional history, in Salvador, A., ed., Gulf of Mexico Basin: Geological Society of America, The Geology of North America.

Galloway, W.E., and Brown, L.F., Jr., 1973, Depositional systems and shelf-slope relations on cratonic basin margin, uppermost Pennsylvanian of north-central Texas: American Association of Petroleum Geologists Bulletin, v. 57, p. 1185-1218.

Galloway, W.E., Dingus, W.F., and Paige, R.E., 1988, Depositional framework and genesis of Wilcox submarine canyon systems, northwest Gulf Coast: American Association of Petroleum Geologists Bulletin, v. 72, p. 187-188.

Galloway, W.E., and Hobday, D.K., 1983, Terrigenous clastic depositional systems: New York, Springer-Verlag, 423 p.

Galloway, W.E., Hobday, D.K., and Magara, K., 1982, Frio Formation of Texas Gulf Coastal Plain: Depositional systems, structural framework, and hydrocarbon distribution: American Association of Petroleum Geologists Bulletin, v. 6, p. 649-688.

Galloway, W.E., Jirik, L.A., Morton, R.A., and DuBar, J.R., 1986, Lower Miocene (Fleming) depositional episode of the Texas Coastal Plain and continental shelf: Structural framework, facies, and hydrocarbon resources: University of Texas at Austin, Bureau of Economic Geology Report of Investigations 122, 78 p.

Gloppen, T.G., and Steel, R.J., 1981, The deposits, internal structure and geometry in six alluvial fan-fan delta bodies (Devonian-Norway)—A study in the significance of bedding sequences in conglomerates: Society of Economic Paleontologists and Mineralogists Special Publication 31, p. 49-70.

Haq, B.U., Hardenbol, J., and Vail, P.R., 1987, Chronology of fluctuating sea levels since the Triassic: Science, v. 235, p. 1156-1166.

Heckel, P.H., 1986, Sea-level curve for Pennsylvanian eustatic marine transgressive-regressive depositional cycles along midcontinent outcrop belt, North America: Geology, v. 14, p. 330-334.

Kocurek, G., 1988, First-order and super bounding surfaces in eolian sequences—Bounding surfaces revisited: Sedimentary Geology (in press).

Kraus, M.J., 1987, Integration of channel and floodplain suites, II. Vertical relations of alluvial paleosols: Journal of Sedimentary Petrology, v. 57, p. 602-612.

Leeder, M.R., 1978, A quantitative stratigraphic model for alluvium, with special reference to channel deposit density and interconnectedness, in Miall, A.D., ed., Fluvial sedimentology: Canadian Society of Petroleum Geologists Memoir 5, p. 587-596.

Loutit, T.S., Hardenbol, J., and Vail, P.R., in press, Condensed sections: The key to age dating and correlation of continental margin sequences, in Wilgus, C., et al., eds., Sea level changes—An integrated approach: Society of Economic Paleontologists and Mineralogists Special Publication 42.

McCave, I.N., 1985, Recent shelf clastic sediments, in Brenchley, P.J., and Williams, B.P.J., eds., Sedimentology—Recent developments and applied aspects: Oxford, Blackwell Scientific Publishers, p. 49-65.

Miall, A.D., 1984, Principles of sedimentary basin analysis: New York, Springer-Verlag, 490 p.

Mitchum, R.M., Jr., 1985, Seismic stratigraphic expression of submarine fans, in Berg, O.R., and Woolverton, D.G., eds., Seismic stratigraphy II: An integrated approach to hydrocarbon exploration: American Association of Petroleum Geologists Memoir 39, p. 117-138.

Mitchum, R.M., Jr., Vail, P.R., and Thompson, S., III, 1977, Seismic stratigraphy and global changes of sea level, part 2: The depositional sequence as a basic unit for stratigraphic analysis: American Association of Petroleum Geologists Memoir 26, p. 53-62.

Morton, R.A., and Price, W.A., 1987, Late Quaternary sea-level fluctuations and sedimentary phases of the Texas coastal plain and shelf: Society of Economic Paleontologists and Mineralogists Special Publication 41, p. 181-198.

Mutti, E., 1985, Turbidite systems and their relations to depositional sequences, in Zuffa, G.G., ed., Provenance of arenites: Boston, D. Reidel, p. 65-93.

Mutti, E., and Normark, W.R., 1987, Comparing examples of modern and ancient turbidite systems: Problems and concepts, *in* Leggett, J.K., and Zuffa, G.G., eds., Marine clastic sedimentology: Concepts and case studies: London, Graham and Trotman Ltd., p. 1-38.

Quinlan, G.M., and Beaumont, C., 1984, Appalachian thrusting, lithospheric flexure, and the Paleozoic stratigraphy of the Eastern Interior of North America: Canadian Journal of Earth Sciences, v. 21, p. 955-973.

Raynolds, R.G.H., and Johnson, G.D., 1985, Rates of Neogene depositional and deformational processes, north-west Himalayan foredeep margin, Pakistan: Geological Society Memoir 10, p. 297-311.

Retallack, G.J., 1983, A paleopedological approach to the interpretation of terrestrial sedimentary rocks: The mid-Tertiary fossil soils of Badlands National Park, South Dakota: Geological Society of America Bulletin, v. 94, p. 823-840.

Schumm, S.A., 1977, The fluvial system: New York, Wiley, 338 p.

Schumm, S.A., 1981, Evolution and response of the fluvial system, sedimentologic implications: Society of Economic Paleontologists and Mineralogists Special Publication 31, p. 19-30.

Shen, H.W., 1979, Some notes on the Yellow River: EOS (American Geophysical Union Transactions), v. 60, p. 545-547.

Stow, D.A.V., 1985, Deep-sea clastics: Where are we and where are we going? *in* Brenchley, P.J., and Williams, B.P.J., eds., Sedimentology—Recent developments and applied aspects: Oxford, Black-well Scientific Publishers, p. 67-93.

Swift, D.J.P., 1968, Coastal erosion and transgressive stratigraphy: Journal of Geology, v. 76, p. 444-456.

Vail, P.R., Mitchum, R.M., and Thompson, S., III, 1977, Seismic stratigraphy and global changes of sea level, part 4, *in* Payton, C.E., ed., Seismic stratigraphy—Applications to hydrocarbon exploration: American Association of Petroleum Geologists Memoir 26, p. 83-97.

Vail, P.R., Hardenbol, J., and Todd, R.G., 1984, Jurassic unconformities, chronostratigraphy, and sea-level changes from seismic stratigraphy and biostratigraphy, *in* Schlee, J.S., ed., Interregional unconformities and hydrocarbon accumulation: American Association of Petroleum Geologists Memoir 36, p. 129-144.

Vita-Finzi, C., 1986, Recent earth movements: New York, Academic Press, 226 p.

Walling, D.E., and Webb, B.W., 1983, Patterns of sediment yield, *in* Gregory, K.J., ed., Background to palaeohydrology: New York, Wiley, p. 69-99.

Weimer, R.J., 1984, Relation of unconformities, tectonics, and sea-level changes, Cretaceous of Western Interior, U.S.A., *in* Schlee, J.S., ed., Interregional unconformities and hydrocarbon accumulation: American Association of Petroleum Geologists Memoir 36, p. 7-35.

Winker, C.D., 1979, Late Pleistocene fluvial-deltaic deposition, Texas coastal plain and shelf [M.A. Thesis]: Austin, Texas, The University of Texas at Austin, 187 p.

Zeller, E.J., 1964, Cycles and psychology: Kansas Geological Survey Bulletin 169, p. 631-636.

28

Characteristics of a Field Data Base for Developing and Evaluating Quantitative Stratigraphic Models

E.J. Anderson and Peter W. Goodwin
Department of Geology, Temple University, Philadelphia, PA 19122 USA

Abstract

Evaluation of quantitative stratigraphic models requires comparison of simulations with consistently derived field data. Because theory influences observations, the assumptions and qualitative models inherent in the construction of a particular data base should be clearly delineated and accommodated in the evaluation process. Comparison of simulations with a mixture of data collected with the biases of different qualitative stratigraphic models will yield inconsistent evaluations. The most valid testing of quantitative models will occur if the assumptions and theory underlying the quantitative models are similar to or compatible with those of the qualitative stratigraphic models applied in collecting the field data.

An extensive field data base has been developed for the Lower Devonian Helderberg Group of Pennsylvania and New York based on the assumption of the PAC hypothesis, a qualitative model of small-scale allogenic stratigraphic accumulation. In this data base all Helderbergian facies, peritidal to deep shelf, are divisible into PACs. PACs, in turn, are grouped into PAC sequences, sets of 2 to 6 PACs, recognized by larger patterns of facies change. PACs and PAC sequences, which have recurrence intervals of 10^4 and 10^5 yr, respectively, are correlative for 100s of kilometers.

Basinwide correlation of PACs and PAC sequences reveals systematic loss of PACs toward basin margins at several stratigraphic horizons in the Helderberg Group. These stratigraphic discontinuities, termed cryptic unconformities, are the products of major sea-level drops accompanied by differential subsidence or uplift and may have a recurrence interval as small as 10^5 yr. Correlation of PACs also permits detailed analysis of small-scale thickness differences resulting from differential subsidence.

If this structure of cycles, stratigraphic discontinuities and thickness variability observed in the Helderberg Group is representative of the stratigraphic record, quantitative two-dimensional simulations of such basin fills should contain the following structural elements: (1) a pervasive cyclic structure at more than one scale in all facies from peritidal to deep shelf; (2) cycles defined by patterns of facies change, rather than by particular lithologies; (3) relatively frequent small-scale cryptic unconformities (recurrence interval on the order of 10^5 yr). These elements of stratigraphic fabric appear to place at least the following constraints on variables used in simulation models: (1) a sediment accumulation variable which simulates aggradation simultaneously throughout the full extent of a cycle rather than progradation from a point source; (2) a fluctuating eustatic variable at more than one scale (in the Milankovitch band) to simulate PACs, PAC sequences and cryptic unconformities;

(3) a differential tectonic variable (uplift and/or subsidence) to account for patterns of erosion at cryptic unconformities and patterns of lateral thickness variability within correlative sets of PACs; and, (4) a lag-time variable which is independent of depth, permitting simulation of strongly asymmetric rock cycles (PACs) in all facies areas.

INTRODUCTION

Quantitative models that simulate the stratigraphic record represent a rapidly developing methodology for evaluating the relative significance of the interdependent variables which may have produced this record. Typically a particular quantitative model is designed with a specific stratigraphic pattern in mind. For example, Read et al. (1986) developed a one-dimensional quantitative model that simulates both meter-scale carbonate cycles representative of "basic peritidal cyclic sequences" and large-scale "basin-to-platform shallowing sequences" (grand cycles). The starting points for producing this quantitative model were recognition of the frequent occurrence of peritidal cycles in the stratigraphic record (e.g., James, 1984) and the definition by Aitken (1978) of grand cycles. Similarly, Goldhammer et al. (1987) developed a one-dimensional simulation model, "Mr. Sediment," to evaluate variables which may have controlled development of the cyclic structure in the Latemar Group of the Dolomites. Matthews and Frohlich (1987) presented a quantitative model that simulates cyclic diagenetic overprint on a carbonate sequence based initially on their experience with diagenesis and sea-level fluctuation in Barbados. These quantitative models are particularly useful in that they allow assessment of the influence of a single variable within a set of variables considered significant in controlling stratigraphic accumulation.

We anticipate that future development of quantitative models for stratigraphic analysis will emphasize simulations of stratigraphic cross sections rather than stratigraphic columns. Such two-dimensional simulations necessarily will encompass a broad facies spectrum rather than a particular sedimentary or diagenetic facies as in the examples cited above. Two-dimensional quantitative stratigraphic models could be developed to serve two distinctive objectives, one primarily sedimentologic, the other stratigraphic. In the first case the object might be to simulate contemporaneous facies patterns and the evolution of those facies patterns between stratigraphic discontinuities (e.g., cycle boundaries). In contrast, the second objective might be simulation of patterns of stratigraphic accumulation at a larger scale (e.g., sequences of cycles). We think that these two distinct objectives require different quantitative models because the processes responsible for intracycle facies patterns are different from those producing sequences of cycles.

Specifically, sedimentologic processes (e.g., current intensity and persistence) that generally can be modeled from modern environments might account for grain size distribution, type of lamination, most bedding and contemporaneous facies patterns. In contrast and at a larger scale, basin fills are characterized by other properties, such as cyclic structure and lateral thickness variability, that are determined by stratigraphic processes (e.g., eustasy or tectonism). These processes are generally longer term, independent of local sedimentary environments and cannot be modeled readily from modern environments.

Both sedimentologic and stratigraphic simulations are dependent on the kind of *qualitative* stratigraphic models assumed. Because quantitative models are attempts to simulate actual sedimentologic or stratigraphic patterns which, in turn, are perceived from the bias of a particular qualitative model, the qualitative models underlying all rock observations must be made explicit. Inevitably, all quantitative models constructed to simulate particular stratigraphic perceptions will include the biases of the underlying qualitative assumptions. We have proposed one such qualitative model, the hypothesis of punctuated aggradational cycles (Goodwin and Anderson, 1985 and in press; Goodwin et al., 1986). This hypothesis states that the stratigraphic record is pervasively cyclic at a meter scale, and that these cycles (PACs) are allogenic time-stratigraphic units. PACs are not defined by specific facies, but are thin basinwide units characterized by internal facies gradations (usually shallowing upward) and sharp boundaries marked by change to deeper disjunct facies.

If the PAC hypothesis is correct, that is, if basinwide meter-scale allogenic cycles are pervasive, it then would be necessary to base quantitative models on a qualitative model which predicts allogenic cycles in a broad spectrum of facies, not just in a particular areally limited paleoenvironment. In this paper we describe specific characteristics of a field data base derived by applying such an allogenic qualitative model to the Helderberg Group. Our intent is to illustrate the details of the stratigraphic framework and to focus on the aspects of that framework which affect the selection of variables in quantitative two-dimensional stratigraphic modeling.

THE HELDERBERGIAN EXAMPLE

Our stratigraphic analyses of the Helderberg Group, Lower Devonian of New York State and Pennsylvania (Anderson et al., 1984 and 1986; Goodwin et al., 1986; Goodwin and Anderson, in press), provide an example of a field data base that encompasses a broad facies spectrum. This data base was constructed by consistent application of a single qualitative stratigraphic theory, the PAC hypothesis (Goodwin and Anderson, 1985). Our reconstruction of the Helderberg Group reveals a stratigraphic framework consisting of PACs and PAC sequences that appears to require a sea-level function containing more than one period of oscillation. PAC correlations, in turn, demonstrate the existence of otherwise hidden, or cryptic, unconformities and also reveal significant lateral thickness changes that seem to require a tectonic explanation (e.g., differential subsidence or uplift). The recurrence intervals of PACs and PAC sequences appear to fit Milankovitch band cyclicity, i.e., 20,000 and 100,000 yr cycles (Goodwin and Anderson, 1985), whereas the hidden unconformities are less frequent but still recur at intervals in the 10^5 yr range.

PACs—The Fundamental Stratigraphic Unit

The first characteristic of the stratigraphic structure of the Helderberg Group is that each section is totally divisible into meter-scale cycles (PACs) irrespective of local facies (Figs. 1 and 2). This means that cycles are recognized in a broad range of subtidal through peritidal facies. In the two illustrated sections (Fig. 2), only two PACs (PAC 26 at Oriskany Falls and PAC 12 at Kingston) are typical peritidal cycles. The great majority of cycles in these two sections are defined in a wide range of subtidal facies.

Figure 1. Helderberg outcrop belt with locations of the two stratigraphic sections (Kingston and Oriskany Falls). S = Syracuse and A = Albany, New York.

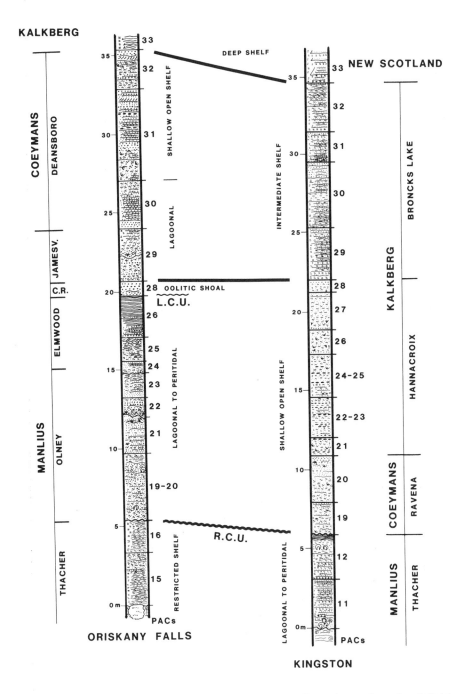

Figure 2. Correlated columnar sections, Helderberg Group (formations and members), Oriskany Falls and Kingston, New York. Numbering at the right side of the column indicates PAC correlations. Local (L.C.U.) and regional (R.C.U.) cryptic unconformities are defined at surfaces where PACs are missing. Facies grade from shallower to deeper from Oriskany Falls to Kingston in correlative PACs. Major allogenic events indicated by solid lines, interpreted from major regional facies changes, are keys to correlation. Symbol explanations shown on facing page.

PAC boundaries are recognized in these subtidal facies by a diverse set of criteria including continuity of facies within cycles and facies dislocations at cycle boundaries. Additional criteria include variation in argillaceous content, vertical patterns in bedding thickness, textural changes, appearance and disappearance of key faunal elements, and integration of patterns of facies change with patterns at nearby localities. For example, PAC 15 at Oriskany Falls and PACs 11 and 30 at Kingston become thicker bedded, coarser textured, less argillaceous and lighter in color from base to top. In another example, the appearance of gypidulid brachiopods at the base of PAC 30 and their disappearance at the base of PAC 33 at Oriskany Falls represent faunal responses to the stratigraphic events that produced these PAC boundaries.

Two stratigraphic conclusions are drawn concerning individual meter-scale cycles in the Helderberg Group. First, cycles or PACs are recognized by patterns of facies change in whatever general facies they occur. They are not dependent on the presence of specific facies components in the sense that peritidal shallowing-upward cycles are defined (James, 1984). Second, cycles and their boundaries cut across major facies. For example, at Oriskany Falls PAC 26 is a peritidal cycle, whereas at Kingston it is represented by open shelf facies.

Finally, an analysis of facies thickness within a single PAC contributes information on the stratigraphic dynamics of cycle accumulation. In particular, accommodation space for stratigraphic accumulation is related both to events that produce cycle boundaries (e.g., rapid sea-level rises) and to processes that continue through the cycle's accumulation (e.g., continuous subsidence or sea-level rise). For example, thick sequences of depth-specific facies, such as the 2 m of cryptalgal laminites at the top of PAC 26 at Oriskany Falls, appear to require continuous addition of accommodation space during the accumulation of the PAC.

PAC Sequences

PACs appear to be grouped into sets of two to six related PACs separated by PAC boundaries of unusually large facies change. While analysis of this cyclic structural element in the Helderberg Group is not complete, possible examples of such sequences include PACs 22 to 26 and PACs 28 to 32. The PAC 22 to 26 sequence is distinguished by large facies changes at its boundaries, including a local cryptic unconformity at the top (Fig. 2), and internally by a general shallowing pattern. Analogous patterns have been described by Goldhammer et al. (1986) from the Latemar Group, Strasser (1988) from Purbeckian carbonates, and Goodwin and Anderson (1985) from elsewhere in the Helderberg Group. The common occurrence of a PAC sequence pattern seems to require a hierarchy of sea-level functions. The stratigraphic structure of the Helderberg Group is, in this way, in accord with observations from other stratigraphic intervals that indicate that simulation models need to incorporate a compound eustatic curve of Milankovitch type (see Goldhammer et al., 1987; Strasser, 1988).

Cryptic Unconformities

Cryptic or hidden unconformities are revealed by basinwide PAC correlations (Goodwin and Anderson, in press). These unconformities generally are recognized by progressive loss of PACs toward a basin margin due to erosion and/or nondeposition. Because major cryptic unconformities may be basinwide in extent, and are not limited to the tectonically active portions of the basin, they were apparently produced by major sea-level falls. Variable durations of hiatus and vacuity (Wheeler, 1958; Sloss, 1984) along a single unconformity indicate that differential tectonism also played a role in producing these surfaces. Although we have not yet completely mapped the stratigraphic occurrences of cryptic unconformities throughout the Helderberg Group, our preliminary observations indicate a relatively low recurrence interval, perhaps as low as 10^5 yr. In this respect these surfaces are similar to the basin margin unconformities in the Lower Chesapeake Group described by Kidwell (1984). The relatively low recurrence interval (10^5 yr) suggests a comparison with boundary surfaces of major cyclothems of the Midcontinent (Busch and Rollins, 1984; Heckel, 1986; Busch and West, 1987). Further mapping of cryptic unconformities in the Helderberg Group should reveal the relationship between these surfaces and the boundaries of PAC sequences.

A major regional cryptic unconformity occurs at the boundary between the Thacher and the overlying Olney or Ravena Members (Fig. 2). Between Oriskany Falls and Kingston, New York, PACs are progressively lost at this surface (Goodwin et al., 1986; Goodwin and Anderson, in press). At least five PACs (13 to 17) are interpreted as missing in an erosional vacuity at Kingston. This surface of unconformity is covered first by PAC 19 in the localities in Figure 2, but to the west in the Syracuse area, the surface is first onlapped by PAC 21.

A minor local cryptic unconformity is recorded between PACs 26 and 28 in central New York by the loss of PAC 27 between the illustrated localities (Fig. 2). This smaller discontinuity is interpreted as a hiatus in deposition produced by a minor sea-level drop and the loss of a cycle at the western basin margin. There is no evidence of progressive erosional loss of cycles (vacuity) at the unconformity surface. Such unconformities may not be recognizable by field evidence at individual localities or by biostratigraphy. They appear to be identifiable only by correlating and mapping meter-scale cycles. However, they are a basic part of the stratigraphic structure that quantitative models must simulate.

VARIABLES FOR QUANTITATIVE SIMULATIONS

Analysis of the cyclic and discontinuous fabric of the Helderberg Group provides insight about variables that need to be incorporated in two-dimensional quantitative stratigraphic models. Of principal importance is the allogenic origin of this framework as well as its lateral persistence through a variety of facies. To simulate such a stratigraphic fabric the most important variables are sediment accumulation, lag time, eustasy and subsidence (e.g., Read et al., 1986; Burton et al., 1987; Goldhammer et al., 1987).

Sediment Accumulation

MODE Our observations suggest that distribution of sediment by progradation from a point source is *not* the principal mode of accumulation within individual PACs. Rather sediment accumulates largely by aggradation throughout the full extent of the basin, albeit at very different rates and in different absolute amounts in different coeval environments. For example, there is no evidence or logic to suggest that accumulation of the shelf facies of PAC 28 at Kingston was in any way dependent on accumulation of the oolitic shoal facies in the same PAC at Oriskany Falls, 150 km to the west (Fig 2). Instead very distinct facies developed independently in these places as a function of different initial water depths and position in the basin. Locally, in peritidal environments progradation certainly occurred, likely in many directions, as in island development in Florida Bay today. However, the general mode of accumulation throughout the full lateral extent of each PAC must have been nearly simultaneous aggradation.

If other stratigraphic sequences (carbonate and siliciclastic) are as pervasively cyclic at a meter scale as the Helderberg Group, it seems unlikely that quantitative models that employ progradation from a point source as the principal means of distributing sediment in the modeled cross section will have any utility in simulating basinwide accumulation patterns. Rather, modelers need to consider sedimentary accumulation in each environment as a separate entity, recognizing that interdependencies are probably limited to immediately contiguous environments.

RATE   What controls the rate of sediment accumulation, or more precisely stratigraphic accumulation, in paleoenvironments is problematic, in that the relative roles of stratigraphic and sedimentologic processes are not fully understood. Sadler (1981, p. 578) suggested that although short-term sedimentation rates may be both very high and highly variable, "accumulation rate trends are convergent" on low rates, (<<1m/kyr), "and become largely coincident . . . at time spans longer than 1000 years, suggesting a common large scale controlling influence." In other words, Sadler concluded that stratigraphic accumulation rates for time spans equivalent to those involved in the deposition of a PAC (10^4 yr) are largely independent of local environments and the sedimentologic processes that operate wihtin them. Instead the controlling process may be larger scale isostatic or tectonic subsidence.

Because stratigraphic processes (e.g., eustasy, tectonism, isostasy) may play a significant role in determining sediment accumulation rates even during the deposition of a single PAC, it is probably inaccurate to simply utilize modern sedimentation rates (e.g., 0.4 to 1.1 m/kyr; Schlager, 1981) in environments producing facies comparable to those within a PAC to construct a sediment accumulation variable. Rather, it seems to us that stratigraphic accumulation rates must be derived principally from stratigraphic data, such as the thickness and changes in thickness within a single cycle throughout a sedimentary basin and lateral patterns in cumulative thickness for larger sets of cycles.

Comparison of stratigraphic thickness within single PACs in the Helderberg Group indicates that some PACs are as much as two or three times as thin in offshore facies as in onshore facies. However this pattern is not consistent. When limited by available accommodation space onshore PACs are thin, and offshore PACs may be thicker than their onshore equivalents when they are associated with progradational sequences of PACs. Variability in accumulation rates within single PACs is thus high. Processes controlling this variability are not only environmental but also stratigraphic.

Sadler's (1981) concept of converging accumulation rates for different facies and a large-scale controlling factor is an even better model when sequences of cycles (i.e., representing 10^5 to 10^6 yr) are simulated. In the Helderberg Group, long-term sediment accumulation rates (i.e., averaged over several cycles) converge on low values as suggested by Sadler for long time periods, and in some cases appear to be largely independent of facies.

For example, PACs 19 to 28 in the two illustrated sections (Fig. 2) are within a meter of the same cumulative total thickness (15 to 16 m). This relationship holds despite differences in facies and availability of accommodation space. At the western locality the thickness of this interval was determined by subsidence as deposition began and ended at or near sea level. However deposition at the eastern locality was not constrained by accommodation space as this area was continuously subtidal after transgression of the basal unconformity. At an even larger scale (the PAC 19 to 32 interval), 14 cycles have a cumulative thickness of 29 to 30 m. If each of these lithic cycles represents the 20,000 yr Milankovitch precessional cycle, the long-term accumulation rate was about 0.1 m/kyr, a value in excellent agreement with Sadler's calculations. We conclude that sediment accumulation rates are stratigraphically controlled, converge on low values and can be determined best by thickness analysis of cycles.

Lag Time

Sharp PAC boundaries separating disjunct facies suggest that sediment accumulation was discontinuous, negligible during punctuation (transgressive) events and at a maximum during periods of relatively stable base level between events. In modeling peritidal cycles, Read et al. (1986) simulated this pattern by introducing a variable of lag time, an estimate of the time interval between flooding of an emergent shelf and resumption of sediment production. Others (e.g., Goldhammer et al., 1987) have handled this variable as a depth-dependent function in which sediment production, which was terminated by accumulation to sea level, resumes when subsidence or eustatic rise brings the surface of sediment accumulation to a certain water depth.

The pervasiveness of the PAC motif across a broad facies spectrum suggests that these treatments of lag time are suspect. Sharp contacts, nondeposition and disjunct facies cannot be depth-related phenomena because all PACs, regardless of depth of water, exhibit the same motif. Also there is no evidence of emergence at most PAC boundaries in the Helderberg Group, implying that something other than lack of accommodation space caused cessation of deposition. Perhaps lag time is not a separate variable, but is, instead, an effect of the rapid base-level rise resulting from combining subsidence and the shortest period eustasy variable. At the inflection point of a eustasy curve where the rate of sea-level rise is at a maximum, the added influence of subsidence might produce rates of base-level rise sufficient to permit little or no deposition, even with constant production rates, until the rate of rise slows.

However, if rates of relative sea-level rise needed to simulate the PAC motif are geologically unreasonable, it may be necessary to introduce a time-related variable that implies disruption of sediment production or supply by the eustatic event. If this type of disruption is a critical process, then a sediment accumulation variable might be modeled as an inverse function of the rate of base-level change. A variable designed in this manner would result in lowest rates of sediment accumulation at the inflection point of eustatic rise and another period of lowered accumulation rates during the time of maximum sea-level fall. PAC boundaries would be produced by this variable during sea-level rises and intracycle discontinuities might be produced by sea-level drops which do not cause exposure. However, caution is recommended in the design of the lag-time variable because it is not connected with certainty to any directly observable process.

Eustasy

The pervasiveness of PACs, their grouping into PAC sequences and the frequent occurrence of cryptic unconformities suggest that sea-level oscillations with two or three different periods (and possibly different amplitudes) need to be incorporated in quantitative simulations. Estimates of periodicity from the Helderberg Group (Goodwin and Anderson, 1985) point to either the precession or obliquity oscillations (10^4 yr) as the forcing mechanism for PAC boundaries and to the combined effects of these short-term oscillations and the longer term eccentricity cycle (100,000 yr) as the cause of unusually large sea-level rises at PAC sequence boundaries. Cryptic unconformities may possibly be the result of the additional effects of the longer term (400,000 yr) eccentricity cycle combined with tectonism.

Relative magnitudes of facies changes at these three types of surfaces suggest that PAC boundaries are produced by relatively low-amplitude oscillations, PAC sequence boundaries by intermediate amplitude events and cryptic unconformities by even higher amplitude eustatic fluctuations. Normal PAC boundaries separate environmentally disjunct facies but do not represent events which significantly altered environmental patterns across the basin. PAC sequence boundaries represent events which produced major facies changes throughout the basin (Fig. 2). Cryptic unconformities represent widespread exposure following major sea-level falls.

The magnitude of punctuation events (sea-level rises) is a function of the amplitude of the individual Milankovitch cycles as well as their combined effects augmented by subsidence. Large relative rises of sea level should occur when the different orbital cycles are in phase. Smaller sea-level rises would be produced by the short-period cycles during times in which the effects of the long- and short-period Milankovitch cycles are not in phase (Goodwin and Anderson, 1985; Goldhammer et al., 1987).

Subsidence

Stratigraphic patterns in the Helderberg Group suggest that subsidence is continuous and gradual as well as variable in rate with respect to time and position in the basin. Although theoretically possible, it does not seem probable that episodic subsidence is the mechanism producing PAC boundaries. Basinwide episodic subsidence at this scale and frequency has not been documented by structural geologists. Rather it seems that PAC boundaries are produced by eustatic rises coupled with gradual subsidence. Evidence for continued gradual subsidence is found in over-thickened facies in the upper portions of PACs. For example, cryptalgal laminites, which form in the high intertidal zone, are occasionally as much as three meters thick, indicating that accumulation kept pace with gradual continued eustatic rise or subsidence. Lateral variation in thickness within the same facies in a single PAC may require differential subsidence rather than eustasy to explain the over-thickened facies.

At a larger scale, vertical and lateral variation in subsidence rates can be documented from detailed analysis of the fine-scale, time-stratigraphic framework of PACs and cryptic unconformities. For example, differences in thickness between paleoisotopographic surfaces (PAC boundaries on top of widespread facies deposited at sea level) can be attributed to differential subsidence. Significant thickness differences within tens of meters of section and over tens of kilometers of distance indicate that subsidence varies considerably over short distances and relatively short time periods, perhaps tens of thousands of years (Anderson et al., 1986). Two-dimensional quantitative simulations will need to accommodate these observations by treating subsidence as a very flexible variable rather than by assuming a constant subsidence rate over great distance and longer periods of time.

In the same vein, cryptic unconformities, which mark major reorganizations of pale-ogeography, also indicate relatively frequent (10^5 yr) changes in subsidence rates on a basinwide scale. For example, at the Thacher–Ravena regional cryptic unconformity (Fig. 2) there is a significant change in tectonic subsidence rates and patterns (Goodwin and Anderson, in press). Beneath the unconformity, Manlius strata across New York State exclusively contain facies deposited within a few meters of sea level, suggesting generally low and similar subsidence rate throughout the whole outcrop belt which transects most of the basin in New York State. Following erosion and inundation of the erosion surface, sediment accumulation in eastern New York (e.g., Kingston column, Fig 2) never again reached sea level; instead an episodic progression to markedly deeper facies of the Kalkberg and New Scotland Formations was initiated. In central New York (Oriskany Falls column, Fig. 2) sea-level facies recurred briefly (e.g., Upper Elmwood, PAC 26) but were limited in geographic and temporal occurrence. These patterns seem to indicate a general increase in subsidence rates as well as marked lateral differential subsidence related to increased tectonism associated with the active eastern side of the basin.

CONCLUSIONS

In this paper a case is presented that a critical element in developing new *quantitative* stratigraphic models for analysis of cyclic stratigraphy is a field data base constructed with consistent application of a *qualitative* cyclic model. In an example of such a data base, the Helderberg Group described with assumption of the PAC hypothesis, three key stratigraphic elements appear: meter-scale allogenic cycles (PACs) which cut across facies; groups of two to six meter-scale cycles delineated by major facies changes (PAC sequences); and cryptic unconformities revealed by basinwide correlation and mapping of PACs. This kind of stratigraphic structure is important in quantitative modeling because it places constraints on variables such as eustasy, stratigraphic accumulation and subsidence and because it places an emphasis on basinwide allogenic processes as the fundamental controls on stratigraphic accumulation.

ACKNOWLEDGMENTS

This research was supported by National Science Foundation Grants EAR–8107690 and EAR–8305900. We are grateful to Tim Cross, Ashton Embry and Robley Matthews for helpful reviews of the manuscript.

REFERENCES CITED

Aitken, J.D., 1978, Revised models for depositional grand cycles, Cambrian of the southern Rocky Mountains, Canada: Bulletin of Canadian Petroleum Geology, v. 26, p. 515-542.

Anderson, E.J., Goodwin, P.W., and Goodmann, P.T., 1986, Reconstruction of patterns of differential subsidence using an episodic stratigraphic model, in Allen, P.A., and Homewood, P., eds., Foreland basins: International Association of Sedimentologists Special Publication 8, p. 437-443.

Anderson, E.J., Goodwin, P.W., and Sobieski, T.H., 1984, Episodic accumulation and the origin of formation boundaries in the Helderberg Group of New York State: Geology, v. 12, p. 120-123.

Burton, R., Kendall, C.G.St.C., and Lerche I., 1987, Out of our depth: On the impossibility of fathoming eustasy from the stratigraphic record: Earth-Science Reviews, v. 24, p. 237-277.

Busch, R.M., and Rollins, H.B., 1984, Correlation of Carboniferous strata using a hierarchy of transgressive-regressive cycles: Geology, v. 12. p. 471-474.

Busch, R.M., and West, R.R., 1987, Hierarchal genetic stratigraphy: A framework for paleoceanography: Paleoceanography, v. 2, p. 141-164.

Goldhammer, R.K., Dunn, P.A., and Hardie, L.A., 1987, High frequency glacio-eustatic sealevel oscillations with Milankovitch characteristics recorded in Middle Triassic platform carbonates in northern Italy: American Journal of Science, v. 287, p. 853-892.

Goodwin, P.W., and Anderson, E.J., 1985, Punctuated aggradational cycles: A general hypothesis of episodic stratigraphic accumulation: Journal of Geology, v. 93, p. 515-533.

Goodwin, P.W., and Anderson, E.J., in press, Episodic development of Helderbergian paleogeography, New York State, Appalachian Basin, in Embry, A., and McMillan, N.J., eds., Second International Symposium on the Devonian System, Calgary, Alberta, Canada.

Goodwin, P.W., Anderson, E.J., Goodman, W.M., and Saraka, L.J., 1986, Punctuated aggradational cycles: Implications for stratigraphic analysis, in Arthur, M.A., and Garrison, R.E., eds., Milankovitch cycles through time: Paleoceanography, v. 1., p. 417-429.

Heckel, P.H., 1986, Sea-level curve for Pennsylvanian eustatic marine transgressive-regressive cycles along midcontinent outcrop belt, North America: Geology, v. 14, p. 330-334.

James, N.P., 1984, Shallowing-upward sequences in carbonates. in Walker, R.G., ed., Facies models: Geosciences Canada Reprint Series 1, p. 213-228.

Kidwell, S.M., 1984, Outcrop features and origin of basin margin unconformities in the lower Chesapeake Group (Miocene), Atlantic coastal plain, in Schlee, J.S., Interregional unconformities and hydrocarbon accumulation: American Association of Petroleum Geologists Memoir 36, p. 37-58.

Matthews, R.K., and Frohlich, C., 1987, Forward modeling of bank-margin carbonate diagenesis: Geology, v. 15, p. 673-676.

Read J.F., Grotzinger, J.P., Bova, J.A., and Koerschner, W.F., 1986, Models for generation of carbonate cycles: Geology, v. 14., p. 107-110.

Sadler, P.M., 1981, Sediment accumulation rates and the completeness of stratigraphic sections: Journal of Geology, v. 89, p. 569-584.

Schlager, W., 1981, The paradox of drowned reefs and carbonate platforms: Geological Society of America Bulletin, v. 92, p. 197-211.

Sloss, L.L., 1984, Comparative anatomy of cratonic unconformities, in Schlee, J.S., Interregional unconformities and hydrocarbon accumulation: American Association of Petroleum Geologists Memoir 36, p. 7-36.

Strasser, A., 1988, Shallowing-upward sequences in Purbeckian peritidal carbonates (lowermost Cretaceous, Swiss and French Jura Mountains): Sedimentology, v. 35, p. 369-383.

Wheeler, H.E., 1958, Time stratigraphy: American Association of Petroleum Geologists Bulletin, v. 42, p. 1047-1063.

29

A TECTONIC ORIGIN FOR THIRD-ORDER DEPOSITIONAL SEQUENCES IN EXTENSIONAL BASINS—IMPLICATIONS FOR BASIN MODELING

Ashton F. Embry
Geological Survey of Canada, 3303 - 33rd St. NW, Calgary, Alberta, Canada T2L 2A7

ABSTRACT

Third-order depositional sequences characterize the stratigraphic succession of many extensional basins, and numerous sequences seemingly have intercontinental synchroneity. These sequences are usually tens to hundreds of meters thick and span time intervals of two to eight million years. Seven main variables appear to control the development of depositional sequences: (1) sediment supply, (2) subsidence due to compaction of sediments, (3) thermal subsidence or uplift due to lithospheric temperature changes, (4) subsidence due to isostatic compensation of sediment and water loads, (5) glacio-eustasy, (6) tectono-eustasy, and (7) mechanical subsidence or uplift due to changes in lithospheric horizontal stress fields. The last variable often has been ignored but observations on the Mesozoic depositional sequences of the Sverdrup Basin, Arctic Canada, combined with the theoretical and empirical tectonic models of Cloetingh and associates, indicate that it is probably the forcing function for the observed third-order cyclicity.

Quantitative stratigraphic models designed to simulate stratigraphic successions of third-order depositional sequences must consider all seven of the above variables. Failure to do so will result in simulated sections that have little basis in reality. Determination of realistic values for all seven variables represents a major challenge for quantitative modelers. Unique values for most of the component variables will be very difficult, if not impossible, to determine. Reasonable values may be obtained on a trial and error basis from basins around the world in combination with qualitative observations of stratigraphic relationships and adherence to the laws of physics.

INTRODUCTION

Quantitative modeling of the development and evolution of extensional basins began over a decade ago and early models emphasized the thermal-mechanical properties of the lithosphere, compaction of basin fill, paleobathymetry, and long term eustasy (McKenzie, 1978; Keen and Barrett, 1981; Steckler and Watts, 1981; Watts et al., 1982; Beaumont et al., 1982). In general these models simulated the overall basin fill reasonably well, and in some cases generated stratigraphic geometries and relationships compatible with observations (e.g., Watts and Thorne, 1984).

Simultaneous with the development of these quantitative models, qualitative studies of basin stratigraphy, often employing regional seismic reflection data, revealed that the stratigraphic succession of many basins is divisible into stratigraphic units termed third-order depositional sequences (Vail et al., 1977; Vail et al., 1984). Mitchum et al. (1977, p. 53) defined a depositional sequence as "a stratigraphic unit composed of a relatively conformable succession of genetically related strata and bounded at its top and base by unconformities or their correlative conformities," and indicated that most pre-Quaternary seismic sequences represent a time span on the order of 1 to 10 m.y.

Recently, quantitative models of basin fill have attempted to simulate basin stratigraphy characterized by successive depositional sequences (e.g., Burton et al., 1987). These studies recognized that the main input variables for such models are rates of sediment supply, subsidence and eustatic change, with rate of subsidence usually divided into thermal, mechanical and compaction components. The problems associated with quantifying each of these variables were discussed by Burton et al. (1987). They demonstrated that it is impossible to derive a unique eustatic curve from the stratigraphic record. Thus modelers must make assumptions regarding the relative contribution of thermo-mechanical subsidence versus that of eustatic change to the development of a basin for any given time step. Such assumptions must be based on qualitative studies of basin stratigraphy which provide insight into the relative importance of these three main variables.

A key question that presently confronts basin modelers is "which variable exhibits a significant change in magnitude every two to eight million years to initiate a new cycle of deposition—that is, which variable is the main forcing function for the origin of third-order depositional sequences?" A consensus answer to this question has not yet been reached. It is generally agreed that rate of sediment supply is not the critical variable for the origin of sequences because it cannot account for the subaerial unconformities that bound the sequences on basin margins. The main point of contention is whether tectonic subsidence or eustatic change is the main forcing function. Vail et al. (1977), and more recently Haq et al. (1987) and Vail et al. (1987), strongly favor eustatic change in this argument. Their views have gained widespread acceptance and most quantitative modelers take this approach (e.g., Burton et al., 1987). However, a number of workers maintain that tectonic subsidence is the key variable controlling the development of depositional sequences (Bally, 1982; Watts, 1982; Cloetingh et al., 1985; Lambeck et al., 1987).

In summary it is important for quantitative basin modeling to know if tectonics or eustasy is the main forcing function for the origin of third-order sequences and to learn how each forcing function may be expressed in the stratigraphic record. This problem is addressed in this paper, and the conclusions reached are based on observations on the stratigraphic relationships of third-order depositional sequences that constitute the Mesozoic fill of the Sverdrup Basin, Arctic Canada (Fig. 1).

MESOZOIC DEPOSITIONAL SEQUENCES OF THE SVERDRUP BASIN

The Sverdrup Basin was the main depocenter in the Canadian Arctic Archipelago from Carboniferous to early Tertiary during which time up to 13,000 m of strata accumulated in the basin (Balkwill, 1978). Stephenson et al. (1987) recognized four main phases of tectonic

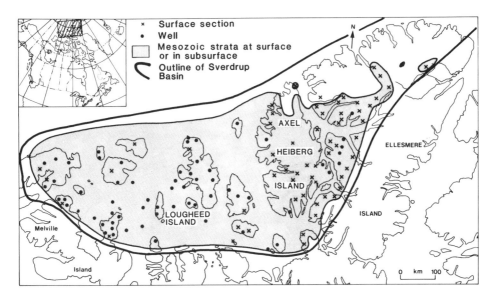

Figure 1. Distribution of Mesozoic strata in the Sverdrup Basin of Arctic Canada and available control points.

development for the basin: Carboniferous-Early Permian rifting, Early Permian-earliest Cretaceous thermal subsidence, Early Cretaceous rifting, and Late Cretaceous-early Tertiary thermal subsidence. The last phase was complicated by foreland subsidence related to the early stages of the Eurekan Orogeny. The basin was deformed and uplifted in late Eocene to Oligocene during the end of the orogeny.

The Mesozoic succession of the basin is up to 9,000 m thick and has been divided into thirty, unconformity-bounded depositional sequences (Embry, in press, a). Each sequence spans a time interval between four and eight m.y. and thus represents a third-order cycle. Abundant outcrops in the eastern portion of the basin combined with well and seismic data from the central and western portions of the basin provide excellent control for determining the stratigraphic relationships of the sequences.

Stratigraphic relationships that characterize a typical Mesozoic sequence of the Sverdrup Basin are illustrated in Figure 2. On the basin margins subaerial unconformities form the sequence boundaries. A thin, transgressive unit, commonly consisting of calcareous sandstone, glauconitic sandstone, arenaceous limestone, or oolitic ironstone, overlies the basal unconformity and is capped by a submarine unconformity (hiatal surface, *sensu* Frazier, 1974). Overlying the submarine unconformity is a thick, regressive succession of shale and siltstone that coarsens-upward into sandstone of shallow shelf and strand plain origin. A subaerial unconformity caps the sandstone and forms the upper boundary of the sequence. The subaerial unconformities and the sandstone unit disappear basinward (Fig. 2) and the sequence boundaries become conformities that separate regressive strata from overlying transgressive beds. In the central portion of the basin transgressive deposits are very thin or absent and submarine unconformities commonly form the sequence boundaries.

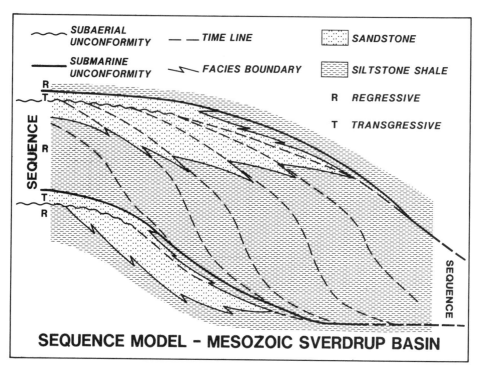

Figure 2. Stratigraphy of a depositional sequence. On the basin margins subaerial unconformities form the boundaries of a sequence whereas submarine unconformities or conformities (dashed time line) form the sequence boundaries in the more central parts of the basin.

The stratigraphic relationships portrayed on Figure 2 are a slight oversimplification because only two significant stratigraphic surfaces, the subaerial unconformity and the submarine unconformity, are illustrated. Actually up to four distinct stratigraphic surfaces can be recognized at or just above a sequence boundary on the flanks of the basin (Fig. 3).

1. The subaerial unconformity is an erosional surface cut by fluvial or weathering processes during the regressive phase of a sequence. By definition this surface forms the boundary of a sequence. The surface is sometimes difficult to recognize with confidence because nonmarine strata which overlie the unconformity can arguably be assigned to the uppermost regressive portion of the underlying sequence. This problem has been discussed by Demarest and Kraft (1987).

2. The shoreface unconformity or ravinement surface is a scour surface cut by wave action in the plunge zone of the shoreface during transgression (Demarest and Kraft, 1987). This surface sometimes overlies nonmarine strata which are the initial deposits on top of the subaerial unconformity. However it commonly completely erodes these basal nonmarine deposits and forms the sequence boundary (Fig. 3). This relationship is present at most marginal localities of the Sverdrup Basin where nonmarine strata occur only in local topographic lows on the subaerial unconformity.

3. The submarine unconformity is a regional scour surface cut by marine currents (e.g., storm surge) in the offshore area where sedimentation rates during transgression are so low that net erosion occurs over an extended period of time. A thin lag deposit of coarse material derived from the underlying strata may locally overlie the submarine unconformity.

4. The downlap surface (Haq et al., 1987) is a nonerosive sedimentary contact of offshore shale and siltstone on the basal lag deposit. These fine clastics are the initial regressive deposits. In many cases a basal lag bed is absent and the downlap surface coincides with the submarine unconformity (Fig. 3).

The subaerial unconformity and ravinement surface disappear basinward and the sequence boundary is placed, by definition, at the stratigraphic horizon which correlates with the unconformity. For the Sverdrup sequences this horizon is placed at the boundary between regressive and overlying transgressive strata (Figs. 2 and 3), a position that often must be defined subjectively.

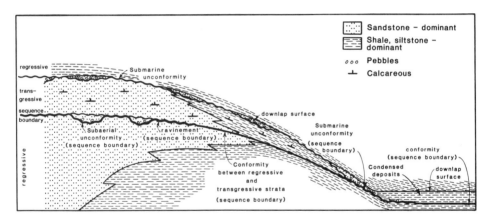

Figure 3. Stratigraphic surfaces associated with a sequence boundary. Up to four distinct stratigraphic surfaces are recognizable at or near a sequence boundary: subaerial unconformity, ravinement surface, submarine unconformity and downlap surface. The nature of the sequence boundary (distinguished by thicker line work) varies along its extent and includes a subaerial unconformity, ravinement surface, and submarine unconformity.

In the central part of the basin, transgressive strata are absent except for a basal lag or thin chemical deposit (e.g., ironstone) which are referred to as condensed deposits by Haq et al. (1987). Here the sequence boundary is placed at the basal contact of this unit which can be either a submarine unconformity (Fig. 3) or a conformity (no erosion). These various stratigraphic surfaces are illustrated in photographs and on well-log cross sections in Embry (in press a, b) and Embry and Podruski (in press).

In summary, the nature of a sequence boundary varies considerably over its extent from a subaerial unconformity or a ravinement surface on the basin margins to a conformity or submarine unconformity in more seaward localities (Fig. 3). Note that the submarine unconformity commonly occurs above the sequence boundary and only is coincident with that boundary where it rests directly on regressive strata of the underlying sequence. The type of sequence boundary that develops at any given locality is dependent on the paleogeographic setting of the locality at the time when the sequence is initiated by a relative sea-level rise. In areas of net erosion (shoreface, outer shelf, upper slope) the sequence boundary will be an unconformity (ravinement surface, submarine unconformity), whereas in areas of net sedimentation (inner shelf, lower slope, basin) the sequence boundary will be a conformity.

For most of the sequences the subaerial and submarine unconformities are separated from each other by a few tens of meters or less of transgressive strata and thus are not distinguishable on seismic sections. The interval between the subaerial unconformity and the downlap surface is commonly represented by a single seismic reflector that can be interpreted erroneously as a single, through-going unconformity. As illustrated on Figures 2 and 3 the subaerial unconformity (or ravinement) and the submarine unconformity do not join to form a continuous unconformity surface. Rather these two very different types of unconformities interleave on the basin margins and are separated by a thin transgressive interval. It is perhaps this blurring of the stratigraphic record on seismic sections that has led to the interpretation of a single unconformity surface that extends from the basin margin to the basin center (e.g., Vail et al., 1977, p. 78, Fig. 13a).

Mesozoic depositional sequences of the Sverdrup Basin differ from those described by Vail et al. (1977) and Haq et al. (1987) in a few ways. These include:

1. Onlapping nonmarine strata that characterize the basal portion of many of the sequences illustrated by Vail et al. (1977) and Haq et al. (1987) are absent in all but one of the Sverdrup sequences.
2. Transgressive strata are marine in origin, usually very thin (<5 m), and apparently represent a short time interval (<1 m.y.). This contrasts with the transgressive portions of sequences described by Vail et al. (1977) and Haq et al. (1987) which often contain substantial nonmarine to paludal strata, are thick (>25 m), and represent time spans in excess of 1 m.y.
3. Subaerial unconformities are restricted to the margins of the Sverdrup Basin and extend only a short distance into the basin. Many of the subaerial unconformities that bound the sequences described by Vail et al. (1977, 1984) extend far into the basin, often beyond the shelf break.

Despite these differences, most of the Mesozoic sequences of the Sverdrup Basin are time equivalent to third-order cycles on the Haq et al. (1987) chart. Thus the Sverdrup cycles have an intercontinental extent and this factor must be considered in any interpretation of their origin. I explored the possible origins of the Triassic depositional sequences (T-R cycles) of the Sverdrup Basin in a manuscript submitted in 1985 (Embry, in press, b) and offered two interpretations: (1) alternating tectonic uplift and subsidence against a background of variable sediment influx and long-term (>50 m.y.) eustatic change; and, (2) short term (<10 m.y.) eustatic rise and fall against a background of variable sediment supply and gradually dissipating thermal subsidence.

The second interpretation, which advocated eustasy as the forcing function, was favored because of the widespread synchroneity of sequence boundaries. At that time there was no known tectonic model which could account for widespread, synchronous, short-term tectonic events.

In an abrupt about-face, in an abstract (Embry, 1986) submitted a year after the Embry (in press, b) manuscript, I proposed that tectonism was the main factor for the origin of the Sverdrup depositional cycles. I had recognized for some time that the Sverdrup sequences possessed a number of features that suggested tectonic control on their origin. These included: (1) the sediment source area often varies greatly from one sequence to the next; (2) the sedimentary regime of the basin commonly changed drastically and abruptly across a sequence boundary; (3) faults terminate at sequence boundaries; (4) significant changes in subsidence and uplift patterns within the basin occurred across sequence boundaries; and, (5) there were significant differences in the magnitude of the extent of some of the subaerial unconformities recognized on the slowly subsiding margins of the Sverdrup Basin and time equivalent ones recognized by Vail et al. (1977, 1984) in areas of high subsidence.

Previous to 1986 I rationalized these features, documented by Embry (in press, a), either in terms of eustasy or "swept them under the proverbial carpet" because only eustasy seemingly could explain the intercontinental extent of the sequences. In 1985 Cloetingh and co-workers published the first of a series of papers which provided a tectonic explanation for global, third-order cyclicity (Cloetingh et al., 1985; Cloetingh, 1986; Lambeck et al., 1987; Cloetingh and Kooi, this volume). They proposed that changes in the magnitude of horizontal lithospheric stresses related to plate movement could cause widespread relative sea-level changes with a frequency of a few million years. In light of this tectonic hypothesis the Sverdrup sequences, with their various attributes—troublesome to explain by eustasy—became more readily explained by tectonics.

TECTONIC ORIGIN FOR MESOZOIC SEQUENCES OF THE SVERDRUP BASIN

As outlined above, the Mesozoic depositional sequences of the Sverdrup Basin exhibit a number of features that suggest tectonic control. The abrupt changes of sedimentary source areas and sedimentary regimes from one sequence to the next are indicative of variable tectonic uplift and subsidence of land areas surrounding the basin. Also these same tectonic movements are expressed within the basin as changes in patterns of uplift and subsidence across sequence boundaries, and by the termination of faults at sequence boundaries.

Finally the large differences in magnitudes of relative sea-level falls between the Sverdrup Basin and those recorded in other areas of the world indicate a tectonic origin of sequences and sequence boundaries. For example the late Anisian and latest Ladinian (Middle Triassic) unconformities, which are third-order cycle boundaries on the Haq et al. (1987) chart, occur in the Triassic succession of northern Italy (R. Goldhammer, personal communication, 1988). The Anisian-Ladinian succession in this area, which consists of shallow water carbonates, is about 600 m thick (Bosellini and Rossi, 1974; Goldhammer et al., 1987) indicating the area was one of relatively high subsidence (6 cm/kyr). Stratigraphic and sedimentologic studies of these strata indicate relative sea-level falls of at least 100 m for each of the unconformities (R. Goldhammer, personal communication, 1988). Given the high subsidence rate of the area, these relative falls would translate into even larger eustatic falls if the main cause of these two global unconformities (Haq et al., 1987) was eustasy. Furthermore in an area of slower subsidence, such as the margin of the Sverdrup basin where the Middle Triassic succession is 200 m thick (equivalent to a subsidence rate of about

2 cm/kyr), such large eustatic falls should have resulted in major subaerial unconformities that extend far into the basin. Although these unconformities are present in the Sverdrup Basin (Embry, in press, a, b) supporting their intercontinental nature, they are confined to the basin margins and record relative sea-level falls of only a few tens of meters. This is the exact opposite of what would be expected if eustasy was the main variable controlling relative sea-level changes and hence third-order cyclicity.

Similar comparisons can be made between the Jurassic relative sea-level falls of the Sverdrup Basin and equivalent events of much higher magnitude in the North Sea (Vail et al., 1984). The concurrent development of large relative sea-level falls in areas of overall high subsidence (e.g., Italy, North Sea) with small relative sea-level falls in areas of low subsidence (e.g., Sverdrup Basin margin) is incompatible with a eustatic explanation for the events. However, such observations fit well with a tectonic origin for the unconformities because it would not be surprising if large uplifts occurred in tectonically active areas and small uplifts characterized tectonically quiescent areas.

The tectonic model proposed by Cloetingh and co-workers (Cloetingh et al., 1985; Lambeck et al., 1987) "in which changes in sea-level relative to land are produced by an interaction between fluctuating horizontal stress fields in the lithosphere and basin subsidence resulting from lithospheric cooling and sediment loading" (Lambeck et al., 1987, p. 259) explains the various features of the Sverdrup Basin sequences discussed above, including their intercontinental extent. This tectonic model predicts increased subsidence (i.e., relative sea-level rise) on a basin margin when horizontal tensional stresses are increased (or compressional stresses decreased), and uplift of the margin when tensional stresses are reduced (or compressional stresses increased). The model further predicts that relative sea-level changes would be up to 100 m in magnitude, occur as frequently as every few million years, be intercontinental and perhaps global in extent, and exhibit considerable variation in magnitude. All of these expectations are compatible with the observations from the Sverdrup sequences. Thus the "Cloetingh Tectonic Model" is favored as the primary forcing function for the origin of third-order sequences.

INPUT VARIABLES FOR QUANTITATIVE STRATIGRAPHIC MODELING

Quantitative stratigraphic models that simulate basin stratigraphy characterized by third-order cycles must consider rates of: sediment supply, eustatic change, thermal-mechanical subsidence, and sediment compaction. Quantifying rates of sediment supply and sediment compaction does not appear to present insurmountable problems, and reasonable estimates can be made and varied as the situation demands (e.g., Burton et al., 1987). However, the problems of disentangling rates of eustatic change from rates of thermal-mechanical subsidence are more difficult. Burton et al. (1987) unequivocally stated that it is impossible to determine absolute values for eustasy from the stratigraphic record. They indicated that quantitative models must employ a number of relative sea-level curves, each of which is a unique combination of eustatic and tectonic effects.

Relative sea-level curves for extensional basins may be formed by adding the following component curves: (1) thermal subsidence or uplift due to lithospheric temperature changes; (2) subsidence due to isostatic compensation of sediment and water loads; (3) mechanical subsidence or uplift caused by changes in the horizontal stress field in the lithosphere; 4) glacio-eustasy related to the transfer of water between the land and the sea; and (5) tectono-eustasy related to the changing volume of the ocean basins. There is broad agreement regarding the general shapes of the first two curves, although further refinements are required (e.g., Stephenson, this volume). This study strongly supports the contention that mechanical subsidence or uplift related to changes in horizontal stresses must be included in quantitative models. Unfortunately, published attempts at quantitative stratigraphic modeling (e.g., Burton et al., 1987) appear to have ignored this important variable. The amplitude of this curve will vary greatly from a few tens of meters in tectonically quiet areas to 100 m in tectonically active areas. Periodicity would vary from a few to perhaps 20 m.y.

Glacio-eustatic changes have probably operated throughout most of geologic time (see Goldhammer et al., 1987) and known periodicities are the Milankovitch cycles of approximately 20,000, 40,000 and 100,000 years. Longer periodicities of glacio-eustatic cycles may exist but there is no known physical process to account for such cycles. The amplitude of a glacio-eustatic cycle varies greatly depending on whether continental glaciers are present (hundreds of meters) or not (<10 m).

A tectono-eustatic component curve can be estimated for the last 150 m.y. (Pitman, 1978). Estimates of such a curve for this time interval indicate amplitudes of a few hundred meters and periodicities in excess of 50 m.y. (Pitman and Golovchenko, 1983). Worsley et al. (1984) argued that tectono-eustasy differs significantly from one phase of a 430 million-year-long Wilson Cycle to another. Because the tectono-eustatic curve for the last 150 m.y. is representative of only the "opening phase" of the Wilson Cycle it would not be applicable to time intervals that cover the "closing," "Pangea," and "rift" phases of the Cycle. The amplitude and periodicity of tectono-eustatic curves for these phases are poorly known and represent a problem area for the quantitative modeling of pre-Upper Jurassic successions.

SUMMARY AND CONCLUSIONS

A quantitative stratigraphic model designed to simulate the stratigraphy of an extensional basin characterized by third-order cyclicity should consider at least seven variables: (1) sediment supply, (2) subsidence due to compaction of sediments, (3) thermal subsidence or uplift due to lithospheric temperature changes, (4) subsidence due to isostatic compensation of sediment and water loads, (5) mechanical subsidence or uplift due to changes in the horizontal stress field, (6) glacio-eustatic variations, (7) tectono-eustatic changes. The fifth variable is ignored in most quantitative models. However, observations on the Mesozoic sequences of the Sverdrup Basin, in combination with the theoretical and empirical studies of Cloetingh and associates, indicate that it is an important factor and is probably the forcing function for the origin of third-order cyclicity.

Determining reasonable input curves for each of the seven variables represents a substantial challenge for quantitative modelers, but must be done if a realistic stratigraphic succession is to be generated and the model used with confidence. The input curves selected should have amplitudes and periodicities compatible with both known physical processes and qualitative stratigraphic observations from basins throughout the world. Each attempt at modeling a basin has the potential to provide insights into the dynamic variables that shape the stratigraphic record. Thus it would appear that quantitative dynamic stratigraphy holds great promise for increasing our understanding of the processes responsible for the complex stratigraphic mosaics which confront us.

ACKNOWLEDGMENTS

The Geological Survey of Canada supported this study through the Frontier Geoscience Program and allowed publication of the results. Polar Continental Shelf Project provided much logistical support during the field seasons. Discussions on the subject of tectonism versus eustasy with Randell Stephenson, Sierd Cloetingh, Jim Dixon and Bob Goldhammer have been very enlightening and enjoyable. Jim Dixon, Randell Stephenson, Pete Goodwin and Bill Galloway critically read the manuscript and their constructive remarks improved my manuscript. I am especially indebted to Tim Cross for inviting me to participate in the QDS meeting and for a thorough job of editing this paper. Billie Chiang typed the manuscript with speed and accuracy, and Elspeth Snow and Paul Wozniak kindly drafted the figures. Geological Survey of Canada Contribution No. 14988.

REFERENCES CITED

Balkwill, H.R., 1978, Evolution of Sverdrup Basin Arctic Canada: American Association of Petroleum Geologists Bulletin, v. 62, p. 1004-1028.

Bally, A.W., 1982, Musings over sedimentary basin evolution: Philosophical Transactions of the Royal Society of London, Series A, v. 305, p. 325-338.

Beaumont, C., Keen, C.E., and Boutilier, R., 1982, On the evolution of rifted continental margins: Comparison of models and observations for the Nova Scotia margin: Geophysical Journal of the Royal Astronomical Society, v. 70, p. 667-715.

Bosellini, A., and Rossi, D., 1974, Triassic carbonate buildups of the Dolomites, northern Italy, *in* Laporte, L.F., ed., Reefs in time and space: Society of Economic Paleontologists and Mineralogists Special Publication 18, p. 209-233.

Burton, R., Kendall, C.G., and Lerche, I., 1987, Out of our depth: On the impossibility of fathoming eustasy from the stratigraphic record: Earth Science Reviews, v. 24, p. 237-277.

Cloetingh, S., 1986, Intraplate stresses: A new tectonic mechanism for fluctuations of relative sea level: Geology, v. 14, p. 617-621.

Cloetingh, S., McQueen, H., and Lambeck, K., 1985, On a tectonic mechanism for regional sea level variations: Earth and Planetary Science Letters, v. 75, p. 157-166.

Embry, A.F., 1986, Mesozoic depositional sequences: Anatomy and origin: Geological Society of America Abstracts with Programs, v. 18, p. 594.

Embry, A.F., in press (a), Mesozoic history of the Arctic Islands, *in* Trettin, H.P., ed., Innuitian Orogen and Arctic Platform: Canada and Greenland: Geological Survey of Canada, Geology of Canada 3.

Embry, A.F., in press (b), Triassic sea-level changes: Evidence from the Canadian Arctic Archipelago, *in* Wilgus, C., et al., eds., Sea level changes—An integrated approach: Society of Economic Paleontologists and Mineralogists Special Publication 42, in press.

Embry, A.F., and Podruski, J.A., 1988, Third-order depositional sequences of the Mesozoic succession of Sverdrup Basin, *in* James, D.P., and Leckie, D.A., eds., Sequences, stratigraphy, sedimentology: Surface and subsurface: Canadian Society of Petroleum Geologists Memoir 15, p 73-84.

Frazier, D.E., 1974, Depositional episodes: Their relationship to the Quaternary stratigraphic framework of the northwestern portion of the Gulf basin: The University of Texas at Austin, Bureau of Economic Geology Geological Circular 71-1, 27 p.

Goldhammer, R.K., Dunn, P.A., and Hardie, L.A., 1987, High frequency glacio-eustatic sea-level oscillations with Milankovich characteristics recorded in Middle Triassic platform carbonates in northern Italy: American Journal of Science, v. 287, p. 853-892.

Haq, B., Hardenbol, J., and Vail, P.R., 1987, Chronology of fluctuating sea level since the Triassic (250 million years to present): Science, v. 235, p. 1156-1167.

Keen, C.E., and Barrett, D.L., 1981, Thinned and subsided continental crust on the rifted margin of eastern Canada; crustal structure, thermal evolution and subsidence history: Geophysical Journal of the Royal Astronomical Society, v. 65, p. 443-465.

Lambeck, K.S., Cloetingh, S., and McQueen, H., 1987, Intraplate stresses and apparent changes in sea level: The basins of northwestern Europe, in Beaumont, C., and Tankard, A., eds., Sedimentary basins and basin forming mechanisms: Canadian Society of Petroleum Geologists Memoir 12, p. 259-268.

McKenzie, D., 1978, Some remarks on the development of sedimentary basins: Earth and Planetary Science Letters, v. 40, p. 25-32.

Mitchum, R.M., Vail, P.R., and Thompson, S., III, 1977, Seismic stratigraphy and global changes of sea level, Part 2: The depositional sequence as a basic unit for stratigraphic analysis, in Payton, C.E., ed., Seismic Stratigraphy—Applications to hydrocarbon exploration: American Association of Petroleum Geologists Memoir 26, p. 53-62.

Pitman, W.C., 1978, Relationship between eustacy and stratigraphic sequences of passive margins: Geological Society of American Bulletin, v. 89, p. 1389-1403.

Pitman, W.C., and Golovchenko, X., 1983, The effect of sea level change on the shelf edge and slope of passive margins, in Stanley, D.J., ed., The shelf break: Critical interface on continental margins: Society of Economic Paleontologists and Mineralogists Special Publication 33, p. 41-58.

Steckler, M.S., and Watts, A.B., 1981, Subsidence history and tectonic evolution of Atlantic-type continental margins: American Geophysical Union Geodynamics Series, v. 6, p. 184-196.

Stephenson, R.A., Embry, A.F., Nakiboglu, S.M., and Hastaoglu, M.A., 1987, Rift-initiated Permian to Early Cretaceous subsidence of the Sverdrup Basin, in Beaumont, C., and Tankard, A., eds., Sedimentary basins and basin-forming mechanisms: Canadian Society of Petroleum Geologists Memoir 12, p. 213-231.

Vail, P.R., Mitchum, R.M., Todd, R.G., Widmier, J.M., Thompson, S., III, Sangree, J.B., Bubb, J.N., and Hatlelid, W.G., 1977, Seismic stratigraphy and global changes in sea level, in Payton, C.E., ed., Seismic stratigraphy—Applications to hydrocarbon exploration: American Association of Petroleum Geologists Memoir 26, p. 49-212.

Vail, P.R., Hardenbol, J., and Todd, R.G., 1984, Jurassic unconformities, chronostratigraphy and sea level changes from seismic stratigraphy and biostratigraphy, in Schlee, J.S., ed., Interregional unconformities and hydrocarbon exploration: American Association of Petroleum Geologists Memoir 36, p. 129-144.

Vail, P.R., Van Wagoner, J.C., Mitchum, R., and Posamentier, H.W., 1987, Seismic stratigraphy interpretation using sequence stratigraphy, in Bally, A.W., ed., Atlas of seismic stratigraphy, v. 1: American Association of Petroleum Geologists Studies in Geology 27, p. 1-14.

Watts, A.B., 1982, Tectonic subsidence, flexure and global changes of sea level: Nature, v. 297, p. 469-474.

Watts, A.B., Karner, G.D., and Steckler, M.S., 1982, Lithospheric flexure and the evolution of sedimentary basins: Philosophical Transactions of the Royal Society of London, Series A, v. 305, p. 249-281.

Watts, A.B., and Thorne, J., 1984, Tectonics, global changes in sea level, and their relationship to stratigraphical sequences at the U.S. Atlantic continental margin: Marine and Petroleum Geology, v. 1, p. 319-339.

Worsley, T.R., Nance, D., and Moody, J.B., 1984, Global tectonics and eustasy for the past 2 billion years: Marine Geology, v. 58, p. 373-400.

30

Comments on Sedimentary-Stratigraphic Verification of Some Geodynamic Basin Models: Example from a Cratonic and Associated Foreland Basin

George deV. Klein
Department of Geology, University of Illinois at Urbana-Champaign, 245 Natural History Building, 1301 W. Green St., Urbana, IL, 61801-2999 USA

Abstract

Quantitative tectonic subsidence analysis utilizing backstripping and two-dimensional thermal modeling of cratonic and adjacent foreland basins shows remarkable parallelism between subsidence mechanisms and sedimentation patterns. The Illinois basin formed by initial rifting (with associated nonmarine and coastal clastics), followed by thermal subsidence (with associated carbonate ramps and shelves). Fault controlled mechanical subsidence accompanied thermal subsidence to control Cambro-Ordovician cyclic stratigraphic patterns. Later Paleozoic subsidence in the Illinois basin was caused by broad flexural subsidence associated with Alleghenian-Hercynian collision which yoked these basins with the Appalachian basin. This flexural subsidence is associated with deposition of cyclothems. Overlapping eustatic change and foreland basinal deepening controlled differentiation of a continuum of cyclothem styles from eustatic dominated (Kansas) to foreland tectonic dominated (Appalachian) types. Ordovician and Devonian flexure in the Appalachian basin influenced Illinois basin sedimentation to a minor degree because continental crust beneath the Appalachian basin was less rigid, thus favoring development of deeper, narrower foreland basins. These findings show that acceptable stratigraphic verification of geodynamic basin models is attainable at basin scales and long temporal scales, but is less reliable at the outcrop scale shorter temporal scales.

Introduction

During the past decade, the geophysical community has proposed a variety of geodynamic models for basin formation and subsidence. These models involve both single (McKenzie, 1978) and multilayered stretching of the crust and mantle (Hellinger and Sclater, 1983; White and McKenzie, 1988), associated thermal subsidence in both vertical (McKenzie, 1978) and lateral mode (Sawyer et al., 1987), and flexure (Watts et al., 1982). Synthetic stratigraphies have been calculated from both thermal and flexural subsidence models (Watts et al., 1982;

Quantitative Dynamic Stratigraphy (1989), T.A. Cross, ed., Prentice Hall, p. 503–517.

503

White and McKenzie, 1988), and used to explain a variety of tectonic controls on sea-level change (Cloetingh, 1986; Watts et al., 1982; White and McKenzie, 1988).

To the writer's knowledge, only one study (Quinlan and Beaumont, 1984) made a preliminary attempt to establish whether basin modeling and subsidence analysis can be used to verify and explain known stratigraphic relations. The case they presented involved periodic Paleozoic collisional thrust loading of the Atlantic margin of North America which caused repeated flexural bending and development of the Appalachian basin. Development of the Appalachian foreland basin caused uplift of arches between it and adjacent embayed cratonic basins such as the Illinois and Michigan basins. Unconformities within the stratigraphic record on these arches appear to fit the modeled flexural deformation in the Appalachian basin. Moreover, they proposed that the oval-shaped plan of the present Illinois and Michigan basins owes its origin to flexural processes caused by collisional events along the Atlantic margin of North America. One consequence of Quinlan and Beaumont's (1984) study is that tectonic processes occurring in one basin influence stratigraphic and sedimentary processes beyond that basin's boundaries into adjacent highs and basins.

Comparative stratigraphy of the Illinois and central Appalachian basins (Heidlauf et al., 1986; Klein, 1987; Klein and Hsui, 1987; Willard and Klein, in review) shows a parallelism of depositional trends, but of different duration, that appear to fit tectonic subsidence models. Both basins began with a rifting history (Heidlauf et al., 1986; Fichter and Diecchio, 1986) involving deposition of nonmarine and marine clastic sediments in response to supercontinent breakup. Both show nearly coeval thermal subsidence with associated carbonate deposition. Thermal subsidence, with associated carbonate deposition, followed rifting. The thermal subsidence in the Illinois basin (Heidlauf et al., 1986) involved both a 60 m.y. period of linear thermal subsidence (defined by $t^{1/2}$) followed a 110 m.y. period of asymptotic thermal subsidence. In contrast, thermal subsidence in the central Appalachian basin was halted by compressional foreland basin tectonics associated with the Taconic (480-435 Ma) and Acadian (380-340 Ma) orogenies (Glover et al., 1983). Thus mid-Paleozoic stratigraphy of the central Appalachian basin differs from the Illinois basin. During the Alleghenian-Hercynian orogeny (330-230 Ma), stratigraphies of the Illinois and central Appalachian basins again were in parallel, because the two basins were yoked (Quinlan and Beaumont, 1984).

This paper examines whether the stratigraphic record can provide verification of geodynamic models of basin subsidence. This paper also addresses whether basin forming processes deduced from geodynamic modeling can explain certain patterns of sedimentation. In particular, I focus on how changes in foreland basin subsidence in the central Appalachian basin contributed not only to its late Paleozoic yoking with the Illinois basin, but also to the origin of the Pennsylvanian cyclothems which are coeval with this yoking event.

ILLINOIS BASIN SUMMARY

Quantitative tectonic subsidence history of the Illinois basin (Fig. l) was examined in detail by Heidlauf et al. (1986) and summarized by Klein and Hsui (1987). In their tectonic subsidence analysis, Heidlauf et al. (1986) used a two-dimensional lateral heat flow model (cf. Sawyer et al., 1987) which incorporated an Airy isostatic model and backstripping procedures of Sclater and Christie (1980). Additional modeling assumptions included: porosity decreases with depth, sediment accumulation rates kept pace with subsidence rate,

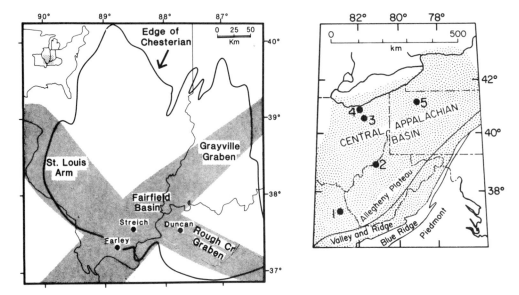

Figure 1. Maps showing location of wells used in tectonic subsidence analysis in the Illinois basin (left) and the central Appalachian basin (right) as per Heidlauf et al. (1986) and Willard and Klein (in review). 1 = Ashland No. 1 Well; 2 = Wood Co. Deep Well; 3 = Armstrong No. 8 Well; 4 = Krause No. 1 Well; 5 = Well No. 3737

the mass of rock and the water column above the depth of isostatic compensation is constant everywhere in the basin, and sediment contained no radioactive elements and were characterized by a zero coefficient of thermal expansion. Their model assumed also that extension was nonuniform (cf. Hellinger and Sclater, 1983) and that the physical processes within the basin can be considered independently during the short discrete time steps into which the stratigraphy of the Illinois basin was divided for tectonic subsidence analysis. Because of computer program limitations with respect to time intervals and scaling, only the older portion of the recovered well data was used for tectonic subsidence analysis; the remaining tectonic subsidence trends were determined by splicing a second subsidence curve for the younger section in one case (Streich well), and determination of sediment accumulation rates (Figs. 2 and 3).

The Illinois basin is a large oval-shaped basin in which 6000 m of sediment accumulated during Paleozoic time. Initially, sediments accumulated in a south facing, open-marine embayment formed by rifting (McGinnis, 1970; McGinnis et al., 1976). The three-arm rift system that underlies the basin is demarcated by magnetic and gravity anomalies and earthquake epicenters (Braile et al., 1982, 1986; Hildenbrand, 1985; McGinnis, 1970; McGinnis et al., 1976). The southern end of this embayment was closed by uplift of the Pascola arch between Pennsylvanian and Cretaceous time (Marcher, 1961; Stearns, 1957; Bethke, 1985). Hypothesized yoking of the central Appalachian and Illinois basins, as predicted from modeling by Quinlan and Beaumont (1984), suggests the Pascola arch probably was uplifted during Pennsylvanian time. This prediction is consistent with development of a major paleoceanographic sill in southern Illinois during late Mississippian time (Treworgy, 1985).

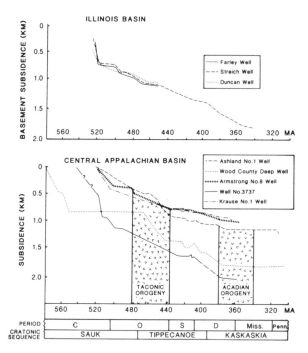

Figure 2. Tectonic subsidence curves for Illinois basin (upper) and central Appalachian basin (lower). Redrawn to common scale from Heidlauf et al., 1986, and Willard and Klein, in review. Curves for Farley and Duncan wells in Illinois basin obtained from lower Paleozoic only because of limitations of computer program. Curves for Streich well represents splicing of two subsidence curves as per Heidlauf et al. (1986; their Figs. 2 and 5).

Tectonic subsidence analysis of three deep wells in the basin depocenter (Fig. 1) demonstrated that the basin formed by rapid, initial mechanical subsidence associated with rifting. This was followed by thermal subsidence which began around 520 Ma (Heidlauf et al., 1986; see Figs. 2 and 3). By comparing reconstructed and calculated basement elevations through time, observed subsidence is consistent with a thermal model with $t^{1/2}$ for a 60 m.y. period, followed by longer-term (110 m.y.) asymptotic thermal subsidence (Heidlauf et al., 1986). The asymptotic thermal trend may have been coupled also with subsidence caused by excess mass (cf. DeRito et al., 1983) as igneous intrusions cooled (Klein and Hsui, 1987). Thermal subsidence was followed by flexural subsidence (Figs. 2 and 3) in response to thrust loading in the Appalachians during the Alleghenian-Hercynian orogeny (Quinlan and Beaumont, 1984; Klein and Hsui, 1987). Such flexural subsidence is expressed as a post-late Mississippian increase in slope of the tectonic subsidence curve of the Streich well (Fig. 2) and increased sediment accumulation rates (Fig. 3) in the basin (Heidlauf et al., 1986). This tectonic stage was followed by late Pennsylvanian and Permian intrusion of alnoites, dated by Zartman et al. (1967), which were derived by partial melting and are suggestive of resurgent rifting (Lewis, 1987, and 1987 personal communication). During its entire history

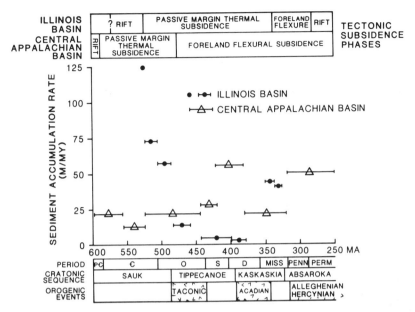

Figure 3. Comparison of sediment accumulation rates with time and tectonic phases for Illinois basin (as per Heidlauf et al., 1986) and central Appalachian basin (as per Willard and Klein, in review).

including the present day, reactivation of faulting was associated with the rift phase of the Illinois basin (Davis, 1987). The evolution of the Illinois basin (Fig. 4) demonstrates a close correlation of facies style, sediment accumulation rates (Fig. 3) and tectonic history (Heidlauf et al., 1986; Klein and Hsui, 1987). The Illinois basin shows coeval trends in thermal subsidence with the Michigan and Williston basins (Klein and Hsui, 1987).

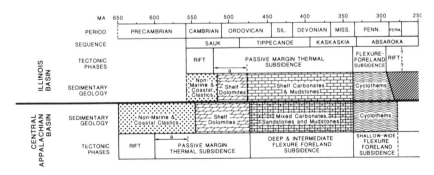

Figure 4. Correlation diagram of Illinois basin and central Appalachian basin comparing tectonic subsidence phases with sedimentary geology (adapted from Heidlauf et al., 1986, their Fig. 6; and Willard and Klein, in review, their Fig. 8). Thermal subsidence during interval marked by letter "a" represents period of linear thermal subsidence, whereas remainder of thermal subsidence phase is asymptotic thermal subsidence.

Central Appalachian Basin Summary

The central Appalachian basin is a long linear basin that is oriented northeast-southwest parallel to the orogenic grain of the Appalachian foldbelt and contains up to 19.2 km of sediment (Colton, 1970). This basin evolved with a complex history of basin formation and sedimentary geology (Colton, 1970; Hatcher, 1972; Rodgers, 1983; Fichter and Diecchio, 1986; Tankard, 1986; Willard and Klein, in review, their Table 1). Initially the central Appalachian basin was part of a late Precambrian supercontinent which evolved into a Cambrian passive margin by rifting, extension and associated thermal subsidence (Bond et al., 1984; Fichter and Diecchio, 1986). Plate convergence started a compressional phase causing collision of North America with island arc terrains and microcontinents (Hatcher, 1972; Rodgers, 1983; Fichter and Diecchio, 1986). Such compressional tectonics dominated the Appalachian basin region throughout most of Paleozoic time, with three major accretionary events causing foreland basin development during the Taconic (480-435 Ma), Acadian (380-340 Ma) and Alleghenian-Hercynian (330-230 Ma) orogenies (Glover et al., 1983). In short, the Appalachian basin shows a complex history starting with initial rifting and thermal subsidence, followed by successive foreland basins. The sedimentary-stratigraphic history of the central Appalachian basin parallels these tectonics events (Colton, 1970; Fichter and Diecchio, 1986; Simpson and Sunberg, 1986; Willard and Klein, in review).

Willard and Klein (in review) completed a tectonic subsidence analysis of the central Appalachian basin for the time interval 580-320 Ma utilizing data from five deep wells (Fig. 1). They used the same tectonic subsidence program as Heidlauf et al. (1986) with the same model assumptions reviewed in the previous section. Figure 2 summarizes tectonic subsidence data from these wells (Willard and Klein, in review). Initially the central Appalachian basin underwent mechanical, rapid subsidence caused by early rifting, followed by thermal subsidence. This thermal subsidence was confirmed by comparing the elevation of basement with $t^{1/2}$ and a short-term linear trend was observed (Willard and Klein, in review, their Fig. 7); no subsequent asymptotic trend was observed, however. Starting with the Acadian orogeny, foreland flexural subsidence was observed primarily from the nonlinear nature of $t^{1/2}$ (Willard and Klein, in review), and this flexural subsidence persisted until the Late Permian. Although the Appalachian region experienced both a post-Taconic and a post-Acadian rift-drift cycle, tectonic subsidence analysis (Willard and Klein, in review) failed to show recurrence of thermal subsidence within the central Appalachian basin.

In the central Appalachian basin significant changes in subsidence rate through time are attributed to rapid subsidence during the Taconic orogeny and considerably slower subsidence during the Acadian orogeny (Fig. 5). A similar change in subsidence rate was inferred previously by Tankard (1986) from stratigraphy. Subsidence rate was even less during the Alleghenian-Hercynian orogeny (Tankard, 1986). This temporal decrease in subsidence rate is attributed to progressive increase of rigidity of continental crust below the Appalachian basin by overthrusting; consequently, flexural deformation decreased and wavelength increased (Tankard, 1986; Willard and Klein, in review). Thus during the Taconic and Acadian orogenies, the central Appalachian basin was narrower and deeper than during the Alleghenian-Hercynian orogeny when this basin was shallower and much wider because crustal rigidity increased from middle to late Paleozoic time. Taconic strata consist of deltaic and slope facies confined mostly to the basin; Acadian sedimentation was dominantly deltaic

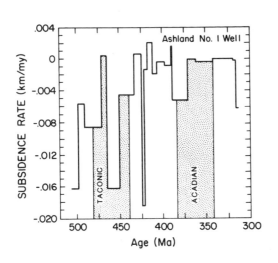
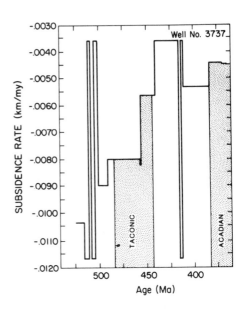

Figure 5. Comparison of subsidence rate through time, Ashland No. 1 well (left) and well 3737 (right), central Appalachian basin. Redrawn from Willard and Klein, in review.

with associated black shales that extended onto the eastern interior (Ettensohn, 1985; Ettensohn and Elam, 1985); and Alleghenian-Hercynian sedimentation ranged from non-marine to shallow shelf. The deltaic and shallow shelf deposits were organized into coal-bearing cyclothems which extended from the central Appalachian basin to the Illinois and Michigan basins. The evolution of the central Appalachian basin (Fig. 4) shows a close correlation of facies styles, sediment accumulation rate (Fig. 3) and tectonic subsidence history (Willard and Klein, in review) although initial rifting and subsequent thermal subsidence occurred earlier than in the Illinois basin.

YOKED DEPOSITIONAL PATTERNS IN THE ILLINOIS AND CENTRAL APPALACHIAN BASINS AND THE NORTHERN MIDCONTINENT

I have shown that the comparative stratigraphies of the Illinois and central Appalachian basins were parallel during initial rifting and thermal subsidence phases of basin evolution, and again during Alleghenian-Hercynian yoking of the two basins with coeval development of coal-bearing cyclothems (Fig. 4). Foreland flexural subsidence during the middle Paleozoic in the central Appalachian basin caused development of a mixed clastic and carbonate assemblage, whereas in the Illinois basin carbonate deposition continued with minor interruption (Fig. 4). Shale was deposited in the Illinois basin coeval with the Taconic and Acadian orogenies (Heidlauf et al., 1986, their Fig. 6) probably as a consequence of laterally extensive turbid layer flow fed by sediment from the Taconic and Acadian orogens.

During the Alleghenian-Hercynian orogeny, large scale stratigraphic continuity was established between the Illinois and central Appalachian basins and the northern midcontinent with deposition of coal-bearing cyclothems (Fig. 4). The origin of these cyclothems is controversial (see review by Heckel, 1984), and earlier workers proposed at least two dominant mechanisms including tectonic processes (Weller, 1930, 1956, 1959), eustatic processes of unidentified origin (Moore, 1931, 1950), or climatically controlled, glacial-eustatic changes (Wanless and Shepard, 1936; Wheeler and Murray, 1957). Recently, the controversy was revived by Tankard (1986) and Heckel (1986). Tankard (1986), working primarily with the nonmarine dominated cyclothems of the Appalachians (or Appalachian-type cyclothems), proposed that they are the product of episodic foreland thrust faulting caused by collision of different microcontinents and North America (Hatcher 1972; Lefort, 1983; Lefort and Van der Voo, 1981; Van der Voo, 1983, 1988). In this view, episodic thrust loading caused repeated flexural subsidence. With each flexural subsidence event, basins were underfilled—favoring marine transgression, but as sediment yield from uplifted orogens increased the basins overfilled slowly and swampy conditions gave way to later nonmarine deposition. Each thrust event repeated deposition of cyclothems. Heckel (1986) suggested that northern midcontinent Pennsylvanian (Kansas-type) cyclothems were domi-nated by eustatic changes. He considered the black shale members (core shales) to represent the relatively deepest water environment associated with both a rise in sea level and expansion of the oxygen minimum zone (cf. Arthur et al., 1987; Schlanger et al., 1987). Periodicity of Kansas-type cyclothems fits the Milankovitch band in response to glacially driven eustasy (Heckel, 1986).

Heckel (1986) and Tankard (1986) each isolated one major end-member process of cyclothem formation, but failed to recognize that combined global and regional tectonic events are causing cyclothem formation. The Alleghenian-Hercynian orogeny is a regional component of a global orogenic cycle that led to the formation of the supercontinent Pangea. Times of supercontinent development, such as the Pennsylvanian, effect climatic regimes in such a way that continent-wide glaciers form (Fischer, 1984; Worsley et al., 1984; Veevers and Powell, 1987). Glaciation was caused both by supercontinent growth and uplift in response to excess heat underneath the continents (Anderson, 1982; Worsley et al., 1984) and supercontinent position in colder and wetter polar and subpolar latitudes. Thus, glacial-eustatic processes identified by Heckel (1986) are an indirect response to global supercontinent collisional tectonics and are preserved best in Kansas-type cyclothems which developed in stable terrains far removed from foreland flexural subsidence and uplift. The Appalachian-type cyclothems, however, appear to be formed in direct response to foreland basin flexural subsidence because they accumulated adjacent to the collision zone, thus masking glacial eustasy. These two end members of Pennsylvanian cyclothem deposition (Fig. 6) involve both eustatic processes and foreland flexural processes corresponding to the Kansas-type and Appalachian-type cyclothems of North America, respectively. The Illinois-type cyclothem, as designated by Weller (1930), represents a transitional stage of nearly equal influence of both glacially driven eustatic change and moderate flexural subsidence. Thus it is the remarkable coincidence of a combination of supercontinent development, concomitant glaciation, associated episodic thrust loading and foreland basin subsidence of small magnitude on progressively more rigid crust that caused the development of Pennsylvanian coal-bearing cyclothems of North America over a broad region.

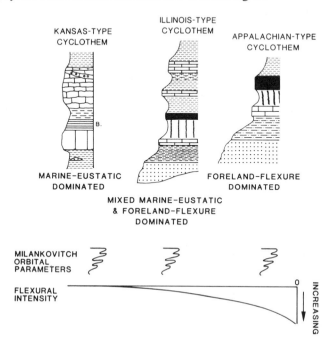

Figure 6. Comparison of classic types of North American coal-bearing cyclothems as per Heckel (1986) and Weller (1958) showing relative significance of the role of end-member eustatic processes and foreland flexural basin forming processes. (Stipple = sandstone; dot/dash pattern = siltstone; brickwork pattern = limestone; horizontal lines with "B" = black or "core" shale; dashes = gray shale; black = coal; vertical lines = underclay).

OTHER SEDIMENTARY RESPONSES TO BASIN SUBSIDENCE MODELS

Examination of Figure 4 shows a consistent association of sedimentary facies and processes with tectonic phases in both the Illinois basin (cf. Klein and Hsui, 1987) and the central Appalachian basin. The tectonic phases were established from application of tectonic subsidence analysis incorporating modeling assumptions and methods of McKenzie (1978), Sclater and Christie (1980), Sawyer et al. (1987) and Watts et al. (1982). The effect of rapid mechanical, fault-controlled initial subsidence is coeval with deposition of both nonmarine and coastal clastic sedimentary rocks. This relationship is to be expected because the effect of rapid fault-controlled mechanical subsidence is to create large accommodation space and basinal relief. Both permit increased rates of erosion of the rift shoulders, the sediment from which accumulates as thick basinal nonmarine clastics. As relief is reduced both by the net effect of erosion of the rift shoulders and deposition of sediments within the basin, rates of sediment accumulation tend to become lower (Hellinger and Sclater, 1983). Moreover, during the latest stages of rifting, thermal influences increase and basin subsidence develops over a wider areas (Hellinger and Sclater, 1983; White and McKenzie, 1988) permitting marine deposition to occur. Some of this marine clastic deposition persisted into the earlier stage of a passive margin type of thermal subsidence. These clastics persisted in part because during the transition from mechanical to thermal subsidence within basins, the two subsidence processes overlap, rather than being abrupt as suggested from tectonic subsidence models (Watso, 1988). These findings suggest that on a basin scale and a longer-term time scale, the stratigraphic record can be correlated to tectonic subsidence phases in both these basins as well as used to calibrate and verify tectonic subsidence modeling.

Overlap of mechanical and thermal basin subsidence can be documented from the Illinois basin (Fig. 2) and appears to influence stratigraphic relationships observed in the platform area represented by the Wisconsin arch, northwest of the Illinois basin. Recently, Watso (1988) confirmed that the Upper Cambrian of Wisconsin is organized into two main facies cycles as proposed on lithologic grounds by Ostrom (1978). The basal zone of these cycles are 510 and 478 Ma, which are identical in age (Watso, 1988) to the start of two major positive deviations observed in subsidence curves from the Streich and Farley wells (Heidlauf et al., 1986) in the Illinois basin (Fig. 2). These positive deviations on the tectonic subsidence curves (Fig. 2) appear to represent late stages of superimposed mechanical subsidence during a phase of linear thermal subsidence. These two mechanical subsidence events appear to cause repeated deposition of Cambro-Ordovician facies and lithologic cycles (Watso, 1988; Ostrom, 1978). Tectonic subsidence analysis of a deep well on the Wisconsin arch by Watso (1988) showed an early short (10 m.y.) period of moderate subsidence followed by slow mechanical subsidence during Late Cambrian and Early Ordovician time. Thus uplift by flexure or other processes could not be documented on the Wisconsin arch. It appears that mechanical subsidence within the Illinois basin, rather than uplift on the Wisconsin arch influenced cyclic sedimentation on the adjacent platform areas northwest of the Illinois basin.

Another sedimentary process observed to be coeval with tectonic subsidence processes is the occurrence of extensive dolomitization during passive margin thermal subsidence. Heidlauf et al. (1986) were the first to observe this relationship in the Illinois basin, but Figure 4 shows that an identical relationship exists also in the central Appalachian basin. Similarly, Barrett (1987) reported that dolomitization in the Smackover Limestone (Jurassic) of the Gulf coastal plain is coeval with a passive margin thermal subsidence stage. Like Heidlauf et al. (1983), neither Barrett (1987) nor I offer an explanation for these findings. Clearly, this problem remains open, but it demonstrates that a possible solution to understanding the origin of this coincidental relationship probably lies in a more detailed analysis of geodynamic modeling and associated stratigraphic and petrologic study on a smaller spatial and temporal scale. These observations suggest that smaller-scale stratigraphic, sedimentological, and petrological features occurring over shorter-term time scales do not permit verification of basinal geodynamic models at this time.

Summary and Conclusions

Comparison of both tectonic subsidence analysis and stratigraphy of the Illinois and central Appalachian basins shows that both basins were initiated by late Precambrian to Cambrian rifting processes followed by thermal subsidence, and that both ended during the late Paleozoic with a period of yoked flexural foreland subsidence. During initial rifting, depositional patterns were dominantly nonmarine to coastal and shelf clastics, followed by carbonate bank deposition during thermal subsidence. In the Illinois basin, middle Paleozoic thermal subsidence reached an asymptotic stage, whereas the central Appalachian basin experienced two episodes of middle Paleozoic flexural subsidence in response to collisional tectonics. Middle Paleozoic deposition in the Illinois basin consisted dominantly of a carbonate sequence with minor quantities of shale deposition derived from the Acadian and Taconic orogens, whereas sedimentation in the central Appalachian basin during flexural subsidence of both the Acadian and Taconic orogenies consisted of mixed carbonate and deltaic-shelf systems. During late Paleozoic time, cyclothem deposition dominated both basins because they were yoked by foreland flexural subsidence and tectonics. Deposition of broad midcontinent Cambro-Ordovician stratigraphic cycles in the Wisconsin arch correlate to two fault-controlled mechanical subsidence events superimposed on linear thermal subsidence in the Illinois basin.

The origin of the Pennsylvanian cyclothems is attributed to a remarkable set of concurrent events and processes related to supercontinent accretion. As supercontinents accreted and were uplifted in response to excess heat under the continents in polar and subpolar latitudes, glacial-eustatic changes caused rapid facies shifts in stable interiors far removed from the orogenic belts of plate margins, forming Kansas-type cyclothems. Closer to the margins where collision tectonics occurred, periodic thrust loading caused repeated flexural subsidence in foreland basins. Such repeated flexural subsidence favored an alternation of underfilling of basins with marine transgressive deposits followed by an overfilling by coastal sediments, coals, and nonmarine fluvial facies as uplift rates in the orogen increased. These repeated tectonic processes formed Appalachian-type cyclothems. Glacial-eustatic change was superimposed on this flexural process. Thus it is the coupling of foreland flexural tectonics in response to repeated thrust loading and the genetically related growth of supercontinents in polar regions which triggered polar glaciations, that caused the development of the Pennsylvanian cyclothems of North America.

This paper has demonstrated that stratigraphic and sedimentary verification of tectonic subsidence models is possible if both stratigraphic and sedimentological data are integrated into larger units of long-term temporal and large spatial scales. Tectonic subsidence models separate basin subsidence processes into distinct longer-term phases which existed over a large area. Integrated sedimentary packages of lithofacies, environmentally controlled facies, and trends in sediment accumulation rates on a basin-wide scale fit remarkably well with tectonic subsidence stages (Figs. 3 and 4) suggesting that the temporal changes of sedimentary patterns is dependent on subsidence mechanism. This finding confirms that the stratigraphic record can be used to verify model utility on a basin-wide scale and at temporal resolution ranging from approximately 15 to 30 m.y.. It appears that the smallest level of temporal resolution of current geodynamic basin modeling is of the magnitude of a second-order sea-level change as per Vail et al. (1977). Current controversy over the tectonic role on second-order sea-level change (Cloetingh, 1986; Watts et al., 1982; White and McKenzie, 1988) shows that this level of resolution requires considerable improvement (cf. Burton et al., 1987). Attempts to verify quantitative basin modeling at the scale of outcrops, facies, diagenetic events or processes appear to be at its infancy and represent one of many challenges for future research in quantitative dynamic stratigraphy.

ACKNOWLEDGMENTS

Research support came from two grants from the University of Illinois Research Board. A.T. Hsui is thanked for his guidance in the tectonic subsidence analysis summarized in the text and elsewhere. Discussions with D.T. Heidlauf, A.T. Hsui, D.R. Kolata, and Debra A. Willard proved helpful. Teresa E. Jordan and J.A. Nunn are thanked for their cogent and constructive reviews of an earlier manuscript version. I thank Mary L. Barrett for calling to my attention the timing of Jurassic dolomitization in the Gulf Coastal Plain with thermal subsidence events, and for providing me with a copy of her Ph.D. dissertation.

REFERENCES CITED

Anderson, D.L., 1982, Hot spots, polar wandering, Mesozoic convection and the geoid: Nature, v. 297, p. 391-393.

Arthur, M.A., Schlanger, S.O., and Jenkyns, H.C., 1987, The Cenomanian-Turonian oceanic anoxic event, II. Palaeoceanographic controls on organic-matter production and preservation: in Brooks, J., and Fleet, A.J., eds., Marine petroleum source rocks: Geological Society of London Special Publication 26, p. 401-420.

Barrett, M.L., 1987, The dolomitization and diagenesis of the Jurassic Smackover Formation, southwest Alabama [Ph.D. thesis]: Baltimore, Maryland, Johns Hopkins University, 362 p.

Bethke, C.M., 1985, A numerical model of compaction-driven groundwater flow and heat transfer and its application to the paleohydrology of intracratonic sedimentary basins: Journal of Geophysical Research, v. 90, p. 6817-6828.

Bond, G.D., Nickeson, P.A., and Kominz, M.A., 1984, Breakup of a supercontinent between 625 Ma and 555 Ma: New evidence and implication for continental histories: Earth and Planetary Science Letters, v. 70, p. 325-345.

Braile, L.W., Keller, G.R., Hinze, W.J., and Lidiak, E.G., 1982, An ancient rift complex and its relation to contemporary seismicity in the New Madrid seismic zone: Tectonics, v. 1, p. 225-237.

Braile, L.W., Hinze, W.J., Keller, G.R., Lidiak, E.G., and Sexton, J.L., 1986, Tectonic development of the New Madrid rift complex, Mississippi embayment, North America: Tectonophysics, v. 131, p. 1-21.

Burton, R., Kendall, C.G.St.C., and Lerche, I., 1987, Out of our depth: On the impossibility of fathoming eustasy from the stratigraphic record: Earth Science Reviews, v. 24, p. 237-277.

Cloetingh, S., 1986, Intraplate stresses: A new tectonic mechanism for fluctuations of relative sea level: Geology, v. 14, p. 617-620.

Colton, G.W., 1970, The Appalachian basin—Its depositional sequences and their geologic relationships, in Fischer, G.W., Pettijohn, F.J., Reed, J.C. Jr., and Weaver, K.N., eds., Studies of Appalachian geology: Central and southern: New York, Wiley, p. 5-47.

Davis, H.G., 1987, Pre-Mississippian hydrocarbon potential of Illinois basin: American Association of Petroleum Geologists Bulletin, v. 71, p. 546-547.

DeRito, R.F., Cozzarelli, F.A., and Hodge, D.S., 1983, Mechanism of subsidence of ancient cratonic rift basins: Tectonophysics, v. 94, p. 141-168. Ettensohn, F.G., 1985, The Catskill delta complex and the Acadian orogeny: A model, *in* Woodrow, D.L., ed., The Catskill Delta: Geological Society of America Special Paper 201, p. 39-49.

Ettensohn, F.R., and Elam, T.R., 1985, Defining the nature and location of a Late Devonian-early Mississippian pycnocline in eastern Kentucky: Geological Society of America Bulletin, v. 96, p. 1313-1321.

Fichter, L.S., and Diecchio, R.J., 1986, Stratigraphic model for timing the opening of the proto-Atlantic ocean in northern Virginia: Geology, v. 14, p. 307-309.

Fischer, A.G., 1984, Two Phanerozoic supercycles, *in* Berggren, W.A., and Van Couvering, J.A., eds., Catastrophes in earth history—The new uniformitarianism: Princeton, New Jersey, Princeton University Press, p. 129-150.

Glover, L, III, Speer, J.A., Russell, G.S., and Farrar, S.D., 1983, Ages of regional metamorphism and ductile deformation in the central and southern Appalachians: Lithos, v. 16, p. 223-245.

Hatcher, R.D., Jr., 1972, Developmental model for the southern Appalachians: Geological Society of America Bulletin, v. 83, p. 2735-2760.

Heckel, P.H., 1984, Changing concepts of midcontinent Pennsylvanian cyclothems, North America: Compte Rendu Neuvieme Congres International de Stratigraphie et de Geologie du Carbonifere, Carbondale, Illinois, Southern Illinois University Press, v. 3, p. 535-553.

Heckel, P.H., 1986, Sea-level curve for Pennsylvanian eustatic marine transgressive-regressive depositional cycles along midcontinent outcrop belt, North America: Geology, v. 14, p. 330-334.

Heidlauf, D.T., Hsui, A.T., and Klein, G.deV., 1986, Tectonic subsidence analysis of the Illinois basin: Journal of Geology, v. 94, p. 779-794.

Hellinger, S.J., and Sclater, J.G., 1983, Some comments on two-layer extensional models for the evolution of sedimentary basins: Journal of Geophysical Research, v. 88, p. 8251-8269.

Hildenbrand, T.G., 1985, Rift structure of the northern Mississippi embayment from the analysis of gravity and magnetic data: Journal of Geophysical Research, v. 90, p. 225-245.

Klein, G.deV., 1987, Current aspects of basin analysis: Sedimentary Geology, v. 50, p. 95-118.

Klein, G.deV., and Hsui, A.T., 1987, Origin of cratonic basins: Geology, v. 15, p. 1094-1098.

Lefort, J.P., 1983, A new geophysical criterion to correlate the Acadian and Hercynian orogenies of western Europe and eastern North America: *in* Hatcher, R.D., Jr., Williams, H., and Zietz, I., eds., Contributions to the tectonics and geophysics of mountain chains: Geological Society of America Memoir 158, p. 3-18.

Lefort, J.P., and Van der Voo, R., 1981, A kinematic model for the collision and complete suturing between Gondwanaland and Laurussia in the Carboniferous: Journal of Geology, v. 89, p. 537-551.

Lewis, R., 1987, Introducing the Wauboukigou alnoite province, USA, or ultramafic intrusives associated with Permian rifting of the northern Mississippi embayment, New Madrid rift complex: Illinois State Geological Survey Seminar, April, 1987, p. 1.

Marcher, M.V., 1961, The Tuscaloosa Gravel in Tennessee and its relation to the structural development of the Mississippi embayment syncline: U.S. Geological Survey Professional Paper 424, p., B90-B93.

McGinnis, L.D., 1970, Tectonics and the gravity field in the continental interior: Journal of Geophysical Research, v. 75, p. 317-331.

McGinnis, L.D., Heigold, C.P., Ervin, C.P., and Heidi, M., 1976, The gravity field and tectonics of Illinois: Illinois State Geological Survey Circular 494, 24 p.

McKenzie, D., 1978, Some remarks on the development of sedimentary basins: Earth and Planetary Sciences Letters, v. 40, p. 25-32.

Moore, R.C., 1931, Pennsylvanian cycles in the northern midcontinent region: Illinois State Geological Survey Bulletin 60, p. 247-257.

Moore, R.C., 1950, Late Paleozoic cyclic sedimentation in central United States: 18th International Geological Congress Report, Part 4, p. 5-16.

Ostrom, M.E., 1978, Stratigraphic relations of lower Paleozoic rocks of Wisconsin: Wisconsin Geological and Natural History Survey Field Trip Guidebook No. 3, p. 3-22.

Quinlan, G.M., and Beaumont, C., 1984, Appalachian thrusting, lithospheric flexure, and the Paleozoic stratigraphy of the eastern interior of North America: Canadian Journal of Earth Sciences, v. 21, p. 973-996.

Rodgers, J., 1983, The life history of a mountain range—The Appalachians, in Hsü, K.J., ed., Mountain building processes: London, Academic Press, p. 229-241.

Sawyer, D.S., Hsui, A.T., and Toksöz, M.N., 1987, Extension, subsidence and thermal evolution of the Los Angeles basin—A two dimensional model: Tectonophysics, v. 133, p. 15-32.

Sclater, J.G., and Christie, P.A.F., 1980, Continental stretching: An explanation of the post-mid-Cretaceous subsidence of the central North Sea basin: Journal of Geophysical Research, v. 85, p. 3711-3739.

Schlanger, S.O., Arthur, M.A., Jenkyns, H.C., and Scholle, P.A., 1987, The Cenomanian-Turonian oceanic anoxic event, I. Stratigraphy and distribution of organic carbon-rich beds and the marine delta ^{13}C excursion, in Brook, J., and Fleet, A.J., eds., 1987, Marine petroleum source rocks: Geological Society of London Special Publication 26, p. 371-399.

Simpson, E.L., and Sundberg, F.A., 1986, Early Cambrian age for synrift deposits of the Chilhowee Group of southwestern Virginia: Geology, v. 15, p. 123-126.

Stearns, R.G., 1957, Cretaceous, Paleocene, and lower Eocene geologic history of the northern Mississippi embayment: Geological Society of America Bulletin, v. 68, p. 1345-1360.

Tankard, A.J., 1986, On the depositional response to thrusting and lithospheric flexure: Examples from the Appalachian and Rocky Mountain basins, in Allen, P.A., and Homewood, P., eds., Foreland basins: International Association of Sedimentologists Special Publication 8, p. 369-392.

Treworgy, J.D., 1985, Stratigraphy and depositional setting of the Chesterian (Mississippian) Fraileys/Big Clifty and Haney Formations in the Illinois basin (Ph.D. thesis): Urbana, Illinois, University of Illinois at Urbana-Champaign, 202 p.

Vail, P.R., Mitchum, R.M., Jr, and Thompson, S., III, 1977, Global cycles of relative changes of sea level, in Payton, C.E., ed., Seismic stratigraphy—Applications to hydrocarbon exploration: American Association of Petroleum Geologists Memoir 26, p. 83-98.

Van der Voo, R., 1983, A plate-tectonics model for the Paleozoic assembly of Pangea based on paleomagnetic data, in Hatcher, R.D., Jr., Williams, H., and Zietz, I., eds., Contributions to the tectonics and geophysics of mountain chains: Geological Society of America Memoir 158, p. 19-24.

Van der Voo, R., 1988, Paleozoic paleogeography of North America, Gondwana, and intervening displaced terranes: Comparison of paleomagnetism with paleoclimatology and biogeographical patterns: Geological Society of America Bulletin, v. 100, p. 311-324.

Veevers, J.J., and Powell, C.McA., 1987, Late Paleozoic glacial episodes in Gondwanaland reflected in transgressive-regressive depositional sequences in Euramerica: Geological Society of America Bulletin, v. 98, p. 475-487.

Wanless, H.R., and Shepard, F.P., 1936, Sea level and climatic changes related to late Paleozoic cycles: Geological Society of America Bulletin, v. 47, p. 1177-1206.

Watso, D.C., 1988, The effect of tectonic subsidence on sedimentation processes during deposition of four Late Cambrian formations, upper middle west, USA (M.S. thesis): Urbana, Illinois, University of Illinois at Urbana-Champaign, 260 p.

Watts, A.B., Karner, G.D., and Steckler, M.S., 1982, Lithospheric flexure and the evolution of sedimentary basins: Royal Society of London Philosophical Transactions, Series A, p. 249-281.

Weller, J.M., 1930, Cyclic sedimentation of the Pennsylvanian Period and its significance: Journal of Geology, v. 38, p. 97-135.

Weller, J.M., 1956, Argument for diastrophic control of late Paleozoic cyclothems: American Association of Petroleum Geologists Bulletin, v. 40, p. 17-50.

Weller, J.M., 1958, Cyclothems and larger sedimentary cycles of the Pennsylvanian: Journal of Geology, v. 66, p. 195-207.

Wheeler, H.E., and Murray, H.H., 1957, Base-level control patterns in cyclothemic sedimentation: American Association of Petroleum Geologists Bulletin, v. 41, p. 1985-2011.

White, N., and McKenzie, D., 1988, Formation of the "steer's head" geometry of sedimentary basins by differential stretching of the crust and mantle: Geology, v. 16, p. 250-253.

Willard, D.A, and Klein, G.deV., in review, Tectonic subsidence analysis of the central Appalachian basin: Journal of Geology.

Worsley, T.B., Nance, R.D., and Moody, J.B., 1984, Global tectonics and eustasy for the past 2 billion years: Marine Geology, v. 58, p. 373-400.

Zartman, R.E., Brock, M.R., Heyl, A.V., and Thomas, H.H., 1967, K-Ar and Rb-Ar ages of some alkalic intrusive rocks from central and eastern United States: American Journal of Science, v. 265, p. 858-870.

31

Time in Quantitative Stratigraphy

F.M. Gradstein [1], F.P. Agterberg [2] and M.A. D'Iorio [3]

[1]Atlantic Geoscience Centre, Bedford Institute of Oceanography, P.O. Box 1006,
 Dartmouth, Nova Scotia, Canada B2Y 4A2
[2]Geological Survey of Canada, 601 Booth Street, Ottawa, Ontario, Canada K1A 0E8
[3]Department of Geology, University of Ottawa, Ottawa, Ontario, Canada K1N 6N5

Abstract

Biostratigraphic zonations based on the Ranking and Scaling (RASC) method display both the likely order and the most likely distance in relative time between successive fossil events. If, for some of the ordered and scaled events also an age is known in millions of years, the interfossil distances may be converted in a linear time scale. Larger interfossil distances between stratigraphically successive events or zones of events reflect hiatusses or breaks in this scale. Such hiatusses tie closely to basin wide tectonic events.

The use of ranking and scaling for construction and correlation of time scales in basin analysis is illustrated with examples from the Mesozoic and Cenozoic continental margins of the NW Atlantic. Data sets include foraminifers, ostracods and palynomorphs, either separate or combined. On the Grand Banks of Newfoundland, early Tithonian, early Valanginian and early Aptian were times of significant tectonic and sedimentary changes.

Introduction

Time scales that link many stratigraphically successive events to each other in a linear fashion are essential to the earth sciences. This study deals with the construction and application of such time scales from regional biostratigraphic zonations, using the RASC (ranking and scaling) probabilistic sequencing method and the CASC (correlation and scaling in time) method.

Quantitative processing of fossil occurrence data in stratigraphic sections has a history of at least two decades. Theory and methods only have come of age in the last ten years, in large measure due to the stimulation of international cooperative projects. Older methods include adaptation of conventional multivariate clustering techniques and the semi-objective composite standard method using graphic correlation. More recently, deterministic and probabilistic sequencing techniques have been developed that analyze all stratigraphic order relationships of all fossil events in all stratigraphic sections simultaneously; these are more objective. A recent review of methods and applications is in Gradstein et al. (1985) and Agterberg and Gradstein (1988).

Quantitative Dynamic Stratigraphy (1989), T.A. Cross, ed., Prentice Hall, p. 519–542.

Probabilistic methodology views individual sequences of fossil events from different stratigraphic sections as random deviations from a true solution. Sources of uncertainty occur in both the order and in the spacing of fossil events in relative time. A fossil event is the occurrence of a taxon in its time context, derived from its position in a rock sequence. For stratigraphic purposes we apply the techniques only to certain events, such as first occurrences (appearance, entry), or last occurrences (disappearance, exit), in relative time. The RASC technique is a probabilistic sequencing method that calculates the most likely order and position along an interval scale of a large number of fossil events in multiple stratigraphic sections. Such a data set may contain the last occurrence of about 250 taxa in 25 or more wells or outcrops which involves 1500 or more events. Distance (separation) in relative time is inversely proportional to the frequency of cross-over from well to well for pairs of events. Suppose that a detailed scaled optimum sequence has been calculated in a sedimentary basin. Each event in this sequence has a known interfossil distance in relative time to its neighbors above and below. If, for some of the ordered and scaled events, an age also is known in millions of years, the interfossil distances in relative time can be used to interpolate the scaled optimum sequence in linear time. Large interfossil distances in relative time between two stratigraphically successive clusters of events in a scaled optimum sequence correspond to hiatuses or breaks in the biochronology. Such hiatuses may tie closely to basin-wide tectonic or global sea level changes.

This paper illustrates the use of ranking and scaling for constructing local biochronologies (time scales) in basin analysis. We use examples from the Mesozoic and Cenozoic of the Northwestern Atlantic margin, although the methods and their implications are applicable to any sedimentary basin.

RASC Method for Ranking and Scaling

The RASC computer program for ranking and scaling of stratigraphic events was published originally in Agterberg and Nel (1982a, b). In ranking, the frequency that each event occurs above, within, or below all other events is recorded. The final arrangement of all events in the optimum sequence is based on rank and reflects the most likely (average) order in relative time. Biostratigraphic events cluster along a relative time scale when they change position frequently with respect to one another in stratigraphic sections. This feature is used to calculate cross-over frequencies for all pairs of events that occur in all sections. Higher cross-over frequencies correspond to smaller stratigraphic distances between events. Detailed explanations, further developments of the method, and examples of application are presented in Gradstein et al. (1985). This section explains the scaling method by applying it to simple artificial examples: RASC uses a simple transformation to change cross-over frequencies into distances.

In Figure 1, observed occurrences of two biostratigraphic events (A and B) in 12 vertical stratigraphic sections are compared. An additional event (C) is considered in Artificial Example 4. As a rule, biostratigraphic events are observed only in a subset of the total number of stratigraphic sections (n) in a study region. In Artificial Example 1, n = 12 but A occurs only in n_A = 5 and B in n_B = 6 sections. The number of sections with both A and B present, $n_{A,B}$ = 2, is even smaller. In these two sections, A occurs stratigraphically above B. This relation can be expressed as n_{AB} = 2 and n_{BA} = 0, where AB represents A above B and BA

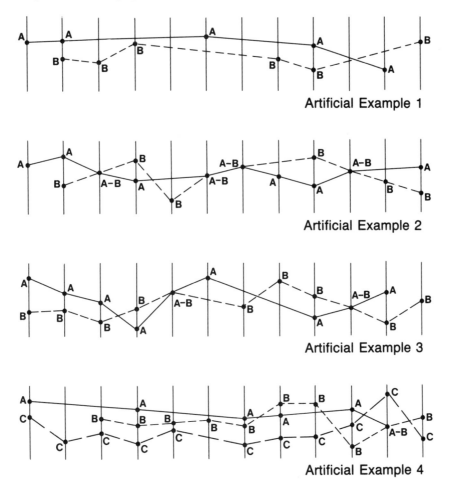

Figure 1. Graphical illustration of RASC method for ranking and scaling of stratigraphic events in many stratigraphic sections (shown as vertical lines). Ranking provides optimum sequences AB (A stratigraphically above B) in Examples 1 and 3; A-B (undecided) in Example 2; and ABC in Example 4. Scaling gives distance estimates of intervals between successive events along a linear (RASC) scale. The distance between A and B is estimated as (1) 1.28, (2) 0.00, (3) 0.32 and (4) 0.36 for Artifical Examples 1,2,3 and 4, respectively.

is B above A. In the other examples of Figure 1, A-B denotes that A and B occur at identical stratigraphic positions; n_{A-B} is the frequency with which this occurs (e.g., $n_{A-B} = 4$ in Artificial Example 2).

   Three threshold parameters have to be set at the beginning of a RASC run: k_c, m_{c1} and m_{c2} with k_c m_{c2} m_{c1}. The critical value kc indicates that an event will be used for computing only if it occurs in at least k_c sections. If we would set $k_c = 6$ in Artifical Example 1, the event A would not be used for ranking and scaling. The parameters m_{c1} and m_{c2} control the minimum number of pairs of events to be used for computing optimum sequences in ranking and scaling, respectively. If $m_{c1} = 1$ and $m_{c2} = 4$ in Artificial Example 1 (with $k_c \geq 5$), A and B would be compared for ranking but not for scaling. In general, if k_c and m_{c2} are increased, statistical precision of results is improved but fewer events are considered.

Several methods of ranking can be used, including presorting and matrix permutation on scores of the relative order of all pairs of events in all sections (Agterberg and Nel, 1982a). These methods produce a simple answer for the examples of Figure 1. If $n_{AB} > n_{BA}$, as in Artificial Examples 1 and 3, the ranking result is AB. The optimum sequence for the fourth example is ABC, and is "undecided" for Artificial Example 2 because $n_{AB} = n_{BA}$.

The scaling technique is conceptually more complex than the ranking technique. Using the frequencies n_{AB}, n_{BA}, n_{A-B} and $n_{A,B}$, a single relative frequency $p_{AB} = (n_{AB} + 0.5n_{A-B}) / n_{A,B}$ is computed. We define $p_{BA} = 1 - p_{AB}$. The principle of scaling is that the frequency p_{AB} is transformed into $z_{AB} = \phi^{-1}$, where p_{AB} is an estimate of the interval between mean positions of A and B along a distance scale (RASC scale), and ϕ represents a fractile of the normal distribution in standard form. If $p_{AB} = 0.5$, it follows that $z_{AB} = 0$ for the situation that A and B coincide along the RASC scale. If $p_{AB} = 1$, there would be certainty for A above B. However, if A and B are relatively close in the sections $p_{AB} = 1$ is replaced by a probability which is less than 1 and $z_{AB} = q$. In Artificial Example 1, $n_{A,B} = 2$ with $p_{AB} = 1$. If this relation were used in conjunction with other frequencies (e.g., for "indirect" estimation discussed later), we would choose $p_{AB} = 0.90$ with $q = 1.282$. The "default" value in RASC is $q = 1.645$ for $p = 0.95$.

The transformation ϕ^{-1} can be approximated by the linear transformation $z'_{AB} = 2.93 \, (p_{AB} - 0.5)$ as illustrated in Table 1. This illustrates that $z = z' = 0$ for $p = 0.5$ when unable to decide whether A should be above or below B in the optimum sequence as in Artificial Example 2. In Artificial Example 3, $p_{AB} = 5/8$, which yields $z_{AB} = 0.319$ and $z'_{AB} = 0.366$. In Artificial Example 4, $p_{AB} = 3.5/5$ which is slightly greater than $5/8$ in Example 3. The resulting distance $z'_{AB} = 0.59$ ($z'_{AB} = 0.52$) is also slightly greater.

For Example 4, $p_{AC} = 5/6$ with $z'_{AC} = 0.98$ ($z_{AC} = 0.97$), and $p_{BC} = 7/9$ with $z'_{BC} = 0.59$ ($z_{BC} = 0.77$). These three estimates of distance are not mutually consistent. For example, $z'_{AB.C} = z'_{AC} - z'_{BC} = 0.29$ provides an indirect estimate of the distance between A and B which differs from the direct estimate $z'_{AB} = 0.59$. This type of inconsistency is ascribed to small sample sizes and may be eliminated by averaging; e.g., $z'_{AB} = 0.5 \, (z'_{AB} + z'_{AB.C}) = 0.38$ which is close to $O(z,-)_{AB} = 0.36$. Especially when many indirect distance estimates occur, such averages are more precise than direct distance estimates. In RASC, the averaging process is refined by considering differences in sample sizes. For example, $p = 1.5/4$ for $n = 4$ is less precise than $p = 4.5/12$ for $n = 12$ although their z-values are the same. The second z-value is given more weight in the calculations because it is based on a larger sample.

The linear transformation was introduced here to illustrate the concept of scaling. In practice it is better to use the normal distribution as in the RASC method, because the linear transformation implies that the frequency density function of the interval between two events along the linear scale of the scaled optimum sequence is uniform. This, in turn, would mean that frequency density functions of individual events along this scale would show different shapes depending on the value of z'; e.g., for $z'_{AB} = 0$, A and B would show U-shaped density functions with local minima at their mean locations. It is more realistic to assume that individual species possess density functions with maxima at or near their mean values. The mode and mean coincide for the normal (Gaussian) curve model used in RASC. Decrease in

Table 1 Inconsistency frequencies (p) are transformed in RASC by using fractiles (z) of normal distribution in standard form. For p = 0.00, q (see text) is used. Linear transformation z' = 2.93 (p-0.50) gives values close to z.

p	z	z^1
0.00	-q	-2.930
0.05	-1.645	-1.319
0.10	-1.282	-1.172
0.20	-0.842	-0.879
0.30	-0.524	-0.586
0.40	-0.253	-0.293
0.50	0.000	0.000
0.60	0.253	0.293
0.70	0.524	0.586
0.80	0.842	0.879
0.90	1.282	1.172
0.95	1.645	1.319
1.00	q	2.930

density away from the mode could be different for different species. Also, for the same species it could be different in different stratigraphic positions. Agterberg and D'Iorio (in press) developed a method for separately estimating the frequency distributions of biostratigraphic events. They showed that a modification of the RASC computer program, which allows different variances for different events along the RASC scale, yields approximately the same scaled optimum sequence as ordinary RASC. However, the modified RASC method helps to identify small-variance events which are potentially useful as markers in a region.

As a result of scaling, all stratigraphic events, including some of those that occur only in one or a few sections (Unique Event Option, see Agterberg and Nel, 1982b), are positioned along the linear RASC scale. Extensive applications of the RASC method for biostratigraphic zonations of foraminifers, nannofossils and palynomorphs of Mesozoic and Cenozoic age (Gradstein et al., 1985; Williamson, 1987; D'Iorio, 1986) demonstrated that the optimum sequence reflects the average order of the events in time. Thus, RASC may be used for high resolution "point–to–point" correlation, after the individual (well) sequence record has been normalized for sampling and other errors (see CASC method). The scaled optimum sequence is a succession of biostratigraphic zones that can be calibrated to standard chronostratigraphy and geochronology.

CASC Method for Correlation and Scaling in Time

Biostratigraphic zonations express the general order and temporal separation of successive fossil ranges or range end points (events). Zonations are a principal tool in geological correlation of strata with comparable fossil content. The principle of the CASC method for automated geological correlation of the zonation obtained with RASC is illustrated in Figure 2. CASC is a multistep processing method that first plots the (scaled) optimum sequence of events against their level in a well section. The line of correlation is a best-fitting cubic spline-curve. The amount of smoothing is controlled by the smoothing factor (also see next section) which can be selected in a subjective manner or determined statistically. The observed positions of events in the individual sequence can now be updated according to the sequence of events in the optimum sequence, as graphically shown in the lower half of Figure 2 for an event A. This event may not have been observed in the well.

In the second processing step, the depth of events are plotted against their level and a best cubic spline is fitted to the data. This second step requires only a small smoothing factor, as depth and level of events are "parallel" (observed) measures. Next, the most likely depth of the events can be calculated as shown graphically in the upper half of Figure 2 for event A. With this correlation technique the expected depth of the scaled optimum sequence events can be calculated, which facilitates the tracing of the successive biozones from one section to the next. The CASC method is described in detail by Gradstein et al. (1985).

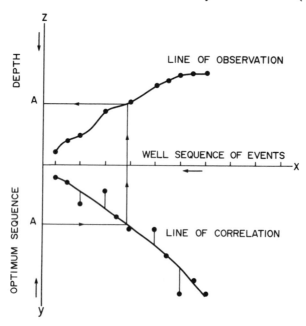

Figure 2. Graphical illustration of CASC method for correlation and scaling in time. y - optimum sequence of events; x - observed sequence of events in one well; z - depth scale for the events in the well. The most likely position of the optimum sequence events in the well is found by projecting these events via the two best fit lines (line of correlation and line of observation) onto the (well) depth scale, as shown for an event A. The best fit lines are calculated with smoothed cubic splines.

If the absolute age is known for a subgroup of events in the scaled optimum sequence, the RASC scale can be transformed into a numerical time scale (in Ma). The method is identical to the one shown in the lower half of Figure 2, except that the well sequence of events (x) is replaced by a linear scale in units of millions of years. The subsets of events in the scaled optimum sequence for which an absolute age is known are located along this scale. The best fitting spline curve of age versus RASC distance is used to calculate the age in Ma units for all events in the RASC sequence.

In the CASC method of time conversion, age-depth curves for individual wells are provided which can be used for multi-well correlation with error bars (Fig. 3). These error bars are projections along the depth axis of error bars initially estimated in time. Estimates of initial intervals for age ± one standard deviation were projected assuming either constant (Fig. 3A) or variable (Fig. 3B) sediment accumulation rates for these intervals. Approximately two-thirds of observed events should fall within the error bars on Figure 3.

The original CASC computer program is interactive and requires the user to decide on the best choice of age-depth curves in a subjective manner. More recently, an objective method for this purpose has also been developed, using cross-validation as discussed below.

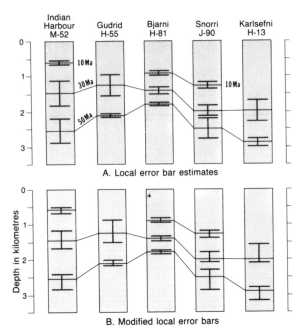

Figure 3. Example of CASC multi-well comparison for 5 wells along the northwest Atlantic margin (after Gradstein et al., 1985, Figure 8, p. 327). Subjective Smoothing Factor (SF) intially estimated along age axis in age-depth diagram for each well was projected along depth axis by assuming constant (Fig. 3A) and variable (Fig. 3B) sediment accumulation rate, respectively. Approximately two thirds of observed biostratigraphic events fall on error bars. Recently, a cross-validation method was developed for estimating optimum SF (see Table 2). Also, separate error bars can be estimated to evaluate uncertainties of estimated age-depth curve and observed biostratigraphic events (see Figure 4 and Table 3).

Computer Simulation Experiment for Age-Depth Spline-Curve Fitting with Error Analysis

During development of the RASC/CASC procedures, three criteria were employed for evaluation: (1) the methods require a firm stratigraphic foundation; (2) they should be logically coherent mathematically and statistically; and (3) the computer program should be efficient and user-friendly. The methods are tested for the first criterion by systematically comparing RASC zonations with subjective zonations, and by stratigraphers' evaluation of computer outputs. Computer simulation experiments are helpful in evaluating those aspects of the method related to the second criterion. Such experiments were performed on the RASC program (Harper, 1984; Gradstein et al., 1985) and are now underway for CASC.

Results for one of these experiments are displayed in Figure 4 and Tables 2 and 3. A theoretical age-depth curve $T = 9.155 + 0.685x + 0.165x^2 - 0.005x^3$ ($x = 100m$) is shown in Figure 4A. Twenty-one random normal numbers (zero mean, unit variance) of regularly spaced points (labelled $i = 1, 2, \ldots, 21$ in Table 3) were added in order to simulate observations, O_i, of biostratigraphic events in a hypothetical Cenozoic basin. In Reinsch–DeBoor spline-fitting, as in our computer program called SPLIN, the smoothing factor (SF) fully determines the shape of the best-fitting spline-curve. In SPLIN, the square of SF is equal to the averaged squared deviation $(O_i - S_i)^2$ between observations O_i and their corresponding values S_i on the spline-curve S.

Table 2 Cross-validation value (CVi) is minimum for optimum smoothing factor (SFi) within interval between minimum (SF0) and maximum (SF10). For further explanation see text.

i	SF_i	CV_i
0 (minimum)	0.828	3.607
1	0.897	3.314
2	0.967	2.968
3	1.036	2.565
4	1.106	2.219
5 (optimum)	1.175	2.131
6	1.245	2.192
7	1.314	2.290
8	1.384	2.395
9	1.453	2.445
10 (maximum)	1.523	2.448

An optimum value of SF can be found by cross-validation (cf, Agterberg and Gradstein, 1988) as illustrated in Table 2. In general, SF_{min} SF_{opt} SF_{max} where the minimum smoothing factor (SF_{min}) represents the least smooth spline curve whose age increases monotonically with depth. The maximum smoothing factor (SF_{max}) corresponds to the best-fitting straight line. The optimum smoothing factor (SF_{opt}) shows minimum cross-validation value within the interval between SF_{min} and SF_{max}. The spline-fitting function in SPLIN is a modified version of DeBoor's (1978) FORTRAN program which uses the secant method, returning a smoothing factor that is usually slightly greater than an input SF. For this reason, 1.170 was

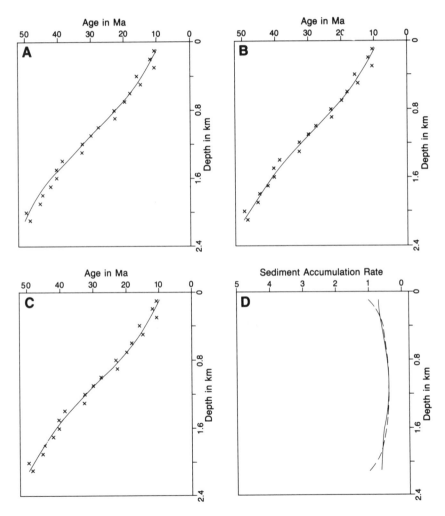

Figure 4. Computer simulation experiment. Random normal deviates were added to theoretical curve (A). Cross-validated smoothing spline (B) was approximated by its Jackknife estimate (C). First derivative of spline-curve in Figure 4B gave sediment accumulation rate curve of Figure 4D (solid line) which is compared with first derivative of theoretical curve (broken line).

selected from Table 2 as input for the cross-validated smoothing spline with SF = 1.174 shown in Figure 4B. Obviously S in Figure 4B is close to T in Figure 4A. In practice, T is not known and confidence intervals on S were constructed to evaluate the difference between S and (unknown) T, and to check residuals $(O_i\text{-}S_i)$ for possible outliers. This problem has been studied by Wahba (1983) and Eubank (1984) who have shown that the column vector §S of values S_i is related to column vector O for O_i by §S = HO where the hat matrix H shares properties similar to the hat matrix in regression analysis. Several methods have been developed for obtaining H in explicit form. This allows estimation of the following two variance-covariance matrices: Var $(S_i\text{-}T_i)$ = $\sigma^2 H$, and Var $(O_i\text{-}S_i)$ = σ^2 (I-H) where I is the identity matrix.

In this paper, we use the following procedure for obtaining the diagonal elements h_{ii} of H. S was closely approximated by its Jackknife spline-curve Q with values Q_i at the observation points (see Figure 4c and Table 3). The Jackknife method provides variances $s^2(Q_i)$ of the values Q_i which are approximately equal to $\sigma^2 h_{ii}$ being the diagonal elements Var$(S_i\text{-}T_i)$. Our smoothing factor SF = 1.174 estimates σ. The diagonal elements of Var $(O_i\text{-}S_i)$ also can be estimated because $O_i \cdot$ Si. These are written as $s^2(E_{1i})$ in Table 3. Approximate 95% and 99% confidence intervals are obtained by multiplying $s(E_{1i})$ by 2 and 3, respectively. As expected, no outliers are indicated for the random normal numbers used in our computer simulation experiments. Because $s(E_{1i})$ applies to observations used for estimating the spline-curve S, Table 3 also shows residuals $E_{2i} = R_{2i} - Q_i$ for new observations R_{2i} obtained by adding 21 other random numbers to T_i. These new observations have wider confidence belts with widths controlled by $s(E_{2i})$ = SFr$(1+h_{ii})$ also shown in Table 3. This second type of confidence belt would, for example, apply to test inspected outliers not used for calculating the smoothing spline.

The three standard deviations $s(Q_i)$, $s(E_{1i})$ and $s(E_{2i})$ can be projected onto the depth axis by using the sediment accumulation rate curve (Fig. 4D). This type of curve is more difficult to estimate than the age-depth curve as indicated by larger discrepancies in Figure 4D.

Cenozoic Zonation and Isochron Correlation, Labrador Shelf and Grand Banks

RASC was applied to an integrated data set of last occurrence events of 437 Cenozoic foraminifer and palynomorph taxa in 23 Grand Banks and Labrador offshore wells. The selected zonation only uses taxa that occur in 7 or more wells; this reduces the number of taxa to 85, comprising 47 foraminifers and 38 palynomorphs. The final scaled optimum sequence is shown in Figure 5 (D'Iorio, 1987). In order to enhance the zonation with taxa that are considered potentially of chronostratigraphic value but which occur in fewer wells than the threshold (k_c) selected, the RASC method allows introduction of special or unique events (UE). Ten events were selected that occur in less than 7 wells, and include *Globorotalia inflata* and *G. crassiformis* (Pliocene), *Neogloboquadrina atlantica* (Pliocene), *Asterigerina gurichi* (peak occurrence; middle-late Miocene), *Globorotalia scitula praescitula* (Middle Miocene), *Globigerinoides primordius* (Early Miocene), *Acarinina broedermannii* (early Middle Eocene), *Planorotalites chapmani* (Paleocene/Eocene boundary), and *Subbotina triloculinoides* and *S. pseudobulloides* (Danian).

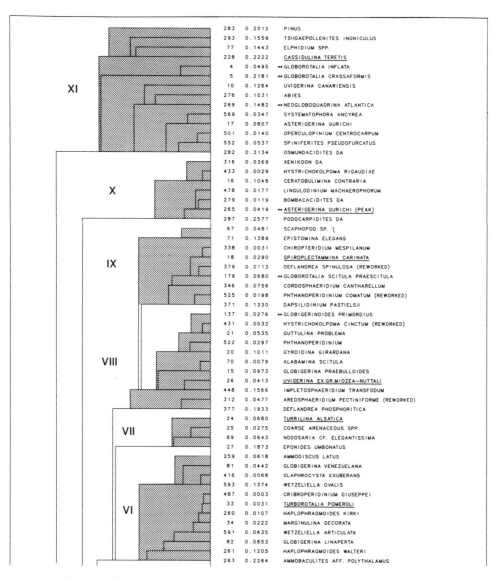

283	0.2012	PINUS
293	0.1559	TSUGAEPOLLENITES INGNICULUS
77	0.1443	ELPHIDIUM SPP.
228	0.2222	CASSIDULINA TERETIS
4	0.0495	**GLOBOROTALIA INFLATA
5	0.2181	**GLOBOROTALIA CRASSAFORMIS
10	0.1264	UVIGERINA CANARIENSIS
276	0.1031	ABIES
269	0.1482	**NEOGLOBOQUADRINA ATLANTICA
569	0.0347	SYSTEMATOPHORA ANCYREA
17	0.0807	ASTERIGERINA GURICHI
501	0.0140	OPERCULOPINIUM CENTROCARPUM
552	0.0537	SPINIFERITES PSEUDOFURCATUS
282	0.3134	OSMUNDACIDITES DA
316	0.0369	XENIKOON DA
433	0.0029	HYSTRICHOKOLPOMA RIGAUDIAE
16	0.1048	CERATOBULIMINA CONTRARIA
478	0.0177	LINGULODINIUM MACHAEROPHORUM
279	0.0119	BOMBACACIDITES DA
265	0.0419	**ASTERIGERINA GURICHI (PEAK)
287	0.2577	PODOCARPIDITES DA
67	0.0481	SCAPHOPOD SP.
71	0.1389	EPISTOMINA ELEGANS
338	0.0031	CHIROPTERIDIUM MESPILANUM
18	0.0290	SPIROPLECTAMMINA CARINATA
379	0.0113	DEFLANDREA SPINULOSA (REWORKED)
179	0.0680	**GLOBOROTALIA SCITULA PRAESCITULA
346	0.0756	CORDOSPHAERIDIUM CANTHARELLUM
525	0.0198	PHTHANOPERIDINIUM COMATUM (REWORKED)
371	0.1330	DAPSILIDINIUM PASTIELSII
137	0.0276	**GLOBIGERINOIDES PRIMORDIUS
431	0.0032	HYSTRICHOKOLPOMA CINCTUM (REWORKED)
21	0.0535	GUTTULINA PROBLEMA
522	0.0297	PHTHANOPERIDINIUM
20	0.1011	GYROIDINA GIRARDANA
70	0.0079	ALABAMINA SCITULA
15	0.0973	GLOBIGERINA PRAEBULLOIDES
26	0.0413	UVIGERINA EX.GR.MIOZEA–NUTTALI
448	0.1566	IMPLETOSPHAERIDIUM TRANSFODUM
312	0.0477	AREOSPHAERIDIUM PECTINIFORME (REWORKED)
377	0.1933	DEFLANDREA PHOSPHORITICA
24	0.0680	TURRILINA ALSATICA
25	0.0275	COARSE ARENACEOUS SPP.
69	0.0643	NODOSARIA CF. ELEGANTISSIMA
27	0.1873	EPONIDES UMBONATUS
259	0.0618	AMMODISCUS LATUS
81	0.0442	GLOBIGERINA VENEZUELANA
416	0.0068	GLAPHROCYSTA EXUBERANS
593	0.1374	WETZELIELLA OVALIS
487	0.0003	CRIBROPERIDINIUM GIUSEPPEI
33	0.0031	TURBOROTALIA POMEROLI
260	0.0107	HAPLOPHRAGMOIDES KIRKI
34	0.0222	MARGINULINA DECORATA
591	0.0635	WETZELIELLA ARTICULATA
82	0.0853	GLOBIGERINA LINAPERTA
261	0.1205	HAPLOPHRAGMOIDES WALTERI
263	0.2264	AMMOBACULITES AFF. POLYTHALAMUS

Figure 5. High-resolution interval zonation based on the scaled optimum sequence of Cenozoic foraminifers and palynomorphs in 23 wells, Grand Banks-Labrador Shelf. Zones I through XI of Paleocene through Pliocene/Pleistocene age are shaded and nominate taxa are underlined. Ages are in Table 4.

(*continued*)

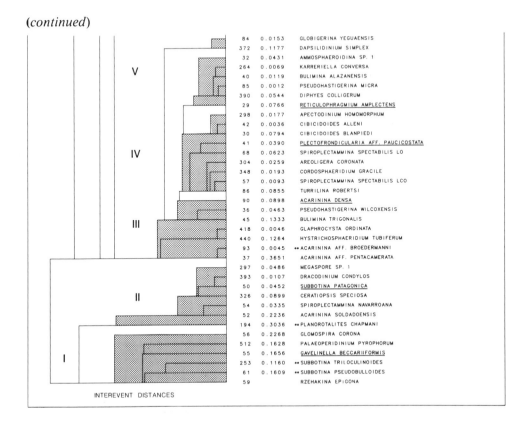

84	0.0153	GLOBIGERINA YEGUAENSIS
372	0.1177	DAPSILIDINIUM SIMPLEX
32	0.0431	AMMOSPHAEROIDINA SP. 1
264	0.0069	KARRERIELLA CONVERSA
40	0.0119	BULIMINA ALAZANENSIS
85	0.0012	PSEUDOHASTIGERINA MICRA
390	0.0544	DIPHYES COLLIGERUM
29	0.0766	RETICULOPHRAGMIUM AMPLECTENS
298	0.0177	APECTODINIUM HOMOMORPHUM
42	0.0036	CIBICIDOIDES ALLENI
30	0.0794	CIBICIDOIDES BLANPIEDI
41	0.0390	PLECTOFRONDICULARIA AFF. PAUCICOSTATA
68	0.0623	SPIROPLECTAMMINA SPECTABILIS LO
304	0.0259	AREOLIGERA CORONATA
348	0.0193	CORDOSPHAERIDIUM GRACILE
57	0.0093	SPIROPLECTAMMINA SPECTABILIS LCO
86	0.0855	TURRILINA ROBERTSI
90	0.0898	ACARININA DENSA
36	0.0463	PSEUDOHASTIGERINA WILCOXENSIS
45	0.1333	BULIMINA TRIGONALIS
418	0.0046	GLAPHROCYSTA ORDINATA
440	0.1264	HYSTRICHOSPHAERIDIUM TUBIFERUM
93	0.0045	**ACARININA AFF. BROEDERMANNI
37	0.3651	ACARININA AFF. PENTACAMERATA
297	0.0486	MEGASPORE SP. 1
393	0.0107	DRACODINIUM CONDYLOS
50	0.0452	SUBBOTINA PATAGONICA
326	0.0899	CERATIOPSIS SPECIOSA
54	0.0335	SPIROPLECTAMMINA NAVARROANA
52	0.2236	ACARININA SOLDADOENSIS
194	0.3036	**PLANOROTALITES CHAPMANI
56	0.2268	GLOMOSPIRA CORONA
512	0.1628	PALAEOPERIDINIUM PYROPHORUM
55	0.1656	GAVELINELLA BECCARIIFORMIS
253	0.1160	**SUBBOTINA TRILOCULINOIDES
61	0.1609	**SUBBOTINA PSEUDOBULLOIDES
59		RZEHAKINA EPIGONA

INTEREVENT DISTANCES

i	T_i	R_i	O_i	S_i	Q_i	$s(Q_i)$	$s(E_{1i})$	E_{1i}	$s(E_{2i})$	E_{2i}
1	10.00	0.54	10.54	9.69	9.79	1.02	0.59	0.75	1.55	-0.48
2	11.15	0.63	11.77	11.14	11.23	0.62	1.00	0.54	1.33	-0.57
3	12.56	-1.99	10.57	12.66	12.64	0.57	1.02	-2.08	1.31	-0.91
4	14.22	1.66	15.87	14.29	14.03	0.71	0.97	1.85	1.37	-1.10
5	16.08	-1.22	14.86	16.05	15.59	0.84	0.82	-0.73	1.44	-0.46
6	18.13	-0.04	18.08	17.97	17.39	0.90	0.76	0.69	1.48	1.40
7	20.32	-0.56	19.76	20.03	19.46	0.90	0.75	0.30	1.48	1.12
8	22.63	0.36	23.00	22.24	21.82	0.87	0.79	1.17	1.46	1.62
9	25.04	-2.45	22.59	24.58	24.40	0.78	0.88	-1.81	1.41	1.71
10	27.50	0.16	27.67	27.04	27.04	0.59	1.02	0.62	1.31	-0.15
11	30.00	-0.15	29.85	29.55	29.73	0.45	1.08	0.12	1.26	0.73
12	32.50	0.13	32.63	32.04	32.39	0.51	1.06	0.24	1.28	-1.51
13	34.96	-2.33	32.63	34.47	34.90	0.70	0.94	-2.27	1.37	0.42
14	37.37	1.30	38.66	36.81	37.17	0.91	0.75	1.49	1.48	-0.56
15	39.68	0.75	40.43	38.96	39.32	0.94	0.70	1.21	1.51	1.72
16	41.87	-1.51	40.36	40.93	41.36	0.79	0.87	-1.00	1.41	-0.85
17	43.92	-1.74	42.18	42.78	43.21	0.59	1.02	-1.04	1.31	1.47
18	45.79	-1.10	44.68	44.54	44.87	0.45	1.08	-0.19	1.26	1.82
19	47.44	-2.04	45.40	46.26	46.42	0.44	1.09	-1.02	1.25	0.76
20	48.86	0.67	49.52	47.93	47.95	0.55	1.04	1.57	1.29	0.43
21	50.00	-1.52	48.48	49.55	49.58	0.78	0.88	-1.10	1.41	0.26

Table 3 Random normal deviates (R_i) were added to theoretical values (T_i) on curve of Figure 3A to give observed values O_i. Cross-validated smoothing spline values (S_i) on curve of Figure 3B were approximated by their Jackknife estimates (Q_i) of which standard deviations $s(Q_i)$ could be computed. Standard deviations $s(E_{1i})$ and $s(E_{2i})$ are for residuals of data used (E_{1i}) and not used (E_{2i}) for estimating Q_i, respectively.

Eleven interval zones stand out (Fig. 5), from Paleocene through Pliocene/Pleistocene age, with the characteristic taxa listed stratigraphically in order of average disappearance. Table 4 lists the age of each zone in this zonation which is the most detailed one available for the region. It is interesting to observe that the original RASC zonation based on foraminifers only (see Gradstein et al., 1985) is virtually unchanged upon insertion of the palynomorph events. The zones are more apparent because they contain more events, and large distances between some adjacent interval zones may be reduced as, for example, between zones VI and VII and VIII when compared to the original zonation.

Table 4 Names and ages of interval zones identified in Figure 5.

Zone	Age of Zone	Name of Marker Event
I	Paleocene	*Gavelinella beccariiformis*
II	Early Eocene	*Subbotina patagonica*
III	early Middle Eocene	*Acarinina densa*
IV	late Middle Eocene	*Plectofrondicularia* aff. *paucicostata*
V	Late Eocene	*Reticulophragmium amplectens*
VI	Late Eocene	*Turborotalia pomeroli*
VII	Oligocene	*Turrilina alsatica*
VIII	Early Miocene	*Uvigerina* ex. gr. *miozea-nuttali*
IX	Middle Miocene	*Spiroplectamina carinata*
X	middle Late Miocene	*Asterigerina qurichi*
XI	Pliocene-Pleistocene	*Cassidulina teretis*

To use the RASC scaled optimum sequence for stratigraphic correlation, the arbitrary RASC scale of relative event distances can be converted to a numerical time scale. This produces a RASC biochronology and allows geochronologic correlations with the CASC method. The events used as chronologic markers are obtained from the literature and are listed in Table 5. These ages are plotted against the RASC distances of the events to obtain the age/RASC distance function and cubic spline fitting is used to assign ages to all events in the scaled sequence. The advantage of this fitting technique is that it creates a detailed local time scale.

Five wells were selected for correlation of isochrons: Karlsefni, Bjarni, Indian Harbour, Dominion and Osprey. These wells constitute a geographically representative cross section of the region, over a distance of more than 1000 km. The most likely age-depth estimates for

Table 5 Marker events used in CASC

Foraminiferal markers		Dinoflagellate, spore & pollen markers	
Event #	Age[1]	Event #	Age[1]
4 [2,5]	3.5		
5 [2,5]	3.5		
269 [2,5]	3.5		
		552 [3]	9.5
17 [2]	11		
		569 [4]	11
265 [2,5]	12		
15 [2]	17		
179 [2]	17		
		346 [3]	19
26 [2]	20		
137 [2,5]	20		
		377 [3]	24
24 [2]	30		
259 [2]	37		
		390 [3]	37
		487 [2]	37
33 [2]	38		
82 [2]	38		
85 [2]	38		
		591 [3]	38
29 [2]	40		
		298 [3]	41
86 [2]	49		
90 [2]	49		
		393 [3]	50
57 [2]	52		
93 [2]	52		
		326 [3]	53
50 [2]	55		
194 [2,5]	57		
55 [2]	58		
56 [2]	58		
		512 [3]	59
61 [2,5]	63		
253 [2,5]	63		

[1] Time scale from Berggren et al. (1985)
[2] Gradstein and Agterberg (1982)
[3] Williams et al. (1985)
[4] Harland (1978)
[5] Events that do not meet the minimum occurrence threshold

each well were obtained through the technique described in the two previous sections. Figure 6 is a spline-fitted plot of age versus depth in the Indian Harbour well. The spline function yields the age-depth values for geological correlation.

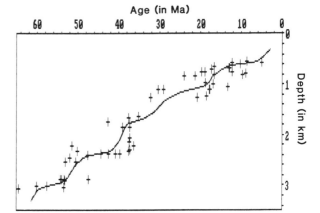

Figure 6. Spline-fitted plot of age versus depth for the Cenlozoic section of the Indian Harbour well, Labrador Shelf. The points that make up the spline curve are used as age-depth estimates in a regional isochron correlation (Fig. 7) involving this well.

Two types of correlations were made in this application of CASC, one traces isochrons and the second zone-boundary events. The isochron correlation of the five wells (Fig. 7a) shows that the stratigraphic thickness of each time interval is maintained from one well to the next, with the major exception of the 35 to 45 Ma interval in the Indian Harbour well (#3) which is thicker than in the other wells. This change in thickness is due to a higher Eocene sedimentation rate. The Osprey (#5) well is shallower than the others and its section generally is thinner. Moreover, the 55 Ma isochron is absent in this section. The absence of the 55 Ma isochron means that no point on the spline curve showed an ordinate of 55 Ma, and does not necessarily imply than no sections older than 45 Ma are present in the well. Noisy data in the oldest part of the well caused the spline curve in the age versus level plot to be too steep near the bottom, and age estimates for the corresponding events were slightly underestimated.

The second type of correlation, using zone-boundary events, is shown in Figure 7b. For this purpose, the original RASC optimum sequence was divided into eleven biozones. Several zone boundary events selected from this RASC biozonation scheme were correlated using their calculated (CASC) depths in the five wells. Events 67 (*Scaphopod* sp. 1) and 24 (*Turrilina alsatica*) delineate an interval ranging from the Late Oligocene to the Middle Miocene. This interval is thinning slightly towards the south. Events 264 (*Karreriella conversa*) to 33 (*Turborotalia pomeroli*) and events 33 to 24 mark Middle to Late Eocene and Late Eocene zones, respectively. The older zone indicates a greater sedimentation rate in the Indian Harbour well in contrast with the Late Eocene zone which possesses relatively equal

Figure 7 A. Correlation of isochrons (in Ma) in five Labrador Shelf/Grand Banks wells, i.e. from left (north) to right (south): Karlsefni, Bjarni, Indian Harbour, Dominion and Osprey. The isochrons are interpolated at 10 Ma intervals from 55 Ma to 5 Ma. Modified local error bars (cf. Figure 3B) are shown for first four wells. **B.** Correlation of event positions in the wells. The events chosen are found at or near zone boundaries (Fig. 5). The original observed positions are shown as crosses.

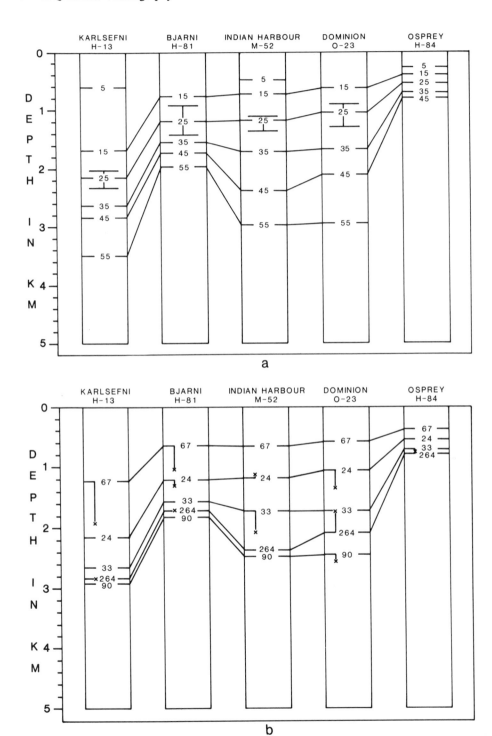

thicknesses in all wells. Finally, the zone bounded by events 264 and 90 (*Acarinina densa*) is not observed in the Osprey well. As previously explained for the 55 Ma isochron, this absence is probably due to noisy data in the oldest section of this well.

MESOZOIC BIOCHRONOLOGY, SEDIMENTARY SEQUENCES AND BURIAL HISTORY

Williamson (1987) studied the stratigraphy and depositional history of the shallow marine Upper Jurassic and Lower Cretaceous deposits near the Hibernia oil field, northern Grand Banks. This application of ranking and scaling used the highest occurrences of 116 taxa of foraminifers and ostracods in 13 exploration wells. In the computed scaled optimum sequence, which consists of 54 taxa, each event occurs in 4 or more wells and each pair of events occurs in at least 3 wells. Expansion of the number of wells from 13 to 25, and the use of higher thresholds leaves virtually the same scaled optimum sequence, an indication that the results are stable.

Figure 8 is the RASC scaled optimum sequence with chronostratigraphically useful interval zones highlighted by shading. Nine unique events (shown with two asterisks), were added for age calibration. Eleven interval zones stand out, numbered XI (Kimmeridgian) through I (Cenomanian). This zonation considerably refines stratigraphic resolution previously available. Good correspondence exists between the average position of the disappearance levels of taxa in the wells and the upper part of stratigraphic ranges reported in European literature. Some longer ranging taxa reported in this literature have shorter ranges on the Grand Banks, as is the case with *L. nodosa* (no. 10) and *D. gradata* (no. 111). In addition, *N. varsoviensis* (no. 64) occurs in younger strata on the Grand Banks than reported elsewhere. The tight clustering of events in the Albian zones III and II reflects considerable uncertainty on their exact disappearance levels. For example, in standard zonations *P. buxtorfi* is considered to disappear later than *R. ticinensis*, but in the Grand Banks zonation the order is reversed. Because *R. ticinensis* is rare, this event may be incompletely sampled, and *P. buxtorfi* is associated with other taxa of "older but less certain" connotation. Zones II and III therefore have a strong overlap in age.

It is likely that large distances between successive interval zones in the scaled optimum sequence are caused by major sedimentary changes or tectonic breaks that separate the majority of events below from those occurring above them. Figure 8 reveals such large interfossil distances between zones X (Tithonian) and IX (Valanginian) and between zones VI (Barremian) and IV (Aptian). The lower of the two breaks is associated with the deltaic and unfossiliferous Hibernia reservoir sands, overlain by a condensed arenaceous limestone or sandstone sequence, termed Catalina sands, and associated with seismic marker B (marker horizon event 162 in Figure 8).

Figure 8. Eleven-fold interval zonation, based on the scaled optimum sequence from ranking and scaling (RASC), using the highest occurrences of 116 taxa of foraminifers and ostracods in 13 wells, northern Grand Banks (after Williamson, 1987). The 54 taxa in the probabilistic zonation occur in a minimum of 4 wells. Values along y-axis are distances between events in relative time and expressed in dendrogram format. Large distances between fossil events correspond to basin wide tectonic or paleoceanographic changes (see Figure 9).

DENDROGRAM

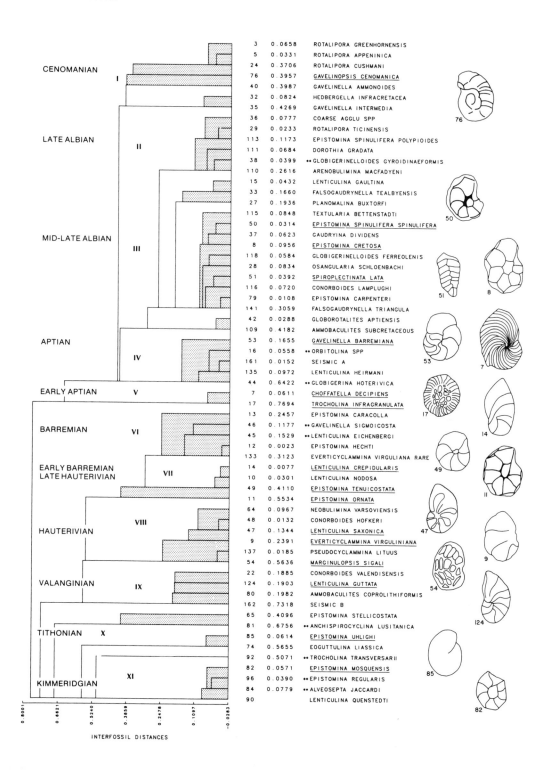

3	0.0658	ROTALIPORA GREENHORNENSIS
5	0.0331	ROTALIPORA APPENINICA
24	0.3706	ROTALIPORA CUSHMANI
76	0.3957	GAVELINOPSIS CENOMANICA
40	0.3987	GAVELINELLA AMMONOIDES
32	0.0824	HEDBERGELLA INFRACRETACEA
35	0.4269	GAVELINELLA INTERMEDIA
36	0.0777	COARSE AGGLU SPP
29	0.0233	ROTALIPORA TICINENSIS
113	0.1173	EPISTOMINA SPINULIFERA POLYPIOIDES
111	0.0684	DOROTHIA GRADATA
38	0.0399	••GLOBIGERINELLOIDES GYROIDINAEFORMIS
110	0.2616	ARENOBULIMINA MACFADYENI
15	0.0432	LENTICULINA GAULTINA
33	0.1660	FALSOGAUDRYNELLA TEALBYENSIS
27	0.1936	PLANOMALINA BUXTORFI
115	0.0848	TEXTULARIA BETTENSTADTI
50	0.0314	EPISTOMINA SPINULIFERA SPINULIFERA
37	0.0623	GAUDRYINA DIVIDENS
8	0.0956	EPISTOMINA CRETOSA
118	0.0584	GLOBIGERINELLOIDES FERREOLENIS
28	0.0834	OSANGULARIA SCHLOENBACHI
51	0.0392	SPIROPLECTINATA LATA
116	0.0720	CONORBOIDES LAMPLUGHI
79	0.0108	EPISTOMINA CARPENTERI
141	0.3059	FALSOGAUDRYNELLA TRIANGULA
42	0.0288	GLOBOROTALITES APTIENSIS
109	0.4182	AMMOBACULITES SUBCRETACEOUS
53	0.1655	GAVELINELLA BARREMIANA
16	0.0558	••ORBITOLINA SPP
161	0.0152	SEISMIC A
135	0.0972	LENTICULINA HEIRMANI
44	0.6422	••GLOBIGERINA HOTERIVICA
7	0.0611	CHOFFATELLA DECIPIENS
17	0.7694	TROCHOLINA INFRAGRANULATA
13	0.2457	EPISTOMINA CARACOLLA
46	0.1177	••GAVELINELLA SIGMOICOSTA
45	0.1529	••LENTICULINA EICHENBERGI
12	0.0023	EPISTOMINA HECHTI
133	0.3123	EVERTICYCLAMMINA VIRGULIANA RARE
14	0.0077	LENTICULINA CREPIDULARIS
10	0.0301	LENTICULINA NODOSA
49	0.4110	EPISTOMINA TENUICOSTATA
11	0.5534	EPISTOMINA ORNATA
64	0.0967	NEOBULIMINA VARSOVIENSIS
48	0.0132	CONORBOIDES HOFKERI
47	0.1344	LENTICULINA SAXONICA
9	0.2391	EVERTICYCLAMMINA VIRGULINIANA
137	0.0185	PSEUDOCYCLAMMINA LITUUS
54	0.5636	MARGINULOPSIS SIGALI
22	0.1885	CONORBOIDES VALENDISENSIS
124	0.1903	LENTICULINA GUTTATA
80	0.1982	AMMOBACULITES COPROLITHIFORMIS
162	0.7318	SEISMIC B
65	0.4096	EPISTOMINA STELLICOSTATA
81	0.6756	••ANCHISPIROCYCLINA LUSITANICA
85	0.0614	EPISTOMINA UHLIGHI
74	0.5655	EOGUTTULINA LIASSICA
92	0.5071	••TROCHOLINA TRANSVERSARII
82	0.0571	EPISTOMINA MOSQUENSIS
96	0.0390	••EPISTOMINA REGULARIS
84	0.0779	••ALVEOSEPTA JACCARDI
90		LENTICULINA QUENSTEDTI

CENOMANIAN   I

LATE ALBIAN   II

MID-LATE ALBIAN   III

APTIAN   IV

EARLY APTIAN   V

BARREMIAN   VI

EARLY BARREMIAN LATE HAUTERIVIAN   VII

HAUTERIVIAN   VIII

VALANGINIAN   IX

TITHONIAN   X

KIMMERIDGIAN   XI

0.8001   0.6621   0.5240   0.3859   0.2478   0.1097   -0.0283

INTERFOSSIL DISTANCES

The upper of the two large breaks is associated with RASC zones VI to V, and represents the pre-Aptian regional unconformity. The smaller break between RASC zones V to IV, below seismic marker A (event 161 in Figure 8), probably ties to the Avalon reservoir sands. These events are associated with the onset of sea-floor spreading between Portugal and Grand Banks.

Because it is possible to estimate numerical geological ages for a subset of the events in the scaled optimum sequence, the distance scale can be converted to a linear biochronology (time scale). The RASC distances in Figure 8 were scaled in linear time using last appearance datums for 13 events, denoted by arrows in Figure 9. The geological time scale used follows

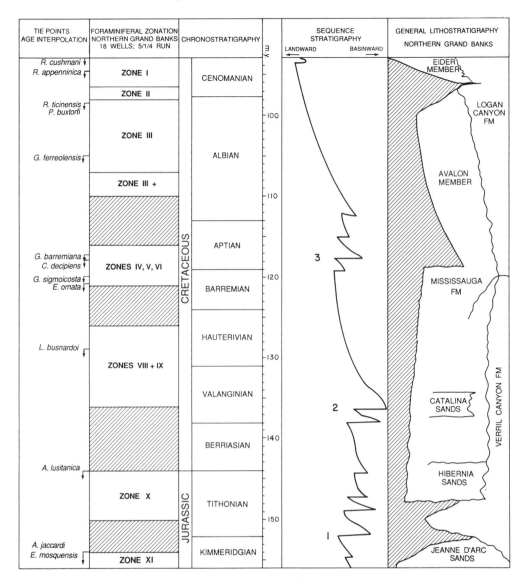

the Decade of North American Geology standard scale (Kent and Gradstein, 1985). We are now able to interpolate the ages in millions of years for all events in RASC zones XI through I. This regional foraminiferal time scale (or interpolated biochronology) extends from 154 to 93 Ma. The zonation is shown in Figure 9, which also gives the Exxon-based "global" sequence stratigraphy (Haq et al., 1987) calibrated to the standard geological time scale and the generalized lithostratigraphic scheme compiled for the northern Grand Banks. It is clear that a fair agreement exists of "missing intervals" between the zones, the principal basinward shifts of the strand line due to apparent sea-level drop, hiatuses, and the development of lowstand wedges which provide the hydrocarbon reservoir. The "missing intervals" between zones are reflected in the large interfossil distances and simply indicate that the majority of events below a sequence boundary are stratigraphically separated from those above, the result of dramatic change in biofacies and lithofacies or of tectonically induced hiatuses. On the northern Grand Banks, early Tithonian, Early Valanginian and Early Aptian were times of significant changes that appear related to "eustatic events."

Another route to the interpretation of the abrupt breaks in faunal pattern is through analysis of burial history. For four representative Grand Banks wells, burial rates were calculated and Figure 10 shows resulting plots of sediment plus overlying water (paleo waterdepth) with respect to time (Stam, et al., 1987). The thickness (depth) of the paleo water column for successive time increments was estimated from paleoecologic interpretations of foraminifer and ostracod assemblages. Sediment thicknesses were decompacted using a conversion of downhole sonic velocity to porosity for shale, sandstone, siltstone and limestone based on the program DEPOR (Stam et al., 1987). The Jurassic and Lower Cretaceous water depth was largely shallow neritic and contributed less than 4% to the burial depth below (paleo) sea level, corrected for first-order eustatic trends.

For comparison, the burial trends of two long stratigraphic outcrop sections were added as observed in Portugal, which was adjacent to the Grand Banks in Jurassic and Early Cretaceous time. These two sections are at Cape Espichel, approximately 60 km southwest of Lisbon, and near Montejunto, approximately 100 km north of Lisbon.

The cumulative burial trends show that over 50% of the sediment was deposited between 160 and 110 Ma, which represents part of the syn-rift phase of these margins. Sea-floor spreading started around 120 Ma ago. On average, burial rates in excess of 150 m/10^6yr changed abruptly to ±40 m/10^6yr near 150 Ma and to ±10 m/10^6yr near 120 Ma. Since the abrupt facies changes and RASC "hiatuses" literally straddle both the "150" and "120" Ma tectonic events, the relationship is probably causal. The sudden decreases in subsidence may have caused the strand line to shift seaward. If these tectonic events are the result of sudden intra-plate compression (*sensu* Cloetingh, 1988), such an apparent sea-level change may have been caused by basin-margin uplift.

Figure 9. Relation between the Mesozoic RASC zonation for the northern Grand Banks scaled in linear time, global sequences stratigraphy and lithostratigraphy for the northern Grand Banks. Alignment between the zones, apparent sealevel falls and the deposition of sands is a crucial result. To the left are age calibration tiepoints, used to convert the cumulative distance scale of the scaled optimum sequence in Figure 1 to linear time.

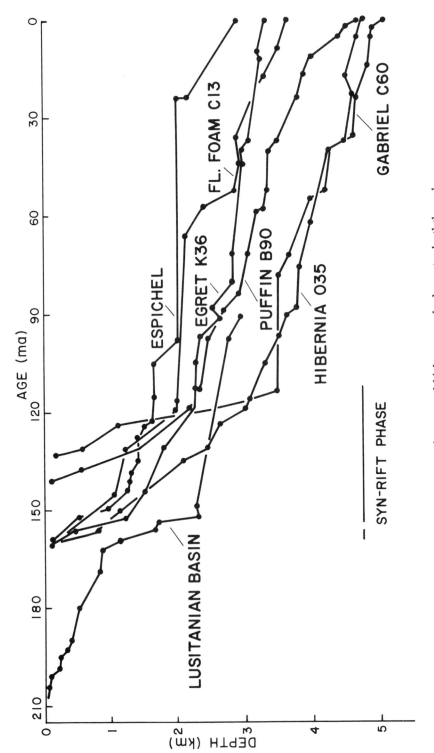

Figure 10. Burial curves using restored thicknesses and paleo water depth through time for 4 Grand Banks wells and two thick outcrop sections, Lusitanian Basin, Portugal. Abrupt facies changes and RASC "hiatusses" of Figure 9 literally straddle both the "150" and "120" Ma tectonic events.

ACKNOWLEDGMENTS

This study benefitted from discussions and manuscript reviews by M. Fearon, M. Kaminski, G. Klein, D. McAlpine, and D. Williams. N. Koziel kindly typed the versions of this text. This is Geological Survey of Canada Contribution No. 50887.

REFERENCES CITED

Agterberg, F.P., and D'Iorio, M.A., in press, Frequency distributions of highest occurrences of Cenozoic Foraminifera along the northwestern Atlantic Margin: Proceedings, Fourth South American COGEODATA conferences, Ouro Preto, Brazil, November, 1987.

Agterberg, F.P., and Gradstein, F.M., 1988, Recent developments in quantitative stratigraphy: Earth-Science Reviews, v. 25, p. 1-73.

Agterberg, F.P., and Nel, L.D., 1982a, Algorithms for the ranking of stratigraphic events: Computers and Geosciences, v. 8, p. 69-90.

Agterberg, F.P., and Nel, L.D., 1982b. Algorithms for the scaling of stratigraphic events: Computers and Geosciences, v. 8, p. 163-189.

Berggren, W.A., Kent, D.V., Flynn, J.J., and Van Couvering, J.A., 1985, Cenozoic geochronology: Geological Society of America Bulletin, v. 96, p. 1401-1418.

Cloetingh, S., 1988, Intra-plate stresses: A new element in basin analysis, in Kleinspehn, K.L., and Paola, C., eds., New perspectives in basin analysis: New York, Springer Verlag, p. 205-230.

De Boor, C., 1978, A practical guide to splines: New York, Springer Verlag, 392 p.

D'Iorio, M.A., 1986, Integration of foraminiferal and dinoflagellate data sets in quantitative stratigraphy of the Labrador Shelf and Grand Banks: Bulletin of Canadian Petroleum Geologists, v. 34, p. 277-283.

D'Iorio, M.A., 1987, Quantitative biostratigraphic analysis of the Cenozoic of 23 Canadian Atlantic offshore wells: The Compass, v. 64, no. 4, p. 264-277.

Eubank, R.L., 1984, The hat matrix for smoothing splines: Statistics and Probability Letters, v. 2, p. 9-14.

Gradstein, F.M., and Agterberg, F.P., 1982, Models of Cenozoic foraminiferal stratigraphy, northwestern Atlantic margin, in Cubitt, J., and Reyment, R., eds., Quantitative stratigraphic correlation: New York, John Wiley & Sons, p. 119-170.

Gradstein, F.M., Agterberg, F.P., Brower, J.C., and Schwarzacher, W., 1985, Quantitative stratigraphy: Unesco and Reidel Publishing Co., 598 p.

Gradstein, F.M., Kaminski, M.A., and Berggren, W.A., 1988, Cenozoic foraminiferal biostratigraphy of the Central North Seam, in Rögl, F., and Gradstein, F.M., eds., Proceedings, II workshop agglutinated benthic foraminifera, Vienna, 1986, p. 97-108.

Haq, B.U., Hardenbol, J., and Vail, P.R., 1987, Chronology of fluctuating sea levels since the Triassic: Science, v. 235, p. 1156-1166.

Harland, R., 1978, Distribution of biostratigraphically diagnostic dinoflagellate cysts and miospores from the northwest European continental shelf and adjacent areas: Trondheim, Continental Shelf Institute Publication 100, p. 7-17.

Harper, C.W., Jr., 1984, A FORTRAN IV program for comparing ranking algorithms in quantitative biostratigraphy: Computers and Geosciences, v. 10, p. 3-29.

Kent, D.V., and Gradstein, F.M., 1985, A Jurassic and Cretaceous geochronology: Geological Society of America Bulletin, v. 96, p. 1419-1427.

Stam, B., Gradstein, F.M., Lloyd, P., and Gillis, D., 1987, Algorithms for porosity and subsidence history: Computers and Geosciences, v. 13, p. 312-349.

Wahba, G., 1983, Bayesian "confidence intervals" for the cross-validated smoothing spline: Journal of the Royal Statistical Society, v. B45, p. 133-150.

Williams, G.L., and Bujak, J.P., 1985, Mesozoic and Cenozoic dinoflagellates, *in* Bolli, H., Saunders, J., and Perch-Nielsen, C., eds., Plankton stratigraphy: Oxford, Cambridge University Press, p 847-964.

Williamson, M.A., 1987, Quantitative biozonation of the Late Jurassic and Early Cretaceous of the East Newfoundland Basin: Micropaleontology, v. 33, p. 37-65.

32

Selected Approaches of Chemical Stratigraphy to Time-Scale Resolution and Quantitative Dynamic Stratigraphy

Douglas F. Williams
Department of Geological Sciences, University of South Carolina, Columbia, SC 29208 USA

Abstract

Chemical measures of sediments and individual sedimentary components offer powerful, quantitative information on sedimentation processes and diagenetic histories of basin fill. Chemical stratigraphy, when integrated with biostratigraphy and seismic stratigraphy, provides some of the essential components for Quantitative Dynamic Stratigraphy. Potential chemical signals can be derived from stable isotopic, organic geochemical and trace-elemental analyses of biogenic and nonbiogenic sedimentary components. Quantitative signal processing techniques are available for defining, comparing and predicting chemical stratigraphic signals in both time and frequency domains. Present limitations to the application of chemical stratigraphy are: the speed and precision with which chemical stratigraphic data can be generated: uncertainties in the structural, diagenetic and stratigraphic contexts for interpreting the data; and the level of resolution within current biostratigraphic, radiometric, paleomagnetic and seismic frameworks for calibrating the chemical signal(s) to time. Five case histories, based on both proprietary and public-domain data, are presented to illustrate the relevance of chemical stratigraphy to problems of stratigraphic correlation and time-scale resolution.

Introduction

By its very nature, the discipline of stratigraphy is dynamic, whether one uses qualitative or quantitative terminology to describe and relate strata to the geologic column or absolute time scale. What is considered "dynamic" and "quantitative" will depend to a large extent on an individual's background or viewpoint. One perspective was offered in the opening talk of a recent AAPG research conference on "chemostratigraphy" where F. Nietsche was quoted: "It is an illusion that something is known when we possess a mathematical formula for it. It is quantified and described, nothing more." Description and quantification of any process or phenomenon are just the initial steps in the scientific search for understanding. An opposing view to the Nietsche quotation is expressed in a quotation attributed to Lord Kelvin (Hsu, 1986, p. 111): "...when you can measure what you are speaking about, and express it in numbers, you know something about it; but when you cannot express it in numbers, your knowledge is of a meagre and unsatisfactory kind; it may be the beginning of knowledge, but you have scarcely, in your thoughts, advanced to the state of Science, whatever the matter

Quantitative Dynamic Stratigraphy (1989), T.A. Cross, ed., Prentice Hall, p. 543–565.

may be." Quantification, be it mathematical or otherwise, has obvious advantages over qualitative description in that others can arrive, independently and objectively, at the same description and in the process evaluate the path taken to arrive there. Likewise, quantitative description enhances the possibility that advances in understanding will be forthcoming.

Rationale of this Paper

One of the premises of this paper is that quantification is essential for understanding stratigraphic processes objectively and accurately. Another central premise is that geochemistry can provide quantifiable information essential for Quantitative Dynamic Stratigraphy (QDS). Too often, however, geochemistry and stratigraphy are regarded as disciplines with little in common. This perception may be due to the fact that many geochemists have been interested primarily in the thermodynamic and/or diagenetic processes responsible for sedimentary components, in and of themselves, without regard to the stratigraphic context of the geochemical data. On the other hand, some stratigraphers either don't appreciate the relevance of geochemistry or regard diagenesis as a complicated, even mysterious, process which may overprint the stratigraphic signal of interest and which is best left to the geochemists to understand.

One of the objectives of this contribution is to illustrate the relevance of chemical stratigraphy to stratigraphy in general and QDS in particular. I offer a few examples from our laboratory and from the literature which illustrate the linkage between geochemistry and stratigraphy, especially in light of efforts to develop and promote Quantitative Dynamic Stratigraphy.

Chemical stratigraphy, or chemostratigraphy, is herein defined as the use of chemical signatures in a stratigraphic context. Chemical signatures may be contained in sediments, rocks and/or individual rock components. Very often these chemical signals provide important information about litho-, bio- and chronostratigraphic processes (Fig. 1). Chemical

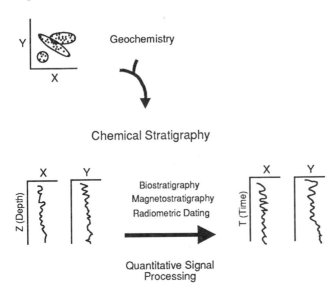

Figure 1. Schematic representation of stratigraphic signals often displayed by chemical data and the development of a chronostratigraphic framework with chemical stratigraphy when integrated with biostratigraphy, paleomagnetic stratigraphy, radiometric dating and quantitative signal processing of the chemical signals in the depth, time and frequency domains.

signals can be used either for correlation purposes or for determining temporal changes in primary depositional, climatic-oceanographic or diagenetic processes. Many geochemical parameters display considerable character when plotted stratigraphically instead of in the more traditional x-y format (Fig. 1). Chemical stratigraphic signals compared against depth (or relative time) often display events, trends or shifts which can be used to correlate depositional cycles, paleoclimatic-paleoceanographic changes or changing diagenetic patterns (Holser and Kaplan, 1966; Veizer and Hoefs, 1976; Allen and Matthews, 1977; Holser et al., 1979; Scholle and Arthur, 1980; Claypool et al., 1980; Veizer et al., 1980). Furthermore, chemical stratigraphy can sometimes be used to achieve a high level of resolution in an absolute time stratigraphic framework (Fig. 1), as in depositional systems of the deep sea and continental margins (Shackleton and Opdyke, 1973; Trainor and Williams, 1987; Williams et al., 1988).

Criteria for Linking Chemical Stratigraphy and QDS

Three basic criteria must be met for any particular parameter, technique or strategy to be incorporated into Quantitative Dynamic Stratigraphy. The parameter must be measurable (quantifiable), dynamic (changing), and a common feature of strata (the stratigraphic record). Chemical stratigraphy meets these basic criteria in many depositional environments. For example, many types of chemical signals are recorded and measurable in most stratigraphic sections. The potential exists, therefore, to develop multiple chemical stratigraphies with different chemical signals which reflect different processes in the sediments under study. Depending on the depositional system, chemical signals can be derived from stable isotopic, organic geochemical and trace-elemental analyses of biogenic and nonbiogenic sedimentary components. Measurements of such signals can be made precisely, sometimes rapidly, using laboratory instrumentation and techniques which typically are routine and well documented. Chemical stratigraphy of marine or terrestrial sediments can supply important temporal information about the evolution of the ocean environment or about the depositional and diagenetic histories of sedimentary basins. The ocean and terrestrial environments are certainly dynamic, thereby imparting equally dynamic chemical signals with different time constants or wavelengths (frequencies), depending on whether the signals are biotically or abiotically derived.

   To derive a meaningful chemical stratigraphy in a sedimentary section, one must first select the appropriate chemical parameter(s) to measure. Possible diagenetic influences on that parameter must be considered and tests made to determine the extent to which the chemical signal is primary or secondary. If micro- or macrofossils are the sedimentary components being analyzed as the chemical recorders, possible vital effects, taxonomy, taphonomy and habitat preferences must be evaluated. In sedimentary rocks, petrography is especially critical. The use of trace-elemental analyses has proven to be important for handling subtle diagenetic processes in some stable isotope investigations (Veizer, 1983). Determination of a chemical signal in the context of other geochemical, geophysical, lithological, structural or petrographic information permits the chemical stratigrapher to determine which aspects of the geochemical data are primary and secondary. One can then select the most appropriate primary or secondary signal for use in QDS models. In this context primary chemical signals are regarded as chemical stratigraphic signals due to the original depositional or climatic-oceanographic processes responsible for their origin. Secondary chemical signals are those due to diagenesis or other post-depositional processes which may obscure in some way or eliminate the primary signals.

To achieve the goal of placing chemical data into a chronostratigraphic framework, chemical stratigraphic data must be integrated and calibrated properly with biostratigraphy, paleomagnetic stratigraphy, radiometric dating, seismic stratigraphy and sequence stratigraphy. Of course all these desirable frameworks are not always available in all depositional systems or stratigraphic sections. One must also consider the best age model for the chronostratigraphy since considerable differences exist between various authors (Haq et al., 1987; Gradstein et al., 1988).

Thus, chemical stratigraphy, when integrated with other stratigraphic approaches (e.g., biostratigraphic, lithostratigraphic, paleomagnetostratigraphic, petrographic, seismostratigraphic), provides some of the essential components for numerical modeling of sedimentation processes. Limitations are the rapidity and precision with which chemical stratigraphic data can be generated, the structural and stratigraphic contexts for interpreting the data, and the biostratigraphic, radiometric, paleomagnetic and seismic resolution available for calibrating the chemical signal(s) to time. Standard quantitative signal processing techniques are available for defining, comparing and predicting the completeness of the signals in the time and frequency domains. Another important consideration is that no one investigator or laboratory can derive these frameworks, making multidisciplinary and collaborative studies necessary.

In this contribution, five case histories are used to illustrate the relevance of chemical stratigraphy to problems of stratigraphic correlation and time-scale resolution on a variety of temporal and spatial scales. These examples show that chemical stratigraphy offers stratigraphic and time-scale resolution which very often exceeds the resolution possible from some biostratigraphic and seismic stratigraphic frameworks. Chemical stratigraphic signals in an absolute time frame also provide essential information for numerical modeling of contemporaneous (depositional) and post-depositional (diagenetic) processes acting in basins.

CASE HISTORY 1: CHRONOSTRATIGRAPHY OF CONTINENTAL MARGIN SEDIMENTS USING STABLE ISOTOPIC COMPOSITIONS OF FORAMINIFERA

Chronostratigraphic frameworks for precisely correlating Plio-Pleistocene exploration wells from the northwestern margin of the Gulf of Mexico are now established (Williams and Trainor, 1986; Trainor and Williams, 1987). The chemical signals are provided by the $^{18}O/$ ^{16}O and $^{13}/^{12}C$ ratios of foraminifera (Fig. 2A, B). Decades of oxygen isotopic studies, and more recently carbon isotope studies, of fairly continuous deep-sea sediment sections from every major ocean basin have firmly demonstrated that the isotopic compositions of benthic and planktonic foraminifera are important recorders of global changes in ocean chemistry and atmospheric carbon dioxide (Emiliani, 1966; Duplessy, 1978; Shackleton and Opdyke, 1973; Shackleton et al., 1983; Ruddiman et al., 1986, 1987; Williams et al., 1988a). Changes in ocean chemistry and atmospheric CO_2 during the Pleistocene have occurred quasiperiodically with frequencies strongly suggestive of a linkage with earth-sun orbital geometries as predicted by Milankovitch in 1941 (Hays et al., 1976; Imbrie et al., 1984; Shackleton and Pisias, 1985). Isotopic fractionation due to metabolic processes and depth habitat also are fairly well known although many investigators continue to explore these and other effects on the stratigraphic and paleoceanographic record (Williams et al., 1977; Graham et al., 1981; Williams et al., 1988b). These oxygen and carbon isotopic signals are stable in that they do not undergo detectable radioactive decay and the potential effects of diagenesis are understood fairly well.

The $^{18}O/^{16}O$ and $^{13}/^{12}C$ records shown in Figure 2 (A, B) offer two independent but corroborative chemical stratigraphies in sections containing foraminifera. The current age model for the last 4.5 million years of the Plio-Pleistocene (Fig. 2A, B) is based on linear interpolation of detailed isotope records between biostratigraphic and paleomagnetostratigraphic horizons (Williams et al., 1988a). Figure 3 illustrates how the $^{18}O/^{16}O$ chemical stratigraphy, in an absolute time frame, provides information about sedimentation rate variations, hiatuses in sedimentation (perhaps unconformities?), and age, all essential information for describing and modeling depositional processes in offshore Plio-Pleistocene exploration basins. The same type of reconstructions are possible with carbon isotope chemical stratigraphy.

The next step is to integrate this high-resolution chronostratigraphic information (Fig. 3) into QDS and basin analysis models using seismic, well log and other information. Dynamic modeling of sedimentary basins and structural trends, like the Flexure Trend in the offshore Gulf of Mexico, is possible at a temporal resolution of several 10^5 years. Such models offer exciting new possibilities for exploration. Our work to date in the northwestern Gulf of Mexico has shown that isotopic signals in foraminifera are stable diagenetically into the latest Miocene at burial depths shallower than 5000 to 6000 m. It remains to be seen if diagenesis interferes with the development of similar chemical stratigraphies for Miocene sections, which have been buried deeply for longer periods of time.

CASE HISTORY 2: CHEMICAL STRATIGRAPHY OF CONTINENTAL MARGIN SEDIMENTS USING DETRITAL, AUTHIGENIC AND/OR DIAGENETIC CARBONATE

In many depositional environments, especially those dominated by siliciclastic sediments, the biogenic components represented by calcareous foraminifera are either not present or not preserved throughout the sedimentary section. In Plio-Pleistocene sections of the northwestern Gulf of Mexico or offshore Trinidad, for example, sediments tens to hundreds of meters thick are often devoid of calcareous foraminifera. Such circumstances complicate, if not frustrate, attempts to develop a chemical chronostratigraphy using foraminifera (Case History 1).

However, alternative chemical stratigraphic information is often retrievable using other sedimentary components. Often this alternate chemical information can be obtained more easily because generation of the data does not require isolation of foraminiferal tests from sediment, knowledge about microfossil taxonomy, or in some cases, sophisticated instrumentation like an isotope ratio mass spectrometer. For example, Figure 4 shows a composite $\delta^{13}C$ signal of carbonate from the fine silt-sized fraction (<63 um) of well cuttings from several exploration wells of the northwestern Gulf of Mexico (Williams and Trainor, 1987). The fine silt-sized fraction (herein referred to as the fine fraction) can be saved routinely during the washing of cuttings samples prior to the search for foraminiferal tests from the

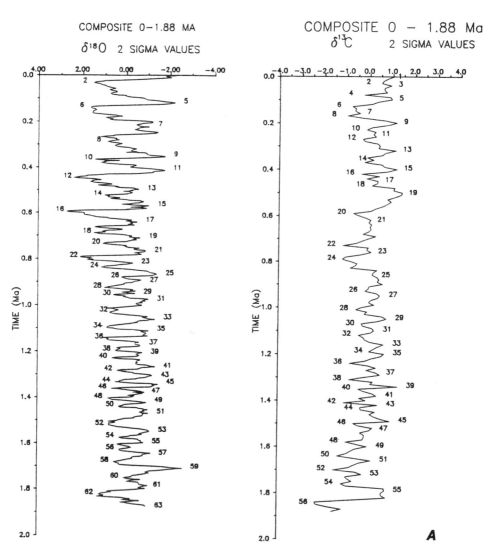

Figure 2A. Provisional oxygen and carbon stable isotope chronostratigraphy for the Plio-Pleistocene (A, 0 to 1.88 Ma; B, 1.88 to 4.5 Ma) based on isotope stages recorded by deep-sea foraminifera which have been calibrated to time (million years before present, Ma) using biostratigraphy and paleomagnetic stratigraphy (Trainor and Williams, in prep.).

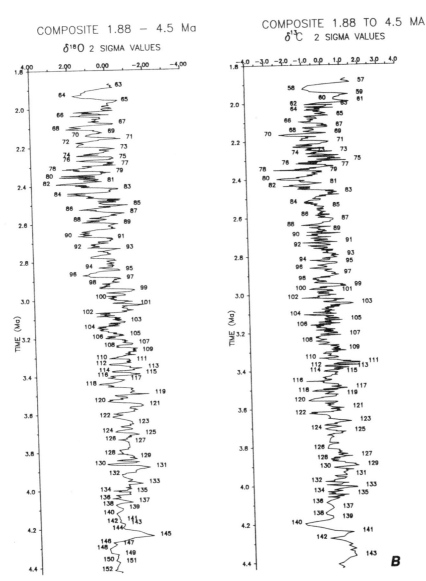

Figure 2B. Provisional oxygen and carbon stable isotope chronostratigraphy for the Plio-Pleistocene (A, 0 to 1.88 Ma; B, 1.88 to 4.5 Ma) based on isotope stages recorded by deep-sea foraminifera which have been calibrated to time (million years before present, Ma) using biostratigraphy and paleomagnetic stratigraphy (Trainor and Williams, in prep.).

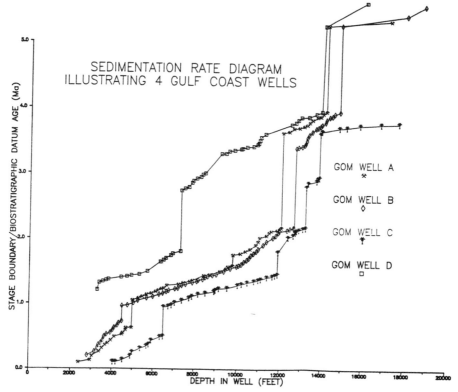

Figure 3. Sedimentation rate diagrams, uncorrected for compaction, of four Plio-Pleistocene exploration wells from the northwestern Gulf of Mexico based on the application of the chemical chronostratigraphic framework in Figure 2A, B to the foraminiferal isotope records of the wells (proprietary data, Isotope Stratigraphy Group).

coarse fraction (>63 um). The isotopic composition of the carbonate contained in the fine fraction is determined using the same basic protocol as for foraminifera (Williams and Trainor, 1987). As Figure 4 illustrates, the $\delta^{13}C$ signals of the foraminifera and fine-fraction carbonate are entirely different. The $\delta^{13}C$ range of the fine-fraction carbonate is >25 per mil as compared with 2-3 per mil in the foraminifera. The carbonate carrying the $\delta^{13}C$ signal may be detrital, biogenic, authigenic, diagenetic, or a combination of these origins. We know from studies of a number of wells from the Green Canyon, Ewing Banks and South Timbalier areas, offshore Louisiana, that certain events and shifts in the fine-fraction $\delta^{13}C$ signal correlate stratigraphically (Williams and Trainor, 1987). This conclusion is based on detailed chemical stratigraphy from foraminiferal isotope records, integrated with planktonic foraminiferal and calcareous nannofossil biostratigraphy, as described in Case History 1 (Figs. 2A, B; 3). At present, the proprietary nature of the exploration wells prevents the release of the well locations and data bases. In addition, it is not possible to describe what proportion of detrital, biogenic, authigenic or diagenetic carbonate is driving the $\delta^{13}C$ events and shifts shown in Figure 4, although calcite is the carbonate phase carrying the signal.

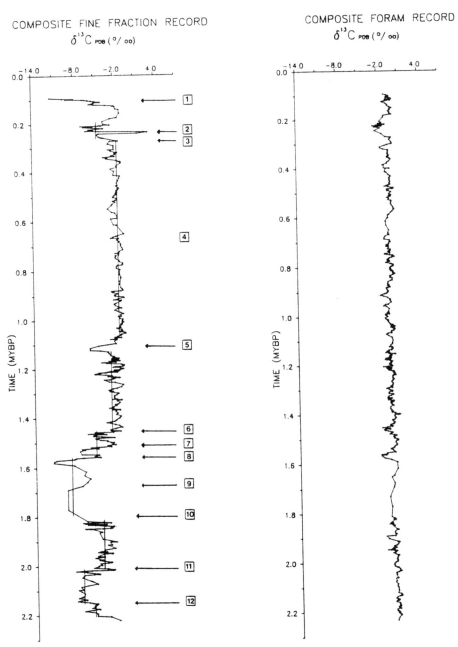

COMPOSITE FINE FRACTION RECORD
$\delta^{13}C$ PDB (°/ oo)

COMPOSITE FORAM RECORD
$\delta^{13}C$ PDB (°/ oo)

Figure 4. The chemical stratigraphic signal provided by the carbon isotopic composition of the carbonate contained in the fine-fraction sediment (<63 microns) of several Plio-Pleistocene exploration wells of the northwestern Gulf of Mexico (from Williams and Trainor, 1987). The composite record is calibrated to time (Ma) using the biostratigraphy and companion foraminiferal chronostratigraphies available for the same wells (proprietary data, Isotope Stratigraphy Group).

Notwithstanding these limitations, however, Case History 2 illustrates the nature of the fine-fraction $\delta^{13}C$ signal as another chemical stratigraphy. This signal can be used either in place of, or in conjunction with, a foraminiferal-based chronostratigraphy (Figs. 2A, B; 3) to correlate Plio-Pleistocene exploration sections in areas of the northwestern Gulf of Mexico.

To determine additional chemical stratigraphic information from some of the same Plio-Pleistocene exploration sections illustrated in Case Histories 1 and 2, the percentage of total carbonate in two exploration wells from the upper slope offshore Gulf of Mexico was measured using a simple gasiometric technique. The technique is modified from that of Hulsemann (1966). A visual comparison of the chemical stratigraphic signal provided by percent carbonate in these two wells (Fig. 5) suggests correlations of major carbonate fluctuations A-A', B-B', C-C', D-D' and E-E'. The timing of these changes, as well as the overall shift from relatively carbonate-poor (<5% $CaCO_3$) to carbonate-rich (as much as 20% $CaCO_3$) contents in the two wells (Fig. 5), is in close agreement with the biostratigraphy and three independent chemical stratigraphies in the same wells: the foraminiferal $\delta^{18}O$ and $\delta^{13}C$ chronostratigraphies and $\delta^{13}C$ fine-fraction chemical stratigraphy. The lithostratigraphic pattern observed in the total carbonate records of these and other wells from the northwestern Gulf of Mexico reflects large- and small-scale changes in the balance between siliciclastic input and carbonate production along the Plio-Pleistocene margin of the northern Gulf of Mexico (Williams et al., in press).

CARBONATE CHEMICAL STRATIGRAPHY

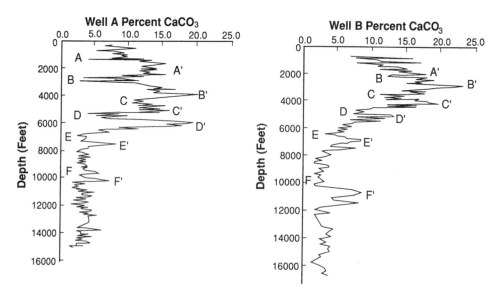

Figure 5. The chemical stratigraphic signal provided by the percentage of calcium carbonate with subsurface depth in two Plio-Pleistocene exploration wells from the northwestern Gulf of Mexico. Indicated stratigraphic correlations are in agreement with the biostratigraphy, companion foraminiferal chronostratigraphies and fine-fraction isotope records available for the same wells (proprietary data, Isotope Stratigraphy Group).

CASE HISTORY 3: CHEMICAL STRATIGRAPHY OF PALEOZOIC SILICICLASTIC ROCKS OF THE DELAWARE BASIN USING STABLE ISOTOPIC COMPOSITIONS OF MICRITE CEMENTS

Can a chemical stratigraphy be developed using the stable isotopic composition of lithified rock components, like micrite cements, even in carbonate-poor depositional environments such as the Permian siliciclastic rocks of the Delaware basin, West Texas? Figure 6 shows the "$\delta^{18}O$" stratigraphic signal through a subsurface section of the Cherry Canyon Formation of the Delaware Mountain Group (Ahr and Berg, 1982). The range of "$\delta^{18}O$" values is greater than 100 per mil. These positive "$\delta^{18}O$" values are not from the primary chemical signal of the calcite cements, because no known distillation or fractionation process exists to produce such positive isotopic values. Instead the anomalous values are due to a contaminant gas phase, most likely nitrogen dioxide of masses 46, 47 and 48 (Mucciarone, 1988). The NO_2 of mass 46 contributes to and sometimes overwhelms the target $^{12}C^{16}O^{18}O$ mass 46 signal. The NO_2 mass 46 contribution is greatest, and the apparent "$\delta^{18}O$" values most positive, in samples which are calcite-poor and clay-rich. Furthermore, the NO_2 mass 46 signal is not widespread in the Brushy Canyon sections of similar lithology studied to date (Mucciarone et al., 1986; 1987). The positive "$\delta^{18}O$" signal appears to exhibit three large cycles in the subsurface section of the Cherry Canyon formation (Fig. 6). It is postulated that the contaminant NO_2 gas is the result of ammonia in clays, which is most likely produced during the phosphoric acid extraction of carbon dioxide gas from the carbonate contained in the whole-rock sample. The CO_2 gas can be purified of oxides, like NO_2 gas of masses 46, 47 and 48, by passing the gas sample through a hot catalytic reduction furnace. As shown in Figure 7, this procedure yields no $\delta^{18}O$ values of the micrite cements more negative than 0‰(PDB) (Mucciarone, 1988).

At this time we do not know the exact origin of the NO_2 mass 46 signal but we have observed anomalously positive "$\delta^{18}O$" values in other depositional environments (Mucciarone et al., 1987). We also do not understand the significance of the NO_2 signal in terms of its relationship with either deposition of particular clay minerals or post-depositional alteration and emplacement of ammonia. The stratigraphic integrity of either the apparent "$\delta^{18}O$" signals from NO_2 plus CO_2 or, true $\delta^{18}O$ signals from carbonate-derived CO_2 (Figs. 6, 7), awaits better petrographic control than is currently available for these sections.

CASE HISTORY 4: CHEMICAL (MOLECULAR) STRATIGRAPHY USING BIOGEOCHEMICAL COMPONENTS OF SEDIMENTS

Biochemical or molecular stratigraphy is a relatively new but rapidly growing form of chemical stratigraphy. Vast numbers of organic compounds are present in sedimentary rocks as chemical fossils (Mackenzie et al., 1982). Some of these complex compounds are source-specific, widely distributed and fairly stable on geologic time scales (Farrimond et al., 1985). Organic compounds such as steroids, lipids and ketones, may provide important information about the stratigraphic record and the origin and diagenetic history of organic constituents of sedimentary basins. For example, the chemical behavior of lipids may be linked to levels of oxic and anoxic conditions present during sedimentation (Didyk et al., 1978).

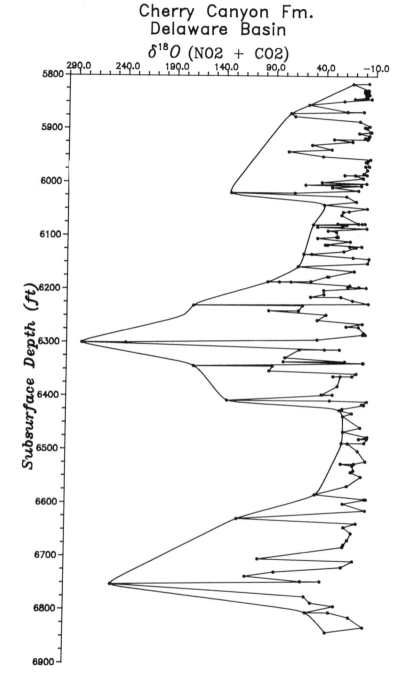

Figure 6. The chemical stratigraphic signal provided by the apparent oxygen-18 signal in whole-rock analyses of subsurface core taken from the Permian Cherry Canyon Formation, Delaware basin, West Texas (unpublished data, Isotope Stratigraphy Group).

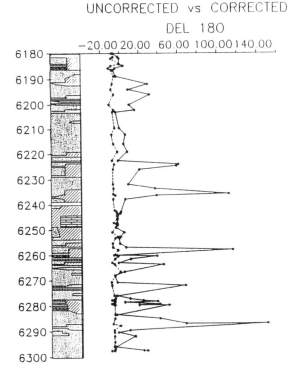

Figure 7. Comparison of the chemical stratigraphic signals of the apparent d¹⁸O signals in a subsurface section of the Cherry Canyon Formation. The d¹⁸O values are uncorrected and corrected for the contribution of nitrogen dioxide gas of mass 46 to carbon dioxide of mass 46 from micritic calcite-phosphoric acid reaction (Mucciarone, 1988).

This case history is chosen from the recent literature to illustrate the potential for obtaining chronostratigraphic and paleoclimatic information from molecular stratigraphic records. One such molecular stratigraphy provided by an alkenone unsaturation index, U_{37} (Fig. 8; Brassell et al., 1985, 1986). Long-chain alkenones are specific constituents of ubiquitous marine phytoplankton, primarily coccolithophorids (Volkman et al., 1980) and the unsaturation of alkenones is temperature dependent (Brassell et al., 1986). The correlation of the U_{37} index and the $\delta^{18}O$ record of foraminifera, in combination with preliminary spectral analysis of the U_{37} record showing significant power at frequencies similar to the Milankovitch orbital parameters (Brassell et al., 1985), suggests that molecular stratigraphy will receive more attention in future applications of chemical stratigraphy.

CASE HISTORY 5: CHEMICAL STRATIGRAPHY AND CHRONOSTRATIGRAPHY USING STRONTIUM ISOTOPIC COMPOSITION OF MARINE CARBONATES

One of the most promising chemical stratigraphies being developed is that provided by the stable strontium isotope ratio ($^{87}Sr/^{86}Sr$) of marine carbonates (Figs. 9, 10). Sr is one of the principal trace elements incorporated in marine carbonates from seawater precipitation. The $^{87}Sr/^{86}Sr$ of seawater is determined by the mass balance involving submarine hydrothermal activity, riverine influxes of weathering products from different sources with different ages and $^{87}Sr/^{86}Sr$ ratios—young volcanic rocks, old continental sialic rocks and Phanerozoic

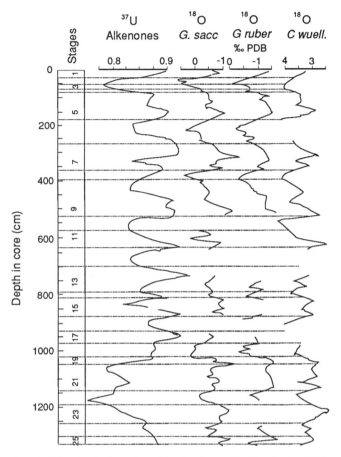

Figure 8. The chemical (molecular) stratigraphic signal provided for the late Pleistocene by the the alkenone saturation index U_{37} and the oxygen-18 values of foraminifera from a deep-sea core (Meteor 16415-2) collected from the North Atlantic (from Brassell et al., 1986).

carbonate rocks (Faure et al., 1965; Spooner, 1976; Faure 1986), and submarine recycling of carbonates (Baker et al., 1982). Given the long residence time of strontium in the oceans (5 x 10^6 yr) relative to the much shorter mixing time of the oceans (approximately 10^3 yr), analyses of seawater and modern marine carbonates show that the oceans are homogeneous with respect to $^{87}Sr/^{86}Sr$.

Importantly, modern unaltered carbonates faithfully reflect the oceanic $^{87}Sr/^{86}Sr$, and thus, it should be possible to use the $^{87}Sr/^{86}Sr$ composition of well-preserved ancient marine carbonates to determine the $^{87}Sr/^{86}Sr$ composition of seawater through geologic time (Peterson et al., 1970). Indeed, early work demonstrated that large-scale changes in the $^{87}Sr/^{86}Sr$ ratio of seawater have occurred in the Phanerozoic (Peterson et al., 1970; Dasch and Biscaye, 1971; Veizer and Compston, 1974; Faure et al., 1978; Starinsky et al., 1980; Jorgensen and Larsen, 1980) and even Precambrian (Veizer and Compston, 1976; Veizer et al., 1983).

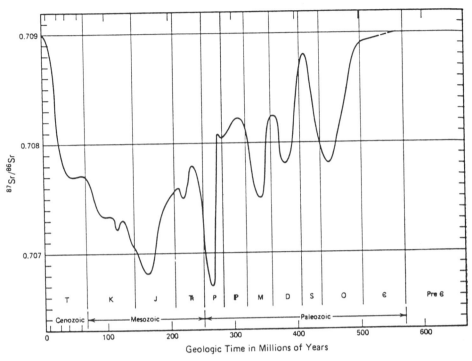

Figure 9. The chemical stratigraphic signal provided by large-scale changes in the $^{87}Sr/^{86}Sr$ ratio of seawater in the Phanerozoic as recorded by well preserved marine carbonates (Burke et al., 1982, as illustrated in Faure, 1986). Some temporal changes in $^{87}Sr/^{86}Sr$ have occurred in a slow, near monotonic fashion (i.e., from mid-late Jurassic to the Pleistocene). Other $^{87}Sr/^{86}Sr$ fluctuations occurred very rapidly (i.e., in the early Devonian, early and late Mississippian, early to middle Permian, and across the Permo-Triassic boundary).

It is fair to say, however, that excitement in the application of $^{87}Sr/^{86}Sr$ as a chemical stratigraphic tool for correlation and age-dating of marine carbonates really developed momentum with the study spanning the Phanerozoic by Burke et al. (1982; Fig. 9). This $^{87}Sr/^{86}Sr$ record revealed that large changes had occurred in a slow, near monotonic fashion from mid-Late Jurassic to the Pleistocene. Such a trend offers excellent potential for age-dating of sedimentary sequences. Other $^{87}Sr/^{86}Sr$ fluctuations occurred very rapidly, like those seen in the Early Devonian, early and late Mississippian, Early to middle Permian, and across the Permo-Triassic boundary (Fig. 9). This landmark study of over 700 marine carbonates was followed shortly after by more detailed studies of the Cenozoic molluscs, foraminifera and chalks by DePaolo and Ingram (1985), Renard (1985) and Hess et al. (1986; Fig. 10). These studies began to confirm the high-resolution character of the Cenozoic $^{87}Sr/^{86}Sr$ record.

Like many other areas of geochemistry, this increased interest in $^{87}Sr/^{86}Sr$ variations paralleled technological improvements in instrumentation, in this case, thermal-ionization mass spectrometers. Relative to the $^{87}Sr/^{86}Sr$ of seawater and modern/Holocene carbonates

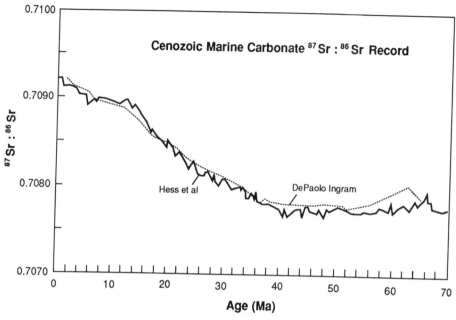

Figure 10. The chemical (chrono)stratigraphic signal provided by detailed changes in the $^{87}Sr/^{86}Sr$ ratio of Cenozoic molluscs, foraminifera and chalks (DePaolo and Ingram, 1985; Hess et al., 1986).

(0.70906 to 0.709234; Faure, 1982; DePaolo and Ingram, 1985; Hess et al., 1986), precision measurements to the 5th and 6th decimal place are needed to resolve time-dependent changes in the $^{87}Sr/^{86}Sr$ of Phanerozoic marine carbonates (Figs. 9, 10). The chemistry involved in the $^{87}Sr/^{86}Sr$ analytical procedures are slightly more complicated than those for the lighter stable isotopes of oxygen and carbon, but this minor disadvantage is balanced by the fact that chemostratigraphic correlations are often possible with fewer analyses. In addition, the $^{87}Sr/^{86}Sr$ of marine carbonates often is less sensitive to diagenesis, to a first approximation, than the $^{18}O/^{16}O$ and $^{13}C/^{12}C$ ratios (Elderfield and Gieskes, 1982; Baker et al., 1982).

$^{87}Sr/^{86}Sr$ chemical stratigraphy has another advantage over our current capabilities with $^{18}O/^{16}O$ and $^{13}C/^{12}C$ chemical stratigraphies in that it is less dependent on the presence of calcareous fossils. For example, it is quite possible that $^{87}Sr/^{86}Sr$ chemical stratigraphy can be extended to noncarbonate sediment fractions (Dasch, 1969; Boger and Faure, 1974; 1976), phosphatic fish debris in deep-sea red clay sequences, and authigenic sedimentary components such as manganese and other metal oxides, pyrite, phosphorites and evaporites.

WHERE SHOULD WE GO FROM HERE?: IMPROVEMENTS IN AND A STRATEGY FOR QUANTITATIVE CHEMICAL STRATIGRAPHY

To use the chemical stratigraphies like those illustrated in Case Histories 1-5 in QDS models, numerous standard procedures are available for treating quantitatively the chemical signals in time and frequency domains (Williams et al., 1988c). In all of the case histories, quantitative approaches are now called for in order to enhance the confidence of stratigraphic correlations and chronostratigraphic interpretations. A strategy for achieving a more quantifiable (better?) analysis of chemical stratigraphic data, and calibration to time, is the subject of a recent book (Williams et al., 1988c). The need for a quantitative strategy is particularly illustrated by isotopic signals recorded in fossil foraminifera (Fig. 2A, B). A primary control on the Plio-Pleistocene $\delta^{18}O$ signal is solar insolation variations as described by and named after Milankovitch (1941; see Hays et al., 1976). The signals are certainly dynamic, in that they vary in a quasiperiodic fashion with time (age) as a function of global changes in ocean chemistry. Data-dependent methodologies are required to define more precisely the timing of these signals and the frequencies at which these changes occur in geologic time.

Behavior of the $\delta^{18}O$ signal at frequencies predicted by Milankovitch suggests that time resolution can be enhanced by "tuning" the chronology of the signal to fit the predicted frequencies of obliquity and/or precession (Imbrie et al., 1984). It is unclear, however, whether the linkage between global climate and the insolation variables of obliquity (power at 41,000 years) and precession (power at 21,000 and 19,000 years) is entirely direct and linear (Kominz and Pisias, 1980; Imbrie and Imbrie, 1980; Berger, 1984; McKenna et al., 1988; Williams et al., 1988a; 1988c). A significant unexplained observation is the fact that eccentricity (a period of approximately 100,000 yr) is a dominant component of the power spectrum of the Plio-Pleistocene $\delta^{18}O$ signal and yet, eccentricity is not considered a principal climatic variable in the Milankovitch (1941) hypothesis (Kominz and Pisias, 1979). Figure 11 shows that eccentricity is also a significant component of the deep-sea $\delta^{13}C$ record. Other frequencies common to the $\delta^{18}O$ and $\delta^{13}C$ signals of the last 4.5 Ma (Fig. 2A, B) are possible harmonics of the orbital frequencies, but others may be fundamentals not predicted by Milankovitch (1941). The choice of which frequency(s) to "tune" the $\delta^{18}O$ and $\delta^{13}C$ signals becomes critical for chronostratigraphic age models and further refinements of exploration models (Fig. 3).

With regard to chemical stratigraphies of Case History 2 (% carbonate and fine-fraction $\delta^{13}C$), investigations must now be conducted to determine the multiple origins of the carbonate in the fine-fraction. Further geochemical and sediment fabric studies including SEM, microprobe and trace-elemental analyses are needed to determine the primary, authigenic and diagenetic origins of the fine-fraction carbonate material of the northwestern margin of the Gulf of Mexico.

Even given the proprietary nature of the exploration well data in Case Histories 1 and 2, I believe the chemical stratigraphies illustrated in Figures 4 and 5 provide interesting, and perhaps important, information about the depositional and diagenetic history of Plio-Pleistocene sedimentation on the northwestern Gulf of Mexico. Chronostratigraphic information may even be retrievable from these alternative chemical stratigraphies by calibrating these chemical signals to time using well-established biostratigraphy and foraminifera-based chronostratigraphy (Williams and Trainor, 1986; Trainor and Williams, 1987). Combining the fine-fraction isotope and total carbonate signals with foraminiferal or other chemical signals offers the potential for Multiple Integrated Chemical Stratigraphy (MICS). MICS combined with standard quantitative signal processing techniques offers the potential for achieving a Quantitative Dynamic Stratigraphy of this continental margin.

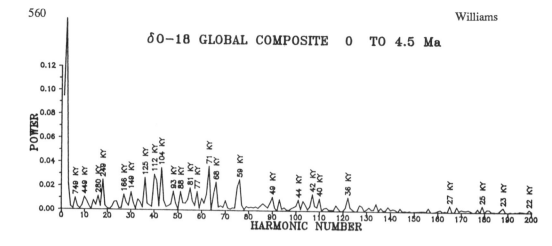

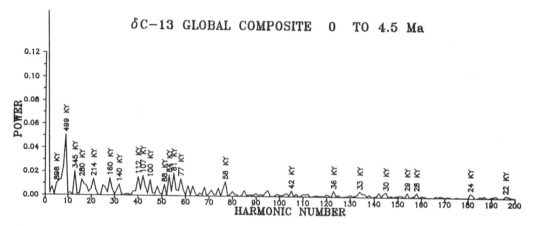

Figure 11. Smoothed power spectra of the composite deep-sea oxygen-18 (A) and carbon-13 (B) records of the last 4.5 Ma shown in Figure 2A, B (Trainor and Williams, in prep).

We also know that some of the large negative $\delta^{13}C$ anomalies or excursions are not stratigraphic in the sense that they are site specific. The very negative $\delta^{13}C$ values (–10 to –50 per mil PDB) strongly suggests some level of authigenic interaction between the fine-fraction calcite and hydrocarbons. The site-specific negative anomalies are related to the present or past occurrence of hydrocarbons within the sediment. Work is continuing to determine how the fine-fraction isotope signal provides insights into the depositional and diagenetic history of the sedimentary sections by mathematically describing the variables that are driving the $\delta^{13}C$ signal, with the optimistic and hopeful objective of using the fine-fraction signal more knowledgeably in exploration.

In studies like that of the Delaware basin siliciclastic sections (Case History 3), experiments are needed to determine the source(s) of the NO_2 or NO_x compounds that is suspected to be ammonia on the basis of our current studies (Mucciarone, 1988). Work also is needed to integrate more petrographic information with the geochemical data. It should be noted that positive "$\delta^{18}O$" anomalies from NO_2 have been observed in carbonate-poor sediments from other depositional environments like the Miocene Monterey Formation and the Plio-Pleistocene fine-fraction of the northwestern Gulf of Mexico (Mucciarone et al., 1987).

With regard to the $^{87}Sr/^{86}Sr$ record through geologic time, definitive experiments need to be conducted on the diagenetic behavior of $^{87}Sr/^{86}Sr$ in different sedimentary components and on distinguishing potential localized from global variations in the $^{87}Sr/^{86}Sr$ of ancient seawater. These experiments will be especially important in efforts to improve the chronostratigraphic fidelity and applicability of the carbonate $^{87}Sr/^{86}Sr$ record. Even though some portions of the late Phanerozoic $^{87}Sr/^{86}Sr$ record appear suitable for chronostratigraphic interpretations (Fig. 9), particularly from the Middle Jurassic to Pleistocene where steady increases in the ratio occur from less than 0.70600 to modern values around 0.70900, temporal resolution is currently limited to between 0.5 to 1 m.y. in the Cenozoic (Fig. 10). Further work is needed in the Paleocene-Eocene and Middle to Late Miocene portions of the Cenozoic to determine if short-term fluctuations of the $^{87}Sr/^{86}Sr$ can be resolved and exploited. Improvements in biostratigraphic and paleomagnetic frameworks also will improve understanding of the high-resolution character of the marine $^{87}Sr/^{86}Sr$ record. Integrating $^{87}Sr/^{86}Sr$ chemical stratigraphy with other chemical stratigraphies will lead to improvements in biostratigraphic zonations.

The importance of stable strontium isotopes in chemical stratigraphy goes far beyond marine carbonates, although little direct evidence currently exists to support this claim. As improvements continue to be made in the analytical procedures for measuring strontium isotopes and in the availability of high-precision heavy-isotope mass spectrometers, strontium isotopes will continue to play a major role in chemical stratigraphic and diagenetic studies of sedimentary rocks.

SUMMARY REMARKS

Many other inorganic and organic chemical parameters are available for chemical stratigraphy that have not been discussed in this contribution: trace elements (Veiser, 1983); the $^{87}Sr/^{86}Sr$ of noncarbonate components; the stable isotope ratios of sulfur ($^{34}S/^{32}S$), nitrogen ($^{15}N/^{14}N$) and carbon $^{13}C/^{12}C$ in organic materials; the $^{18}O/^{16}O$ in phosphatic components (Kolodny et al., 1983; Shemish et al., 1983), including icthyolith debris; the $^{18}O/^{16}O$ and D/H of silica and cherts (Knauth and Epstein, 1975; 1976); and of course, the wide variety of molecular compounds which have only begun to be examined in stratigraphic sections (Mackenzie et al., 1982).

It cannot be overemphasized that no one chemical stratigraphy offers a "cure-all" for unravelling complex depositional and post-depositional histories. Instead, multiple integrated chemical stratigraphies (MICS) are possible. MICS offers the opportunity to monitor, in quantitative terms, multiple components of a depositional system. An integrated and iterative approach is the most ideal whereby available petrographic, structural and lithological information is integrated with geochemistry, preferably in a quantitative manner, using an array of signal processing techniques (Williams et al., 1988) to achieve the best and potentially most objective solution(s) to dynamic processes in stratigraphy.

ACKNOWLEDGMENTS

I thank the organizers of the workshop on "Quantitative Dynamic Stratigraphy" for inviting this contribution, and Tim Cross for his efforts in running a very exciting workshop and putting together this volume. I also thank George deVries Klein, Ian Lerche, Howard Spero and Tim Cross for criticism of this manuscript. Our work on chemical stratigraphy at the University of South Carolina would not have been possible without the dedicated efforts of the members of the Isotope Stratigraphy Group, the support of the College of Science and Math, and the support and encouragement of Amerada-Hess, Amoco, ARCO, Chevron, CONOCO, Elf-Aquitaine, Exxon, Gulf Oil, Mobil, Marathon, Phillips Petroleum, Standard Oil, Shell Development, Sun, Texaco, Transco, and UNOCAL.

REFERENCES CITED

Ahr, W.M., and Berg, R.R., 1982, Deepwater evaporites in the Bell Canyon Formation, Delaware basin, west Texas, *in* Handford, C.R., Loucks R.G., and Davies, G.R., eds., Depositional and diagenetic spectra of evaporites—A core Workshop: Society of Economic Paleontologist and Mineralogists Core Workshop 3, p. 305-395.

Allen, J.R., and Matthews, R.K., 1977, Carbon and oxygen isotopes as diagenetic and stratigraphic tools: Surface and subsurface data, Barbados, West Indies: Geology, v. 5, p. 16-20.

Baker, P.A., Gieskes, J.M., and Elderfield, H., 1982. Diagenesis of carbonates in deep-sea sediments— Evidence from Sr/Ca ratios and interstitial dissolved Sr^{2+}: Journal of Sedimentary Petrology, v. 52, p. 71-82.

Berger, A.L., 1984, Accuracy and frequency stability of the Earth's orbital elements during the Quaternary, *in* Berger, A.L., Imbrie, J., Hays, J., Kukla, G., and Saltzman, B., eds., Milankovitch and climate: Understanding the response to astronomical forcing: NATO ASI Series C, Mathematical and Physical Sciences, Dordrecht, Netherlands, Reidel, v. 126, p. 3-39.

Boger, P.D., and Faure, G., 1974, Strontium isotope stratigraphy of a Red Sea core: Geology, v. 2, p. 181-183.

Boger, P.D., and Faure, G., 1976, Systematic variations of sialic and volcanic detritus in piston cores from the Red Sea: Geochimica et Cosmochimica Acta, v. 40, p. 731-742.

Brassell, S.C., Brereton, R.G., Eglinton, G., Grimalt, J., Liebezeit, G., Marlowe, I.T., Pflaumann, U., and Sarnthein, M., 1985, Palaeoclimatic signals recognized by chemometric treatment of molecular stratigraphic data: Advances in Organic Geochemistry, v. 10, p. 649-660.

Brassell, S.C., Eglinton, G., Marlowe, I.T., Sarnthein, M., and Pflaumann, U., 1986, Molecular stratigraphy—A new tool for climatic assessment: Nature, v. 290, p. 693-696.

Burke, W.H., Denison, R.E., Hetherington, E.A., Koepnick, R.B., Nelson, N.F., and Otto, J.B., 1982, Variation of seawater $^{87}Sr/^{86}Sr$ throughout Phanerozoic time: Geology, v. 10, p. 516-519.

Claypool, G.E., Holser, W.T., Kaplan, I.R., Sakai, H., and Zak, I., 1980, The age curves of sulfur and oxygen isotopes in marine sulfate and their mutual interpretations: Chemical Geology, v. 28, p. 199-260.

Dasch, J.J., 1969, Strontium isotopes in weathering profiles, deep-sea sediments and sedimentary rocks: Geochimica et Cosmochimica Acta, v. 33, p. 1521-1552.

Dasch, E.J., and Biscaye, P.E., 1971, Isotopic composition of strontium in Cretaceous to Recent pelagic foraminifera: Earth and Planetary Science Letters, v. 11, p. 201-204.

DePaolo, D.J., and Ingram, B.L., 1985, High resolution stratigraphy with strontium isotopes: Science, v. 227, p. 938-941.

Didyk, B.M., Simoneit, B.R.T., Brassell, S.C., and Eglinton, G., 1978, Organic geochemical indicators of palaeoenvironmental conditions of sedimentation: Nature, v. 272, p. 216-222.

Duplessy, J.-C., 1978, Isotope studies, *in* Gribben, J., ed., Climatic Change: Cambridge, Cambridge University Press, p. 46-67.

Elderfield, H. and Gieskes, J.M., 1982, Sr isotopes in interstitial waters of marine sediments from Deep-Sea Drilling Project cores: Nature, v. 300, p. 493-497.

Eglinton, G., 1973, Chemical fossils: A combined organic geochemical and environmental approach, *in* Swain, T., ed., Chemistry in evolution and systematics: I.U.P.A.C. Symposium, Strausbourg, Butterworth, p. 611-633.

Emiliani, C., 1966, Paleotemperature analysis of Caribbean cores P6304-8 and P6304-9 and a generalized temperature curve for the past 425,000 years: Journal of Geology, v. 74, p. 109-126.

Farrimond, P., Eglinton, G., and Brassell, S.C., 1985, Alkenones in Cretaceous black shales, Blake-Bahama basin, western North Atlantic: Advances in Organic Geochemistry, v. 10, p. 897-903.

Faure, G., 1982, The marine-strontium geochronometer, *in* Odin, G.S., ed., Numerical dating in stratigraphy: Chinchester, UK, John Wiley & Sons, p. 73-79.

Faure, G., 1986, Principles of isotope geology (second edition): New York, John Wiley & Sons, 589 p.

Faure, G., Hurley, P.M., and Powell, J.L., 1965, The isotopic composition of strontium in surface water from the North Atlantic Ocean: Geochimica et Cosmochimica Acta, v. 29, p. 209-220.

Faure, G., Assereto, R., and Tremba, E.L., 1978, Strontium isotope composition of marine carbonates of Middle Triassic to Early Jurassic age, Lombardic Alps, Italy: Sedimentology, v. 25, p. 523-543.

Gamble, R., Trainor, D.M., and Corbin, J., 1988, Calcium carbonate preservation on the margin of the northern Gulf of Mexico: Geological Society of America Abstracts with Programs (in press).

Graham, D.W., Corliss, B.H., Bender, M.L. and Keigwin, L.D., 1981, Carbon and oxygen isotopic disequilibria of Recent deep-sea benthic foraminifera: Marine Micropaleontology, v. 6, p. 483-497.

Gradstein, F.M., Agterberg, F.P., Aubry, M.-P., Berggren, W.A., Flynn, J.J., Hewitt, R., Kent, D.V., Klitgord, K.D., Miller, K.G., Obradovich, J., Ogg, J.G., Prothero, D.R., and Westermann, G.E.G., 1988, Sea level history [Technical Comments on Haq et al., 1987]: Science, v. 241, p. 599-601.

Haq, B.U., Hardenbol, J., and Vail, P.R., 1987, Chronology of fluctuating sea levels since the Triassic: Science, v. 235, p. 1156-1167.

Hays, J.D., Imbrie, J. and Shackleton, N.J., 1976, Variations in the Earth's orbit: Pacemaker of the Ice Ages: Science, v. 194, p. 1121-1132.

Hess, J., Bender, M.L., and Schilling, J.-G., 1986, Evolution of the ratio of strontium-87 to strontium-86 from the Cretaceous to present: Science, v. 231, p. 979-984.

Holser, W.T. and Kaplan, I.R., 1966, Isotope geochemistry of sedimentary sulfates: Chemical Geology, v. 1, p. 93-135.

Holser, W.T., Kaplan, I.R., Sakai, H., and Zak, I., 1979, Isotope geochemistry of oxygen in the sedimentary sulfate cycle: Chemical Geology, v. 25, p. 1-18.

Hsu, K.J., 1986, The great dying: New York, Ballentine Books, 288 p.

Hulsemann, J., 1966, On the routine analysis of carbonates in unconsolidated sediments: Journal of Sedimentary Petrology, v. 36, p. 622-625.

Imbrie, J. and Imbrie, J.Z., 1980, Modeling the climatic response to orbital variations: Science, v. 207, p. 943-953.

Imbrie, J., Hays, J., Martinson, D.G., McIntyre, A., Mix, A.C., Morley, J.J., Pisias, N.G., Prell, W.L., and Shackleton, N.J., 1984, *in* Berger, A.L., Imbrie, J., Hays, J., Kukla, G., and Saltzman, B., eds., Milankovitch and climate: Understanding the response to astronomical forcing: NATO ASI Series C, Mathematical and Physical Sciences, Dordrecht, Reidel, v. 126, p. 269-305.

Jorgensen, N.O., and Larsen, O., 1980, The strontium isotopic composition of Maastrichtian and Danian chalk: Bulletin Geological Society of Denmark, v. 28, p. 127-129.

King, P.B., 1942, Permian of west Texas and southeastern New Mexico: American Association of Petroleum Geologists Bulletin, v. 26, p. 535-763.

Kolodny, Y., Luz, B., and Navon, O., 1983, Oxygen isotope variations in phosphate of biogenic apatites, I. Fish bone apatite—Rechecking the rules of the game: Earth and Planetary Science Letters, v. 64, p. 398-404.

Kominz, M.A., Heath, G.R., Ju, T.-L., and Pisias, N.G., 1979, Brunhes time scales and the interpretation of climatic change: Earth and Planetary Science Letters, v. 45, p. 394-410.

Kominz, M.A., and Pisias, N.G., 1979, Pleistocene climate: Deterministic or stochastic?: Science, v. 204, p. 171-173.

Mackenzie, A.S., Brassell, S.C., Eglinton, G., and Maxwell, J.R., 1982, Chemical Fossils: The geological fate of steroids: Science, v. 217, p. 491-504.

McKenna, T. E., Lerche, I., Williams, D.F., and Full, W.E., 1988, Quantitative techniques in isotope chronostratigraphy: Paleogeography, Paleoclimatology, Paleoecology, v. 64, p. 241-264.

Milankovitch, M.M., 1941, Canon of insolation and the ice-age problem: Beograd, Koninglich Serbische Akademie, 484 p. (English translation by Israel Program for Scientific Translation and published for the U.S. Department of Commerce and National Science Foundation).

Mucciarone, D.A., Williams, D.F., and Bouma, A.H., 1986, Stable isotope evidence for depositional cycles and diagenetic patterns in Brushy Canyon and Cherry Canyon Formations, Delaware basin, West Texas: American Association of Petroleum Geologists Bulletin, v. 70, p. 624.

Mucciarone, D.A., Trainor, D.M., and Williams, D.F., 1987, Whole-rock stable isotope geochemistry for correlation and diagenetic interpretation: Delaware basin and Monterey Formation examples: American Association of Petroleum Geologists Bulletin, v. 71, p. 596.

Mucciarone, D.A., 1988, The effect of anomalous $\delta^{18}O$ values on stratigraphic interpretations: Delaware basin, West Texas [M.S. thesis]: Columbia, South Carolina, University of South Carolina, 145 p.

Peterson, Z.E., Hedge, C.E., and Tourtelot, H.A., 1970, Isotopic composition of strontium in seawater throughout Phanerozoic time: Geochimica et Cosmochimica Acta, v. 34, p. 105-120.

Renard, M., 1985, Géochimie des Carbonates Pélagiques [Ph.D. Thesis]: Paris, France, University of Paris, Documents du BRGM n° 85, 650 p.

Ruddiman, W.F., Raymo, M., and McIntyre, A., 1986, Matuyama 41,000-year cycles: North Atlantic Ocean and northern hemisphere ice sheets: Earth and Planetary Science Letters, v. 80, p. 117-129.

Ruddiman, W.F., Raymo, M., and McIntyre, A., 1987, Paleoenvironmental results from North Atlantic sites 607 and 609, in Ruddiman, W.F., Kidd, R., et al., eds., Initial Reports of the Deep Sea Drilling Project: Washington, D.C., U.S. Government Printing Office, v. 92, p. 855-878.

Scholle, P., and Arthur, M.A., 1980, Carbon isotope fluctuations in Cretaceous pelagic limestones: Potential stratigraphic and petroleum exploration tool: American Association of Petroleum Geologists Bulletin, v. 64, p. 67-87.

Shackleton, N.J., and Opdyke, N.D., 1973, Oxygen isotope and paleomagnetic stratigraphy of equatorial Pacific core V28-238: Oxygen isotope temperatures and ice volumes on a 10^5 year scale: Quaternary Research, v. 3, p. 39-55.

Shackleton, N.J., Hall, M.A., Line, J., and Shuxi, C., 1983, Carbon isotope data in core V19-30 confirm reduced carbon dioxide concentration of ice age atmosphere: Nature, v. 306, p. 319-322.

Shackleton, N.J., and Pisias, N.G., 1985, Atmospheric carbon dioxide, orbital forcing and climate, in Sundquist, E.T., and Broecker, W.S., eds., The carbon cycle and atmospheric CO_2: Archean to present: American Geophysical Union Monograph 32, p. 303-317.

Shemish, A., Kolodny, Y., and Luz, B., 1983, Oxygen isotope variations in phosphate of biogenic apatites, II. Phosphorite rocks: Earth and Planetary Science Letters, v. 64, p. 405-416.

Spooner, E.T.C., 1976, Strontium isotopic composition of seawater and seawater-oceanic crust interaction: Earth and Planetary Science Letters, v. 31, p. 167-174.

Starinsky, A.M., Bielsky, M., Lazer, B., Wakshal, E., and Steinitz, S., 1980, Marine $^{87}Sr/^{86}Sr$ ratios from the Jurassic to Pleistocene: Evidence from groundwaters in Israel: Earth and Planetary Science Letters, v. 47, p. 75-80.

Trainor, D.M., and Williams, D.F., 1987, Isotope chronostratigraphy: High resolution stratigraphic correlations in deep-water exploration tracts of the northern Gulf of Mexico: Gulf Coast Association of Geological Societies Transactions, v. 37, p. 247-254.

Veiser, J., 1983, Chemical diagenesis of carbonates: Theory and application of trace element technique, *in* Arthur, M.A., Anderson, T. F., Kaplan, I.R., Veiser, J., and Land, L.S., eds., Stable isotopes in sedimentary geology: Society of Economic Paleontologists and Mineralogists Short Course 10, p. 3-1 – 3-100.

Veizer, J., and Compston, W., 1974, $^{87}Sr/^{86}Sr$ composition of seawater during the Phanerozoic: Geochimica et Cosmochimica Acta, v. 38, p. 1461-1484.

Veizer, J., and Compston, W., 1976, $^{87}Sr/^{86}Sr$ in Precambrian carbonates as an index of crustal evolution: Geochimica et Cosmochimica Acta, v. 40, p. 905-914.

Veizer, J., Compston, W., Clauer, N., and Schidlowski, M., 1983, $^{87}Sr/^{86}Sr$ in late Proterozoic carbonates: Evidence for a "mantle" event at ~900 Ma ago: Geochimica et Cosmochimica Acta, v. 47, p. 295-302.

Veizer, J., and Hoefs, J., 1976, The nature of O^{18}/O^{16} and C^{13}/C^{12} secular trends in sedimentary carbonate rocks: Geochimica et Cosmochimica Acta, v. 40, p. 1387-1395.

Veizer, J., Holser, W.T., and Wilgus, C.K., 1980, Correlation of $^{13}C/^{12}C$ and $^{34}S/^{32}S$ secular variations: Geochimica et Cosmochimica Acta, v. 44, p. 579-587.

Volkman, J.K., Eglinton, G., Corner, E.D.S., and Sargent, J.R., 1980, Novel unsaturated straight-chain C_{37}-C_{39} methyl and ethyl ketones in marine sediments and a coccolithophore *Emiliania huxleyi*, *in* Douglas, A.G., and Maxwell, J.R., eds., Advances in Organic Geochemistry: Oxford, Pergamon Press, p. 219-227.

Williams, D.F., Lerche, I., and Full, W.E., 1988c, Isotope chronostratigraphy: Theory and methods: London, Academic Press, 345 p.

Williams, D.F., Sommer, M.A., and Bender, M.L., 1977, Carbon isotopic compositions of Recent planktonic foraminifera of the Indian Ocean: Earth and Planetary Science Letters, v. 36, p. 391-403.

Williams, D.F., Thunell, R.C., Tappa, E., Rio, D., and Spovieri, I., 1988a, Chronology of the Pleistocene oxygen isotope record: 0-1.88 million years before present: Paleogeography, Paleoclimatology, Paleoecology, v. 64, p. 221-240.

Williams, D.F., Ehrlich, R., Healy-Williams, N., and Gary, A.C., 1988, Shape and isotopic differences between conspecific foraminiferal morphotypes and resolution of paleoceanographic events: Paleogeography, Paleoclimatology, Paleoecology, v. 64, p. 153-162.

Williams, D.F., and Trainor, D.M., 1986, Application of isotope chronostratigraphy in the northern Gulf of Mexico: Gulf Coast Association of Geological Societies Transactions, v. 36, p. 589-600.

Williams, D.F., and Trainor, D.M., 1987, Carbon isotope signals for chemical stratigraphy and hydrocarbon exploration in the northern Gulf of Mexico: Gulf Coast Association of Geological Societies Transactions, v. 37, p. 287-293.

Williams, D.F., Trainor, D.M., Gamble, R., Guilderson, T., and Corbin, J., 1988, Carbonate patterns Gulf of Mexico: Gulf Coast Association of Geological Societies Transactions (in press).

33

CORRELATION OF K-BENTONITE BEDS BY CHEMICAL FINGERPRINTING USING MULTIVARIATE STATISTICS

Warren D. Huff [1] and *Dennis R. Kolata* [2]

[1]Department of Geology, University of Cincinnati, Cincinnati, OH 45221 USA

[2]Illinois State Geological Survey, 615 E. Peabody Dr., Champaign, IL 61820 USA

ABSTRACT

Ordovician K-bentonite beds can be correlated along a 900 km transect in the Mississippi Valley by chemical fingerprinting using discriminant function analysis, a multivariate statistical method of pattern recognition. Statistical modeling provides a criterion by which subtle but persistent between-bed differences in chemical composition can be shown to be greater than within-bed differences, thus permitting individual beds to be distinguished from one another on a regional scale. Assumptions in the model include the equality of covariance matrices for all groups, random selection of samples, and the likely membership of unknown samples in one of the model subgroups. Geological assumptions are that K-bentonite chemistry reflects original ash composition and that individual beds retain their chemical identity over long distances. The validity of these assumptions is upheld by cross-validation and the "jackknife" method of sample selection. Maximum advantage of the discriminant model is achieved when combined with well-documented lithostratigraphic and biostratigraphic field evidence.

INTRODUCTION

Late Ordovician K-bentonite (altered volcanic ash) beds are distributed over $1.3 \times 10^6 \, km^2$ in eastern North America (Fig. 1) within a succession of siliciclastic and carbonate rocks representing basinal, ramp and platform environments. Potential use of the beds as regional time surfaces has been known since the work of Sardeson (1924), Nelson (1922), and others who recognized the geologically instantaneous nature of deposition of the ashes. Despite numerous attempts to trace them regionally (e.g., Kay, 1935), positive identification of individual K-bentonite beds has been precluded by the lack of persistent, macroscopically or microscopically identifiable features unique to each bed. K-bentonite beds have been correlated locally on the basis of gross stratigraphic position and relation to established biostratigraphic zones (Huffman, 1945; Milici and Smith, 1969). In some areas, they can be recognized in the subsurface on certain geophysical logs where enclosing strata are characterized by contrasting log response (Cable and Beardsley, 1984). In areas of complex stratigraphic and structural settings, for example in the Appalachian orogenic belt, it has not been possible to trace individual K-bentonite beds with certainty.

Recent work has shown that individual K-bentonite beds can be identified by means of unique chemical fingerprints (Huff, 1983; Cullen-Lollis and Huff, 1986; Kolata et al., 1986, 1987). K-bentonites can now be correlated on a regional scale across major facies boundaries. In addition, examination of the trace elements provides insight into the original magmatic composition, tectonic setting and subsequent alteration of the air-fall ash. This paper illustrates the application of multivariate statistical techniques to the solution of a stratigraphic problem and examines the underlying assumptions and limitations of statistical modeling when used to support chemical stratigraphy. It is based on the correlation of four K-bentonite beds in the upper Mississippi Valley reported by Kolata et al. (1986).

Figure 1. Distribution of Ordovician K-bentonite beds (shaded area) in the eastern United States and Canada (from Kolata et al., 1986).

CORRELATION OF K-BENTONITE BEDS

At least fifteen K-bentonite beds occur in the Galena Group (Upper Ordovician) of the upper Mississippi Valley region. This study concentrates on the chemical correlation of four of the beds, the Deicke, Millbrig, Elkport and Dickeyville K-bentonites (Willman and Kolata, 1978) in the basal part of the Galena Group. These beds occur in numerous widely distributed outcrops where stratigraphic control is very good. The question posed was whether each of the four beds has a distinct chemical fingerprint that can be recognized over wide areas.

Twenty-five Deicke, 24 Millbrig, 6 Elkport, and 5 Dickeyville samples collected at 39 localities in southeastern Minnesota, northeastern Iowa, and southwestern Wisconsin were designated as a control set to test whether the beds have a distinctive chemical signature. Whole-rock samples were analyzed for 26 elements by X-ray fluorescence spectroscopy and instrumental neutron activation analysis.

Differences in elemental abundance among the four beds were apparent in bivariate plots of a few of the elements, as illustrated in Figures 2 and 3. Plots like these, however, do not always give clear, unequivocal clusters of data points for each bed, and it is awkward to assimilate information from several plots at the same time. Hence, a method of evaluating several of the elements together was needed to determine if a distinctive chemical signature for each bed could be defined.

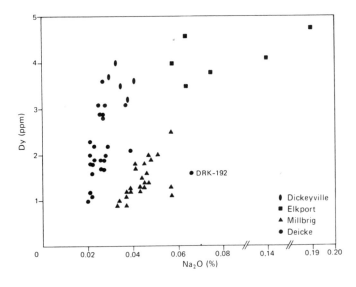

Figure 2. Distribution by bed of sodium and dysprosium in the control group samples of the Galena Group K-bentonites (from Kolata et al., 1986).

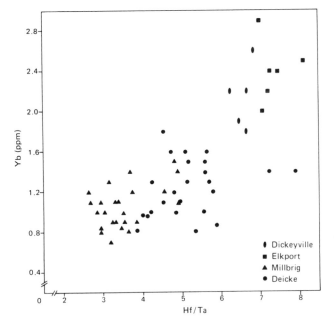

Figure 3. Distribution by bed of the hafnium/tantalum ratio versus ytterbium in the control group samples of the Galena Group K-bentonites (from Kolata et al., 1986).

APPLICATION OF THE STATISTICAL PROCEDURE

A multivariate statistical method, discriminant function analysis, was employed to analyze the 26 variables (elements) and four groups (beds). The discriminant analysis method was selected because it seeks to statistically distinguish between two or more groups of samples using a set of variables that are thought to differ between groups (Klecka, 1981). The mathematical objective is to weight and linearly combine the discriminating variables so that the groups are forced to be as statistically distinct as possible. This is illustrated in Figure 4 for a two-variable, two-group case. In this example, the analysis consists of finding a transform which gives the minimum ratio of the difference between the two-group multivariate means to the multivariate variance within the two groups (Davis, 1986). The linear discriminant function gives an adequate separation of the two groups (Fig. 4). A univariate analysis of such a problem would examine each variable independently, ignoring its redundancy with or its effects upon the others. The discriminant function is especially useful in detecting subtle combinations of variables which, when considered together, result in a larger group difference than any of the variables considered alone. If that difference is significant, then the computed discriminant functions can be used to assign unknown samples to one of the groups.

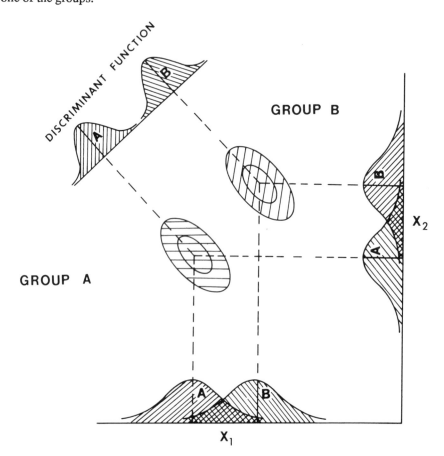

When there are large numbers of variables of unequal or uncertain discriminating power, it usually is necessary to use a stepwise algorithm for selecting variables to include in the discriminant functions. The stepwise procedure begins by selecting the single best-discriminating variable, according to a separation criterion, described below. The second variable selected is the one that, in combination with the first, best improves the value of the discrimination criterion. The third and subsequent variables are similarly selected according to their ability to contribute to further discrimination. Discriminant analysis thus offers a weight of evidence criterion that cannot be observed by measuring any of the individual variables in the model, and thereby extends the limitations of conventional empirical analysis by maximizing the advantage of variable assemblages.

The stepwise procedure was used in the analysis of the K-bentonite beds. The criteria by which independent variables are chosen for the stepwise analysis is based upon their individual discriminating ability as expressed by the multivariate F ratio, which is the ratio of the mean between-group variance to mean within-group variance as determined by analysis of variance. To achieve an optimal ranking of variables we employed the Wilks' lambda method, which minimizes the overall Wilks' lambda (or maximizes the multivariate F ratio) after each new variable is entered. The Wilks' lambda method is one of several separation criteria described in the literature (Klecka, 1981). Lambda is the ratio of unexplained variance to total variance; a measure of the degree to which the discriminant function explains the observed variance. Initial ranking before step 0 is based on the F-ratio criterion. F scores are recalculated after each step and variables are selected on the basis of their ability to account for the remaining variance. Thus variables which, by themselves, possess low F ratios and have little discriminatory value can add significantly to the discriminant model when combined systematically with other variables.

Three types of information useful for stratigraphic correlation result from discriminant analysis. First is a list of discriminant function scores computed for each sample. The reader is referred to standard texts on multivariate statistics for a discussion of the equations involved (e.g., Tatsuoka, 1971; Harris, 1975). Second, a classification table is produced which indicates what proportion of the samples are correctly classified, and whether there is any evidence of systematic misclassification of specific samples. Thus, even if all samples in the initial population have known bed membership *a priori,* they will not necessarily be correctly classified if sufficient separation between groups has not been achieved in the discriminant analysis. Third, the two most effective discriminant functions are used to construct a territorial map showing group centroids and the territorial boundaries that separate them (e.g., see Fig. 5). Samples with unknown affinities may be assigned to one of the recognized groups by plotting their first and second discriminant function scores on the map.

Figure 4. Graphical representation of the linear discriminant function. Two groups, A and B, overlap with respect to the distribution of variables X1 and X2 in two-dimensional space. In N-dimensional multivariate space the discriminant function is defined by the coordinates of the axis along which maximum separation of the two groups occurs (after Davis, 1986).

RESULTS

Discriminant function analysis was performed on the chemical data from the 60 K-bentonite samples in the control set (Kolata et al., 1986). The number of discriminant functions calculated is equivalent to the number of variables entered, or to one less than the number of groups (i.e., K-bentonite beds), whichever is smaller. Table 1 summarizes the results and lists the three functions, their eigenvalues and a corresponding canonical correlation coefficient. The latter is a measure of the function's ability to discriminate among the groups. Values of the functions as calculated at the group means also are given. They may be thought of as defining point coordinates within a three-dimensional orthogonal grid. The eigenvalues, a measure of the relative amount of variance among the group of elements accounted for by each function, indicate that the third function is relatively small compared to the first two and contributes relatively little to the discriminant analysis. The first two functions include 95.9% of the variance accounted for by the model. Moreover, the canonical correlation coefficients associated with the functions show that the first two discriminant functions are each very highly correlated with the groups and the third is somewhat less correlated.

Table 1 Properties of the discriminant functions for four K-bentonite beds.

Function	Eigenvalue	Percent of Variance	Canonical Correlation
1	16.65	61.9	0.97
2	9.16	34.0	0.95
3	1.10	4.1	0.72
		100.0	

Group	Function 1	Function 2	Function 3
1 (Deicke)	-3.28	2.45	0.06
2 (Millbrig)	-0.03	-3.55	0.17
3 (Elkport)	9.25	3.00	1.59
4 (Dickeyville)	5.47	1.17	-3.05

(Data from Kolata et al., 1986)

The correlations show that the functions, especially the first two, effectively separate the four beds. Different elements are important in each of the three functions. The order of importance of the elements to the discriminant model is Eu, Sm, Sc, Ti, Zr, Th, Hf, Lu, Ta, Dy, Tb, Fe, Co, Na, and Sb. The 60 samples were back-classified using the discriminant functions. All classified correctly in their respective groups, indicating the discriminant functions are successful in achieving group (bed) separation.

Two other less-biased tests also were used to estimate the ability of the discriminant function model to classify the control-set samples correctly (Norusis, 1985, p. 87). The first is cross-validation, in which the control set of 60 samples was randomly split into two parts. The first half of the samples was used to derive the discriminant model, and the second half used as unknowns to test the model. Then the second half was used to derive the discriminant functions, and the first half of the samples used as unknowns to test them. In both cases, all Deicke, Millbrig and Elkport samples classified into their correct groups. Three of the five Dickeyville samples were misclassified as Elkport or Millbrig; however, the misclassification is not considered significant since a random split of five samples will not yield a statistically valid subset size for cross-validation. In the second test, the "jackknife" method was employed. One sample was left out of the control set and the discriminant function analysis was performed on the remaining 59 samples. The omitted sample was classified on that model. This was done for samples selected randomly from the control set. Each time the omitted sample was correctly classified by the discriminant model derived the from remaining set. Thus, it is possible to identify a unique chemical fingerprint for each of the K-bentonite beds within the control group area. Figure 5 shows the territorial map constructed using the first two canonical discriminant functions and the positions of the control-set samples plotted on the map.

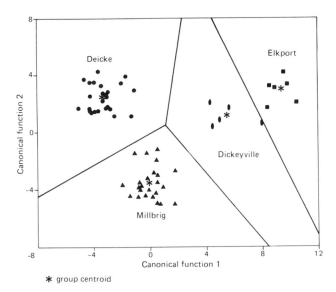

Figure 5. Territorial map constructed from the two major canonical discriminant functions calculated for 15 elements in the 60 control group samples. Asterisks mark the group centroids or means. Boundaries are defined by the loci of points equidistant from pairs of centroids (from Kolata et al., 1986).

Forty-eight additional samples from localities extending from southeastern Minnesota to Missouri were identified in the field as probably either Deicke or Millbrig. They were tested using the discriminant function coefficients derived in the analysis of the control-set samples. Figure 6 shows the unknown samples plotted on the territorial map. All but two samples classified correctly into their respective beds based on their inferred stratigraphic position. Those two, DRK-236 and DRK-213, contained high concentrations of light rare earth elements and, although considered to be Deicke samples based upon their field relationships, could not be properly classified. However, the persistent agreement between the field identification and chemical fingerprint of 96% of the unknown Deicke and Millbrig samples is convincing evidence that the two beds can be correlated beyond the control group area on the basis of trace-element chemistry. The Elkport and Dickeyville K-bentonite beds, on the other hand, lack sufficient statistical control to be identified with confidence outside the reference area.

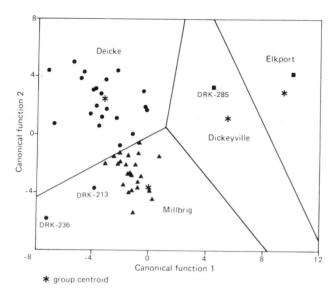

Figure 6. Classification of 48 non-control group samples of Galena Group K-bentonites as illustrated by their scores for the two major canonical functions derived in the discriminant analysis. Asterisks mark the control group centroids or means (from Kolata et al., 1986).

ASSUMPTIONS AND LIMITATIONS

Consideration of the possible misclassification or misidentification of samples requires an understanding of the inherent limitations and assumptions of the discriminant model. Several assumptions are necessary regarding the characteristics of a data set in order that the discriminant analysis be valid (Davis, 1986). First, the multivariate distribution of the variables should be normal. If any of the variables depart significantly from a normal distribution it is likely that the multivariate assumption is incorrect. However, if the variables are normally distributed it does not necessarily follow that the multivariate distribution is normal. Second, it is assumed that the variance-covariance matrices of the groups are equal in size. Most standard discriminant analysis programs contain appropriate statistical tests for these assumptions. Fortunately, the discriminant function is not seriously affected by slight departures from normality or by limited covariance inequality.

Two other assumptions are that an unknown sample may belong with equal probability to any group under consideration, and that the samples in each group are randomly chosen (Davis, 1986). The validity of both assumptions is difficult to demonstrate, particularly if there is significant variation among the population sizes of the various groups. In many geologic situations data are limited by the number of outcrops or drillholes where samples can be collected; thus, sampling is done not by design but by opportunity. The importance of group size can be seen in Figure 7 showing mean values and confidence intervals for the first four elements in the discriminant model. Calculation of the confidence intervals about the means shows a strong correlation with N, the number of samples in each group (bed). The horizontal bar represents the range of values within which the actual or true mean may occur. Those groups with relatively large numbers of samples have means that are more likely to coincide with the true mean for that group than those with small numbers of samples. These are relative expressions of variance, of course, but clearly argue that some minimum sample size is required to optimize the discriminant function. The Mississippi Valley reference group contains 60 samples representing four groups. The preponderance of Deicke and Millbrig samples over Elkport and Dickeyville suggests that Elkport and Dickeyville samples have a higher chance of misclassification.

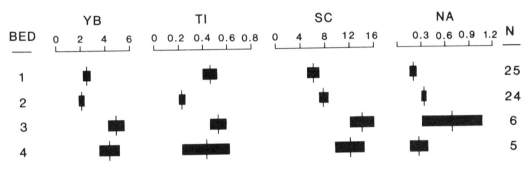

Figure 7. Confidence intervals for sample means calculated for each of four elements in the four-bed model to illustrate the effect of sample size on variance. Bar length indicates the range of values within which the actual or true mean lies, and sample population means are represented by the vertical line in each bar. Thus, for short bars, the probability is high that the sample population means and true means coincide, whereas for longer bars that probability is reduced. Beds 1 through 4 are, respectively, the Deicke, Millbrig, Elkport and Dickeyville K-bentonites. N equals number of samples. Arithmetic scales at the top are in ppm for Yb, Ti and Sc, and percent for Na.

For discriminant function analysis it is important to have data on many elements among which to look for differences between groups. Figure 8 shows the sequential combination of variables against classification output based on the discriminant analysis summary table. With five elements in the analysis, 80% of the cases (samples) can be correctly classified. With ten elements 92% of the cases are correctly classified. Thus the quality of the information provided can be balanced against the cost and availability of sample analysis as part of the particular project design.

The selection of elements to be studied when designing a discriminant analysis experiment depends upon the geological material. In the search for distinguishing chemical characteristics in K-bentonites (altered volcanic ash) it is assumed that the concentrations of those elements that are unique to each bed were similarly different in the separate ash falls. It has been shown in studies of Cenozoic tephra (Randle et al., 1971; Borchardt et al., 1971; Westgate et al., 1977) that the immobile elements are useful in distinguishing between different ash beds. We make the additional assumption that the immobile element concentrations have been preserved or altered consistently in the transformation of volcanic ash to K-bentonite, and further, that the chemical composition of the original ash was constant over the region or varied systematically along a line from source to point of deposition. The validity of these assumptions rests upon the weight of accumulated evidence from studies of volcanic rocks and ashes. When 26 elements are examined and ranked for their ability to distinguish between K-bentonite beds we find that those elements ranked highest in the discriminant analysis are many of the same elements found by others to be geochemically stable in both weathering and low-temperature metamorphic environments (Pearce and Cann, 1973; Winchester and Floyd, 1977; Wood, 1980).

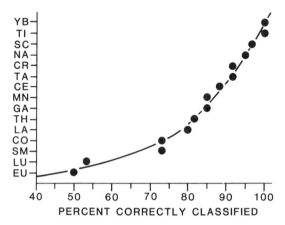

Figure 8. Relationship between the number of elements analyzed and the probability that individual beds can be correctly identified.

CONCLUSIONS

We conclude that the chemical differences between beds are geologically significant. K-bentonite beds of the Galena Group each have unique chemical fingerprints that probably reflect differences in the original composition of the parent volcanic ash. They can be correlated on a regional scale and two of the beds, the Deicke and Millbrig, have been chemically identified from outcrop and subsurface samples between southeastern Minnesota and southeastern Missouri, a distance of approximately 900 km. Discriminant analysis is a powerful tool for the study of chemical differences in the K-bentonites and for the identification of unknown samples. The method is particularly useful when combined with well-documented lithostratigraphic and biostratigraphic field evidence. The value of multivariate analysis lies in the enhanced interpretation of that evidence it provides, and its limitations are a function of appropriate sample sizes and sample distribution.

ACKNOWLEDGMENTS

This study was supported by NSF grants EAR-8025889, EAR-8208480, and EAR-8407018. We thank Roger Stuebing, Duane Moore, Bob Osborne, Lynn Watney and Tim Cross for constructive reviews of the manuscript.

REFERENCES CITED

Borchardt, G.A., Harward, M.E., and Schmitt, R.A., 1971, Correlation of volcanic ash deposits by activation analysis of glass separates: Quaternary Research, v. 1, pp. 247-260.

Cable, M.S., and Beardsley, R.W., 1984, Structural controls on late Cambrian and early Ordovician carbonate sedimentation in eastern Kentucky: American Journal of Science, v. 284, p. 797-823.

Cullen-Lollis, J., and Huff, W.D., 1986, Correlation of Champlainian (Middle Ordovician) K-bentonite beds in central Pennsylvania based on chemical fingerprinting: Journal of Geology, v. 94, p. 865-874.

Davis, J.C., 1986, Statistics and Data Analysis in Geology (second edition): New York, Wiley, 646 p.

Harris, R.J., 1975, A primer of multivariate statistics: New York, Academic Press, 332 p.

Huff, W.D., 1983, Correlation of Middle Ordovician K-bentonites based on chemical fingerprinting: Journal of Geology, v. 91, p. 657-669.

Huffman, G.G., 1945, Middle Ordovician limestones from Lee Country, Virginia, to central Kentucky: Journal of Geology, v. 53, p. 145-174.

Kay, G.M., 1935, Distribution of Ordovician altered volcanic materials and related clays: Geological Society of America Bulletin, v. 46, p. 225-244.

Klecka, W.R., 1981, Discriminant analysis: *in* Nie, N.H., Hull, C.H., Jenkins, J.G., Steinbrenner, K., and Bent, D.H., eds., SPSS: Statistical package for the social sciences (second edition), New York, McGraw-Hill, p. 434-467.

Kolata, D.R., Frost, J.K., and Huff, W.D., 1986, K-bentonites of the Ordovician Decorah Subgroup, upper Mississippi Valley: Correlation by chemical fingerprinting: Illinois State Geological Survey Circular 537, 30 p.

Kolata, D.R., Frost, J.K., and Huff, W.D., 1987, Chemical correlation of K-bentonite beds in the Middle Ordovician Decorah Subgroup, upper Mississippi Valley: Geology, v. 15, pp. 208-211.

Milici, R.C., and Smith, J.W., 1969, Stratigraphy of the Chickamauga Supergroup in its type area: Georgia Geological Survey Bulletin, v. 80, p. 1-35.

Nelson, W.A., 1922, Volcanic ash beds in the Ordovician of Tennessee, Kentucky, and Alabama: Geological Society of America Bulletin, v. 33, p. 605-615.

Norusis, M.J., 1985, SPSS-X advanced statistics guide: SPSS Inc., New York, McGraw-Hill, 505 p.

Pearce, J.A., and Cann, J.R., 1973, Tectonic setting of basin volcanic rocks determined using trace element analyses: Earth and Planetary Science Letters, v. 19, p. 290-300.

Randle, K., Goles, G.G., and Kittleman, L.R., 1971, Geochemical and petrological characterization of ash samples from Cascade Range volcanoes: Quaternary Research, v. 1, p. 261-282.

Sardeson, F.W., 1924, Volcanic ash in Ordovician rocks of Minnesota: Pan-American Geologist, v. 42, p. 45-52.

Tatsuoka, M.M., 1971, Multivariate analysis: Techniques for educational and psychological research: New York, Wiley, 310 p.

Westgate, J.A., Christiansen, E.Q., and Boellstorff, J.D., 1977, Wascana Creek ash (Middle Pleistocene) in southern Saskatchewan: Characterization, source, fission track age, paleomagnetism, and stratigraphic significance: Canadian Journal of Earth Science, v. 14, p. 357-374.

Willman, H.B., and Kolata, D.R., 1978, The Platteville and Galena Groups in northern Illinois: Illinois State Geological Survey Circular 502, 75 p.

Winchester, J.A., and Floyd, P.A., 1977, Geochemical discrimination of different magma series and their differentiation products using immobile elements: Chemical Geology, v. 20, p. 325-343.

Wood, D.A., 1980, The application of a Th-Hf-Ta diagram to problems of tectonomagmatic classification and to establishing the nature of crustal contamination of basaltic lavas of the British Tertiary Volcanic Province: Earth and Planetary Science Letters, v. 50, p. 11-30.

Part V

POTENTIAL APPLICATIONS OF QDS MODELS

34

STRATIGRAPHIC AND GEOCHEMICAL CONTROLS ON THE OCCURRENCE OF ACIDIC MINE WATERS AND PREDICTIVE TECHNOLOGIES

Frank T. Caruccio

Department of Geological Sciences, University of South Carolina, Columbia, South Carolina 29208 USA

ABSTRACT

Acid mine drainage occurs when pyrite-enriched carbonate-deficient strata are exposed to accelerated chemical weathering by land disturbances. Under oxidizing environments, iron sulfides are converted to soluble hydrous metallic sulfate salts that hydrolyze to produce highly acidic metal- and sulfate-enriched drainages. Minor amounts of carbonate dramatically suppress the oxidation process and provide a buffering capacity. Mine drainage quality can be approximated by the simple ratio of acid potential (usually the sulfur content) and alkaline potential (the carbonate content) of the sedimentary sequence that is disturbed. Sophisticated weathering simulations are used to quantitatively assess the acidity and alkalinity that a sample will produce. Results of these simulations can be factored into conceptual or analytical models.

The depositional environment, which controls the form and distribution of both the sulfide and carbonate components within the strata, can be used to identify the potential occurrence of this acid problem on a regional scale. A conceptual model is presented that relates mine-drainage quality to pyrite and carbonate contents of the strata, bacteria activity and ground-water chemistry. The formulation of numerical models, capable of accurately evaluating mine-drainage quality and that can accommodate the complexities of the hydrology, geochemistry and random distribution of acid and alkaline materials within a reconstructed mine site, is identified as a critical research need.

ACID MINE DRAINAGE

Acid mine drainage (AMD), a highly acidic, sulfate- and iron-enriched drainage, may decimate the ecology of recipient streams. Iron disulfides (hereafter collectively termed pyrite) are recognized as the primary cause of acidity due to their spontaneous oxidation and subsequent hydrolysis of oxidation products (Barnes and Romberger, 1968). Although most commonly associated with coal mines, AMD also occurs where pyrite is exposed during other land disturbances, such as metallic sulfide mines, highway road cuts, and subway tunnels. Because AMD can have a deleterious impact on local streams and rivers, accurate premining prediction of drainage quality is of the utmost concern by mine operators and regulatory agencies.

Oxidation of pyrite produces ferrous iron which may be further oxidized to ferric iron by bacteria. In turn, ferric iron can oxidize pyrite and accelerate reactions leading to increased acidity. Singer and Stumm (1970) reported that iron oxidation is the rate-limiting step of the reaction that proceeds slowly under sterile conditions. Iron oxidizing bacteria catalyze the oxidation of ferrous iron. These bacteria are indigenous to aqueous environments with pH values that range from 2.8 to 3.2 (Kleinmann et al., 1981).

The mineralogy of the disturbed material and the capacity of the minerals to produce acidity and alkalinity determines the quality of the drainage. If alkalinity exceeds acidity, the catalyzing actions of bacteria are inhibited, the solubility of Fe^{2+} and Fe^{3+} is reduced, and the acid production reactions are inhibited. On the other hand, if acidity exceeds alkalinity, the solubility of FeS_2 and acid production is increased. The low pH of the geochemical system enhances the catalyzing bacteria, thereby increasing the solubility of Fe^{3+} and the acid load. Once the acid reaction begins, it is self-propagating and all acid-producing mechanisms interact synergistically (Caruccio, 1969).

It often is assumed that the amount of acidity produced is proportional to the concentration of pyrite present; that is, the higher the pyrite (or sulfur) content the greater the potential of the sample to produce acid. However, Caruccio and Parizek (1968) and Caruccio (1969) showed that carbonate-deficient rocks containing pyrite produced acidities that varied more directly with pyrite particle size. They showed that coarse grained pyrite (>50 microns) were the least reactive of the sedimentary forms and remained relatively stable as compared to smaller pyrite particles (<0.25 microns) that rapidly decompose upon exposure to atmosphere. All other factors being equal, samples with a predominance of fine-grained (framboidal) pyrite generate orders of magnitude more acid than samples containing higher percentages of coarse-grained pyrite (Caruccio, 1973).

However, the acid potential of a rock also is dependent upon the carbonate content which, if present, produces an alkaline, buffered and potentially neutralizing leachate (Caruccio, 1969; Caruccio and Geidel, 1980). In contact with calcareous material, the pH of the groundwater regime is buffered to higher values, which effectively suppresses the activity of the iron-oxidizing bacteria. Recent evidence suggests that calcareous material also inhibits the oxidation of pyrite (Caruccio et al., 1981).

Acid-forming reactions have few rate-limiting steps due to the high solubility of the oxidation products. By contrast, the amount of alkalinity produced by calcareous material is constrained by the low solubility of the specific carbonate mineral in water (siderite excluded). Alkalinity values are controlled by the partial pressure of carbon dioxide, the duration of water contact, and the solubility constant of the specific mineral or rock (Geidel, 1980). Under natural conditions acidity can be orders of magnitude greater than alkalinity.

PREDICTIVE FRAMEWORK

Overburden Analyses

Prediction of mine-water quality within a particular geologic setting requires accurate assessment of the system's potential to produce acidity versus its potential to produce alkalinity. The geochemical characteristics of a backfill in a mine also are affected by permeability and particle-size distribution of the backfill, pore-gas composition, ground-water chemistry, frequency of precipitation, mining method, reclamation techniques, and the hydrology of the mine site.

Predictions of mine-drainage quality must be made from overburden analyses performed on rock samples collected from a core or exposure. In turn, the whole-rock analytical data are used to predict the aqueous phase (Sobeck et al., 1978). The rates at which sulfides produce acidity and carbonates produce alkalinity are not similar. Thus, in using overburden analyses to assess mine-drainage quality, the disproportionate kinetics of acid/alkaline release must be evaluated.

Translation of overburden analyses to the prediction of mine-drainage quality requires extrapolation of laboratory results to anticipated field responses. Short of monitoring the internal hydrogeochemical plumbing system of a backfill, the overburden analytical method must first accurately predict the leachate composition of a rock sample expected under a given set of field conditions, then quantify the net acid and alkaline loads produced by each stratum within the overburden and integrate these loads through a mine-drainage predictive model. Should a particular technique fail to accurately assess the leachate composition of a given sample, the model generates erroneous mine-drainage projections. Regardless of the precision and accuracy of a predictive model, errors entered into the model are carried and often amplified as output errors.

Field Setting

In most reclaimed mine sites, the highly permeable nature of the backfill material prevents the water table from mounding to near surface horizons. During periods of low rainfall and spring thaw, water rinses the accumulated weathering products from the rock surfaces. After the wetting front passes through a particular horizon, water retained on particle surfaces by capillary forces forms micro-environments for chemical reactivity leading toward alkalinity or acidity production. The quality of mine drainage is greatly affected by the length of time between flushings and the different rates of the acid/alkaline reactions. Most of the alkalinity is generated during the initial chemical reactions and the reaction rate declines through time. After the infiltrating waters have achieved carbonate-bicarbonate equilibria, additional contact with calcium carbonate does not significantly increase the alkalinity of the system (the maximum value is about 450 mg/l as $CaCO_3$ at $pCO_2 = 10^{-3.5}$). Thus, regardless of the amount of calcium carbonate present in the strata, the solubility of the carbonate, and hence the alkalinity of the system, is limited by the geochemical constraints imposed by equilibrium conditions.

Unlike alkalinity production, the generation of acidity through dissolution of hydrous iron sulfates (the oxidation products of pyrite) in water is limited by pyrite breakdown. In the presence of oxygen, oxidation continues. As in all geochemical systems, a solubility limit will eventually arrest the oxidation reaction. However, iron sulfate and sulfide compounds are an order of magnitude more soluble than carbonate in the system. Because of this high solubility product acidity increases through time.

Calcium Carbonate/Pyrite Interactions

Numerous studies have shown that calcium carbonate can stabilize pyrite as well as produce alkalinity. As early as 1954, Temple and Koehler showed that the stability of sulfur balls is related to the presence of an acid-soluble mineral. Stable, nonoxidizing pyrite samples became reactive when rinsed with dilute hydrochloric acid solutions, presumably removing associated calcareous materials. Caruccio et al. (1981) tested this concept with numerous samples having various calcareous material/pyrite contents and found that the carbonates have a dominant affect.

In summary, the kinetics of acid and base release are not similar and have different solubility constraints. The quantity of either an acid- or base-producing solid phase cannot be related stoichiometrically to anticipated acid loads. And, the occurrence of calcium carbonate, at particular concentrations relative to the amount of pyrite, may serve either as an inhibitor of pyrite oxidation or neutralizer of acidity. Again, the volumes of either component cannot be used directly to determine the anticipated sample response.

PREDICTIVE TECHNOLOGY—MINE-DRAINAGE CHARACTERIZATION

Prediction of mine-drainage quality based exclusively on overburden analyses ignores the multivariate nature of the acid mine-drainage problem and produces erroneous conclusions. Instead, the kinetics of acid or base release, the ground-water geochemistry, the chemistry of recipient streams, and the chemistry of precipitation must be considered. Further, the arrangement of the spoil material in the mine (blending versus segregation), the hydrology of the mine, the evolution of pore-gas chemistry and its effect on the acid and base leachates must be considered as evolving through time.

Table 1 Matrix showing anticipated drainage quality as a function of carbonate/pyrite ratios.

		CARBONATE	
		Absent	Present
P Y R	0.85%	Low Acid Low Ionic Strength	Alkaline Low Ionic Strength
I T E	0.85 - 1.5 %	AMD	Alkaline (?) High Ionic Strength
(wt. %)	1.5%	AMD	Low Acid (?) High Ionic Strength

Carbonate/Pyrite Ratio Model

Caruccio and Geidel (1980) showed that carbonates play a dominant role in controlling acid production. In the presence of calcium-magnesium carbonate pyrite oxidation is retarded, in many situations to the point that sulfate ion (a measure of the pyrite oxidation) is absent in mine drainages. Carbonate beds generally occur over large, laterally pervasive areas, and their presence in cores suggests that large areas are underlain by nonacidic stratigraphic units. As such, the matrix presented in Table 1 may be used to identify AMD occurrences.

Conceptual Model

The factors determining mine-drainage quality include: calcium carbonate content of the strata, ground-water pH, mode of occurrence of the disulfide, and the neutralizing and buffering capacity of the ground water. The conceptual model shown in Figure 1 attempts to organize these factors into a matrix that can be used to qualitatively predict mine-drainage quality.

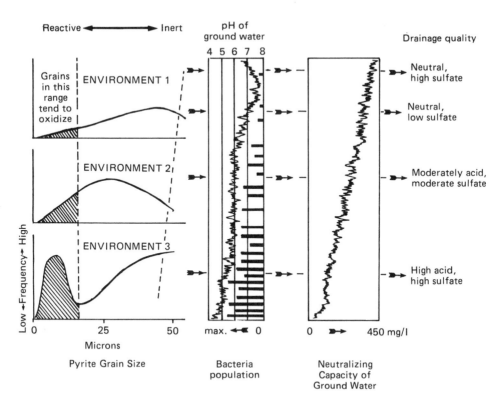

Figure 1. Conceptual model relating drainage quality to the geochemistry of the overburden.

In the Hydrogeochemical Environment I, pyrite is primarily coarse grained and relatively slow reacting, the carbonate buffers the groundwater pH to 6.4, and catalyzing iron bacteria are inhibited. Here, minor amounts of acidity produced by the coarse pyrite are neutralized by the alkalinity, the acid reaction is not catalyzed, and the pyrite-stabilizing carbonates, if present, are preserved. In this environment mine drainages are chemically neutral and have low sulfate concentrations.

At the other end of the spectrum, in Hydrogeochemical Environment III, most of the pyrite is fine grained and can rapidly oxidize to produce acidity. In the absence of carbonate, the pH of the ground water is low (<4) which enhances the viability of the iron-oxidizing bacteria. The lack of carbonate also produces neutral, mildly buffered ground waters (low specific conductance) that are easily degraded by the acidity produced. Mine drainages in this environment have high acidity, sulfate, and specific conductance. Hydrogeochemical Environment II is intermediate to the other two and is characterized by correspondingly intermediate amounts of acidity and sulfate. Neutral, high-sulfate drainages are produced when strongly alkaline ground waters commingle with acid mine drainages.

Quantitative Geochemical Models

Several numerical models to predict mine-drainage quality have been advanced. Some, such as the computer models by Jaynes (1983) and Rogowski et al. (1977), consider the long-term oxidation of pyrite and removal of reaction products but ignore the character of the overburden and do not utilize overburden analyses.

Caruccio and Geidel (1981) presented an algebraic model that makes extensive use of overburden analyses to predict mine-drainage quality. Simulated weathering tests and associated reaction-rate data are used to assess the chemical weathering attributes of strata collected from cores of the overburden. With some assumptions, the minimum acid load produced by the rock sequence is calculated and balanced against the alkalinity available to determine the mine's potential to produce acid waters (Caruccio and Geidel, 1981).

This technique has been field tested in the evaluation of 13 mines in southwestern Pennsylvania and two in West Virginia. Of the 15 mines evaluated, four were predicted to produce nonacid drainages and were given operating permits. Three of these mines were opened and produced alkaline drainages as predicted. The remaining 11 mines were judged to be acid producing. Of these, two were inadvertently permitted and produced acid mine drainage within two months.

Paleoenvironmental Model

The models presented above can be related to the paleoenvironment (the depositional environment) of the strata which serves to enhance the model's predictive capability. Because most strata are laterally continuous throughout the area, results of detailed studies in a small area may be applied to other areas. Marine and brackish water paleoenvironments are identified with acid prone areas, whereas the fresh water paleoenvironment is closely related with nonacid systems. Mapping the paleoenvironment of an area should thus provide a broad indication of the mine-drainage quality of mines sited in the area (Caruccio et al., 1977, Rose et al., 1983).

Using the paleoenvironmental models, however, requires the detailed and accurate mapping of a coal seam's paleoenvironment which, in many cases, is either without a funding base from the company or state agency, or lacks the stratigraphic control necessary to adequately define the depositional environment within the mine site. At best, the conceptual models provide a regional assessment of expected drainage quality and site-specific data are required to confirm the prediction.

SUMMARY AND CONCLUSIONS

Within the Appalachian coal fields, the quality of mine drainage is partly controlled by the relative proportions of carbonate and pyrite occurring in the disturbed strata. Carbonates not only produce alkaline, strongly buffered leachates through dissolution, but have been shown to reduce pyrite oxidation rates as well. Pyrite produces acidity through oxidation reactions and the hydrolysis of the oxidation products. The kinetics of base/acidity release are not similar. The solubility of calcareous material limits the mineral's dissolution and production of alkalinity. In contrast, the oxidation and dissolution of pyrite, and the consequent production of acidic waters are, essentially, unlimited. Thus, within the natural setting, the capacity of an acid-producing rock to produce acid is orders of magnitude greater than the capacity of carbonate to produce alkalinity.

Of the overburden analytical techniques available, simulated weathering (leaching) tests are preferred over whole-rock analyses (acid-base accounting). Leaching tests provide leachate samples and reaction-rate data that are directly related to the chemical weathering attributes of the samples. Acid-base accounting data, on the other hand, must be interpreted to predict the quality of leachate expected from a particular rock sample. Several lines of evidence show that variations in the kinetics of acid/base release and the complex interactions of varying pyrite-carbonate ratios invalidate predictions of leachate quality by the numerical summation of acid-base accounting data.

The kinetic data derived from the leaching tests of acid-producing samples, coupled with some assumptions regarding the production of alkalinity, infiltration capacities, and surface areas of reactivities, can be used to quantify acid loads and are easily incorporated in a conceptual or computer model used to predict mine-drainage quality and evaluate the environmental impact of a proposed mining operation. At this level, the complexities arising from porosity variations within the backfill (or reclaimed area), pore-gas chemistries, hydrologic variations, temporal and spatial distributions of varying rock chemistries and geochemical responses, must be quantitatively evaluated and included in the model. At the time of this writing this is identified as a critical research need.

ACKNOWLEDGMENTS

Appreciation is extended to Eileen Poeter and Sam Romberger for their review of the paper. My special gratitude to Tim Cross for his critical review and constructive editorial comments.

References Cited

Barnes, H.L., and Romberger, S.B., 1968, Chemical aspects of acid mine drainage: Journal Water Pollution Control Federation, v. 40, p. 371-384.

Caruccio, F.T., 1969, Characterization of strip mine drainage, in Ecology and reclamation of drastically disturbed sites: New York, Gordon and Breach Publ., v. 1, p. 193-224.

Caruccio, F.T., 1973, Estimating the acid potential of coal mine refuse: in Chadwick, M., and Goodman, G., eds., The ecology of resource degradation and renewal: Oxford, Blackwell Scientific Publications, p. 197-206.

Caruccio, F.T. and Parizek, R.R., 1968, An evaluation of factors affecting acid mine drainage production and the ground water interactions in selected areas of western Pennsylvania, in Proceedings of the second symposium on coal mine drainage research: Monroeville, Pennsylvania, Bituminous Coal Research, p. 107-151.

Caruccio, F.T., Ferm, J.C., Horne, J., Geidel, G., and Baganz, B., 1977, Paleoenvironment of coal and its relation to drainage quality: Cincinnati, Ohio, Environmental Protection Agency, EPA-600/7-77-067, 107 p.

Caruccio, F.T., and Geidel, G., 1980, The geologic distribution of pyrite and calcareous material and its relationship to overburden sampling: U.S. Bureau of Mines Information Circular IC-8863, p. 2-12.

Caruccio, F.T., and Geidel, G., 1981, Estimating the minimum acid load that can be expected from a coal strip mine, in Proceedings of the 1981 symposium on surface mining hydrology, sedimentology and reclamation: Lexington, Kentucky, University of Kentucky, p. 117-122.

Caruccio, F.T., Geidel, G., and Pelletier, M., 1981, Occurrence and prediction of acid drainages: Journal of Energy Division, American Society of Civil Engineers, v. 107, EY1, p. 167-178.

Geidel, G., 1980, Alkaline and acid production potentials of overburden material, the rate of release: Reclamation Review, v. 2, p. 101-107.

Jaynes, D.B., 1983, Acid mine drainage model—POLS [Ph.D. Thesis]: University Park, Pennsylvania, The Pennsylvania State University, 252 p.

Kleinmann, R.L., Crerar, D.A., and Pacelli, R.R., 1981, Biochemistry of acid mine drainage and a method to control acid formation: Mining Engineering, March, p. 300-305.

Rogowski, A.S., Pionka, H.B., and Broyan, J.G., 1977, Modelling the impact of strip mining and reclamation processes on quality and quantity of mine water: Journal of Environmental Quality, v. 6, p. 237-244.

Rose, A., Williams, E., and Parizek, R., 1983, Predicting potential for acid drainage from coal mines: University Park, Pennsylvania, The Pennsylvania State University, Bulletin of Earth and Mineral Sciences, v. 52, p. 1-4.

Singer, P.C., and Stumm, W., 1970, Acid mine drainage—the rate limiting step: Science, v. 167, p. 1121-1122.

Sobek, A.A., Schuller, W.A., Freeman, J.R., and Smith, R.M., 1978, Field and laboratory methods applicable to overburdens and minesoils: Springfield, Virginia, National Technical Information Service, EPA-600/2-78-054, 203 p.

Temple, K.L., and Koehler, W.A., 1954, Drainage from bituminous coal mines: Morgantown, WV, West Virginia University, Engineering Station Research Bulletin 25, 35 p.

35

THE INFLUENCE OF STRATIGRAPHY IN RESERVOIR SIMULATION

Jack L. Shelton [1] and Timothy A. Cross [2]
[1]Amoco Production Company, P. O. Box 800, Denver, Colorado 80201 USA
[2]Department of Geology and Geological Engineering, Colorado School of Mines,
 Golden, Colorado 80401 USA

ABSTRACT

The most critical item in a reservoir engineering simulation of a petroleum reservoir is often the reservoir description. The reservoir description must identify the spatial relationship of fluid flow units and provide simulator input data averaged over a volume consistent with the size of the individual grid blocks used in the simulator. The geologist and reservoir engineer must work together to ensure that the geological input data represent the fluid flow properties of the reservoir for the particular reservoir recovery process being simulated. This paper summarizes the factors involved in a reservoir simulation and explains the elements of a reservoir description that are most important for the simulation task. References are provided for further reading.

INTRODUCTION

A reservoir simulator is a mathematical description of the geometry and petrophysical properties of a reservoir and the physical and chemical interactions of fluids within it. Reservoir simulation is used to estimate reservoir performance, especially the rates at which fluids are produced from the reservoir and, if applicable, the rates at which fluids can be injected into the reservoir. The purpose of reservoir simulation is to make reservoir management decisions. The most common objective is to economically optimize the depletion of the in-place hydrocarbons.

It is necessary to understand the individual factors that go into a reservoir simulation in order to identify the crucial elements of a reservoir description. The simulation factors can be thought of as: the reservoir description, the recovery process, the mathematical model, the numerical model, the simulator model, history matching, and performance predicting.

The main emphasis of this paper is the reservoir description. This factor will be considered in more detail later, but for now we will define it as a numerical description of: 1) the reservoir boundaries, 2) the rock and fluid characteristics as a function of position within the reservoir, and 3) the fluid saturation and pressure as a function of position and at some initial time. Reservoir engineers need to predict the rate of flow of fluids into and out of the reservoir. Therefore, the emphasis of reservoir description is on how flow units (geological "pipes" having certain permeability and porosity values) connect to provide conduits for fluids to flow at various velocities at various locations in the reservoir.

Quantitative Dynamic Stratigraphy (1989), T.A. Cross, ed., Prentice Hall, p. 589–600.

589

RECOVERY PROCESSES

There are numerous recovery processes; one or more may be viable as a depletion plan for a reservoir. Some of these recovery processes or mechanisms are listed in Table 1 and the following observations can be made about them. First, the reservoir characteristics determine the recovery mechanisms or processes that are operable. For example, gravity drainage would not be a recovery mechanism in a reservoir where lateral flow barriers prevent the downward flow of oil and the upward flow of gas. Waterflooding would not be operable in a reservoir where flow continuity does not exist between wells. Second, the type of recovery process affects the type of information needed in a reservoir description. For example, in miscible flooding a gaseous solvent may be injected along with water. Due to density differences between gas, water and oil, gravity forces move the fluids differentially in the vertical direction, while the fluids also are driven horizontally along pressure gradients between injection and producing wells during production. To simulate such a process, it is necessary to describe both vertical and horizontal components of flow units. The process may or may not prove economically operable depending upon the relationship of vertical to horizontal permeability.

MATHEMATICAL MODELS

The flow of fluids and energy in reservoirs is described by partial differential equations. As an illustration, consider the flow of a single fluid phase in porous media. Conservation of mass dictates that

$$\frac{\partial(\phi\rho)}{\partial t} = -\nabla \cdot p \vec{v} + q$$

where ϕ is porosity, ρ is density, p is pressure and $\vec{v}$ is a velocity vector. According to this continuum equation, at every vanishingly small point volume the rate of change of mass inside the volume (the left side of the equation) is equal to the difference in the flow rates into and out of the volume from all directions $(\nabla \cdot p \vec{v})$ plus a source or generation term q. For this discussion, q represents the rate of flow into the volume from an outside source, namely a well.

To solve the equation, it is necessary to relate velocity to pressure with Darcy's law. In petroleum engineering, Darcy's law is written as

$$v_x = \frac{k}{\mu} \frac{\partial p}{\partial x}$$

where v is velocity, k is permeability, μ is viscosity, and x is the distance over which pressure is measured. It is therefore seen that the permeability k is a crucial element in describing flow. For multiphase flow, there are conservation equations for each phase, and the saturation and relative permeability of each phase are factors in the equations. If components transfer between phases, equations must also describe this process. If heat flow or chemical reactions occur, additional equations are needed.

Table 1 Examples of Recovery Processes

Fluid Expansion
Solution Gas Drive
Gravity Drainage
Water Influx
Gas Injection
Water Flooding
Polymer Flooding
Miscible Flooding
Surfactant Flooding
Steam Flooding
In situ Combustion

NUMERICAL MODELS

The partial differential equations describing a recovery process are complex and usually cannot be solved analytically; rather, they are solved by numerical methods. For example, the partial differential equations that represent a continuum in time and space can be approximated by finite difference equations and solved numerically at discrete points in space and time. Simply stated, the partial differential equations are replaced with finite difference equations.

Solving the partial differential equations would require input data for each infinitesimal point in the reservoir. The finite difference technique requires that input data be provided for each discrete point. The values at each point must be average values related to the volume associated with the point. The concept of averaging data for input into a simulation model will be discussed later.

SIMULATION MODELS

Digital computers are used to solve the finite difference equations. The simulation model comprises the discretization scheme and solution techniques used to solve the finite difference equations for a particular reservoir simulation. It is essentially a computer program and is usually called the reservoir simulator.

HISTORY MATCHING

Reservoir simulation is an approximation process. The reservoir description, process description, and the finite difference equations are never perfect. Hence, the results of reservoir simulations are never exact. The most valuable aids in simulation work are historical data. A reservoir simulator should be able to simulate past performance of the reservoir. When past performance data are extensive, the simulator is "fine tuned" to match the history, often by adjusting the reservoir description. In fact, it is common to obtain a reservoir description by history matching. For example, historical waterflood data are often simulated to obtain a "waterflood history match reservoir description." This reservoir description is then used in a simulation to forecast future waterflood performance or to predict reservoir performance for a different recovery process. It should be emphasized that a history match reservoir description is not unique and is no substitute for geological information.

PERFORMANCE PREDICTIONS

Performance predictions to estimate future performance are the major goal of simulation work. Simulation is also a powerful analytical tool to help understand significant features in a reservoir project. The complexity and number of variables in recovery processes are so great that the mind often cannot form a mental picture of the conflicting and interrelated phenomena occurring in a reservoir. Simulation is an evolutionary process that is best accomplished by the reservoir engineer and geologist working together from inception to completion of the simulation process.

RESERVOIR DESCRIPTION BASICS

The three broad areas of reservoir description in simulation are: 1) establishing the reservoir geometry and boundaries, 2) developing the grid, and 3) entering physical properties into the grid. All of these require information from stratigraphic observations or quantitative models.

Reservoir Geometry and Boundaries

The volume comprising the reservoir must be described in three dimensions. The structure, or external shape and dimensions, of a reservoir is usually definable from seismic or well control. The size and shape of the reservoir determine the general gridding scheme of the simulation. Boundary conditions, such as the flow connections between the reservoir and the outside system, are required. An example would be the degree of fluid communication between the reservoir and an aquifer.

Gridding

Gridding is the most crucial step in simulation because it fixes the texture and character of the reservoir simulation. The more closely the grid approximates the fluid flow properties of the reservoir, the more likely the reservoir simulator will perform accurately. Typically, reservoir simulators contain up to 50,000 grid blocks. Figure 1 shows an example grid for a portion of a reservoir penetrated by several wells. The grid is 16 by 16 blocks areally by 5 blocks deep and defines 1280 grid blocks representing the reservoir. The value assigned to each fluid and petrophysical property within each grid block is considered an average of that property within the volume of the block. The value of a property may vary from block to block. Even if the reservoir is perfectly uniform, there must be a minimum number of grid blocks to provide resolution on the fluid saturation and pressure distributions. For real reservoirs, gridding is much more critical because it must represent the actual stratigraphy of the reservoir. More specifically, the grid must accurately represent the degree of connectivity of the fluid flow units in the reservoir to the wells. The maximum number of grid blocks ultimately is limited by computational resources.

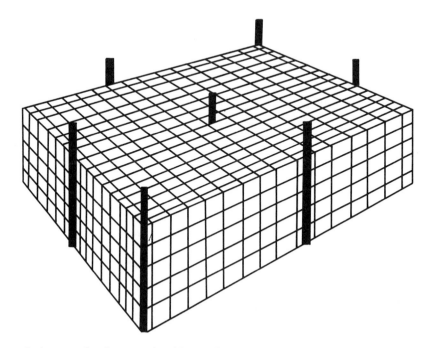

Figure 1. An example of a reservoir grid comprising 16 by 16 by 5, or 1280, grid blocks. The black vertical bars represent wells. For reservoir simulation, values assigned to all fluid and petrophysical properties within each grid block are considered averages of those properties throughout the volume of the block.

Figure 2 is an example of vertical gridding. The illustration shows a four-well cross section of k/φ versus depth, where k/φ represents the flow velocity of fluid in porous media. By correlating similar patterns of k/φ from well to well, the cross section was divided into seven layers. Each layer is a flow unit with fluid communication between the wells. Note that the fluid flow unit third from the top is very thin but has very high velocity. It is identified as a separate flow unit so that the high permeability does not become averaged with overlying and underlying fluid flow layers. The bottom unit has zero flow capacity. In this illustration we might choose to grid the reservoir with seven grid blocks, or layers, in the vertical direction, each of which would have different properties and thicknesses.

Areal gridding must also be provided for simulation. A grid block must be provided for each well and usually several grid blocks are needed between wells. Consultation is needed between the reservoir engineer and the geologist to balance the need for high-resolution areal definition of flow units against the added computation costs for increasing the number of grid blocks.

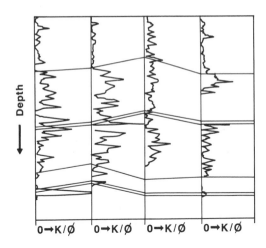

Figure 2. An example of vertical gridding of four wells in a reservoir. Similar fluid flow velocities, represented by k/φ patterns, are correlated from well to well. These k/φ versus depth units represent fluid flow units, and allow the cross section to be divided into seven fluid flow layers of varying flow velocities and thicknesses.

Physical Property Input

Every grid block must contain values for the input data required for the recovery process being simulated. Saturations, pressure, thickness, porosity, permeability, and viscosity always are necessary; additional fluid property data usually are needed. Values for relative permeability, capillary pressure, dispersivity, thermal data, and chemical reaction data may be needed depending on the process.

The values entered into a grid block are averages for the grid block. The volume of a grid block must be taken into account when preparing data for input. Figure 3 is a schematic representation of how a property can vary depending on the volume in which it is measured or averaged. On the microscopic scale, such as seen in thin sections, properties vary widely from pore to pore. On a macroscopic scale, such as the volume scale represented by cores or wireline logs, the variation is not as great with changes in averaging volume. On the megascopic scale of hundreds of feet, even less local variation may occur, but the values of properties within large blocks may vary greatly at the scale of a reservoir. In practice, grid blocks tend to be on the megascopic scale. The challenges in entering data into the grid blocks are obtaining good megascale averages from data at various scales and estimating values for grid blocks that are unsampled.

Several data sources may be available, but well logs and core are the most commonly available and used sources. The the k/φ versus depth (h) data shown in Figure 2 were obtained from porosity and permeability measurements on core plugs. Due to the small averaging volumes, these are macroscopic properties. For the flow units shown in Figure 2, average values of various properties, such as porosity, φ storage capacity, φh, or deliverability, kh, might be computed for each flow unit for each well and entered into the appropriate grid blocks for each well. Arithmetic averages would probably be used. The arithmetic average is always correct for porosity but there is no unique averaging method for permeability. Permeabilities in parallel average arithmetically, but permeabilities in series average geometrically. It is even more complicated because permeability values really are needed in three orthogonal directions. Stratigraphic information is needed for defining the spatial orientation and connectivity of flow paths in a grid block. Average values of permeability derived from sparse well data may be inappropriate input values for unsampled grid blocks. But, at this time, there is no alternative source of information.

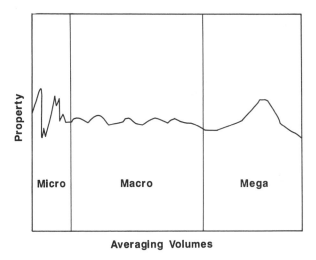

Figure 3. An example of the effect of averaging a particular petrophysical or fluid property throughout different volumes. Microscopic volumes have great ranges in values of a particular property, macroscopic volumes have moderate variations in values, and megascopic volumes have low variations in values but may vary greatly across a an entire reservoir.

An example of the difficulty in obtaining the average permeability of a grid block is illustrated in Figure 4 which is taken from the work of Lake et al. (1986). In this block of cross-stratified eolian sandstone, the arrows show the direction and magnitude of fluid flow with respect to variations in the type and scale of cross stratification. In reality, we have data only where a wellbore exists. The challenge of reservoir engineering and stratigraphy is to develop orthogonal permeability block averages from wellbore data, geologic insight, and quantitative stratigraphic models of appropriate scales and accuracy.

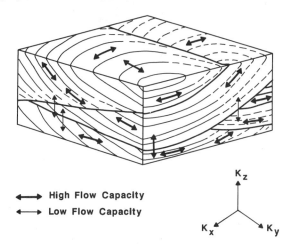

Figure 4. An illustration of the variations in direction and velocity (shown by arrows) of fluid flow within cross-stratified eolian sandstones (from Lake et al., 1986).

Core data usually are available for a few wells at most, but log data often are available for all wells. Logs provide porosity but not permeability data. If core data are available, a correlation between permeability and porosity can sometimes be obtained. An example is shown in Figure 5 where the log of permeability is plotted against porosity. When such a correlation is available, it can be used to develop permeability information from log porosity.

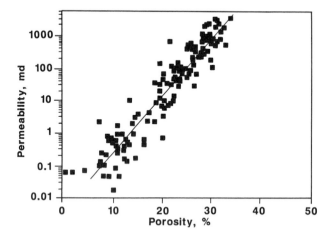

Figure 5. Correlating core porosity with core permeability is one means of estimating permeability from well logs which indirectly measure porosity but not permeability.

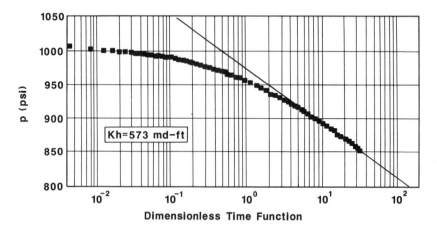

Figure 6. An example of using a pressure falloff test to estimate the average permeability of the stratigraphic units around a well.

Fortunately, some megascopic data are usually available. The two most commonly available types of megascopic data are pressure transient analysis and performance data. Tracer test data, a special case of performance data, occasionally are available. All of these methods require existing wells.

Pressure transient analysis involves perturbing the pressure at one well and measuring the resulting pressure response at one or more wells. This type of analysis measures the kh in the radius of investigation, tens to hundreds of feet, around a well. Figure 6 is an example of data obtained on an injection well during a pressure falloff test, a type of transient analysis. Pressure is plotted against dimensionless time. From the curve slope in certain time regions, the kh can be deduced. Pressure transient data are usually the most accurate permeability data available. Since they represent an average value for the rock volume on the scale of grid blocks, they often are used to calibrate log and core data to an absolute value.

The importance of history matching was discussed earlier. Tracer testing, a special kind of history matching, is accomplished by injecting easily analyzable materials into one or more wells. Produced fluid is analyzed at offsetting producers. Figure 7 is an example of tracer test data. The tracer concentration is plotted against cumulative fluid production. The first sharp peak indicates production from a relatively small, high capacity flow connectivity. The large broad peak represents production from a large, lower flow capacity connectivity. In tracer history matching, an initial k/ϕ distribution is varied in an attempt to match the produced concentrations.

Only a limited number of grid blocks in a reservoir simulator are associated with wells, and for these the data sources are principally from well cores and logs. Entering data into the remaining grid blocks is most simply accomplished by an interpolation scheme. Such a scheme must honor the geologic understanding of the reservoir. For example, where the vertical stratigraphic sequence shows a strong areal correlation, flow units would be interpolated in the areal plane only.

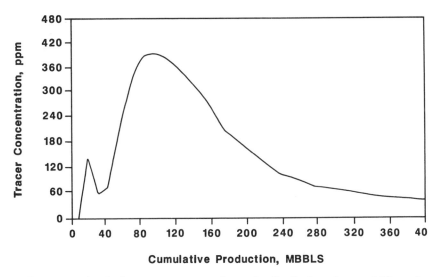

Figure 7. An example of using a tracer test to estimate the distribution of permeability and porosity within one or more fluid flow units between an injection and a production well. The first sharp peak represents production from a relatively small, high capacity flow unit. The large broad peak represents production from a larger flow unit with lower flow capacity.

A recent development for "interpolating" the remaining grid blocks is a stochastic approach using statistically derived spatial distributions of reservoir properties. This approach acknowledges the unknown variability in reservoir properties but recognizes that a degree of order exists in the geological processes that created the reservoir. The key is to somehow estimate or measure the spatial order. This method is expected to improve reservoir descriptions as reservoir engineers and geologists learn to jointly apply geostatistical concepts.

DISCUSSION AND RECOMMENDATIONS

The greatest needs in reservoir simulation are better ways to estimate the spatial distribution of permeability within the reservoir, and to obtain permeability averages over scales representative of grid blocks. A fundamental constraint on the accuracy of a reservoir simulator is the degree of accuracy of the geological information that is entered into the simulator. The geological description of a reservoir must correctly identify fluid flow units, their spatial relationships and their degree of connectedness within the reservoir and among wells. Further, these data must be provided as averages over volumes consistent with the size of individual grid blocks used in the simulator. Typically, dimensions of grid blocks range from several tens to a few hundred feet on a side.

In reservoirs that do not produce primarily from fracture porosity, the stratigraphic architecture, including distributions of facies and diagenetic modifications of original porosities and permeabilities of facies, controls the principal conduit or plumbing system through which fluids flow. Conventionally, sedimentary facies information is collected in a manner designed to understand the depositional and/or early diagenetic processes that operated during the accumulation of particular sedimentologic or stratigraphic units. This information normally is synthesized in the form of facies, process-response, or other models, which are even more general summaries of stratigraphic relationships in a geologic context. Neither facies descriptions nor synthetic models provide information directly usable for reservoir simulation, primarily because they do not specifically describe relations between arrangements of facies in time and space and the ways they would control the fluid flow within the stratigraphic architecture. That is, current methods of obtaining and communicating sedimentologic and stratigraphic information do not also present a quantitative description of the stratigraphic plumbing system. Moreover, most facies, depositional system, or other geologically oriented models fail to provide estimates of their accuracy, ranges in values of included properties, or limits in their predictiveness; such information also would enhance the value of reservoir descriptions for purposes of reservoir simulation.

The development of quantitative stratigraphic models, in combination with existing sedimentologic and stratigraphic models and a new generation of empirical stratigraphic information, appears the most likely avenue by which reservoir descriptions may be improved and applied in reservoir simulators. The following suggestions are intended to focus the attention of sedimentologists and stratigraphers on the types and scales of information that will be required in reservoir engineering applications. Perhaps the most important contribution of quantative stratigraphic models would be to incorporate estimates of predictiveness, accuracy and limits of application of stratigraphic properties simulated by the models. The models should be applicable to the scale of reservoirs and should present stratigraphic information in the form of fluid flow units and their distribution and connectedness in space. New empirical stratigraphic information is required to determine the scales of sedimentologic or stratigraphic heterogeneities that are important in controlling fluid flow and defining fluid flow pathways through strata at the scale of reservoirs. Similarly, the volumes, distributions and arrangements of facies within the larger scale stratigraphic architecture of a variety of depositional environments should be measured and tabulated. Surface and subsurface studies designed to produce this information must also attempt to relate two- or three-dimensional observations to one-dimensional subsurface measurements from a wellbore. These two types of information may then be employed in the processing algorithms of quantitative stratigraphic models that are designed for reservoir description. In order to make existing stratigraphic information more usable, empirical "dictionaries" should be built that translate facies information into some form of flow units. It is likely that this will require collecting new information about fluid movement through specific facies and through lateral and vertical facies associations. Finally, for large or stratigraphically complex reservoirs, more accurate correlation methods are required so that intrareservoir flow pathways and barriers to flow are recognized.

Though not addressed here, geophysical techniques also can be useful in obtaining reservoir descriptions. Surface seismic data can provide estimates of thickness and sometimes can indicate changes in lithological and fluid character. New advances are being made in downhole geophysical methods where sources and receivers are placed in one or more wellbores. Downhole methods have the potential for significantly increasing frequency bandwidth and hence spatial resolution. If resolution on the order of feet can be economically achieved for interwell distances of investigation, improvements in reservoir description may be possible.

ACKNOWLEDGMENTS

We thank Shirley Dutton and Will Schweller for their time in reviewing an earlier draft of this paper, and greatly appreciate their constructive and useful comments. Amoco Production Company is thanked for their permission to publish this review.

SELECTED BIBLIOGRAPHY

Begg, S.H., and Carter, R.R., 1987, Assigning effective values to simulator grid-block parameters in heterogeneous reservoirs: Society of Petroleum Engineers of AIME, 62nd Annual Technical Conference and Exhibitions, SPE 16754.

Haldorsen, H.H., and MacDonald, C.J., 1987, Stochastic modeling of underground reservoir facies (SMURF): Society of Petroleum Engineers of AIME, 62nd Annual Technical Conference and Exhibitions, SPE 16751.

Hewitt, C.H., Morgan, J.T., 1965, The Fry *in situ* combustion test—Reservoir characteristics: Journal of Petroleum Technology, v. 17 (March, 1965), p. 337-312.

Johnson, H.D., and Krol, D.E., 1984, Geological modeling of a heterogeneous sandstone reservoir: Lower Jurassic Statfjord Formation, Brent field: Society of Petroleum Engineers of AIME, 59th Annual Technical Conference and Exhibitions, SPE 13050.

Lake, L.W., and Carroll, H.B., Jr., editors, 1986, Reservoir characterization: New York, Academic Press, 659 p.

Lake, L.W., Scott, A.J., and Kocurek, G.A., 1986, Reservoir characterization for numerical simulation, DOE/BC/10744-8.

Lindquist, S.J., 1988, Practical characterization of eolian reservoirs for development: Nugget Sandstone, Utah–Wyoming thrust belt: Sedimentary Geology, v. 56, p. 1-25.

Matheson, G., Beucher, H., de Fouquet, C., and Galli, A., 1987, Conditional simulations of the geometry of fluvio-deltaic reservoirs: Society of Petroleum Engineers of AIME, 62nd Annual Technical Conference and Exhibitions, SPE 16753.

Miall, A.D., 1989, Can there be life after facies models? The development of a framework for the quantitative descritption of complex, three-dimensional facies architectures, in Cross, T.A., Quantitative Dynamic Stratigraphy: New Jersey, Prentice Hall (this volume).

Pryor, W.A., 1973, Permeability-porosity patterns and variations in some Holocene sand bodies: American Association of Petroleum Geologists Bulletin, v. 57, p. 162-189.

Pryor, W.A., and Fulton, K., 1978, Geometry of reservoir-type sandbodies in the Holocence Rio Grande delta and comparison with ancient reservoir analogs: Society of Petroleum Engineers of AIME, 5th Symposium on Improved Methods for Oil Recovery, SPE 7045.

Tillman, R.W., and Weber, K.J., editors, 1987, Reservoir sedimentology: Society of Economic Paleontologists and Mineralogists Special Publication 40, 357 p.

Weber, K.J., Eijpe, R., Leijnse, D., and Moens, C., 1972, Permeability distribution in a Holocene distributary channel-fill near Leerdam (The Netherlands): Geologie en Mijnbouw, v. 51, p. 53-62.

Weber, K.J., 1982, Influence of common sedimentary structures on fluid flow in reservoir models: Journal of Petroleum Technology, v. 34 (March, 1982), p. 665-672.

36

Can there be Life after Facies Models? The Development of a Framework for the Quantitative Description of Complex, Three-Dimensional Facies Architectures

Andrew D. Miall
Geology Department, University of Toronto, Toronto, Ontario, Canada M5S 1A1

Abstract

Conventional facies modeling techniques are based primarily on Walther's Law and the interpretation of vertical profiles. The kinds of data currently produced by sedimentologists using this methodology are unsuitable for quantitative modeling of depositional systems and the analysis of reservoir heterogeneities because they are essentially one-dimensional descriptions. Further progress in modeling and reservoir description will require systematic methods of three-dimensional facies documentation.

Clastic sedimentary bodies can be subdivided into three-dimensional depositional elements using a hierarchy of bounding surfaces and the technique of architectural element analysis. Bounding surfaces can be ranked according to their genetic and time significance. They may have distinctive shapes and facies associations. Eolian bounding surfaces and a hierarchy of genetic units have been defined by Brookfield and Kocurek, fluvial bounding surfaces and architectural elements have been described by Miall, and the scales of depositional elements in turbidite systems have been described by Mutti and Normark. The hierarchy of depositional elements and bounding surfaces for fluvial and turbidite systems show many similarities.

Field work is in progress to document and quantify the variability in fluvial systems in a variety of Jurassic and Cretaceous units in the Colorado Plateau, where superb exposures permit complete measurement and description of major channel sheets and compound bars (macroforms) hundreds of meters to a few kilometers across. Scaling of the depositional elements is being carried out to build a better data base for the modeling by reservoir engineers of fluid flow behavior in heterogeneous clastic reservoirs, and should lead to a more thorough understanding of medium time scale (10^0 to 10^5 years) depositional processes, as a basis for quantitative synthesis of depositional systems. It is suggested that similar hierarchies of genetic units exist in all channelized clastic systems, and that the same techniques therefore have wide application.

Introduction

Conventional techniques of sedimentological analysis and facies modeling place primary emphasis on the use of Walther's Law and the interpretation of vertical profiles, particularly for the study of clastic sediments (Walker, 1984; Reading, 1986). Although these have proved to be powerful techniques, and have led to much improved stratigraphic and paleogeographic syntheses, they have not provided an adequate basis for quantitative modeling of depositional systems (such as the numerical simulations of other geological processes described elsewhere in this book). Nor have the data found much application in the one area where detailed sedimentological information might be expected to have made its most useful contribution: the documentation of heterogeneities in petroleum reservoirs for the purpose of improved petroleum production, particularly during the design of enhanced recovery programs.

The extensive literature based on current facies model techniques and the appearance of several definitive textbooks on the subject might have created the impression that facies modeling is virtually a completed task, that we have at last reached the "mopping up" phase, that was so prematurely described in the case of turbidites by Walker (1973). However, the lack of success in the two areas noted above suggests that this is not the case. The key to further progress will be an evolution in our methods from the essentially one-dimensional approach of vertical profile analysis to a fully quantitative description of facies in three dimensions. An examination of some possible approaches to this objective is the main purpose of this paper.

If quantitative models of clastic depositional systems are to be attempted, a better understanding will be required of the types of depositional units or elements that compose clastic sequences. For example, it will be essential to describe their dimensions, scales, lithofacies compositions, length of time required for accumulation, preservation potential, and their distributions in time and space.

An immediate practical application of these data would be to improve and recast facies information in a form usable by reservoir engineers and geologists in modeling fluid flow in clastic reservoirs. It has been estimated that complex internal architectures are responsible for intrareservoir stratigraphic entrapment of an average of 30 percent of the original oil in place, amounting to as much as 100 billion barrels of movable oil in the United States alone (Tyler et al., 1984). A better understanding of these architectural complexities would facilitate improved primary production and would increase the success rate of enhanced recovery projects.

Two interrelated ideas might be used to develop a systematic description of facies distributions and their petrophysical properties in three dimensions, ultimately for application in reservoir simulation (Miall, 1988c). The first is the concept of *architectural scale*. Deposits consist of assemblages of lithofacies and structures over a wide range of physical scales, from the individual small-scale ripple mark to the assemblage produced by an entire depositional system. Recent work, particularly in eolian and fluvial environments, suggests that it is possible to formalize a hierarchy of scales. Depositional units at each physical scale originate in response to processes occurring over a particular time scale, and are physically separable from each other by a hierarchy of internal bounding surfaces. The second is the concept of the *architectural element*. An architectural element is a lithosome characterized by its geometry, facies composition, and scale, and is the depositional product of a particular process or suite of processes occurring within a depositional system.

ARCHITECTURAL SCALE AND THE BOUNDING SURFACE HIERARCHY

As noted by Allen (1983, p. 249):

> "The idea that sandstone bodies are divisible internally into 'packets' of genetically related strata by an hierarchically ordered set of bedding contacts has been exploited sedimentologically for many years, although not always in an explicit manner. For example, McKee and Weir (1953) distinguished the hierarchy of the stratum, the set of strata, and the coset of sets of strata, bedding contacts being used implicitly to separate these entities."

Allen (1966) showed that flow fields in such environments as rivers and deltas could be classified into a hierarchical order. His hierarchy was designed as an aid to interpreting variance in paleocurrent data collected over various areal scales, from the individual bed to large outcrops or outcrop groups. The hierarchy consists of five categories, small-scale ripples, large-scale ripples, dunes, channels, and the "integrated system," meaning the sum of the variances over the four scales. Miall (1974) added the scale of the entire river system to this idea, and compiled some data illustrating the validity of the concept.

Brookfield (1977) discussed the concept of an eolian bedform hierarchy, and tabulated the characteristics of four orders of "aeolian bedform elements:" draas, dunes, aerodynamic ripples and impact ripples. These four orders of bedform elements are deposited synchronously and are superimposed on each other. Brookfield showed that this superimposition resulted in the formation of three types of internal bounding surface. His first-order surfaces are major, laterally extensive, flat-lying or convex-up bedding planes between draas (macroforms, in the terminology of Jackson, 1975). Second-order surfaces are low- to moderately-dipping surfaces bounding sets of cross strata formed by the passage of dunes across draas (mesoforms). Third-order surfaces are reactivation surfaces bounding bundles of laminae within crossbed sets, and are caused by localized changes in wind direction or velocity (mesoforms to microforms).

Brookfield's (1977) development of the relationship between the time duration of a depositional event, the physical scale of the depositional product, and the geometry of the resulting lithosome, was a major step of considerable use in the analysis of eolian deposits. Brookfield (1977), Gradzinski et al. (1979) and Kocurek (1981) showed how these ideas could be applied to the interpretation of ancient eolian deposits. Kocurek (1988) recognized two types of first-order surfaces, the most laterally extensive of which he termed "supersurfaces." Characterization of eolian deposits therefore now requires a four-fold hierarchy of bounding surfaces.

Bounding Surfaces in Fluvial Deposits

Several workers have described ranges in physical scales of depositional elements in fluvial deposits. Williams and Rust (1969) proposed an ordering of the scales of channels and bars in the modern Donjek River, Yukon. Campbell (1976), in an analysis of the Westwater Canyon Member of the Morrison Formation in New Mexico, recognized several scales of fluvial sequence that occurred in tabular channel-fill sandstone bodies of a range of dimensions. Jones and McCabe (1980) described three types of reactivation surface occurring within sets of giant crossbedding, and related them to changes in bedform orientation and to stage changes in the river. A similar type of analysis was performed by Haszeldine (1983a, b) on the bounding surfaces within a Carboniferous sand flat deposit. Bridge and Diemer (1983) and Bridge and Gordon (1985) referred to "major" and "minor" bounding surfaces within Paleozoic fluvial sequences. The major surfaces are typically horizontal and planar or slightly concave up, and enclose tabular sheets representing channel-fill successions.

Allen's (1983) study of the Devonian Brownstones of the Welsh Borders is the most explicit attempt to formalize the concept of a hierarchy of bounding surfaces in fluvial deposits. Allen described three types of bounding surface, and referred to Brookfield's work in eolian strata for comparison. He reversed the order of numbering from that used by Brookfield (1977), such that the surfaces with the highest number are the most laterally extensive. No reason was offered for this reversal, but the result is an open-ended numbering scheme that can readily accommodate developments in our understanding of larger scale depositional units, as discussed below. *First-order contacts*, in Allen's scheme, are set boundaries, in the sense of McKee and Weir (1953). *Second-order contacts* "bound clusters of sedimentation units of the kinds delineated by first-order contacts." They are comparable to the coset boundaries of McKee and Weir (1953) except that more than one type of lithofacies may comprise a cluster. Allen (1983) stated that "these groupings, here termed complexes, comprise sedimentation units that are genetically related by facies and/or paleocurrent direction." Many of the complexes in the Brownstones are macroforms, in the sense defined by Jackson (1975). *Third-order surfaces* are comparable to the major surfaces of Bridge and Diemer (1983). No direct relationship is implied between Allen's three orders of surfaces and those of Brookfield, because of the different hydraulic behavior and depositional patterns of eolian and aqueous currents.

The writer has found it useful to expand Allen's classification to a six-fold hierarchy, to facilitate definition of fluvial macroform architecture, and to include the largest, basin-scale heterogeneities in the classification (Miall, 1988a, b, c; Miall and Turner-Peterson, in preparation). This hierarchy is summarized in Table 1, and compared with the hierarchy of turbidite depositional systems erected by Mutti and Normark (1987). Examples of these scales of depositional elements in one stratigraphic unit are listed in Table 2. Note that these elements represent a range of eleven orders of magnitude in time scale, and at least seven orders of magnitude in size (area). These ranges are serious practical obstacles to quantitative modeling.

First- and second-order surfaces record boundaries within microform and mesoform deposits (Fig. 1). The definition of first-order surfaces is unchanged from Allen (1983). Second-order surfaces are simple coset bounding surfaces, in the sense of McKee and Weir (1953).

Third- and fourth-order surfaces are defined when architectural reconstruction indicates the presence of macroforms, such as point bars or sand flats (Fig. 1). Individual depositional units ("storeys" or "architectural elements") are bounded by surfaces of fourth-order or higher rank.

Fifth-order surfaces are surfaces bounding major sand sheets, such as channel-fill complexes (Fig. 1). They are generally flat to slightly concave-upward, but may be marked by local cut-and-fill relief and by basal lag gravels (third-order surfaces of Allen, 1983;

"major" surfaces of Bridge and Diemer, 1983). Sixth-order surfaces are surfaces defining groups of channels, or paleovalleys. Mappable stratigraphic units such as members or submembers are bounded by sixth-order surfaces (Fig. 1). Future work on allocyclic controls on fluvial sedimentation, such as sea-level change and tectonics, may permit a further refinement of large-scale elements of the hierarchy, with the subdivision of sixth-order surfaces into seventh, eighth and higher orders.

Table 1 Comparison of scale hierarchies in fluvial and turbidite systems.

Turbidite systems (Mutti and Normark, 1987)		Fluvial systems (Miall, 1987)		Stratigraphic rank	Time scale (years)	Bedform hierarchy (Jackson, 1975)
Rank	Examples	Rank	Examples			
1	Basin-fill fan complexes	undef.	Basin-fill fluvial complexes	Group, Supergroup	10^6-10^7	-----
2	Turbidite depo. systems, fans	6	Fluvial depo. systems, fans	Formation, Member	10^5-10^6	-----
3	Fan lobes, channel-levee complexes	5	Major channels	Tongue	10^4-10^5	macroform
4	Individual channel fill	4	Point bar, side bar	Storey	10^2-10^3	macroform
5	Lithofacies, bedding patterns, cut and fill (equivalent	3	Growth increments of macroforms	-----	10^0-10^1	mesoform
	to ranks 1-3 of fluvial systems)	2	Cosets of similar facies	-----	10^{-2}-10^{-1}	mesoform
		1	Lithofacies unit	-----	10^{-5}-10^{-3}	microform

Table 2 The range of scales of depositional units in fluvial sandstones, as illustrated by the Westwater Canyon Member, Morrison Formation, near Gallup, New Mexico (Miall, 1988b; Miall and Turner-Peterson, in preparation).

Rank of bounding surface	Lateral extent of unit	Thickness of unit (m)	Area of unit (ha)	Origin	Subsurface mapping methods
6	200x200 km	0-30	4×10^7	members or sub-members, subtle tectonic control	regional correlation of wireline logs
5	1x10 km	10-20	10^4	sheet sandstone of channel origin	intrafield correlation of wireline logs, 3-D seismic
5	0.25x10 km	10-20	2500	ribbon channel sandstone	mapping difficult except with very close well spacing, 3-D seismic
4	200x200 m	3-10	40	macroform elements (elements LA,DA)	dip of 4th- and 3rd-order surfaces may be recognizable in core.
3	100x100 m	3-10	10	reactivation of macroforms	dip of 4th- and 3rd-order surfaces may be recognizable in core
2	100x100 m	5	10	cosets of similar crossbed facies	facies analysis of core
1	100x100 m	2	10	individual crossbed sets	facies analysis of core

In columns 2, 3 and 4 approximate maximum dimensions are indicated. Element codes are discussed below.

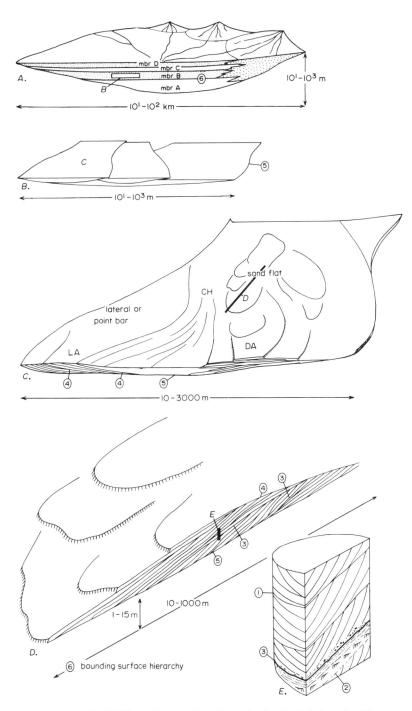

Figure 1. A proposed six-fold bounding surface hierarchy for fluvial deposits. Diagrams A to E represent successive enlargements of part of a fluvial unit, in which bounding surfaces of six distinct types may be recognized. These are discussed in the text (Miall, 1988c).

Fifth- and sixth-order surfaces are potentially the easiest to map in the subsurface because they are laterally extensive and possess essentially simple, flat or gently curved, channelized geometry. Many examples of such mapping have been reported in the literature (e.g., Busch, 1974). Considerable potential now exists for mapping these surfaces with three-dimensional seismic data, as described by Brown (1986). Fourth-, fifth- and sixth-order surfaces may appear very similar to third-order surfaces in core. They are best differentiated by careful stratigraphic correlation between closely spaced cores where well spacing may be a few hundred meters or less.

Identifying and correlating these various bounding surfaces can contribute to unravelling the complexities of a fluvial depositional system. They are likely to be particularly useful in recognizing and documenting macroforms, about which much remains to be learned. Further discussion and illustration of the fluvial bounding surface hierarchy is given by Miall (1988a,b,c).

ARCHITECTURAL ELEMENTS

A succession of clastic strata may be divided into a hierarchy of three-dimensional rock units by bounding surfaces of different scales. This facilitates description, and also makes it easier to visualize the appropriate physical extent and time duration of the processes which controlled sedimentation at each level of the hierarchy. Rock description involves the definition of lithofacies and the recognition of facies assemblages.

Most deposits may be subdivided into several or many types of three-dimensional sediment bodies characterized by distinctive lithofacies assemblages, external geometries and orientations (many of which are macroforms). Allen (1983) coined the term "architectural element" for these depositional units, and Miall (1985) attempted a summary and classification of the current state of knowledge of these elements as they occur in fluvial deposits (Fig. 2).

Two interpretive processes are involved simultaneously in the analysis of outcrops that contain a range of scales of depositional units and bounding surfaces: 1) the definition of the various types and scales of bounding surfaces, and 2) the subdivision of the succession into its constituent lithofacies assemblages, with the recognition and definition of macroforms and any other large features that may be present.

In general the most distinctive characteristic of a macroform is that it consists of genetically related lithofacies with sedimentary structures showing similar orientations, and internal minor bounding surfaces (first- to third-order of the classification given above) that extend from the top to the bottom of the element, indicating that it developed by long-term lateral, oblique or downstream accretion. A macroform is comparable in height to the depth of the channel in which it formed and its width and length is of similar order of magnitude to the width of the channel. However, independent confirmation of these dimensions is difficult in multistorey sandstone bodies, where channel margins are rarely preserved and the storeys commonly have erosional relationships with each other.

The definition of a macroform in any given outcrop is in part an interpretive process. Some types of macroform, such as the lateral accretion deposits that constitute the typical point bar, are by now so well known that their recognition in outcrop would be classified by some workers as a "descriptive" rather than an "interpretive" exercise. Other types of macroform are less well known. The recognition of a macroform may depend in part on the type of bounding surface that encloses it. Conversely, the appropriate classification of a bounding surface may depend on a description of the lithofacies assemblage and geometry of the beds above and below it. For these reasons description, classification and interpretation cannot always be completely separate exercises.

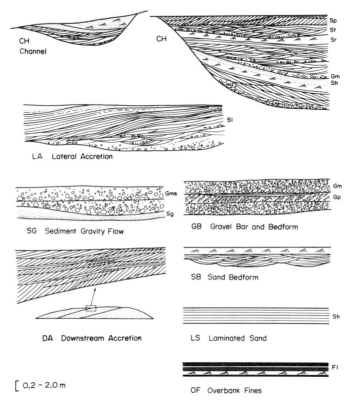

Figure 2. The range of architectural elements in fluvial deposits. Lithofacies codes are those of Miall (1978) (slightly modified from Miall, 1985).

It may be possible to document the range in scales and the potential mappability of the hierarchy of depositional units and their bounding surfaces in any given stratigraphic unit. For example, Table 2 presents the results of work in progress to document the fluvial styles of the Westwater Canyon Member of the Morrison Formation (Jurassic) in northern New Mexico, and Figure 3 illustrates the geometry of some examples of typical architectural elements in this unit (Miall, 1988b; Miall and Turner-Peterson, in preparation). This type of subdivision is of importance in reservoir development, as noted in the next section. However, hierarchies of this type do not have a fixed range of scales. For example, in parts of the Donjek River, Yukon, large compound bars (first-order bars of Williams and Rust, 1969; fourth-order macroforms in the present scheme) are in the order of 200 to 400 m in downstream length. Comparable features in the Brahmaputra River, Bangladesh, are several kilometers in length (Coleman, 1969)—an order of magnitude larger.

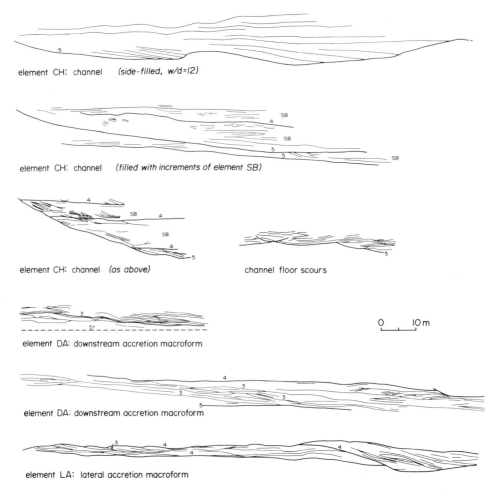

Figure 3. Examples of architectural elements in the Westwater Canyon Member (Morrison Formation), southern San Juan Basin, drawn at natural scale, to illustrate the types of lithosome geometry present in this unit. Bounding surfaces are numbered according to rank. Unnumbered surfaces are of first and second-order rank. Two-letter element codes are illustrated in Figure 2 (from Miall, 1988b; Miall and Turner-Peterson, in preparation).

PRACTICAL APPLICATIONS OF THE CONCEPTS

Quantitative Modeling of Depositional Systems

A central theme of this book is the application of numerical simulation and computer graphic techniques to the modeling of natural geological processes. One of the main purposes of such modeling is to test interpretations and estimates of the rates and scales of physical controls in complex systems with feedback loops and large numbers of unknowns. Modeling of global climates and the lithospheric processes that lead to sedimentary basin formation have, in recent years, demonstrated the immense power of this approach.

Modeling of clastic depositional systems in this way is in its infancy, but several important steps have been achieved in areas applicable to the kinds of complex clastic system discussed in this paper. For example, Rubin (1987) studied the simultaneous migration and interaction of up to three sets of bedforms, and illustrated the varieties of complex cross stratification that result. Bridge, in a series of papers (e.g., Bridge, 1978) modeled the fluvial hydraulics of a meander bend to simulate the stratigraphy of point bars. Leeder (1978) explored rates of fluvial floodplain accretion and channel avulsion and developed some simple equations, which were then used by Bridge and Leeder (1979) to develop two-dimensional models of alluvial stratigraphy. The latter paper demonstrated the importance of compaction and differential subsidence in the evolution of stratigraphic architectures. The independent work of Allen (1978, 1979) along similar lines has also been very influential in the development of architectural concepts. Ross (this volume) explored the large-scale geometry of coastal, paralic and shallow marine systems and their response to sea-level changes.

Within the context of the scale hierarchy discussed here, Rubin's work deals with ranks 1 and 2, Bridge's simulations examine one variety of a rank 4 process, the Bridge and Leeder study and those by Allen are at rank 5, and the work of Ross explores one aspect of rank 6. To date there has been no attempt to carry out a thorough simulation of macroform evolution in a channel system, involving lateral, oblique and downstream accretion (variations of rank 4 geometry) plus episodic growth and erosion (rank 3). And, there has been no attempt to link these processes together into an overall numerical process-response model. To do so would require an extremely complex computer model, which would have to include eleven orders of magnitude of time and seven (or more) orders of magnitude of scale represented by the typical clastic system. The kinds of three-dimensional facies documentation proposed in this paper will be an essential prerequisite for the development and testing of any such computer model.

Quantitative Documentation of Reservoir Heterogeneities

Reservoir development geology makes use of facies and stratigraphic data as input to models of fluid flow (e.g., Shelton and Cross, this volume). Such models are of particular importance in the design of enhanced recovery projects. However, just as in the area of dynamic process modeling discussed in the previous section, the amount of useful quantitative three-dimensional data available to the development geologist is very limited. For this reason the application of sedimentology to development geology and reservoir engineering is underdeveloped, in contrast to the long-standing use of sedimentology in exploration. The growing realization of this fact is reflected by the recent appearance of a special publication devoted to this subject (Tillman and Weber, 1987), and the inclusion of several papers on reservoir sedimentology in a recent volume on reservoir characterization (Lake and Carroll, 1986).

Development geologists use core data to determine porosity and permeability, and they rely on lithostratigraphic correlation to define "flow units" or "lithohydraulic units" for reservoir simulation (Shelton and Cross, this volume). However, such units commonly are assumed to be tabular, and to have uniform porosity and permeability characteristics throughout. In terms of the scale hierarchy discussed in this paper, core data can provide useful information at ranks 1 and 2, whereas subsurface lithostratigraphy typically is used to assess architecture at ranks 5 and 6. There are new developments in subsurface imaging of rank 5 and 6 depositional elements by 3-D seismic (Brown, 1986). The elucidation of complex architectures at an intermediate scale, such as the delineation of isolated sand bodies of rank 5 (channels) and ranks 3 and 4 (point bars and other macroforms) commonly requires more detailed well control than is available.

Can the architectural concepts described here be usefully applied to problems of development geology? It would seem that the first requirement is to recognize the nature of the problem. First, the three-dimensional complexity of clastic strata needs to be more fully understood. Second, up to the present time sedimentologists and petroleum geologists have focused little attention on depositional units of rank 3 and 4, apart from the seemingly ubiquitous point bar (Miall, 1988b). Therefore, there is a pressing need for well-documented case studies of outcrop and subsurface examples. Are such deposits capable of systematic and quantitative classification? Does the concept of the universal architectural element (*sensu* Miall, 1985) have any validity and usefulness? Do these elements have predictable porosity and permeability patterns? Only through acquisition of a great deal more data will these questions be answered.

Can architectural elements, at all ranks, be recognized using the kinds of subsurface data available to exploration and production geologists? As discussed by Miall (1988b) core data, well data and 3-D seismic all may contribute to recognition of architectural elements. The recognition of gently dipping accretion surfaces of ranks 3 and 4 would be a very useful first step, and this should be possible (though far from easy) using all three kinds of data. Success is most likely to be achieved in areas of dense well spacing, particularly where a significant proportion of the wells have been cored. This is commonly the situation that evolves when a field is undergoing a pilot enhanced recovery study. It remains only for the development geologist and management to be aware of the potential that exists in the stratigraphic data base for detailed sedimentological studies, and the useful constraints that such studies may provide during the design of reservoir flow models.

APPLICATION OF THE CONCEPTS TO OTHER DEPOSITS

It is suggested that the quantitative study of all clastic deposits would be facilitated by subdivision using hierarchical classifications based on physical scales, geometries and facies composition. The eolian bounding surface hierarchy of Brookfield (1977) has proved to be an invaluable concept. Ranking of the depositional elements in turbidite systems has recently been introduced by Mutti and Normark (1987), and the classification scheme discussed here is finding application in the study of fluvial deposits. Miall (submitted) summarized the many similarities between the channelized deposits of rivers and submarine fans, and showed how the fluvial classifications could be adapted for use in turbidite systems. Undoubtedly the same principles could be applied to all types of channelized deposits.

By dividing clastic systems into hierarchies of depositional components it will become easier to quantify the scales and the rates of the various processes that contributed to their formation. This will facilitate quantitative stratigraphic modeling and improve our understanding of heterogeneities in clastic reservoirs and ore bodies.

ACKNOWLEDGMENTS

Discussions of bounding surfaces and architectural elements in the field with Larry Middleton and Christine Turner-Peterson, and rigorous reviews of some of my earlier papers on this subject by a variety of colleagues, have helped me to sharpen and clarify my ideas. The present paper was reviewed by Tim Cross, Mark Chapin and Bill Galloway, and my thanks go to them for what are, I hope, some significant improvements.

Financial support of the field work that lead to the ideas described here was provided in part by an Operating Grant from the Natural Sciences and Engineering Research Council (Canada). Acknowledgment is also made to the donors of the Petroleum Research Fund, administered by the American Chemical Society, for partial support of this research.

REFERENCES CITED

Allen, J.R.L., 1966, On bed forms and paleocurrents: Sedimentology, v. 6, p. 153-190.

Allen, J.R.L., 1978, Studies in fluviatile sedimentation: An exploratory quantitative model for the architecture of avulsion-controlled suites: Sedimentary Geology, v. 21, p. 129-147.

Allen, J.R.L., 1979, Studies in fluviatile sedimentation: An elementary geometrical model for the connectedness of avulsion-related channel sand bodies: Sedimentary Geology, v. 24, p. 253-267.

Allen, J.R.L., 1983, Studies in fluviatile sedimentation: Bars, bar complexes and sandstone sheets (low-sinuosity braided streams) in the Brownstones (L. Devonian), Welsh Borders: Sedimentary Geology, v. 33, p. 237-293.

Bridge, J.S., 1978, Palaeohydraulic interpretation using mathematical models of contemporary flow and sedimentation in meandering channels, in Miall, A.D., ed., Fluvial sedimentology: Canadian Society of Petroleum Geologists Memoir 5, p. 723-742.

Bridge, J.S., and Diemer, J.A., 1983, Quantitative interpretation of an evolving ancient river system: Sedimentology, v. 30, p. 599-623.

Bridge, J.S. and Gordon, E.A., 1985, The Catskill magnafacies of New York State, in Flores, R.M., and Harvey, M., eds., Field guide to modern and ancient fluvial systems in the United States: Third International Sedimentology Conference, p. 3-17.

Bridge, J.S. and Leeder, M.R., 1979, A simulation model of alluvial stratigraphy: Sedimentology, v. 26, p. 617-644.

Brookfield, M.E., 1977, The origin of bounding surfaces in ancient aeolian sandstones: Sedimentology, v. 24, p. 303-332.

Brown, A.R., 1986, Interpretation of three-dimensional seismic data: American Association of Petroleum Geologists Memoir 42, 194 p.

Busch, D.A., 1974, Stratigraphic traps in sandstones—Exploration techniques: American Association of Petroleum Geologists Memoir 21, 174 p..

Campbell, C.V., 1976, Reservoir geometry of a fluvial sheet sandstone: American Association of Petroleum Geologists Bulletin, v. 60, p. 1009-1020.

Coleman, J.M., 1969, Brahmaputra River: Channel processes and sedimentation: Sedimentary Geology, v. 3, p. 129-329.

Gradzinski, R., Gagol, J., and Slaczka, A., 1979, The Tumlin Sandstone (Holy Cross Mountains, Central Poland): Lower Triassic deposits of aeolian dunes and interdune areas: Acta Geologica Polonica, v. 29, p. 151-175.

Haszeldine, R.S., 1983a, Fluvial bars reconstructed from a deep, straight channel, Upper Carboniferous coalfield of northeast England: Journal of Sedimentary Petrology, v. 53, p. 1233-1248.

Haszeldine, R.S., 1983b, Descending tabular cross-bed sets and bounding surfaces from a fluvial channel in the Upper Carboniferous coalfield of north-east England,. in Collinson, J.D., and Lewin, J., eds., Modern and ancient fluvial systems: International Association of Sedimentologists Special Publication 6, p. 449-456.

Jackson, R.G., II. 1975, Hierarchical attributes and a unifying model of bed forms composed of cohesionless material and produced by shearing flow: Geological Society America Bulletin, v. 86, p. 1523-1533.

Jones, C.M., and McCabe, P.J., 1980, Erosion surfaces within giant fluvial cross-beds of the Carboniferous in northern England: Journal of Sedimentary Petrology, v. 50, p. 613-620.

Kocurek, G., 1981, Significance of interdune deposits and bounding surfaces in aeolian dune sands: Sedimentology, v. 28, p. 753-780.

Kocurek, G., 1988, First-order and super bounding surfaces in eolian sequences—Bounding surfaces revisited: Sedimentary Geology, v. 56, p. 193-206.

Lake, L.W., and Carroll, H.B., Jr., eds., 1986, Reservoir characterization: Orlando, Academic Press Inc., 659 p.

Leeder, M.R., 1978, A quantitative stratigraphic model for alluvium, with special reference to channel deposit density and interconnectedness, in Miall, A.D., ed., Fluvial sedimentology: Canadian Society of Petroleum Geologists Memoir 5, p. 587-596.

McKee, E.D., and Weir, G.W., 1953, Terminology for stratification and cross-stratification in sedimentary rocks: Geological Society of America Bulletin, v. 64, 381-389.

Miall, A.D., 1974, Paleocurrent analysis of alluvial sediments: A discussion of directional variance and vector magnitude: Journal of Sedimentary Petrology, v. 44, p. 1174-1185.

Miall, A.D., 1985, Architectural-element analysis: A new method of facies analysis applied to fluvial deposits: Earth Science Reviews, v. 22, p. 261-308.

Miall, A.D., 1988a, Architectural elements and bounding surfaces in fluvial deposits: Anatomy of the Kayenta Formation (Lower Jurassic), southwest Colorado: Sedimentary Geology, v. 55, p. 233-262.

Miall, A.D., 1988b, Reservoir heterogeneities in fluvial sandstones: Lessons from outcrop studies: American Association of Petroleum Geologists Bulletin, v. 72, p. 682-697.

Miall, A.D., 1988c, Facies architecture in clastic sedimentary basins, in Paola, C. and Kleinspehn, K., eds., New perspectives in sedimentary basin analysis: New York, Springer-Verlag Inc., p. 67-81.

Miall, A.D., submitted, Architectural elements and bounding surfaces in channelized clastic deposits: notes on comparisons between fluvial and turbidite systems, in Masuda, F., ed., Sedimentary facies in the active plate margin: Tokyo, Terra Scientific Publishing Company.

Miall, A.D., and Turner-Peterson, C., in preparation, Variations in fluvial style in the Westwater Canyon Member, Morrison Formation (Jurassic), San Juan Basin, Colorado Plateau.

Mutti, E. and Normark, W.R., 1987, Comparing examples of modern and ancient turbidite systems: Problems and concepts, *in* Leggett J.K., and Zuffa, G.G., eds., Marine clastic sedimentology: Concepts and case studies: London, Graham and Trotman Limited, p. 1-38.

Reading, H.G., ed., 1986, Sedimentary environments and facies (second edition): Oxford, Blackwell Scientific Publications, 615 p.

Ross, W.C., 1988, Modeling base-level dynamics as a control on basin fill geometries and facies distribution: A conceptual framework, *in* Cross, T.A., ed., Quantitative dynamic stratigraphy: New Jersey, Prentice-Hall.

Rubin, D.M., 1987, Cross-bedding, bedforms, and paleocurrents: Society of Economic Paleontologists and Mineralogists, Concepts in Sedimentology and Paleontology, v. 1, 187 p.

Shelton, J.L., and Cross, T.A., 1988, The influence of stratigraphy in reservoir simulation, *in* Cross, T.A., ed., Quantitative dynamic stratigraphy: New Jersey, Prentice-Hall.

Tillman, R.W., and Weber, K.J., 1987, eds., Reservoir sedimentology: Society of Economic Paleontologists and Mineralogists Special Publication 40, 357 p.

Tyler, N., Galloway, W.E., Garrett, C.M., Jr., and Ewing, T.E., 1984, Oil accumulation, production characteristics, and targets for additional recovery in major oil reservoirs of Texas: Texas Bureau of Economic Geology Geologic Circular 84-2, 31 p.

Walker, R.G., 1973, Mopping up the turbidite mess, *in* Ginsburg, R.N., ed., Evolving concepts in sedimentology: Johns Hopkins University Press, Baltimore, p. 1-37.

Walker, R.G., ed., 1984, Facies models (second edition): Geological Association of Canada, Geoscience Canada Reprint Series No. 1, 317 p.

Williams, P.F., and Rust, B.R., 1969, The sedimentology of a braided river: Journal of Sedimentary Petrology, v. 39, p. 649-679.

INDEX